Analytical Chemistry: Concepts and Applications

Analytical Chemistry: Concepts and Applications

Edited by **Sherley Benson**

NY RESEARCH
P R E S S

New York

Published by NY Research Press,
23 West, 55th Street, Suite 816,
New York, NY 10019, USA
www.nyresearchpress.com

Analytical Chemistry: Concepts and Applications
Edited by Sherley Benson

International Standard Book Number: 978-1-63238-470-6 (Hardback)

Contents

Preface

Every book is initially just a concept; it takes months of research and hard work to give it the final shape in which the readers receive it. In its early stages, this book also went through rigorous reviewing. The notable contributions made by experts from across the globe were first molded into patterned chapters and then arranged in a sensibly sequential manner to bring out the best results.

This book offers information regarding analytical chemistry provided by established academic experts in this field. Analytical chemistry mainly deals with the quantitative as well as qualitative aspects of a substance. It provides extensive knowledge about the structure, composition, as well as the quantity of all the constituents present in a particular matter. Analytical chemistry has been extremely useful since the early days as it helped scientists in separating and identifying various compounds and elements. Modern analytical chemistry has dominated the industry for many years and continues to do so, for example, recent discoveries in this field have made it possible for the pharmaceutical industry to develop more cost-effective ways of treatment. This book outlines the tools, techniques and applications of analytical chemistry in detail. It presents contributions made by international experts that will provide innovative insights into this field.

It has been my immense pleasure to be a part of this project and to contribute my years of learning in such a meaningful form. I would like to take this opportunity to thank all the people who have been associated with the completion of this book at any step.

Editor

Recent advances in high performance poly(lactide): from "green" plasticization to super-tough materials via (reactive) compounding

Georgio Kfoury[1,2], Jean-Marie Raquez[2], Fatima Hassouna[1]*, Jérémy Odent[2], Valérie Toniazzo[1], David Ruch[1] and Philippe Dubois[2]*

[1] Department of Advanced Materials and Structures, Public Research Center Henri Tudor, Hautcharage, Luxembourg
[2] Laboratory of Polymeric and Composite Materials, UMONS Research Institute for Materials Science and Engineering, Center for Innovation and Research in Materials and Polymers, University of Mons, Mons, Belgium

Edited by:
Alfonso Jiménez, University of Alicante, Spain

Reviewed by:
Alberto D'Amore, Second University of Naples, Italy
Michele Galizia, University of Texas at Austin, Italy

*Correspondence:
Fatima Hassouna, Department of Advanced Materials and Structures, Centre de Recherche Public Henri Tudor, Rue Bommel 5, Hautcharage 4940, Luxembourg
e-mail: fatima.hassouna@tudor.lu;
Philippe Dubois, Laboratory of Polymeric and Composite Materials, Center for Innovation and Research in Materials and Polymers, UMONS Research Institute for Materials Science and Engineering, University of Mons, Place du Parc 23, Mons, 7000, Belgium
e-mail: philippe.dubois@ umons.ac.be

Due to its origin from renewable resources, its biodegradability, and recently, its industrial implementation at low costs, poly(lactide) (PLA) is considered as one of the most promising ecological, bio-sourced and biodegradable plastic materials to potentially and increasingly replace traditional petroleum derived polymers in many commodity and engineering applications. Beside its relatively high rigidity [high tensile strength and modulus compared with many common thermoplastics such as poly(ethylene terephthalate) (PET), high impact poly(styrene) (HIPS) and poly(propylene) (PP)], PLA suffers from an inherent brittleness, which can limit its applications especially where mechanical toughness such as plastic deformation at high impact rates or elongation is required. Therefore, the curve plotting stiffness vs. impact resistance and ductility must be shifted to higher values for PLA-based materials, while being preferably fully bio-based and biodegradable upon the application. This review aims to establish a state of the art focused on the recent progresses and preferably economically viable strategies developed in the literature for significantly improve the mechanical performances of PLA. A particular attention is given to plasticization as well as to impact resistance modification of PLA in the case of (reactive) blending PLA-based systems.

Keywords: poly(lactide), (reactive) compounding, mechanical properties, impact resistance, toughening

INTRODUCTION

Over the past decade, there has been a significant research interest on compostable and/or biodegradable polymers in order to alleviate solid waste disposal problems related with petro-based plastics (Lim et al., 2008). These biodegradable polymeric materials are increasingly used today in packaging, agricultural, medical, pharmaceutical, and other areas (Rabetafika et al., 2006; Vroman and Tighzert, 2009). Two main classes of biodegradable polymers can be distinguished (Vroman and Tighzert, 2009) (**Figure 1**):

- *Natural and synthetic biodegradable polymers* produced from feedstocks derived from biological or renewable resources available in large quantities;
- *Synthetic biodegradable polymers* produced from feedstocks derived from non-renewable petroleum resources.

Aliphatic polyesters represent a large part of biodegradable polymers. They are considered as hydrolytically degradable polymers due to the presence in their backbone of hydrolytically sensitive chemical bonds, that is, ester moieties (Li, 1999; Nair

and Laurencin, 2007). There are two routes generally used to chemically develop biodegradable polyesters; step (condensation) polymerization and ring-opening polymerization (ROP) (Nair and Laurencin, 2007). Due to the absence of any by-products released during condensation process, ROP is thereby the most used pathway to prepare biodegradable polyesters. Among them, the most extensively investigated polymers are the poly(α-hydroxyacid)s, which include poly(glycolic acid) and the stereoisomeric copolymers of poly(lactic acid). Due to the commercial and low cost production of high molecular weight polymers using ROP, poly(lactide) (PLA) is one of the most studied candidates (Lim et al., 2008). Indeed, this polymer represents one of the stiffest organic materials with a Young's modulus of ca. 3 GPa, together with good optical and thermal properties [melting temperature (T_m) of ca. 170°C and a glass transition temperature (T_g) of ca. 60°C]. In addition, PLA is directly derived from renewable resources, making it environmentally sustainable in terms of depletion of petroleum resources and CO_2-release. Due to these attributes, PLA holds tremendous promises as an alternative to the ubiquitous petroleum-based materials as shown in **Table 1**. For instance, compared with the general

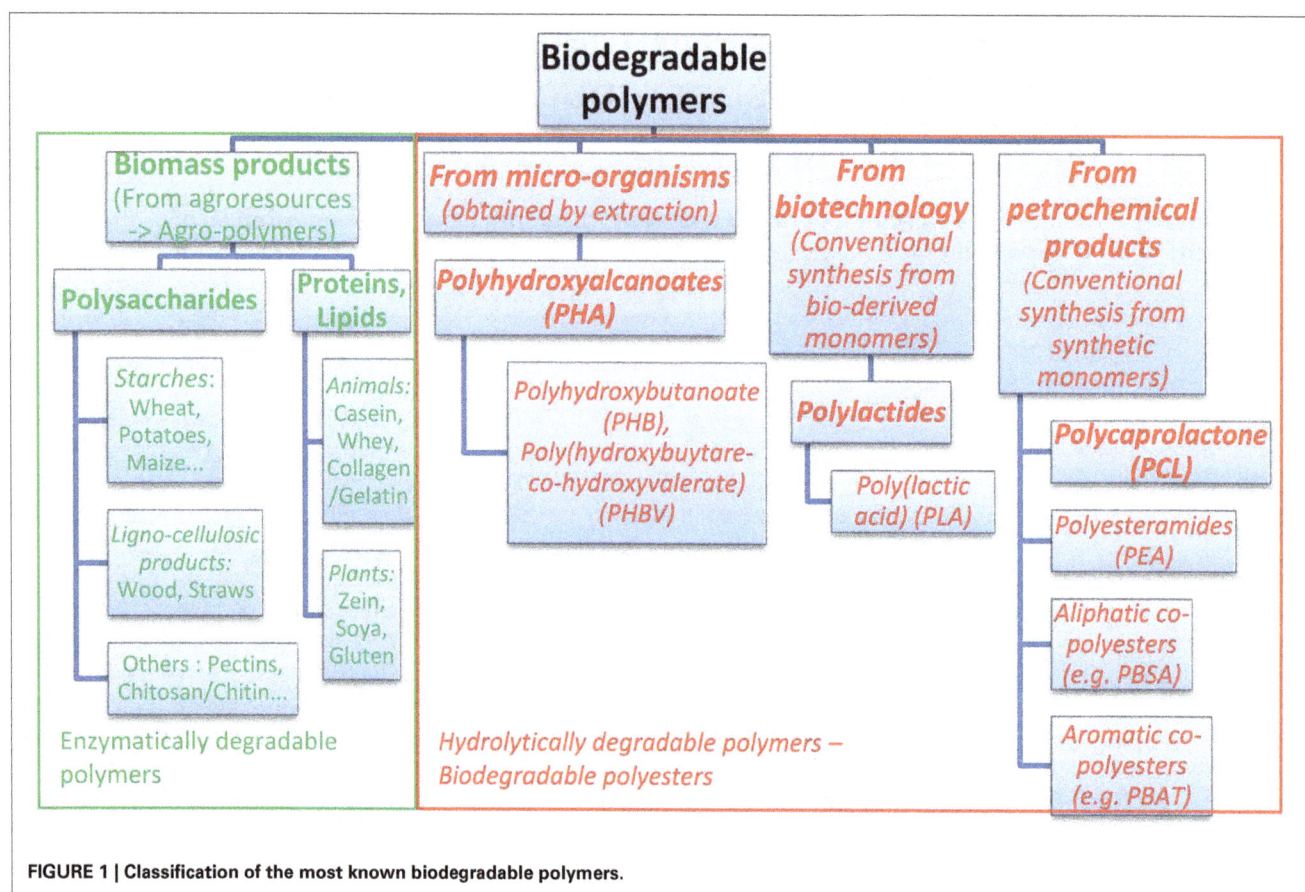

FIGURE 1 | Classification of the most known biodegradable polymers.

Table 1 | PLA mechanical properties compared to those of most common polymers used in commodity applications [Copyright ©(2011) Wiley and Sons; used with permission from Liu and Zhang (2011)].

	PLA	PET	PS	HIPS	PP
T_g (°C)	55–65	75	105	–	10
Tensile strength at break (MPa)	53	54	45	23	31
Tensile modulus (GPa)	3.4	2.8	2.9	2.1	0.9
Elongation at break (%)	6	130	7	45	120
Notched Izod impact strength (J/m)	13	59	27	123	27 (i-PP)
Gardner impact (J)	0.06	0.32	0.51	11.30	0.79
Cost ($/lb)[a]	1–1.5	0.70–0.72	0.99–1.01	1.01–1.03	1.15–1.17

PET, Poly(ethylene terephthalate); PS, Polystyrene; HIPS, High-impact polystyrene; PP, Polypropylene; i-PP, Isotactic polypropylene homopolymer.
[a] Cost cited from "Plastic News," March 31, 2011 except PLA resin.

purpose polystyrene (GPPS), PLA has not only comparable tensile strength and modulus, but also exhibits very similar inherent brittleness (see **Table 1**). However, despite its numerous advantages such as good optical, physical, mechanical properties (high flexural and tensile moduli and strengths), the inherent brittleness significantly impedes its applications in many fields when a high level of mechanical strength is required.

The mechanical resistance of a material is its ability to withstand the application of a sudden load without failure by dissipation of energy of the impact blow. There are two general failure modes, namely *brittle fracture* and *ductile fracture*. While *brittle fracture*, usually resulting of highly concentrated *crazing*, is characterized by a relatively low energy dissipation and a short nearly linear dependence of load–deformation before fracture, a *ductile fracture* is characterized by a high energy dissipation and a large-scale deformation (plastic yielding and plastic flow) (Bucknall, 1978; White, 1984; Argon and Cohen, 1990; Perkins, 1999). A *brittle-ductile transition* is accordingly defined as the point at which the fracture energy increases significantly with a mode of failure passing from brittle fracture to ductile fracture. The importance of this transition zone depends mainly on the strain nature and rate, the temperature gradient, and the specimen geometry (Perkins, 1999). For instance, the same material can exhibit higher brittleness at low temperatures and/or high testing speeds. Mechanical resistance of polymers may be evaluated in terms of the energy absorbed by the specimen during testing by various methods including (Pearson Raymond, 2000):

- *Tensile testing*: The area under the stress strain curve is often used to quantify toughness. However, different stress-strain curve shapes indicating different mechanical behaviors and responses to the impact loading may dissipate the same impact energy.
- *Impact testing*: The energy required to break the sample which is usually entailed by a hammer is measured. The related impact strength is expressed in terms of the difference between the potential energy of the striker before and after the impact. It is generally obtained by dividing the energy required to break the sample by the sample width or cross-sectional area. For impact testing, three different tests are typically performed such as Izod (ASTM D256 where samples are clamped as a cantilever), Charpy (ASTM D6110 unclamped samples are supported at both ends) and Dynstat (DIN 53453 where samples are unclamped at the lower end), which can be done in either a notched or un-notched state.
- Falling weight tests where a projectile propelled onto the specimen or dropped on it under the force of gravity is used to measure the impact energy. Gardner impact tester is a well-known example of this type of instrument which offers the advantage over impact testing method that the fracture shape can be also analyzed.

Typically, like conventional brittle thermoplastics, the reason for brittleness of PLA is strain and stress localizations at its use temperature, which is usually below its glass transition and brittle to ductile transition temperature. Under mechanical loading PLA deforms involving highly localized *crazing mechanism*. As at room temperature its yield stress is superior to the critical stress value for crack formation and propagation, catastrophic damage and break can most likely occur at low deformation and in the elastic zone. The strain-localization can be suppressed namely by compounding the brittle polymer with various softening and toughening agents including plasticizers and rubbery polymers or impact modifiers. However, the most preferred way is to blend PLA with rubbery polymers in order achieve a good toughness-stiffness balance without largely scarifying its glass transition temperature. Like many tough polymer blends, PLA blends can undergo one or a combination of the most known toughening mechanisms, namely *multiple crazing, shear yielding, cavitation and debonding* (Petchwattana et al., 2012). The mechanical energy is therefore transferred to the plastic flow and dissipated through a large volume fraction of material. The energy dissipation mechanisms retard or stop crack initiation and propagation through the polymer, and ultimately result in a material with improved toughness. There are several factors that can influence the amount of toughening, mainly related to the matrix polymer (Kramer, 1983), the rubber phase (type, particle size, concentration, strength and morphology), and the rubber-matrix interfacial interaction. For instance, the correlations between the deformation morphologies (mainly under tensile and impact testing) and the resulting mechanical properties reveal that the blend

compatibility and related morphologies are important factors to influence the toughening mechanisms. The toughening mechanisms can be analyzed through several aspects, including stress whitening, matrix ligament thickness, microstructure evolution under testing, and morphology features of the fracture surface of the impacted sample. For instance, when the matrix ligament thickness is below the critical value, the blends deform to a large extent because of shear yield initiated by stress concentrations and interfacial de-bonding. This may result in the formation of fibers in both tensile and impact samples and the dissipation of a large amount of energy (Han and Huang, 2011).

Many strategies, namely the incorporation of a variety of soft polymers or rubbers, addition of rigid fillers and fibers, and modification of crystalline morphology, have been developed in the literature during the last decades in order to enhance the general toughness of PLA, while maintaining its stiffness-toughness balance acceptable (Anderson et al., 2008; Liu and Zhang, 2011). An optimal toughness balance can be obtained with 10–30% of toughening agents, even if little improvement can be seen by the addition of 5–10% of the latter (Mascia and Xanthos, 1992; Anderson et al., 2008; Liu and Zhang, 2011). In this regard, blending represents an economically viable approach such as plasticization, (reactive) compounding with a variety of flexible/soft polymers or rubbers and the addition of rigid fillers. In this report, an update on the strategies recently developed in the literature to significantly and effectively improve PLA's mechanical properties, will be discussed on its toughening and impact resistance properties.

APPROACHES FOR THE IMPROVEMENT OF PLA'S MECHANICAL PROPERTIES BY MODIFYING ITS INHERENT CRYSTALLINE STRUCTURE

The impact strength of semicrystalline polymers usually varies inversely with the percent crystallinity (Mercier et al., 1965). It is likely that crystallites act as stress concentrators, causing the stress acting on a small volume of the material to be much greater than the average stress applied to the whole sample. As a result, the material breaks at a stress that is less than the expected critical value. Also, crystallites are seen to reduce multiple crazing and shear yielding (Pecorini and Hertzberg, 1993), both energy-dissipative mechanisms of polymer matrices. The size and number of these crystalline structures have a profound influence on impact resistance. It is generally agreed that impact resistance and the brittle to ductile transition temperature are inversely related to spherulite size and morphology which can be tuned by controlling the cooling and drawing rates via thermal and mechanical treatments, respectively (Hammer et al., 1959; Ohlberg et al., 1959; Barish, 1962). This part of the study concerns the PLA matrix itself. In this regard we will report the main approaches that tune up the relationships "physical treatments—crystalline structure—mechanical properties" in order to improve the mechanical properties of PLA-based materials.

THERMAL TREATMENTS—ANNEALING

The effect of annealing treatment on thermal, mechanical and fracture behavior of PLA was investigated. Most of the studies demonstrated that the increase of PLA crystallinity usually leads to; an improvement of its overall mechanical and heat resistance behaviors (Perego et al., 1996; Park et al., 2006; Yu et al., 2008; Nascimento et al., 2010). For instance, (Perego et al., 1996) evidenced that annealed PLA possess higher heat resistance, elastic moduli (tensional and flexural), Izod impact strength. Park et al. (2006) and Nascimento et al. (2010). Annealed PLA under different conditions to obtain several microstructures with varying spherulite size and density. They demonstrated that heat resistance is dramatically improved as crystallinity. Furthermore, the quasi-static fracture toughness of PLA decreases with increase of crystallinity corresponding to decrease of amorphous region; on the other hand, the impact fracture toughness tends to increase with crystallinity. The crack growth behaviors of the PLA specimens having different crystallinity exhibited that under quasi-static loading, disappearance of multiple crazes in the crack-tip region results in the decrease of the fracture toughness with crystallinity. On the contrary, under impact loading, the increase of the fracture toughness with crystallinity is considered to be related to the increase of fibril formation. Finally, for the amorphous PLA, the static toughness was higher than the impact one; mainly owing to extensive multiple craze formation at the static rate. On the contrary, for the crystallized PLA, the impact toughness became larger than the static one due to formation of fibril structure at the impact rate (Gamez-Perez, 2010). Gamez-Perez (2010) applied annealing treatment on two commercial grades of PLA from NatureWorks® (2002D and 4032D) of comparable average molecular weights (M_w) of 212 and 207 kDa, respectively, but they exhibited different optical purities, that is, d-lactic monomer contents of 4.25 and 2%, respectively (Natureworks®, 2005, 2006; Li and Huneault, 2007; Xiao et al., 2009; Carrasco et al., 2010). Annealing the sheets was performed using an oven at 60°C for 20 min, followed by a rapid quenching. The nomenclature employed was "PLA-X" and "PLA-XT" for extruded and thermally treated films, respectively. "X" is set as 96 and 98 for PLAs for a content of 95.75 and 98% l-lactic monomer, respectively. From **Table 2**, it results that the heating at temperatures close to the glass transition temperature (T_g) with the subsequent quenching treatment produces a "de-aging effect," with an increase of the free-volume of polymeric chains, as highlighted by the decrease of the T_g. The increase in the system potential energy was also shown by the disappearance of the endothermic peak at T_g. As a consequence, annealing promotes a brittle-to-ductile change in the fracture behavior of PLA with a decrease of the tensile strength and stiffness and yield stress, regardless the d-lactic isomer content. A shear yielding with a localized neck formation thereby appeared. The fracture parameters, assessed by the EWF method used to characterize the fracture toughness of PLA showed a great enhancement of the toughness after the annealing and quenching treatments. Regarding the influence of the D-lactic isomer content in PLA films, when they were in a glassy stage, no remarkable differences were noticed out in the mechanical properties and fracture behavior. Only when the films were in a de-aged form, the differences in the stiffness of both PLA grades had been revealed. The optical purity, the elastic modulus and the tensile strength were high. However, the deformation to break was still low, only passing from 17% (PLA-98) to 24% (PLA-96).

THERMOMECHANICAL TREATMENTS—SELF-REINFORCING POLYMERIC MATERIALS PROCEDURES (SRPMs)—ALIGNMENT AND ORIENTATION PROCEDURES

Although polymeric composites are referred to as multi-phase or hetero-composites, self-reinforced polymeric materials (SRPMs) are referred to as single-phase or homo-composites because

Table 2 | Effect of some (thermo)mechanical treatments and processing on the thermal and mechanical properties of PLA.

Material	T_g(°C)	T_m(°C)	ΔH_{cc}(J/g)	ΔH_m(J/g)	X_c(%)	Yield stress σ_y (MPa)	Young's modulus E (GPa)	Elongation at break ε_b (%)	Charpy (KJ/m²)	Izod impact (KJ/m²)	References
PLA-96	60	148	–	1	1	56.2 ± 0.7	4.0 ± 0.2	24 ± 5			Gamez-Perez, 2010
PLA-98	61	164	29	31	2	58.4 ± 0.5	4.3 ± 0.1	17 ± 4			
PLA-96T	56	148	–	1	1	47.3 ± 1.1	3.3 ± 0.2	456 ± 100			
PLA-98T	57	165	31	34	3	53.4 ± 0.6	3.5 ± 0.3	422 ± 50			
Un-oriented						47.0	3.65	1.5	12.5	1.6	Grijpma et al., 2002
Oriented (λ = 2.5)						73.3	4.49	48.2	35.9	5.9	
Oriented (λ = 3.4)						66.3	3.74	21.8	No break	52.0	
PLA-I						65.6 ± 1.3	3.7 ± 0.1	4.0 ± 0.8			Carrasco et al., 2010
PLA-EI						65.2 ± 0.9	3.9 ± 0.1	5.4 ± 0.6			
PLA-IA						75.4 ± 0.9	4.1 ± 0.1	2.5 ± 0.2			
PLA-EIA						77.0 ± 1.1	4.1 ± 0.1	3.3 ± 0.3			

I, Injected; IA, Injected then Annealed; EI, Extruded then Annealed; EIA, Extruded then Injected then Annealed.

the same polymer forms both the reinforcing and the matrix phases. The basic concept of self-reinforcement is to create a one-, two- or three-dimensional alignment (1D, 2D, or 3D alignment, respectively) within the matrix to fulfill the role of matrix reinforcement. As a result, the generated structure has to possess a higher stiffness and strength than the matrix as well as to be "well-bonded" to the matrix polymer. Consequently, the stress can be transferred from the "weak" matrix to the "strong" reinforcing structure, according to the "working principle" of all composites. The reinforcing structure can be produced during one (*in situ*) or more processing steps (*ex situ*) (Kmetty et al., 2010). A driving force for SRPMs is the possibility of manufacturing lightweight parts and structures because the density of SRPMs is well-below those of traditional filled polymers, where the "heavier" reinforcements incorporated in the polymeric matrix are of, e.g., glass fibers (density: 2.5–2.9 g.cm^{-3}), carbon fibers (density:1.7–1.9 g.cm^{-3}), basalt fibers (density: 2.7–3.0 g.cm^{-3}), aramid fibers (density: 1.38–1.44 g.cm^{-3}) and/or fillers like talc (density: 2.7–2.8 g.cm^{-3}), chalk (density: 1.1–2.5 g.cm^{-3}) and silica (density: 2.1–2.6 g.cm^{-3}) (Kmetty et al., 2010). Furthermore, the ease of recycling SRPMs must be emphasized when reprocessing via re-melting is targeted. The concepts used to produce SRPMs can be also adapted to biodegradable polymers for improving their property profiles. Reinforcing a PLA matrix by embedding PLA fibers enables to respond the demands for high strength and stiffness required for many applications. The development of high-stiffness and high-strength polymeric fibers is essential to imparting superior mechanical properties for the resulting PLA SRCs (Matabola et al., 2009). The mechanical properties of fibers can be increased via molecular orientation during spinning and drawing (Alcock et al., 2006). The most commonly used methods to produce PLA fibers are melt-spinning and electro-spinning (Mäkelä et al., 2002; Tsuji et al., 2006; Li and Yao, 2008). Significantly improved interfacial bonding can be achieved in materials where both matrix and reinforcing elements have the same chemical structure (Törmälä, 1992). For example, SRCs consisting of oriented PLA fibers surrounded by a PLA matrix have improved strength and rigidity compared to non-reinforced PLA (Tormala et al., 1988; Majola et al., 1992; Wright-Charlesworth et al., 2005).

To control the impact performances, molecular orientation of amorphous poly(D,L-lactide) (PDLLA) chains was carried out through injection moulding techniques at $T < T_g$ or by non-conventional shear controlled orientation by injection moulding (SCORIM) process in which the melt is cooled under oscillating shear conditions. The latter allowed getting oriented PLA-based materials, leading to the elaboration of degradable devices with much improved mechanical properties compared to non-oriented materials (Grijpma et al., 2002). The brittle fracture mechanism of PDLLA via crazing changed from a fragile to a ductile energy dissipation mechanism upon orientation. Consequently, a significant increase in impact strength was obtained. In comparison to the brittle tensile behavior of un-oriented PDLLA, a much more ductile behavior was observed. This increase in toughness was not accompanied by a decrease in tensile strength and stiffness, as it is generally in the case of

plasticization and rubber modification. Due to orientation of the polymer chains in the direction of testing, fibrillation took place during the fracture process. Growing cracks got stopped in the anisotropic structure, and catastrophic failure could be postponed. The mechanical data are summarized in **Table 2**. However, in the perpendicular direction to the orientation, mechanical properties are much poorer and must be taken into account. The effects of operative SCORIM parameters were also investigated. The correlations between processing, morphology and mechanical properties of SCORIM-moulded PLLA were established and compared with conventional injection moulded CIM PLLA (Ghosh et al., 2008). The level of molecular orientation was assessed indirectly by hot recoverable strain HR test. The fracture surface-morphology assessed by optical microscopy and SEM technique showed that, at low mould temperature, the level of molecular orientation increased with shearing time. The SCORIM processing changed the typical heterogeneous skin–core morphology of CIM into a near homogeneous oriented structure. The extent of core-fibrillation increased with shearing time. Under the three-point flexural test, the higher oriented PLLA exhibited dual fractures where the crack initiation started in the skin and transferred to oriented core fraction without decreasing the modulus. At high mould temperatures, the orientation increased steadily with shearing time. However, the level of molecular orientation was lower than the corresponding low mould temperature conditions. The orientation of core-fraction increased steadily with shearing time. Depending on the level of molecular orientation, the SCORIM-processed PLLA products showed four distinct types of fracture surfaces under three-point flexural test: (i) the un-oriented core failed through crazing; (ii) the sub-skins failed either in smooth, rough or fibrillated fracture surfaces depending on the level of molecular orientation; (iii) the less oriented core failed with fibrillation through pronounced plastic deformation; and (iv) the highly oriented skins failed with smooth surface. All the SCORIM-processed PLLA exhibited higher toughness and higher maximum stress compared with conventional injection-moulded PLLA (**Table 2**). The overall increments in maximum stress and toughness were of 134% and 641%, respectively. The increase in maximum stress and toughness were higher in low mould temperatures (30°C) in contrast to high mould temperature temperatures (50°C). Unlike the traditional blending technique, the increments in mechanical performances were achieved without sacrificing the stiffness. The mechanical behavior namely toughness and maximum stress of PLLA processed by SCORIM could be tailored by controlling the melt stage, the in-mould shearing time and the cooling conditions. In another study (Bigg, 2005), biaxial orientation of PLLAs chains by extrusion induced a 5–10-fold increase in elongation and enhanced tensile strength at break, tensile toughness and tensile modulus (**Table 2**). The mechanical processing of PLA (injection and extrusion/injection) as well as annealing of processed materials were studied in order to analyse the variation of its chemical structure, thermal degradation and mechanical properties (Carrasco et al., 2010). Processing of PLA yielded a decrease of its molecular weight and melt-viscosity due to chain hydrolysis. PLA crystal structure was significantly recovered after annealing. The authors also confirmed by proton NMR techniques that the chemical composition of PLA did change after processing, and the proportion of methyl groups from PLA matrix increased, more likely indicating the presence of a different molecular environment. The mechanical behavior was altered as well (**Table 2**). After annealing, the samples showed an increase in Young's modulus (5–11%) and in yield strength (15–18%), which had been explained by the higher degree of crystallinity of annealed materials, with its subsequent decrease in chains mobility. Extruded/injected materials showed a significant increase in elongation at break (32–35% higher), compared to injected materials. It is ascribed to the presence of low molecular-weight chains at high contents, due to hydrolysis reactions in reprocessed materials.

In general, the modification of chain orientation and crystallinity for PLA-based materials can improve its ductility and impact resistance to some extent. Some processing techniques may contribute efficiently to toughening PLA, without compromising its tensile properties. Orientation of chains by injection moulding and especially injection moulding with macroscopic oscillating shear force resulted, for instance, in an enhancement of tensile, Izod and Charpy impact in the orientation direction. In order to increase the crystallinity of PLA blends and therefore tune its mechanical properties, some routes may be considered (Battegazzore et al., 2011):

– *By chain orientation under stress;*
– *By applying thermal treatments (quenching and/or annealing);*
– *By minimizing the amount of the other lactide and mesolactide in the lactide used as the major monomer.* The crystallinity and crystallization rate of PLA decrease as the purity decreases. The crystallization half-time was found to increase by roughly 40% for every 1 wt.% increase in the mesolactide content of the polymerization mixture (Kolstad, 1996). In addition, it is known that a co-monomer content higher of 7 wt% with polymeric chains leads to an amorphous polylactide.;
– *By playing with the moulding conditions, in particular moulding temperature and cooling time.* Even at high L-lactide content, PLA crystallization is typically too slow to develop significant crystallinity unless it is induced by strain like processes used to manufacture bottles. In processes such as injection moulding, where the orientation is limited and the cooling rate is high, it is much more difficult to develop significant crystallinity and therefore formulation or process changes are required.;
– *By adding nucleating agent.*

Nevertheless, these techniques are not very industrially considered because they require increasing the processing time. In addition, studied alone, their influences are usually marginal and the resulting increase of toughness properties is insufficient [but sometimes quite enough because excellent stiffness-toughness balance was achieved in some cases (Gamez-Perez, 2010)] to satisfy the requirement of most applications. However, the combination of these factors with others such as compounding strategies (that will be discussed further) may bring more added-values in terms of the enhancement of PLA's mechanical properties and constitute more prospective routes to improve them.

APPROACHES TO INCORPORATING SOFT COMPONENTS INTO PLA MATRIX VIA COMPOUNDING/BLENDING

Blending polymers is as old as the polymer industry itself. Interestingly, using blending approach, PLA can be readily impact-modified, plasticized, filled, chemically modified and reactive blended and processed like many of other conventional polymers. There are two main ways to improving the ductility and the toughness of PLA materials namely through plasticization or incorporation of soft/rubbery polymers. Plasticization makes possible to achieve improved processing behaviors for polymeric materials, while providing better flexibility in the end-use product. As far as blending is concerned, blending PLA with immiscible polymers produces a new type of polymeric materials with different properties, in which each polymeric partner provide its own feature. Because of their impact-absorbing ability when well-dispersed with the convenient particle size distribution, rubbers should act as stress concentrators at many sites throughout the material. Therefore, they impart great ductility and impact strength to the material, resulting from dissipative micromechanisms initiated by the rubber particles. All of these phenomena are dependent on the deformation, toughening and fracture mechanisms, namely crazing, shear yielding, cavitation, or debonding as mostly reported in the literature (Kambour, 1973; Michler, 1989; Wu, 1990; Könczöl et al., 1992; Ikeda, 1993; Dompas et al., 1994; Lu et al., 1997; O'Connell and McKenna, 2002; Narisawa and Yee, 2006; Bucknall, 2007; Seelig and Van Der Giessen, 2009):

1. *Crazing mechanism* can be initiated in a material when the stress or hydrostatic tension is locally concentrated at a defect which can be a notch, voids, in-homogeneities or rubber particles. Therefore, interpenetrating micro-voids and highly drawn elongated micro-fibrils called *tufts* (usually a fraction of 1 μm in length, depending on the molecular weight of a polymer, several nanometers in diameter, and confined to a small volume of the material), are formed giving rise to macroscopic highly localized zones of plastic dilatational deformation (Kramer and Berger, 1990). Under sufficient mechanical loading, the local stress exceeds a critical value. Thus, the micro-fibrils elongate until breaking and cause the micro-voids growth and coalescence turning into micro-cracks. Crazing mechanism is dilatational in nature and consumes the predominant part of fracture energy in the case of many thermoplastics. Accordingly, crazing is to some extent a precursor to macroscopically brittle failure and is view as a damaging mechanism in the case of brittle polymers when the craze evolution into a micro-crack cannot be arrested. However, when blended with the brittle matrix, the rubbery impact modifier particles have two separate effects but equally important features as a response to load application. They first concentrate locally the stress where craze initiation takes place. The crazes then grow perpendicularly to the maximum applied stress direction. In a second step, the surrounding rubber particles play the role of "craze terminators," preventing the generation of micro-cracks. The result is that a large number of small crazes are formed, in contrast with the small number of large crazes (micro-cracks) within the same polymer

matrix in the absence of rubber particles. This multiple crazing that occurs throughout a comparatively large volume of rubbery-modified material explains the high energy absorption in fracture tests and the extensive stress whitening that accompanies deformation and failure (Perkins, 1999). Some matrices tend to craze because of low entanglement density while high molecular weight is needed to stabilize crazes.

- *Shear yielding mechanism* is highly localized plastic deformation characterized by appearance of oriented shear bands under uniaxial tension at 45°C to the direction of the applied stress. Shear yielding occurs approximately at constant volume while initiation of shear bands is affected by the hydrostatic tension (mean stress). In ductile polymers, shear-yielding is usually the major energy absorbing mechanism. There are also few polymers such as acrylonitrile butadiene styrene (ABS) and rubber-toughened PMMA that exhibit both shear yielding and crazing mechanisms. When the craze initiation stress of the matrix is lower than the yield stress, a polymer will tend to craze; if the yield stress is lower than the craze initiation stress, the matrix will fail by shear-yielding. Mixed crazing and shear yielding tends to occur when the craze initiation stress and the yield stress are comparable or when interactions occur between crazes and shear bands.

- *Cavitation* is void-expansion, which can occur in the matrix (generally coupling with crazing) or initiate inside the rubber particles, which is generally characterized by viewing stress-whitening zones. The essential conditions for void growth is an energy balance between the strain energy relieved by cavitation and the surface energy associated with the generation of a new surface. Cavitation is a precursor to other toughening mechanisms, thereby relieving the hydrostatic strain energy and initiating shear yielding of the matrix. It is assumed that internal rubber cavitation is an instantaneous process, which cannot occur for very small particles (less than 200 nm). In other words, rubber-cavitation mechanism is favored by increasing the particle size within rubber toughening materials or by decreasing the crosslinking density (which can suppress cavitation).

- *De-bonding* is the energy-dissipation due to the interfacial failure. The interface between the phases influences the final blends properties by efficient stress transfer between the two phases. However, interfacial de-bonding can be thought of as a secondary toughening mechanism being more important as a trigger for other induced mechanisms like shear yielding. Accordingly, low interfacial adhesion easily results in premature interfacial failure and hence rapid and catastrophic crack propagation, whereas very strong adhesion is unfavorable for de-bonding and also delays the occurrence of matrix yielding, involving the matrix-particle interface as an important factor that we need to control for optimum energy dissipation.

Toughening mechanisms and competition between both modes of fracture are mainly governed by a variety of factors such as mode of loading, environment, processing conditions, composition and behavior of the matrix, relaxation behavior of the dispersed phase, rubber content, blend morphology, rubber-matrix adhesion, etc. Being a suitable processing technique, reactive extrusion for instance, represents a unique tool to manufacture biodegradable polymers upon different types of reactive modification in a cost-effective polymer processing (Michaeli et al., 1993; Mani et al., 1999). This technique enhances the commercial viability and cost-competitiveness of polymer materials, in order to carry out not only melt blending, but also chemical reactions including polymerization, grafting, compatibilization, branching, functionalization... (Michaeli et al., 1993; Mani et al., 1999). The *in situ* chemical modification of PLA by reactive extrusion has proven to be an effective promising way to elaborate tougher PLA-based materials with improved stiffness-toughness balance compared to neat PLA as it will be detailed later. Here, the forthcoming paragraphs will report the recent investigations on simple plasticization of PLA and blending PLA with rubbery/soft materials.

COMPOUNDING WITH PLASTICIZERS—MISCIBLE TO PARTIALLY MISCIBLE BLENDS

Plasticization is widely used to improve the polymers processability and/or other properties according to specific applications. Plasticizers can act by altering the intermolecular interactions among the host polymer chains to other interactions between the macromolecules and the plasticizer. This promotes conformational changes, resulting in increased mobility of plasticized chains. The Standard ISO427 (1988) define a plasticizer as being as a low or negligible volatility component, which is once incorporated to a plastic material, lowers its softening interval temperature, facilitates its processability and increases its flexibility and ductility. Its behavior can be explained by decreasing the viscosity of the molten plasticized polymer, the glass transition temperature and the elastic modulus of the plasticized materials. The evolution of the elongation at break can be also related to the ductility of a polymer and give information about the plasticization extent of polymers.

To be suitable with PLA, a plasticizer should fulfill the following characteristics (Liu and Zhang, 2011):

- To have an optimum molecular weight and loading level to be miscible with the polymer matrix. Miscibility of plasticizers in a polymer matrix is evaluated by solubility parameters (δ) and magnitude of interaction parameters (χ_T) (Pillin et al., 2006);
- Significantly lower the T_g of PLA and thus enhance tensile toughness;
- Preferably bio-sourced and biodegradable;
- Non-volatile;
- Non-toxic;
- Exhibit minimal even more negligible leaching/migration phase separation from the polymer matrix during ageing.

Many classes of plasticizers were reported by Liu and Zhang and will be discussed in the forthcoming part as follows (Liu and Zhang, 2011):

- Monomeric or small molecule plasticizers;
- Oligomeric and polymeric plasticizers;
- Mixed plasticizers.

In the review, a special emphasis is made on the impact behavior of plasticized PLA, because it has not received enough attention in the literature.

Monomeric or small molecule plasticizers

Many small molecules/monomeric plasticizers have been studied in order to evaluate their plasticization efficiency and their influence on the overall physical properties of PLA (**Table 3**). The optimal plasticizer content has to take into account the molecular weight, solubility δ and interaction parameters χ_T. For instance, most of the studies showed that between 10 and 20 wt.% of plasticizer content in PLA, all studied citrate esters (TEC, TBC, ATEC, ATBC) results in a higher elongation and lower T_g for the as-plasticized PLA materials compared to neat PLA.

Among the monomeric or small molecule plasticizers studied in the literature, lactide monomer (LA) possesses the best plasticization efficiency for PLA. However, due to its low molecular weight compared to the others, lactide tends to migrate toward the PLA surface. Therefore, the toughness of plasticized PLA tends to be reduced with time (Jacobsen and Fritz, 1999). LA can also volatilize during melt processing because of its low boiling point ($\sim$120°C). In terms of good stiffness-toughness balance, Dioctyl adipate (DOA) seems to be the most efficient one by significantly enhancing elongation with a slight depression of tensile modulus (Martino et al., 2009). The plasticizing efficiency of ATBC was higher compared to the others citrate-based plasticizers. Generally, the miscibility of plasticizers with a polymer decreases with increasing molecular weight of the plasticizers. Small molecule plasticizers are usually more efficient than larger ones in order to lower the host polymer's T_g because the mixing entropy is higher in the case of low M_w plasticizers. However, because of their low boiling point, small molecule plasticizers usually evaporate during melt processing (Labrecque et al., 1997; Ljungberg and Wesslén, 2003; Ljungberg et al., 2003; Martino et al., 2009) and have also a strong tendency to migrate toward the surface of the polymeric material (Ljungberg and Wesslén, 2003; Ljungberg et al., 2003, 2005; Martino et al., 2006). The driving force of the migration is ascribed to the enhanced crystallization ability of plasticized samples. Consequently, the ability of PLA to accommodate the plasticizer in the amorphous PLA phase diminishes (Ljungberg and Wesslén, 2002; Ljungberg et al., 2003, 2005; Martino et al., 2006; Pillin et al., 2006). In addition to the loss of the material toughness (plasticized PLA regains part of the brittleness of neat PLA); the plasticizer migration can, for example, contaminate the food or beverage in contact with plasticized PLA in food packaging applications. All monomeric plasticizers should be added in the range of 5–25% (depending on the plasticizer itself) in order to reduce the migration to the maximum, to maintain the optimum balance between tensile modulus, strength and elongation at break and reduce significantly the glass transition temperature of the host polymer. However, monomeric plasticizers cannot fulfill these requirements due to their high tendency to migrate and evaporate. In this regard, researches had been more widely focused on oligomeric and polymeric plasticizers.

Oligomeric and polymeric plasticizers

The common way to reduce plasticizers' migration and evaporation is to increase their molecular weight in such a way to retain their miscibility with the polymer matrix at the same time. In this respect, many researchers have investigated the effect of some oligomeric and polymeric molecules as plasticizers for PLA (**Table 4**).

For 20 wt% plasticizer content, ABA-type block copolymer of PDLLA and PEG400, that is, PDLLA-b- PEG400-b-PDLLA (10/2, molar ratio of D,L-LA monomer to PEG400 used in the feed) (COPO3) and poly(propylene glycol) (PPG720) provide a good stiffness-toughness balance. PPGs 425, 600 and 1000, Glyplast® 206/2 and Glyplast® 206/7 have a better plasticizing efficiency compared to the others. Adipates-based plasticizers are miscible with PLA until a critical concentration reached in function of the molar mass of adipate. A remarkable increase in elongation was achieved when the concentration of plasticizer reached 10 wt%, whereas the decreases in elastic modulus and tensile stress were noted for all the plasticizers investigated. Very recently, it has been shown that PLA can be efficiently plasticized and toughened by melt-blending with poly (1,2-propylene glycol adipate) (PPA) (Zhang et al., 2013b). Thermal and dynamic mechanical analysis revealed that PPA was partially miscible with PLA. In addition, morphological investigation of PLA/PPA blends showed that PPA was compatible with PLA. As a result, with the increase of PPA content (5–25 wt%), the blends showed a decrease in the tensile strength and the Young's modulus (**Table 4**); but the elongation at break and the impact strength dramatically increased due to the plastic deformation. The Izod notched impact strength reached 90 J/m when the PPA amount was of 20 wt%, and even exceeded 100 J/m when PPA amount was of 25 wt%. The plasticization effect of PPA was also highlighted by the lowering of dynamic storage modulus and viscosity in the melt stage of the blends compared with neat PLA. In another recent study, Gui et al. have successfully toughened PLA by melt-blending with poly(ethylene glycol-co-citric acid) (PEGCA) (Gui et al., 2013). The addition of PEGCA to PLA lowered the viscosity and the glass transition temperature of the resulting material. PEGCA was partially miscible in PLA and the blends exhibited a phase-separated morphology. The ductility and toughness of PLA were significantly improved in the presence of PEGCA. Whereas the impact resistance (**Figure 2C**) and the elongation at break (**Figure 2B**) of the blends were remarkably higher than those of neat PLA, the tensile and flexural strength and modulus of the blends (**Figure 2A**) monotonically dropped with increasing PEGCA content.

Hassouna et al. investigated new plasticization ways based on low molecular bio-plasticizers to improve the ductility of PLA. Grafting reactions between anhydride-grafted-PLA (MA-g-PLA) copolymer with hydroxyl-functionalized citrate plasticizer, i.e., tributyl citrate (TbC) (Hassouna et al., 2012), and poly(ethylene glycol) (Hassouna et al., 2011) were carried out through reactive extrusion. All plasticizers drastically decreased the T_g of PLA due to the mobility gained by the polymer chains within the plasticized blends. Regardless the nature of the plasticizer, the elastic modulus and yield stress decrease, while the ultimate strain increases for plasticized PLA. Very recently, we have investigated a novel and efficient pathway to

Table 3 | Plasticization effects on PLA with some monomeric plasticizers [Copyright ©(2011) Wiley and Sons; used with permission from Liu and Zhang (2011)].

Polymer	Molecular weight (g/mol)	Plasticizer	Molecular weight (g/mol)	Plasticizer content (wt%)	Solubility δ (MPa)^0.5	Interaction parameter χT	DSC Tg (°C)	Max. tensile strength (MPa)	Tensile modulus E (MPa)	Tensile elongation at break ε (%)	Notched impact strength (kJ/m²)	References
PLA	-	LA	M_n = 144	1.3	-	-	-	51.7	1993	3	-	Sinclair, 1996
				17.3	-	-	-	15.8	820	288	-	
				19.2	-	-	32–40	29.2	658	536	-	
				25.5	-	-	-	16.8	232	546	-	
PLA	M_w = 137000	-	-		-	-	59.1	51.7	-	7	-	Labrecque et al., 1997
		TEC	M_n = 276	10	19.8	-	42.1	28.1	-	21.3	-	
				20	19.8	-	32.6	12.6	-	382	-	Anderson et al., 2008
				30	19.8	-	-	σ_b = 7.2	-	610	-	Labrecque et al., 1997
		TBC	M_n = 360	10	18.8	-	40.4	22.4	-	6.2	-	
				20	18.8	-	17.6	7.1	-	350	-	
		ATEC	M_n = 318	10	19.6	-	50.8	34.5	-	10	-	
				20	19.6	-	30.0	9.6	-	320	-	
		ATBC	M_n = 402	10	18.7	-	25.4	177	-	2.3	-	Anderson et al., 2008
				15	18.7	-	-	σ_b = 21.3	100	300	-	Labrecque et al., 1997
				20	18.7	-	17.0	9.2	-	420	-	
PLA[a]	M_w = 204450	-	-	-	-	-	58.01	41.69	3364	1.27	-	Sierra et al., 2010
		TBC[a]	M_n = 360	12.44	19.6[d]	-	41.12[b]	38.27	2542	2.14	-	
				12.99		-	37.61[b]	12.44	1248	>100.23	-	
				15.58		-	32.1[b]	6.39	53	>100.16	-	
				22.52		-	21.74[b]	9.84	55	>102.77	-	
PLA[c]	M_w = 204450	-	-	-	-	-	-	41.69	3364	1.27	-	
		TBC[c]	M_n = 360	12.44	9.6[d]	-	39.45	41.95	2301	3.43	-	
				12.99		-	36.97	19.97	1337	>99.98	-	
				15.58		-	30.37	7.67	313	>100.03	-	
				22.52		-	33.95	11.95	159	>100.08	-	
PLA	M_n = 84000	-	-	-	-	-	60	66	3300	1.8	-	Baiardo et al., 2003
		ATBC	M_n = 402	10	-	-	41	50.1	2900	7	-	
				20	-	-	24	23.1	100	298	-	

(Continued)

Table 3 | Continued

Polymer	Molecular weight (g/mol)	Plasticizer	Molecular weight (g/mol)	Plasticizer content (wt%)	Solubility δ (MPa)^0.5	Interaction parameter χT	DSC Tg (°C)	Max. tensile strength (MPa)	Tensile modulus E (MPa)	Tensile elongation at break ε (%)	Notched impact strength (kJ/m²)	References
PLA	M_n = 74500	–	–	–	20.1	–	62	66 ± 2	1020 ± 100	11 ± 3	2.6	Murariu et al., 2008a
		ATBC	M_n = 402	10	19.2	0.46	44	51 ± 1	970 ± 70	11 ± 4	2.4	
				20	19.2	0.46	38	30 ± 1	270 ± 20	317 ± 4	4.6	
		DOA	M_n = 370	10	17.6	1.33	45	29 ± 2	720 ± 90	36 ± 5	2.6	
				20	17.6	1.33	45	21 ± 1	670 ± 120	78 ± 33	28.7	
		GTA	M_n = 218	10	20.1	0.34	48	38 ± 3	760 ± 140	8 ± 2	2.7	
				20	20.1	0.34	29	24 ± 1	10 ± 3	443 ± 13	No Break	
PLA	M_w = 74000	–	–	–	23.1	–	59.2	64.0 ± 1.5	2840 ± 50	3.0 ± 0.3	–	Pillin et al., 2006
		DBS	M_w = 314	10	17.7	3.7	39.9	39.2 ± 4.0	2000 ± 80	2.3 ± 0.2	–	
				20	17.7	3.7	−66.9/26.1	23.1 ± 0.9	430 ± 50	269.0 ± 6.0	–	
		AGM	M_w = 316	10	18.5	1.5	45.8	52.1 ± 4.0	2240 ± 100	32 ± 2.1	–	
				20	18.5	1.5	−65.8/24.3	27.1 ± 3.1	35 ± 5	335.0 ± 2.3	–	
PLA	M_n = 63000	–	–	–	19.93	–	58.2	47 ± 5	2000 ± 200	6 ± 2	–	Martino et al., 2009
		DOA	M_n = 371	10	16.67	–	40.8	27 ± 4	1600 ± 100	259 ± 64	–	
				20	16.67	–	40.1	17 ± 1	1400 ± 100	295 ± 89	–	
PLA	M_n = 81000	–	–	–	–	–	61	52 ± 2	1800 ± 150	6 ± 1	2.6 ± 0.2	Lemmouchi et al., 2009
		TBC	M_n = 360	20	–	–	20	20 ± 1	9 ± 1	320 ± 20	No Break	

[a] Material aged for 10 days.
[b] Material aged for 20 days.
[c] Material aged for 24 days.
[d] (Ljungberg and Wesslén, 2003).

LA, Lactide; TEC, Triethyl citrate; TBC, Tributyl citrate; ATEC, Acetyl triethyl citrate; ATBC, Acetyl tributyl citrate; DOA, Dioctyl adipate (Bis(2-ethyldhexyl) adipate); GTA, Glycerin triacetate (Triacetin); DBS, Dibutyl sebacate; AGM, Acetyl glycerol mono-laurate; σ_b, Tensile strength at break.

Table 4 | Table molecular weight and solubility parameters (δ) of some oligomeric and polymeric plasticizers and their Plasticization effects on PLA [Copyright ©(2011) Wiley and Sons; used with permission from Liu and Zhang (2011)].

Polymer	Molecular weight (g/mol)	Plasticizer	Molecular weight (g/mol)	Plasticizer content (wt%)	Solubility δ (MPa)$^{0.5}$	Interaction parameter χ_T	DSC T_g (°C)	Max. tensile strength (MPa)	Tensile modulus E (MPa)	Tensile elongation at break ε (%)	Notched impact strength (kJ/m^2)	References
PLA	$M_v = 49000$	–	M	0	–	–	58	–	2050 ± 44	9 ± 2	–	Martin and Avérous, 2001
		OLA	–	10	–	–	37	–	1256 ± 38	32 ± 4	–	
				20	–	–	18	–	744 ± 22	200 ± 24	–	
		PEG400	$M_w = 400$	10	–	–	30	–	1488 ± 39	26 ± 5	–	
				20	–	–	12	–	976 ± 31	160 ± 12	–	
		M-PEG	$M_w = 400$	10	–	–	34	–	1571 ± 51	18 ± 2	–	
				20	–	–	21	–	1124 ± 33	142 ± 19	–	
PLA	$M_n = 84000$	–		0	–	–	60	66	3300	1.8	–	Baiardo et al., 2003
		PEG400	$M_w = 400$	10	–	–	23	32.5	1200	140	–	
				20	–	–	19	15.6	500	71	–	
		PEG1500	$M_w = 1500$	10	–	–	42	46.6	2800	5	–	
				20	–	–	20	21.8	600	235	–	
		PEG10K	$M_w = 10000$	10	–	–	42	48.5	2800	2.8	–	
				20	–	–	34	22.1	700	130	–	
PLA	$M_w = 74000$	–		0	23.1	–	59.2	64.0 ± 1.5	2840 ± 50	3.0 ± 0.3	–	Pillin et al., 2006
		PBOH	$M_w = 2100$	10	21.3	2.3	47.6	56.3 ± 1.9	2350 ± 50	3.0 ± 0.1	–	
				20	21.3	2.3	–48.5/30.1	30.2 ± 1.1	350 ± 20	302.5 ± 32.0	–	
		PEG200	$M_w = 200$	10	23.5	0.0	35.8	30.0 ± 4.1	1700 ± 100	2.0 ± 0.6	–	
		PEG400	$M_w = 400$	10	22.5	0.1	37.1	39.0 ± 3.0	1920 ± 53	2.4 ± 0.3	–	
				20	22.5	0.1	–50.2/18.6	16.0 ± 0.3	630 ± 20	21.2 ± 2.3	–	
		PEG1000	$M_w = 1000$	10	21.9	0.5	40.2	39.6 ± 5.0	1970 ± 120	2.7 ± 0.3	–	
				20	21.9	0.5	–62.7/22.4	21.6 ± 0.4	290 ± 50	200.0 ± 12.5	–	
PLA*	$M_w = 160000$	PEG8000	$M_w = 8000$	0	–	–	58	53 ± 2	2200 ± 50	14 ± 1	–	Hu et al., 2003b
				10	–	–	36	23 ± 1	950 ± 30	200 ± 10	–	
				15	–	–	30	16 ± 1	630 ± 20	260 ± 10	–	
				20	–	–	21	5 ± 1	180 ± 20	300 ± 20	–	
		PEG8000[a]		30	–	–	9	–	5 ± 1	500 ± 20	–	
		PEG8000[b]		30	–	–	–	1 ± 0.2	40 ± 5	400 ± 10	–	
		PEG8000[c]		30	–	–	14+	2 ± 0.2	100 ± 10	340 ± 10	–	
		PEG8000[d]		30	–	–	–	7 ± 0.3	220 ± 20	300 ± 20	–	
		PEG8000[e]		30	–	–	22+	9 ± 0.3	370 ± 20	250 ± 10	–	
		PEG8000[f]		30	–	–	27+	9 ± 0.3	400 ± 20	240 ± 20	–	

(Continued)

Table 4 | Continued

Polymer	Molecular weight (g/mol)	Plasticizer	Molecular weight (g/mol)	Plasticizer content (wt%)	Solubility δ (MPa)$^{0.5}$	Interaction parameter χ_T	DSC T_g (°C)	Max. tensile strength (MPa)	Tensile modulus E (MPa)	Tensile elongation at break ε (%)	Notched impact strength (kJ/m²)	References
PLA*	$M_w = 190000$	PEG8000	$M_w = 8000$	0	–	–	60	68 ± 2	2500 ± 200	3 ± 0.5	–	Hu et al., 2003a
		PEG8000		10	–	–	39	26 ± 1	900 ± 50	180 ± 10	–	
		PEG8000		20	–	–	21	4 ± 0.5	150 ± 20	260 ± 20	–	
		PEG8000		30	–	–	12	–	20 ± 2	300 ± 30	–	
		PEG8000[g]		30	–	–	–	2.2 ± 0.3	80 ± 5	250 ± 20	–	
		PEG8000[h]		30	–	–	–	3.8 ± 0.3	140 ± 10	230 ± 20	–	
		PEG8000[i]		30	–	–	–	4.5 ± 0.6	150 ± 10	220 ± 10	–	
		PEG8000[j]		30	–	–	–	5.0 ± 0.2	170 ± 15	170 ± 20	–	
		PEG8000[k]		30	–	–	–	7.0 ± 0.3	240 ± 10	160 ± 10	–	
		PEG8000[l]		30	–	–	–	7.5 ± 0.3	250 ± 10	150 ± 15	–	
PLA	$M_n = 11800$	–		0	–	–	61	62.8	2249	4.5	–	McCarthy and Song, 2002
		PPG720	$M_w = 720$	10	–	–	51	52.2	1820	4.2	–	
		PPG720		20	–	–	38	39.9	1296	260	–	
		PPG640-E	$M_w = 640$	10	–	–	42	50.5	2170	4.2	–	
		PPG640-E		20	–	–	26	28.5	4.4	250	–	
PLA[j]	$M_n = 11800$	–		0	–	–	–	65.6	2338	4.0	–	Kulinski et al., 2006
		PPG720[j]	$M_w = 720$	10	–	–	–	54.9	1827	4.3	–	
		PPG720[j]		20	–	–	–	33.1	1404	260	–	
		PPG640-E[j]	$M_w = 640$	10	–	–	–	54.1	2081	3.9	–	
		PPG640-E[j]		20	–	–	–	28.2	9.9	260	–	
PLA	$M_w = 108000$	–		0	–	–	55.7	41.4 ± 1.5	–	64 ± 42	–	Martino et al., 2009
		PPG425	$M_w = 530$	10	–	–	33.1	21.0 ± 1.5	–	524 ± 66	–	
		PPG1000	$M_w = 1123$	10	–	–	34.0	23.1 ± 0.9	–	473 ± 111	–	
		PPG600	$M_w = 578$	10	19.93	–	31.3	18.5 ± 1.2	–	427 ± 42	–	
PLA	$M_n = 63000$	–		0	–	–	58.2	47 ± 5	2000 ± 200	6 ± 2	–	
		Glyplast® 206/2	$M_n = 1532$	10	21.91	–	39.5	34 ± 2	1600 ± 100	5 ± 1	–	
				20	21.91	–	25.4	25 ± 4	200 ± 100	485 ± 65	–	
		Glyplast® 206/7	$M_n = 2565$	10	22.87	–	42.1	36 ± 2	1700 ± 200	7 ± 5	–	
				20	22.87	–	30.6	28 ± 2	500 ± 100	491 ± 34	–	
		Lapol108	$M_n = 30000-40000$	5	–	–	–	77-84	2160-2313	160-200	–	Lapol, 2009

(Continued)

Table 4 | Continued

Polymer	Molecular weight (g/mol)	Plasticizer	Molecular weight (g/mol)	Plasticizer content (wt%)	Solubility δ (MPa)$^{0.5}$	Interaction parameter χT	DSC T_g (°C)	Max. tensile strength (MPa)	Tensile modulus E (MPa)	Tensile elongation at break ε (%)	Notched impact strength (kJ/m²)	References
		PEG	$M_n = 18500$	10	—	—	—	57–59	1700–1786	180–210	—	Chen et al., 1997
		LA-co-PEG	$M_n = 17600$	11	—	—	—	14.5 ± 2.8	423 ± 20	240 ± 21	—	
				20	—	—	—	24.1 ± 3.1	710 ± 21	204 ± 18	—	
PLA	$M_n = 81000$	—		0	—	—	61	52 ± 2	1800 ± 150	6 ± 1	2.6 ± 0.2	Lemmouchi et al., 2009
		COPO1	$M_n = 650$	20	—	—	29	21 ± 1	790 ± 180	170 ± 10	1.6 ± 0.6	
		COPO2	$M_n = 1000$	20	—	—	26	25 ± 1	300 ± 50	220 ± 20	8.3 ± 2.5	
		COPO3	$M_n = 1050$	20	—	—	36	30 ± 1	1700 ± 100	130 ± 20	1.9 ± 0.6	
		COPO4	$M_n = 1750$	20	—	—	35	24 ± 2	1150 ± 150	170 ± 10	1.9 ± 0.7	
PLA	$M_n = 121400$	PPA	$M_n = 1900$	0			58.6	69.8 ± 3.2	1777 ± 42	6 ± 0.3	~5	Zhang et al., 2013b
				5			49.3	63.2 ± 3.1	1392 ± 34	6 ± 0.5	~5	
				10			40.6	49.6 ± 2.9	1305 ± 29	157 ± 28	<10	
				15			33.3	39.8 ± 2.7	822 ± 26	315 ± 35	<10	
				20			27.0	25.7 ± 2.2	554 ± 21	362 ± 39	~90	
				25			24.3	14.4 ± 1.5	374 ± 17	410 ± 44	>100	

An update of the Table 4 has been made by Kfoury et al. in the frame of this review.

*with low stereoregularity, $^{+}T_g$ measured from E" in DMA

^{a} aging for 6 h.
^{b} aging for 30 h.
^{c} aging for 120 h.
^{d} aging for 500 h.
^{e} aging for 1800 h.
^{f} aging for 2 h.
^{g} aging for 10 h.
^{h} aging for 24 h.
^{i} aging for 120 h.
^{j} aging for 720 h.
^{k} aging for 3000 h.

LA, lactide; OLA, oligomeric lactic acid; PEG, poly(ethylene glycol); M-PEG, poly(ethylene glycol) monolaurate; PPG, polypropylene glycol), the subsequent number represents its nominal molecular weight; PPG-E, epoxy capped poly(propylene glycol); GlyplastVR 206/2 and GlyplastVR 206/7, two kinds of commercial polymeric adipates with molecular weight of 1532 and 2565 g/mol, respectively; COPO1 and COPO2, two kinds of AB-type block copolymers of DLLA and either PEG350 monomethyl ether or PEG750 monomethyl ether, that is, PDLLA-b-PEG350 (10/4, molar ratio of D,L-LA monomer to PEG350 used in the feed) and PDLLA-b-PEG750 (10/4, molar ratio of D,L-LA monomer to PEG750 used in the feed); COPO3, ABA-type block copolymer of PDLLA and PEG400, that is, PDLLA-b- PEG400-b-PDLLA (10/2, molar ratio of D,L-LA monomer to PEG400 used in the feed); COPO4, 3-star(PEG-b-PDLLA) block copolymer (10/1.3, molar ratio of D,L-LA monomer to 3-star-PEG used in the feed); COPO5, 4-star(PEG-b-PDLLA) block copolymer (10/1, molar ratio of D,L-LA monomer to 4-star-PEG used in the feed); PPA, Poly(1,2-propylene glycol adipate).

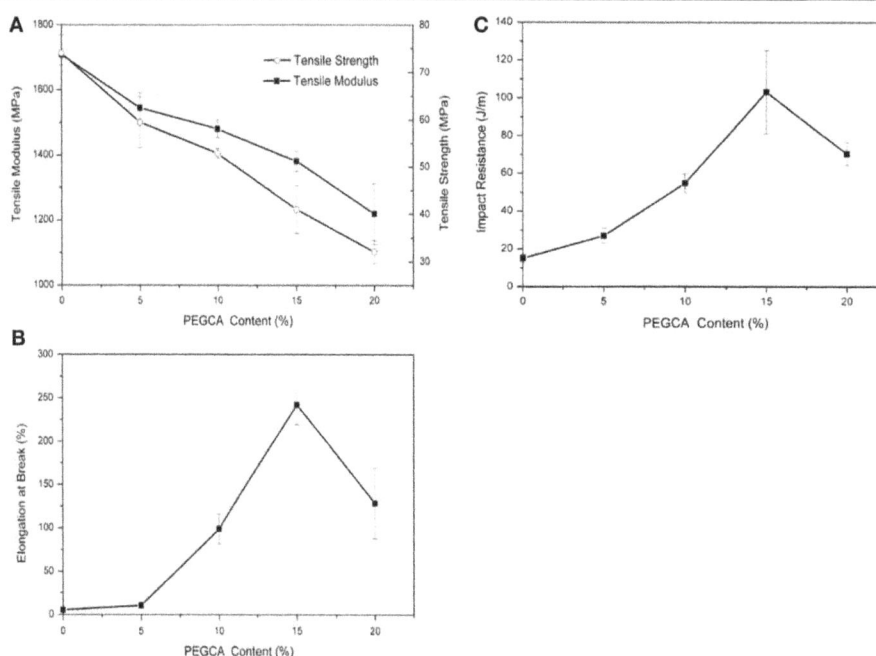

FIGURE 2 | (A) Tensile modulus, tensile strength, **(B)** elongation at break, and **(C)** of neat PLA and PLA/PEGCA blends (Gui et al., 2013, original copyright with kind permission from Springer Science and Business Media).

chemically modify PLA in the presence of "reactive" polyethylene glycol (PEG) derivatives via reactive extrusion (Kfoury et al. *Tunable and durable toughening of polylactide materials via reactive extrusion. Submitted*). In this purpose, polyethylene glycol methyl ether methacrylate (MAPEG) and polyethylene glycol methyl ether acrylate (AcrylPEG) were melt-mixed and extruded with PLA in the absence and in the presence of a free-radical di-tertiary alkyl peroxide, 2,5-dimethyl-2,5-di-(tert-butylperoxy)hexane (Luperox101 or L101). Molecular characterization revealed that in the case of PLA/MAPEG/L101 blends (79.5/20/0.5 wt/wt/wt), about 20% of the initially introduced MAPEG can be grafted onto PLA chains. The remaining fraction (80%) of the plasticizer was a mixture of unreacted/monomeric and "homo-oligomerized" MAPEG. As a result, an efficient plasticization effect was evidenced by a significant lowering of the glass transition temperature (T_g) and storage modulus E' as well as by a drastic increase of the tensile elongation at break of approximately 70 times as compared to neat PLA. More interestingly, in the case of PLA/AcrylPEG/L101 (79.5/20/0.5 wt/wt/wt), up to 65% of the initially introduced AcrylPEG reacted and was grafted onto the PLA chains. The remaining non-grafted AcrylPEG completely homo-oligomerized. As a result, an efficient toughening effect of the resulting materials was reached. This was especially marked by a drastic enhancement of the impact strength, ~36 times, and a significant improvement of the elongation at break, ~63 times.

Lapol®108 is a renewable bioplasticizer of PLA that can be processed using standard processes such as injection moulding, extrusion coating, thermoforming, and cast films (http://www.lapol.net/). It promotes toughness and flexibility without sacrificing modulus, while minimizing the reduction of glass

Table 5 | Comparison of flexural properties of Lapol® HDT blends vs. PLA (unannealed and annealed) (From http://www.lapol.net/).

Flexural properties	Modulus (MPA)	HDT[b] (°C)
PLA	3300	55
Annealed PLA[a]	3800	155
20% Lapol HDT in PLA	3800	165

[a] PLA was annealed for 10 min at 110° C.

[b] Heat deflection temperature is measured using a thermomechanical analyzer using a load of 0.2–0.3 N.

transition temperature. For the lowest plasticizer content (5–10 wt%), the bioplasticizer Lapol®108 seems to be the most convenient one to maintain a good stiffness-toughness balance among this list of investigated plasticizers (**Table 4**). Interestingly, the new Lapol® HDT additive used for increasing the heat deflection temperature of PLA is now available at pilot-production. For many high-performance applications, using PLA requires a high temperature resistance to deformation and deflection, i.e., a heat deflection temperature higher than 100°C. Compounding 20 wt.% of Lapol® HDT with PLA 3001D, 4032D, or 7000D can increase the heat deflection temperature of unannealed PLA from 55°C to about 160°C. This capability greatly expands the potential uses and applications to PLA. This increased heat-performance is achieved without adding inorganic fillers or other additives, although these additions may further enhance some other properties. **Table 5** shows typical flexural properties data for a blend of 20% Lapol® HDT in PLA compared to commercially available neat and annealed PLA.

PLA-based blends containing Lapol® HDT exhibit similar or higher flexural modulus than commercially available PLA (annealed and unannealed). Lapol® HDT may be compounded with an impact modifier to tailor the properties of PLA for specific applications.

Globally, these studies show that oligomeric and polymeric plasticizers are in general less efficient than monomeric ones in order to improve the elongation and reduce the glass transition temperature of resulting blend. However, they have more tendencies to give better stiffness-toughness balance for PLA material compared to small molecule plasticizers. Based on their complementary advantages, the combination of small molecule plasticizers with polymeric or oligomeric ones was also attempted in the literature.

Mixed plasticizers

These mixed plasticizers combine an oligomeric or polymeric plasticizer with a small molecule plasticizer. Therefore, they should lead to a medium level of depression in T_g and more balanced mechanical properties (elongation, modulus and strength) than the individual plasticizers. Some plasticizer combinations were studied. They are reported in **Table 6**.

In general, one can conclude on the behaviors of the plasticizers in PLA and their effect on the properties of the polymer as follows:

- The addition of 10–20% of plasticizers may be a successful way to remarkably reduce T_g and improve PLA flexibility/ductility/tensile elongation at the same time.
- Substantial reductions in tensile strength and modulus are unfortunately unavoidable.
- An excessive incorporation of plasticizer leads to the saturation of the plasticizer in the amorphous phase of PLA, resulting in a migration or phase separation depending on the plasticizer nature.
- Small molecule or monomeric plasticizers are more efficient in order to improve PLA flexibility/ductility/tensile elongation and decrease its T_g, but less efficient on tensile strength and modulus than oligomeric and polymeric plasticizers.
- The higher the molecular weight of the plasticizer, the lower the critical saturation concentration, at which phase separation begin to occur.
- Lower molecular weight PEGs exhibit good miscibility with PLA and result in more efficient reduction of T_g. This can lead to drastic improvement in ductility and/or impact resistance of PLA at low concentrations.
- After ageing for 1 month, the mechanical properties of the plasticized PLA did not change remarkably. This result indicated that PPG and PPG-E could prevent the physical ageing and the embrittlement of PLA.

Table 6 | Summary of effects of some mixed plasticizers on mechanical properties of PLA [Copyright © (2011) Wiley and Sons; used with permission from Liu and Zhang (2011)].

Polymer	Molecular weight (g/mol)	Plasticizer	Molecular weight (g/mol)	Plasticizer content (wt%)	DSC T_g (°C)	Max. tensile strength (MPa)	Tensile modulus E (MPa)	Tensile elongation at break ε (%)	Notched impact strength (kJ/m²)	Reference
PLA	$M_n = 81000$	–	2	0	61	52 ± 2	1800 ± 150	6 ± 1	2.6 ± 0.2	Lemmouchi et al., 2009
		TBC	$M_n = 360$	20	20	20 ± 1	9 ± 1	320 ± 20	No break	
		COPO1	$M_n = 650$	20	29	21 ± 1	790 ± 180	170 ± 10	1.6 ± 0.6	
		COPO2	$M_n = 1000$	20	26	25 ± 1	300 ± 50	220 ± 20	8.3 ± 2.5	
		COPO3	$M_n = 1050$	20	36	30 ± 1	1700 ± 100	130 ± 20	1.9 ± 0.6	
		COPO4	$M_n = 1750$	20	35	24 ± 2	1150 ± 150	170 ± 10	1.9 ± 0.7	
		COPO1/TBC	–	10	42	40 ± 2	2000 ± 110	4 ± 1	2.7 ± 0.2	
				20	24	17 ± 1	9 ± 1	260 ± 20	6.4 ± 1.9	
		COPO2/TBC	–	10	–	27 ± 2	1480 ± 80	140 ± 20	2.4 ± 0.2	
				20	26	24 ± 1	19 ± 5	260 ± 10	No break	
		COPO3/TBC	–	10	–	37 ± 1	1850 ± 200	4 ± 1	–	
				20	25	16 ± 1	16 ± 7	300 ± 20	–	
		COPO4/TBC	–	10	41	39 ± 2	2000 ± 100	4 ± 1	2.5 ± 0.2	
				20	27	22 ± 1	150 ± 65	250 ± 10	5.5 ± 0.8	
		COPO5/TBC	–	10	47	37 ± 1	1950 ± 150	4 ± 1	2.5 ± 0.2	
				20	23	20 ± 1	400 ± 140	260 ± 20	3.8 ± 1.1	

TBC, tributyl citrate; COPO1 and COPO2, two kinds of AB-type diblock copolymers of PDLLA and either PEG350 monomethyl ether or PEG750 monomethyl ether, that is, PDLLA-b-PEG350 (10/4, molar ratio of D,L-LA monomer to PEG used in the feed) and PDLLA-b-PEG750 (10/4, D,L-LA monomer to PEG molar ratio used in the feed); COPO3, ABA-type triblock copolymer of DLLA and PEG400, that is, PDLLA-b-PEG400-b-PDLLA (10/2, molar ratio of D,L-LA monomer to PEG used in the feed); COPO4, 3-star-(PEG-b-PDLLA) block copolymer (10/1.3, molar ratio of D,L-LA monomer to PEG used in the feed); COPO5, 4-star-(PEG-b-PDLLA) block copolymer (10/1, molar ratio of D,L-LA monomer to PEG used in the feed).

– Whilst increasing the molecular weight of the plasticizer can slow down migration rate and thus improve morphological stability of PLA materials during storage, it also decreases its solubility and plasticizing efficiency. Additionally, high-molecular weight plasticizers are keen to phase-separation because of low saturation concentrations of plasticizers.

COMPOUNDING WITH FLEXIBLE/SOFT POLYMERS—PARTIALLY MISCIBLE TO IMMISCIBLE BLENDS—WAYS OF COMPATIBILIZATION

The term "blending" refers to the simple mixing of polymeric materials in the molten state. During the last three decades, polymer blends have become a very important part of the commercialization of polymers because one can tailor blend compositions to meet specific end-use requirements (Baker et al., 2001). Melt-blending polymers is a much more economical and convenient methodology at the industrial scale rather than synthesizing new polymers to achieve the properties unattainable with existing polymers. However, most polymer pairs are immiscible, which can lead to phase-separated materials. The latter has often three inherent problems if the morphology and the interfaces of the blend are not well-controlled: (1) poor dispersion of one polymer phase in the other one; (2) weak interfacial adhesion between the two phases; and (3) instability of immiscible polymer blends (Baker et al., 2001). However, immiscible polymer blends are much more interesting for commercial development because immiscibility allows maintaining the good features of each polymeric component of the blend. One of the most important challenges is thereby to develop compatibilization techniques that allow controlling both the morphology and the interfaces of phase-separated blends. In general, compatibilization in physical blends is tuned by the physical interactions (hydrogen bonds, Van der Waals interactions etc.) between the blend components.

PLA has been blended with various polymers for different purposes, namely for improving its stiffness-toughness balance. A variety of biodegradable and non-biodegradable soft polymers have been used as toughness modifiers for PLA. Recently, it has been shown that new impact modifiers can efficiently strengthen/toughen brittle/stiff PLA, due to their core-shell polymeric structure (a block copolymer). They form a soft or elastomeric block having high compatibility and miscibility with the toughening polymer, and surrounded with a rigid block copolymer, usually having a high compatibility and/or miscibility with the matrix polymer. When the softer component forms a second phase within the stiffer continuous phase, it may act as a stress concentrator, which enables ductile yield and prevents brittle failure (Babcock et al., 2008). At the same time, the core is "locked in" by slight crosslinking and grafting with its shell to avoid phase-separation during blending. Moreover, the adhesion between the two phases, core-shell polymer and polymer matrix, depends strongly on the degree of miscibility of the shell polymer with the matrix, *that is*, whether they are completely miscible, partially miscible, or immiscible. However, a partial miscibility between core-shell modifiers and polymeric matrix is often necessary to obtain blends of desired impact properties. From the literature, multiple crazing initiated from the dispersed rubber phase is recognized to be one of the main mechanisms, which increases the toughness of glassy materials like polylactide-based materials (Ikeda, 1993; Bucknall, 2007; Mahajan and Hartmaier, 2012). Some authors have preferred to blend PLA with biodegradable flexible/soft polymers in order to preserve the overall biodegradability of resulting blends. Some of these blends are in this regard, finding short-term applications, namely packaging and mulch films for agriculture.

Flexible/soft (ε-Caprolactone)-based copolymers

As obtained by ring-opening polymerization of e-caprolactone, poly(e-caprolactone) (PCL) is a biodegradable and flexible/soft polyester with a melting temperature of 60°C and a glass transition temperature of −60°C. Due to the low glass transition temperature, PCL-based materials were considered as interesting impact modifiers. PCL and PLA blends have been extensively investigated over the past years. For instance, Broz et al. investigated the tensile properties of blends made of P(D,L-LA) and PCL at different content in PCL (Broz et al., 2003). Whilst the strain-at-failure decreases monotonically for PCL contents from 0.6 wt%, the modulus and tensile strength increased almost linearly with composition. This was more likely ascribed to some strengthening of the blend interface in this regime. However, DSC and NMR results suggested that PCL was able to crystallize to a certain extent within PCL/P(D,L-LA) blends, indicating that phase-separation was more pronounced under these conditions. However, as shown here, the simple melt blending of PLA and PCL usually results in a marginal toughness improvement because of their poor miscibility (López-Rodríguez et al., 2006). This can be more likely explained by the fact that PCL can readily crystallize within PLA/PCL blends, leading to the more pronounced phase-separation. Accordingly, the simple melt blending of PLA and PCL usually leads to a marginal improvement in toughness because of their immiscibility (López-Rodríguez et al., 2006; Vilay et al., 2009). In this regard, some of us have designed bio-sourced and hydrolytically degradable random copolyesters based on poly(ε-caprolactone) as a soft core component. (Co)polymerization of CL with other lactones affords an elegant way to modulate the thermal and mechanical properties of resulting PCL-based materials. The most interesting feature was that the overall crystallinity of these (co)polyesters decreased with the comonomer content, yielding rubbery-like materials at ambient temperature. When dispersed into glassy materials like PLA, it is well-known that rubbery microdomains can readily absorb the impact energy. In a first study (Odent et al., 2012), 10 wt.% of amorphous poly(ε-caprolactone-*co*-δ-valerolactone) (P[CL-*co*-VL]) random aliphatic copolyesters were thereby synthesized and investigated as biodegradable impact modifiers for commercial PLA using a microcompounder. The use of a high molar mass copolyester (M_n = ca. 60,000 g/mol) with a molar composition of 45/55 mol% (CL/VL) resulted in the optimal improvement in notched Izod impact strength for compression-moulded (vs. injection-moulded) PLA materials ($7 \, kJ/m^2$) compared to $2.5 \, kJ/m^2$ for PLA. According to the author, this improvement in toughness is also related to the mean size ($0.7 \, \mu m$) and size distribution of the dispersed copolymer micro-domains throughout the PLA matrix. In a similar study (Odent et al., 2013b), the random biocopolyester was synthetized and used as impact modifier is poly(ε-caprolactone-*co*-d,l-lactide) (P[CL-*co*-LA]). By varying the comonomer content, a phase inversion was noticed. A control of the affinity between PCL-based impact modifiers and PLA matrix gives access to a mixture of spherical microdomains with similar range of optimum particles diameter (i.e., $0.9\,\mu m$) and nanosized oblong structures, involving a combination of shear yielding and multiple crazing mechanisms. As a result, PLA blended with $10\,wt.\%$ of the CL/LA composition 72/28 mol.% displayed a maximum impact strength of about $11.4\,kJ/m^2$ (**Figure 3**). The mean size of the rubbery micro-domains was $0.9\,\mu m$ in this case.

In the case of brittle polymers, spherical microdomains act as stress reservoirs and initiate crazing upon the microdomains size, i.e., larger microdomains size than $0.5\,\mu m$ are required to nucleate crazing mechanism and enhance fracture energy absorption (Donald and Kramer, 1982; Van Der Wal and Gaymans, 1999). Accordingly, an optimum particle size range (ca. $0.7–0.9\,\mu m$) for PLA toughening was identified by correlating dispersed microdomains size with notched Izod impact strength (Gramlich et al., 2010; Liu et al., 2011). Wu and al. correlated rubber particle diameter with chain structure parameter of the matrix and claimed that the optimum particle size for toughening decreased as the matrix becomes less brittle (Wu, 1990). Kowalczyk et al. reported that rubbery poly[1,4-*cis*-isoprene] microdomains within PLA-based materials initiated crazing at the early stages of deformation, immediately followed by the cavitation phenomena inside rubbery microdomains. This latter promotes further shear yielding for PLA matrix (Kowalczyk and Piorkowska, 2012). More recently, some of us have elaborated ultratough PLA-based materials by synergistically adding PLA, rubber-like poly(ε-caprolactone-*co*-D,L-lactide) copolyester and silica nanoparticles using extrusion techniques (Odent et al., 2013a). A peculiar alteration for the phase-morphology of the rubbery phase within PLA matrix was achieved by co-adding copolyester and silica nanoparticles into

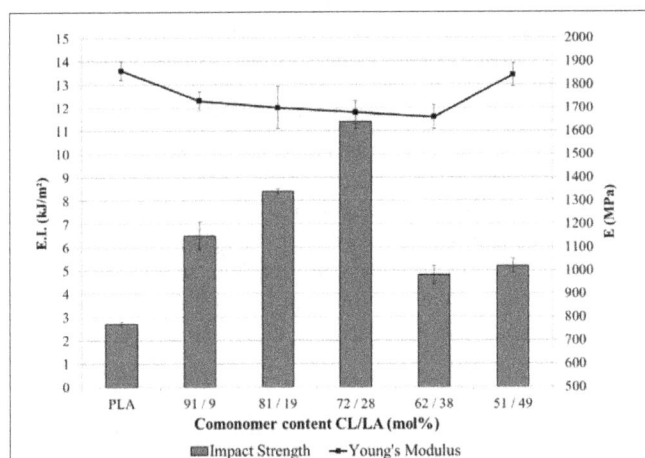

FIGURE 3 | Influence of the LA comonomer content of copolyester on the notched Izod impact strength and Young's modulus of PLA-based materials containing 10 wt.% of P[CL-*co*-LA]. [Reprinted from Odent et al. (2013b) with permission from Elseiver].

PLA matrix. It resulted that regularly obtained spherical nodules convert into almost continuous features after adding nanoparticles in the PLA-based melt-blend. In the latter, an enhancement of 15-fold impact strength was obtained by comparison to unfilled PLA.

The use of small molecule reactive additives during compounding has been demonstrated to be an effective way to improve the compatibility between PLA and PCL. Wang et al. (1998) investigated the tri-phenyl phosphate (TPP) as a catalyst or coupling agent for the preparation of PLA and PCL blends. The addition of 2 phr TPP to PLA/PCL (80/20, w/w) blend during processing resulted in a higher elongation (127 vs. 28%) and tensile modulus (1.0 GPa vs. 0.6 GPa) compared to the binary TPP-free blend. The balance between degradation of molecular weight and the formation of copolymer was believed to govern the final mechanical properties of the blends. Reaction time and molecular weight of PCL used were found to have remarkable effects on mechanical properties of the blends. Higher molecular weight PCL ($M_n = 80,000\,g/mol$) and medium reaction time (15 min) promoted the largest improvement of the ductility. In another study, di-cumyl peroxide (DCP) was used to promote reactive compatibilization of the PLA/PCL blends (Semba et al., 2006). The results showed again that the addition of DCP increased the ductility of the final material. Further addition of DCP beyond the optimum amount had an opposite effect on elongation. AFM observation revealed that the diameter of the dispersed PCL domains decreased with increasing DCP content. The addition of 0.3 phr DCP to the optimum ration PLA/PCL = 70/30 resulted in (1) an impact strength of 2.5 times more than that of neat PLA, (2) an improved blend compatibility, (3) an improved ultimate tensile strain (4) a yield point and ductile behavior under tensile test, and (5) mechanical properties comparable to those of HIPS and ABS. In contrast, the addition of DCP to PLA alone did not alter mechanical properties. It was suggested that DCP caused crosslinking of PLA with PCL and therefore improved interfacial adhesion. Depending on feeding procedure, addition of DCP *via* the splitting feeding method resulted in a higher reverse Izod impact strength than feeding at once through the main hopper (Semba et al., 2007). Lysine tri-isocyanate (LTI) as a reactive compatibilizer improved the compatibility of PLA and PCL, resulting in the reduction of size of PCL spherulites (Takayama and Todo, 2006; Takayama et al., 2006). Impact fracture toughness markedly improved by increasing LTI content, which was attributed to the strengthening structure of the blend as a consequence of crosslinking reactions. The compatibilizing effect of LTI was compared with four other reactive processing agents on the PLA/PCL (80/20, w/w) blends (Harada et al., 2008). The addition of 0.5 phr of each reactive agent resulted in an increase in the un-notched Charpy impact strength in the order of LTI > LDI (lysine diisocyanate) > Duranate TPA-100 [1,3.5-tris(6-isocyanatohexyl)- 1,3,5-triazinane-2,4,6-trione] > Duranate 24A-100 [1,3,5-tris(6-isocyanatohexyl)biuret] > Epiclon 725 (trimethylolpropane triglycidyl ether). It was assumed that the reaction of isocyanates group with both terminal hydroxyl and carboxylic groups of polyesters accounted for improved compatibility at the PLA/PCL interfaces and therefore the enhancement in the physical properties.

Polyhydroxyalkanoates (PHAs) and their copolyesters

Polyhydroxyalkanoates (PHAs) are biodegradable polyesters produced by bacterial fermentation of sugar or lipids (Steinbüchel and Valentin, 1995; Zinn et al., 2001; Poirier, 2002; Noda et al., 2005) when nutrient shortage is present. Since the range of monomers available is impressive within this family, the mechanical properties of PHAs can range from stiff thermoplastics to elastomers dependent on their pendent alkyl chain length (**Scheme 1A**). However, only one grade, i.e., Nodax™, was industrially implemented by Procter and Gamble Co., which correspond to copolymers of 3-hydroxybutyrate with a small amount of 3-hydroxyalkanoate as co-monomer (**Scheme 1B**) (Noda et al., 2004, 2005).

In this regard, Noda et al. (2004) melt-blended PLA with a poly(3-hydroxybutyrate-co-3-hydroxyhexanoate) copolymer, i.e., NodaxH6, containing 5 mol% of 3-hydroxyhexanoate (3-HH) unit. The PLA/NodaxH6 (90/10, w/w) blend exhibited a tensile toughness of 10 times more than that of neat PLA and an elongation superior to 100%. When NodaxH6 content was less than 20 wt % in the blends, its crystallization was largely restricted and thereby NodaxH6 was dispersed as rubbery amorphous droplets within PLA, suggesting that the material was toughened by craze-initiation. Furthermore, it was interesting that the inclusion of these small amounts of PHA did not compromise the optical clarity of PLA itself.

Schreck and Hillmyer investigated the impact toughness of blends of PLLA with a NodaxH6 containing 7 mol.% of 3-HH (Schreck and Hillmyer, 2007). The PLLA/NodaxH6 (85/15, w/w) blend demonstrated a twofold increase in notched Izod impact strength (44 J/m) compared with that of PLLA (22 J/m). Ma et al. toughened PLA by melt-compounding with fully bio-based and bio-compostable poy(β-hydroxybutyrate-co-β-hydroxyvalerate) (PLA/PHBV) with high β-hydroxyvalerate content (40 mol%) (Ma et al., 2013). The blends displayed two separate glass transition temperatures and two separate phases, indicating that the PLA and PHBV were immiscible. The toughness and the ductility of PLA can be effectively improved by incorporation of 10–30 wt% of the PHBV as evidenced by a significant increase in the elongation at break and the impact toughness (**Table 8**). The local deformation mechanism revealed that fibrillation, partial interfacial de-bonding, cavitation and matrix yielding were involved in the toughening mechanism of the PLA/PHBV blends under impact and tensile testing conditions.

Biodegradable poly(butylene adipate) (PBA), poly(butylene succinate) (PBS), and poly(butylene adipate-co-butylene terephthalate) (PBAT)

Poly(butylene adipate-co-terephthalate) (PBAT) is a fully biodegradable aliphatic–aromatic copolyester (**Scheme 1C**), which is commercially available under the trade name of EcoflexVR (BASF Co.).

PBAT has similar thermal properties to those of LDPE, but exhibits higher mechanical properties, more particularly higher flexibility and ductility (elongation > 700%). Even though PLA/PBAT blend are immiscible, PBAT could be dispersed in PLA with an average particle size of about 0.3–0.4 μm without use of compatibilizers in co-rotating twin-screw extruder (Jiang et al., 2006). The mechanical properties of the different PLA/PBAT blends are reported in **Table 8**. It was demonstrated that the de-bonding-induced shear yield was responsible for the remarkable high extensibility of the blends. Because of weak interfacial adhesion in the blends, impact toughness was slightly improved. Interestingly, the PLA/PBAT blends are now being commercially produced by BASF Co. under the trademark EcovioVR for film and extruded foam applications.

To improve the compatibility of PLA/PBAT blends, a random terpolymer of ethylene, acrylate ester, and glycidyl methacrylate (referred as "T-GMA") was investigated as a reactive compatibilizer in melt compounding (Zhang et al., 2009a). Regardless the PLA/PBAT blends composition (70/30, 80/20 or 90/10 wt/wt), the increase of T-GMA content up to 5 wt% resulted in a great improvement of tensile nominal strain at break (**Figure 4A**) and the notched impact strength (**Figure 4B**) to reach more than 150% and 30 kJ/m^2, respectively, approximately two times that of the uncompatibilized binary blends. These results were correlated to the good miscibility and interfacial adhesion between PLA and PBAT, leading to a shear yielding mechanism when increasing the T-GMA content. The authors attributed the better interfacial adhesion to the *in situ* reactive compatibilization phenomena (**Scheme 2A**).

Lin et al. (2012) compatibilized the biodegradable blends poly(lactic acid) (PLA)/poly(butylene adipate-*co*-terephthalate) (PBAT) by *in situ* transesterification using various amounts of tetrabutyl titanate (TBT) as catalyst. The incorporation of 0.5% of TBT into PLA/PBAT blends not only improved their overall mechanical properties as well as gave values of tensile strength, elongation at break and impact strength of 45 MPa, 298% and 9 kJ/m^2 (**Figures 5A,B**), respectively. It was also demonstrated

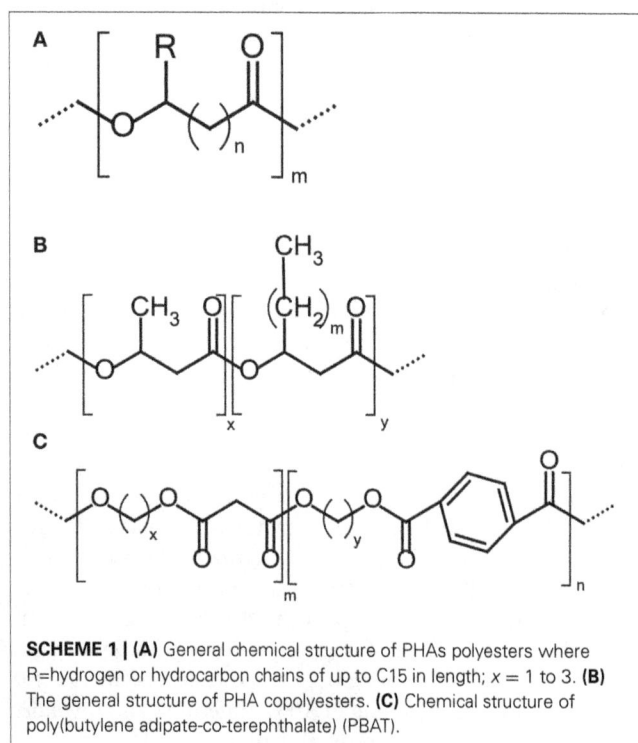

SCHEME 1 | (A) General chemical structure of PHAs polyesters where R=hydrogen or hydrocarbon chains of up to C15 in length; *x* = 1 to 3. **(B)** The general structure of PHA copolyesters. **(C)** Chemical structure of poly(butylene adipate-co-terephthalate) (PBAT).

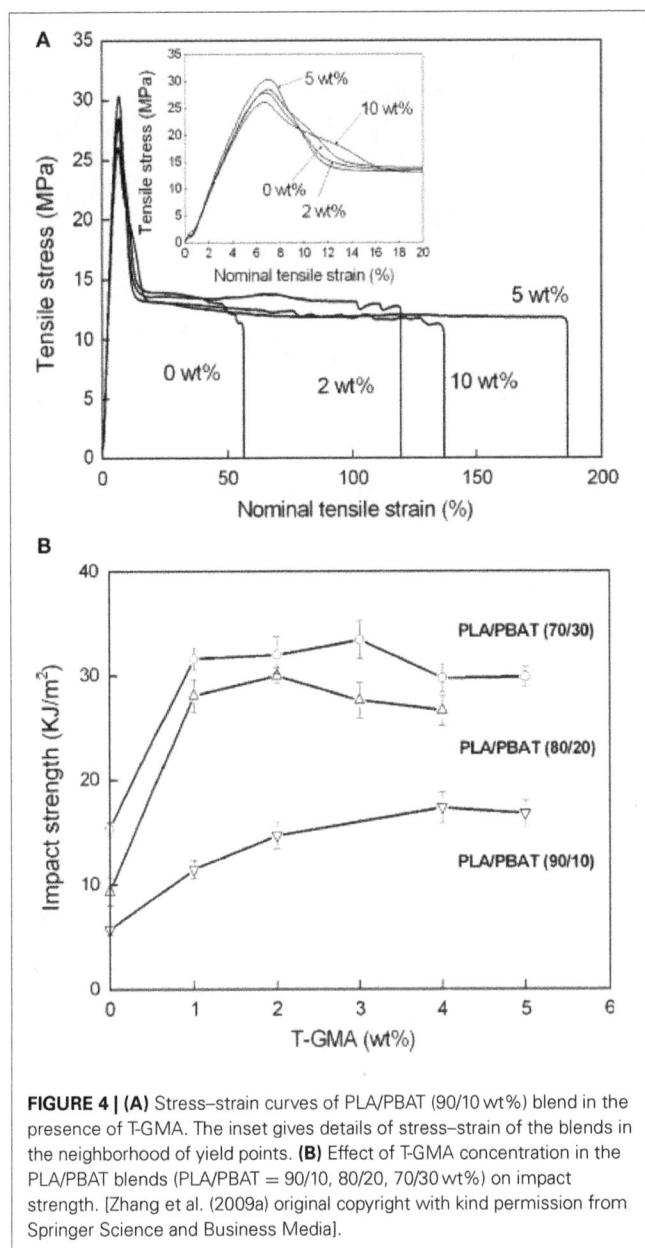

FIGURE 4 | (A) Stress–strain curves of PLA/PBAT (90/10 wt%) blend in the presence of T-GMA. The inset gives details of stress–strain of the blends in the neighborhood of yield points. **(B)** Effect of T-GMA concentration in the PLA/PBAT blends (PLA/PBAT = 90/10, 80/20, 70/30 wt%) on impact strength. [Zhang et al. (2009a) original copyright with kind permission from Springer Science and Business Media].

that the storage modulus of the blends and glass transition temperature (**Figure 5C**) were enhanced compared to the binary blends free of TBT. The SEM micrographs demonstrated that the compatibility between PLA and PBAT was improved *via* transesterification during reactive melt-extrusion. The interfacial debonding and the yielding deformation were the most important mechanisms to improve toughness.

Poly(butylene succinate)s (PBS) and their copolyesters

PLA is immiscible with PBS. In some studies, a third *in situ* reactive component was incorporated to improve compatibility. PBS was melt blended with PLA without compatibilizer and using LDI and LTI as compatibilization agents (Harada et al., 2007). For all PLA/PBS compositions ranging from 90/10 to 80/20 wt.%, only the addition of 0.5 wt.% of LDI or 0.15 wt.% of LTI increased

the elongation at break to more than 150%. Impact strength also increased to reach 50–70 kJ/m^2. For PLA/PBS (80/20 wt/wt) with LTI, the un-notched samples did not break during the impact test. Furthermore, due to the addition of 0.15 phr LTI into the PLA/PBS (90/10, wt/wt) blend, the size of dispersed PBS particles was significantly reduced. Consequently, LTI was an effective reactive processing agent capable of increasing the toughness of the PLA/PBS blends. Similar results were observed by using LTI with PLA/PBSL (Vannaladsaysy et al., 2009) with effectively improved blend compatibility and higher energy dissipation during the initiation and propagation of crack growth. This results in the suppression of spherulite formation of PBSL and the formation of a firm structure made of entanglements between both PLLA and PBSL chains. DCP was also used for *in situ* compatibilization of the PLLA/PBS (80/20 wt/wt) blend (Wang et al., 2009). The uncompatibilized blend showed much higher elongation than PLLA (250 vs. 4%), but only slightly higher notched Izod impact strength (3.7 kJ/m^2 vs.2.5 kJ/m^2 for PLA). Addition of 0.1 phr DCP greatly increased the impact strength of the blend to 30 kJ/m^2. Both strengths and moduli invariably decreased with increasing DCP content. It was found that the addition of DCP led to a reduction in the size of the PBS domains and improved interfacial adhesion between the PLLA and PBS phases. The toughening effect of the blends was considered to be related to the de-bonding initiated shear yielding. In a similar way, blending PLA with other polycondensates like biodegradable poly(butylene adipate) (PBA) and poly(butylene succinate) (PBS) was also investigated on the toughening effect. These results were compared with PBAT/PLA blends. Like PBAT, these (co)polyesters can be readily synthesized by melt-polycondensation (Zhao et al., 2010). As far as blends are concerned, a considerably high elongation at break with a moderate loss of strength was observed for all the blends, regardless the investigated copolyesters. For instance, the elongation at break and the impact strength increase with polyester content, until reaching maximum values (>600% and >35 KJ/m^2, respectively) at a PBA/PBS/PBAT content of 15 and 20%, respectively. The addition of PBA/PBS/PBAT into PLA improves the toughness, but reduces the stiffness of the latter. Moreover, the crystallization ability of PLA blends can be increased by the addition of a small amount of PBS/PBA/PBAT.

Poly(ethylene oxide-b-amide-12) (PEBA) or Pebax®

Biosourced and biodegradable poly(ethylene oxide-*b*-amide-12) (PEBA) was used as a toughening agent for PLA *via* melt compounding. PLA/PEBA blends are an immiscible system with a two-phase morphology (Han et al., 2013). By increasing the PEBA content, the binary blends displayed a marked improvement in toughness. All PLA/PEBA blends showed a clear stress yielding on the stress–strain curves with necking when the PEBA contents varied from 10 to 30%. For the blend with 20% PEBA, the elongation at break markedly increased to 346%, corresponding to a 50-fold increase compared with the elongation at break of neat PLA. The impact strength of the blend was significantly enhanced at 20% (or more) of added PEBA as well. The maximum impact strength reached was of 60.5 kJ/m^2, indicating that a significant toughening effect was achieved (**Table 8**). The phase morphology evolution in the PLA/PEBA blends during

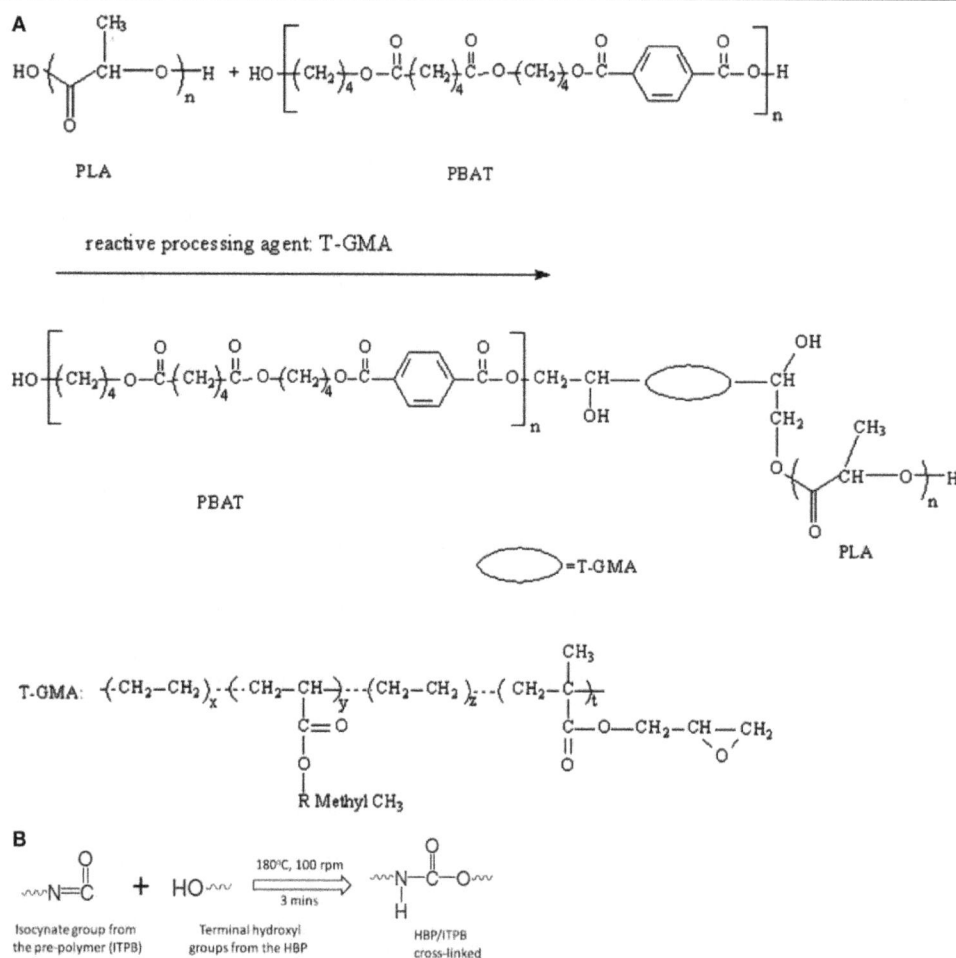

SCHEME 2 | (A) Predicted reaction of PLA, PBAT, and T-GMA [Zhang et al. (2009a), original copyright with kind permission from Springer Science and Business Media]. **(B)** Schematic showing the proposed *in-situ* crosslinking of the terminal hydroxyl groups from HBP or PLA with isocyanate groups from the ITPB during reactive blending in extruder at 180°C [Nyambo et al. (2012), original copyright with kind permission from Springer Science and Business Media].

tensile and impact tests were investigated, and the corresponding toughening mechanism was discussed. Remarkably, a clear shear yielding bands perpendicular to the stretching direction and crack propagation along the tensile direction were observed during the tensile test. Moreover, the obvious plastic deformation in the blend was observed during the impact test. The shear yielding induced energy dissipation and therefore led to the improvement in toughness of the PLA/PEBA blends.

Polyurethane and polyamide elastomers

Feng et al. used a thermoplastic polyurethane (TPU) elastomer with a high strength, toughness and biocompatibility to prepare PLA/TPU blends suitable for a wide range of applications of PLA as general-purpose plastics (Feng and Ye, 2011). The morphological structure and mechanical properties of the PLA/TPU blends indicated that an obvious yield and neck formation was observed for the PLA/TPU blends (**Figure 6A**). The stress–strain curves of the blends exhibited an elastic deformation stress plateau, indicating the transition of PLA from brittle to ductile fracture.

The elongation at break and notched impact strength for the PLA/20 wt% TPU blend reached 350% and 25 kJ/m^2, respectively, without an obvious drop in the tensile strength (**Figure 6B**). The respective T_g's of PLA and TPU in the blends also shifted to intermediate values, suggesting a partially miscible system due to the hydrogen bonding formed between the chains of TPU and PLA. Spherical particles of TPU dispersed homogeneously in the PLA matrix, and the fracture surface presented much roughness. With increasing TPU content, the blends exhibited increasing tough failure thanks to the improved crack initiation resistance and crack propagation resistance. It was evident that the use of TPU greatly improved the toughness of PLA.

In a similar study, Han et al. also investigated the toughness effect of TPU on PLA (Han and Huang, 2011). The study of the blends morphology as a function of TPU contents showed that PLA was incompatible with TPU. The spherical particles dispersed in PLA matrix, and the uniformity decreased with increasing TPU content. There existed long threads among some TPU droplets in blend with 30 wt.% TPU. After addition of

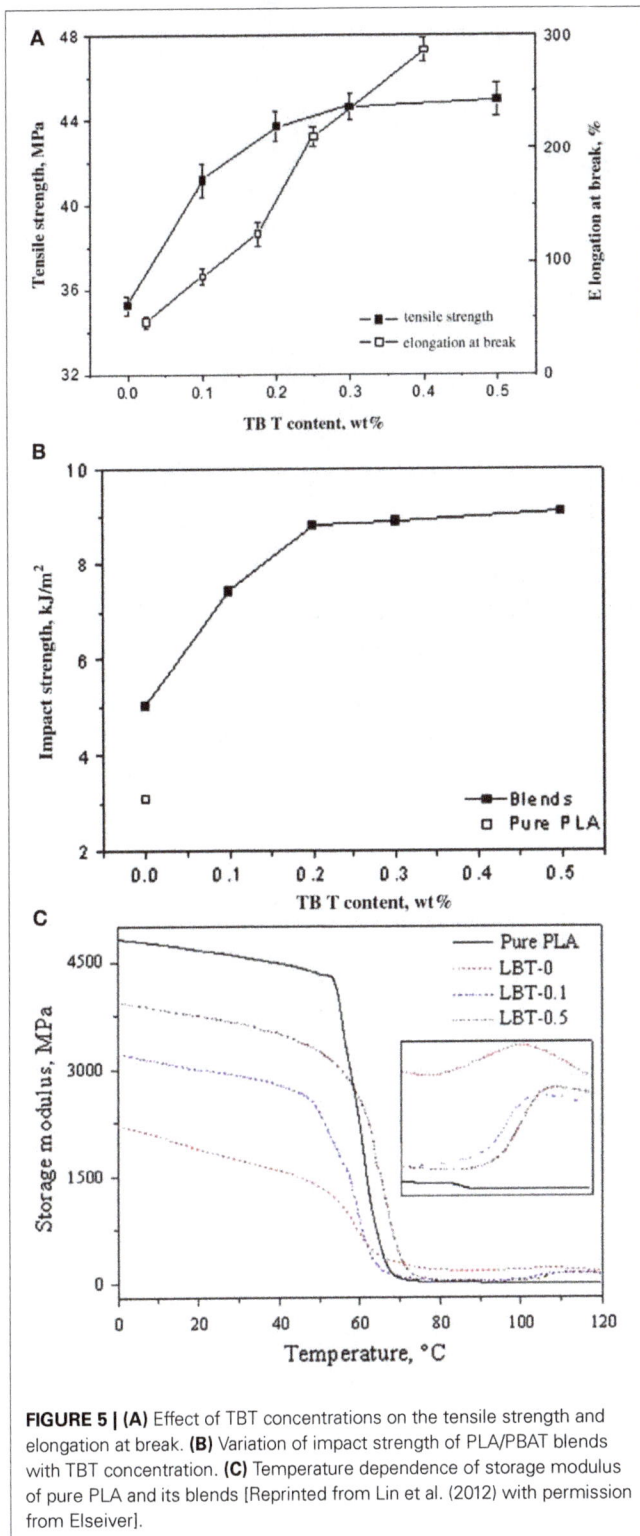

FIGURE 5 | (A) Effect of TBT concentrations on the tensile strength and elongation at break. **(B)** Variation of impact strength of PLA/PBAT blends with TBT concentration. **(C)** Temperature dependence of storage modulus of pure PLA and its blends [Reprinted from Lin et al. (2012) with permission from Elseiver].

the fracture surface of the impacted sample. The matrix ligament thickness of the PLA/TPU blends was below the critical value, and the blends deformed to a large extent because of shear yield initiated by stress-concentrations and interfacial de-bonding. This resulted in the formation of fibers in both tensile and impact samples and the dissipation of a large amount of energy.

Some other blends with polyurethane (PU) and polyamide elastomers (PAE) were elaborated in order to study their mechanical properties (**Table 7**). PLA/PU blends were found to be partially miscible, and PU was dispersed in PLA within domain sizes at the submicrometer scale. It was demonstrated that matrix shear yielding initiated by de-bonding at the matrix/particle interface was considered to be responsible for the improved toughness. Their mechanical properties are reported in **Table 8**.

Dynamic mechanical analysis (DMA) demonstrated a good compatibility between PAE and PLA blends. A gook tu od dispersion of PAE in PLA matrix was shown in SEM images. When the PAE content was fixed to 10%, the tensile strength of blend was similar to that of neat PLA, and the elongation increased significantly to 194.6%. Remarkably, the blends showed a wonderful shape-memory effect. PAE domains act as stress concentrators in system with the stress release locally and lead to energy-dissipation process. This prevents PLA matrix from breaking under high deformation, and lead to the PLA molecular orientation. Consequently, the blends submitted to deformation upon tensile load, and heating up the material reform the shape back to the original shape. **Table 8** lists the obtained results.

Natural rubber

Natural rubber (NR) can be a good impact modifier candidate for PLA because it is derived from renewable resource. However, because of its incompatibility with PLA, it does not provide the desired improvement of PLA toughness. It has also several properties and appearance issues. Interestingly, epoxidized natural rubber (ENR) is more compatible with PLA. The toughness of the final material is dependent on the level of epoxy functions present in the ENR. Bitinis et al. investigated some formulations of natural rubber (NR)–PLA blends (Bitinis et al., 2011). The rubber phase was uniformly dispersed in the continuous PLA matrix with a droplet size ranging from 1.1 to 2.0 μm. The ductility of PLA was significantly improved from 5% for neat PLA to 200% by adding 10 wt.% NR as reported in **Table 8**. At this concentration, the rubber droplets provided an optimum balance between their coalescence and their beneficial effect provided on the material's physical and mechanical behavior, without sacrificing totally the transparency of the material. Moreover, the small molecules contained in the elastomeric phase could be acting as nucleating agent, favoring the crystallization ability of PLA. According to the authors, these materials are, therefore, very promising for industrial applications.

Zhang et al. toughened PLA with epoxidized natural rubber (ENR) by melt-blending in an internal mixer (Zhang et al., 2013a). Whilst the stiffness of the material was slightly reduced, the impact strength of the latter could be improved to 6-fold values as compared to that of pure PLA. Again, the authors attributed this enhancement to a good interfacial adhesion between ENR and PLA.

30 wt.% TPU, the elongation at break of the blend reached about 600% (**Figure 7**), and the samples could not be broken up in the notched Izod impact tests at room temperature. The toughening mechanism was analyzed through three aspects, including the stress whitening, matrix ligament thickness, and observation of

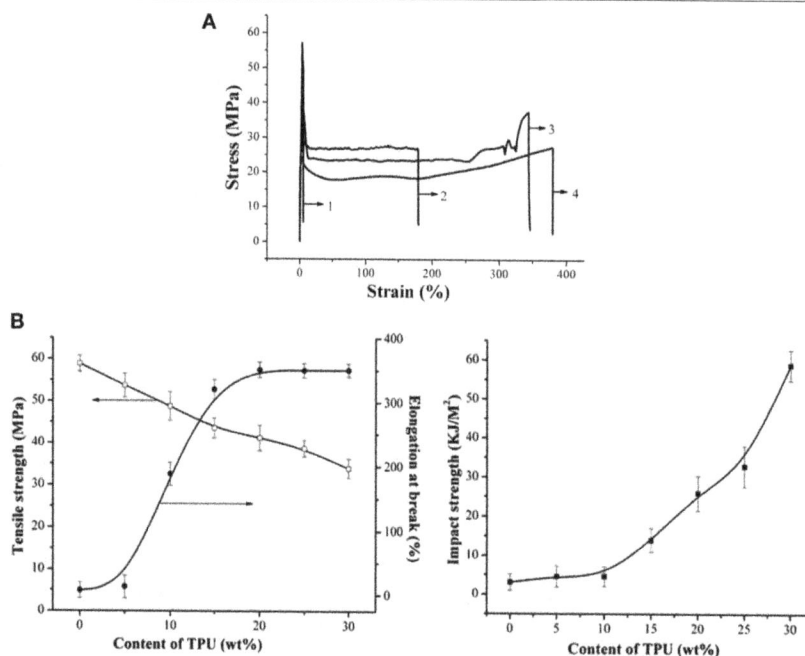

FIGURE 6 | (A) Tensile stress–strain curves of the PLA/TPU blends: (1) PLA/TPU (100/0), (2) PLA/TPU (90/10), (3) PLA/TPU (80/20), and (4) PLA/TPU (70/30). **(B)** Mechanical properties of the PLA/TPU blends as a function of the TPU content. Copyright (2010) Wiley; used with permission from Feng and Ye (2011).

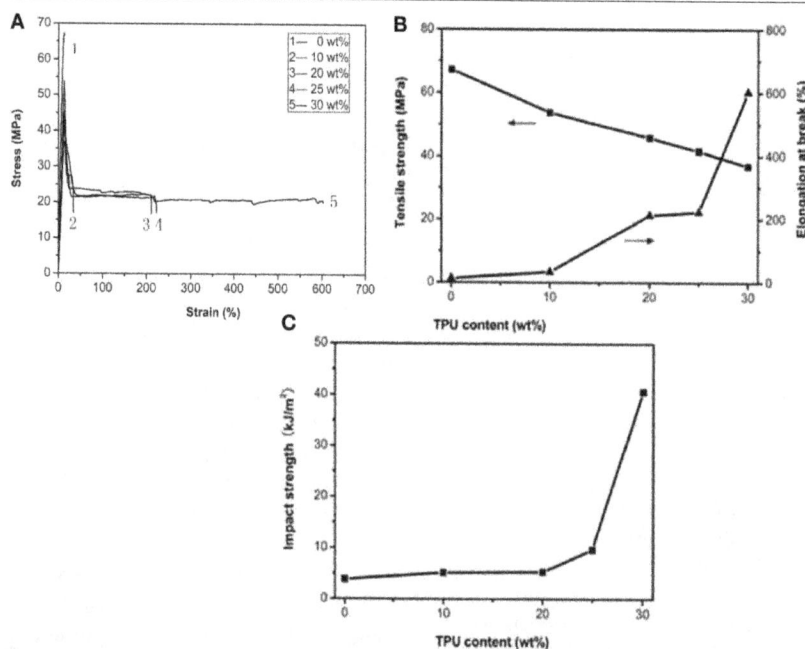

FIGURE 7 | (A) Stress–strain curves, **(B)** Tensile properties, and **(C)** Impact strength of PLA/TPU blends with different TPU contents. Copyright (2011) Wiley; used with permission from Han and Huang (2011).

Ge et al. blended PLA with glycerol monostearate (GMS) (Ge et al., 2013) (**Figure 8**). SEM micrographs of the impact fracture surfaces of PLA/GMS blends had a relatively good separation and this phenomenon was in good agreement with their higher impact strength. The result showed that the addition of GMS enhanced the flexibility of PLA/GMS blends as compared to neat PLA. The impact strength changed from 4.7 kJ/m^2 for neat PLA to 48.2 kJ/m^2 for 70/30 PLA/GMS blend (**Table 8**).

Ma et al. studied the influence of vinyl acetate (VA) content in ethylene-*co*-vinyl acetate copolymer (EVA) rubbers (Levapren®)

Table 7 | Blends with biodegradable elastomers and rubbers.

Elastomer category	Elastomer name	PLA/Elastomer (wt/wt)	Tensile strength (MPa)	Tensile modulus E (GPa)	Tensile elongation ε (%)	Izod impact strength (J/m)	References
–	–	100/0	65	–	4	64 (Unnotched) 27 (Notched)	Li and Shimizu, 2007
Biodegradable PU Elastomer	Thermoplastic Poly(ether)urethane	70/30	31.5	–	363	315 (Unnotched)	
	Pellethane™2102-75A	70/30	–	–	410	769 (Notched)	Natureworks LLC, 2007
–	–	100/0	46.8	1.8	5.1	–	Zhang et al., 2009b
Biodegradable Polyamide Elastomer	Thermoplastic PAE	5	48.1	1.5	161.5	–	
		20	23.4	-	184.6	–	

on toughening mechanisms of PLA-based materials (Ma et al., 2012) (**Figure 9**). They showed that the increase of VA content improves the compatibility between the components of the blend, since PLA is miscible with PVAc (no phase separation). The toughness of the PLA/EVA (80/20 wt/wt) blends firstly increased with VA content up to 50 wt.% and then declined. At high VA content, it resulted the formation of small EVA particles that could not cavitate under impact testing, whereas at low VA content, large EVA particle size was achieved. However, in the latter, there had a weak interfacial adhesion, affecting the toughness of the PLA/EVA blends. As a result, the optimum toughening efficiency of EVA on PLA was obtained at VA content of 50–60 wt.%. The EVA with VA content of 50 wt.% (i.e., EVA50) was selected to study the toughening effect of EVA content in the PLA/EVA blends. Even 5 wt.% EVA50 could already make PLA ductile ($\varepsilon_b \approx$ 300%). However, the notched Izod impact toughness of this blend was not obviously improved due to a strain-rate dependence of the rubber cavitation (Dompas and Groeninckx, 1994; Jansen et al., 1999). Interestingly, the notched Izod impact toughness of the PLA/EVA50 blends was considerably improved in presence of 15 wt.% EVA50. By further increasing the EVA50 content, super-tough PLA/EVA50 blends could be obtained. The reason for brittleness of amorphous polymers is strain-localization, which could be delocalized by the dispersed rubber phase via a (pre)cavitation process. The morphology of PLA/EVA blends could be tuned by the VA content in the EVA copolymers as well as with the EVA content within the blends. The moderate particle size and the low modulus of the non-crosslinked EVA rubber particles are suitable for cavitation in the presence of tri-axial strain/stress [66, (Bucknall and Paul, 2009)]. Consequently, internal EVA rubber cavitation in the PLA matrix occurred under both the impact and tensile testing. Meanwhile, no obvious crazes were observed after deformation. In this regard, internal rubber cavitation in combination with matrix yielding is proposed to be the dominant toughening mechanism for the PLA/EVA blends.

To improve its toughness and crystallization, Zhang et al. (2013c) melt-blended PLA with ethylene/methyl acrylate/glycidyl

methacrylate terpolymer (EGA) containing relatively high-concentration of epoxide groups (8 wt%). Although we cannot exclude any coupling reaction between epoxide groups and end-functionality (hydroxyl) from PLA chains, the addition of EGA accelerated the crystallization rate and increased the final crystallinity of PLA in the blends. Significant enhancements in both toughness and flexibility of PLA were achieved by the incorporation of 20–30 wt% EGA. The impact strength increased from 3 kJ/m² of neat PLA to 60 kJ/m² and the elongation at break increased from 5 to 232% (**Table 11**). The failure mode changed from brittle to ductile fracture of the blend. The phase separated morphology with relatively good interfacial adhesion played an important role in the improvement in crystallization and toughness of the blend.

Petchwattana et al. (2012) utilized ultrafine rubbery particles as toughening agent to reduce the brittleness of PLA. Elastomeric particles of acrylate rubber were added to PLA in the range from 0.1 to 10 wt% (**Figure 10**). Maximum reduction of the flexural modulus and the tensile modulus was achieved by 20 and 45% respectively, when the acrylate rubber content was of 10 wt%. However, under stress, the rubber-modified PLA could be uniaxially deformed to elongation at break of nearly 200%, accounting for an increase by 50 times in comparison to PLA. The toughening efficiency of the ultrafine rubber particles was also reflected through the significant increase in the impact strength by a four-fold factor. Fractographs of the acrylate rubber-modified PLA revealed a plastic deformation and a good dispersion and adhesion of the rubber particles within the PLA matrix. Therefore, they played an important role in dissipating the energy by formation of multiple crazes. The crazing mechanism was found to be the major impact mechanism of the acrylate rubber modified PLA system.

Jiang et al. (2006) and Li and Shimizu (2007) attributed the toughening behavior of the PLA-based blends to debonding at the matrix/particle interface during deformation, which released the hydrostatic stress and facilitated shear yielding to occur.

Table 8 | Mechanical properties of different PLA blends with biodegradable flexible/soft polymers.

Formulation (wt.%)	E_t (MPa)	σ_y (MPa)	ε_b(%)	NIIE (KJ/m^2)			
PLA/PHBV (Ma et al., 2013)					U-NIIE (KJ/m^2)	E_f(GPa)	σ_f (MPa)
100/0		68	4	2.5	16	3.5	109
95/5		62	5	2.7	15	3.4	96
90/10		53	220	3.1	23	3.0	85
80/20		42	230	11	150n.f.	2.5	66
70/30		35	260	10	127n.f.	2.0	51
50/50		15	15	6	41	1.2	21
0/100		9	12	48	45n.f.	0.5	11
PLA/PBAT (Jiang et al., 2006)							
100/0	3400	63	3.7	2.6			
95/5	–	–	~115	–			
80/20	2600	47	>200	4.4			
PLA/PEBA (Han et al., 2013)							
100/0	1170 ± 42	60.0 ± 1.1	6.7 ± 0.4	4.5 ± 0.6			
95/5	1151 ± 75	49.3 ± 1.2	13.7 ± 0.8	7.1 ± 0.3			
90/10	1156 ± 44	46.8 ± 0.4	283 ± 18	7.4 ± 0.4			
85/15	1062 ± 48	42.5 ± 0.9	313 ± 2	9.1 ± 0.4			
80/20	1011 ± 41	42.4 ± 1.0	346 ± 18	39.3 ± 2.2			
70/30	911 ± 54	36.8 ± 0.6	335 ± 5	60.5 ± 1.0			
PLA/PAE (Zhang et al., 2009b)					E_s (MPa)	T_g PAE(°C)	T_g PLA(°C)
100/0	1814	46.8	5.1		2460		79.48
95/5	1517	48.1	161.5		2116	−47.31	77.85
90/10	1633	40.9	194.6		2017	−53.87	75.97
80/20	1240	23.7	184.6		1442	−57.89	74.47
70/30	1050	24.6	367.2		1395	−60.26	73.84
PLA/NR (Bitinis et al., 2011)							
Pristine PLA	2900 ± 100	63.1 ± 1.1	3.3 ± 0.4				
Processed PLA	3100 ± 40	58.0 ± 1.5	5.3 ± 0.7				
95/5	2500 ± 60	50.4 ± 1.6	48 ± 22				
90/10	2000 ± 50	40.1 ± 1.5	200 ± 14				
80/20	1800 ± 80	24.9 ± 0.9	73 ± 45				
PLA/GMS (Ge et al., 2013)							
100/0	1777 ± 42	69.8 ± 3.2	5.7 ± 0.3	4.7 ± 0.2			
95/5	1570 ± 44	44.8 ± 1.3	4.5 ± 0.5	8.1 ± 0.4			
90/10	1200 ± 12	41.9 ± 4.6	7.6 ± 2.4	8.5 ± 0.5			
85/15	1270 ± 36	39.7 ± 1.0	11 ± 5.0	15.5 ± 0.3			
80/20	1210 ± 17	35.1 ± 2.1	9.5 ± 6.5	36.7 ± 0.3			
75/25	1190 ± 24	32.4 ± 1.8	11 ± 3.1	46.1 ± 2.9			
70/30	695 ± 38	29.9 ± 2.6	45 ± 15.8	48.2 ± 4.6			

E_t, Tensile modulus; σ_y, Tensile yield stress; ε_b, Tensile elongation at break; NIE, Notched Izod Impact Energy; U-NIIE, Un-Notched Izod Impact Energy; E_f, Flexural modulus; σ_f, Flexural stress; E_s, Storage modulus; T_g, Glass transition temperature by DMA; n.f., Not (completely) fractured.

Zhao et al. (2013) (**Figure 11**) used a unique ultrafine full-vulcanized powdered ethyl acrylate rubber (EA-UFPR) as toughening modifier for PLA. Largely improved tensile toughness was successfully achieved by the incorporation of only 1 wt% EA-UFPR, while the tensile strength and modulus of the blends were almost the same as pure PLA. The highly efficient toughening of UFPR on PLA could be mainly ascribed to the strong interfacial interaction between PLA and UFPR as well as a good dispersion of UFPR particles in PLA matrix. This induces de-bonding cavitation at the PLA/UFPR interfaces during stretching, leading to an extensive energy dissipation and superior tensile toughness. It should be highlighted that this work provided an effective toughening method to largely improve the mechanical properties of PLA without sacrificing its stiffness, which is very important for the wide application of PLA materials.

Taib et al. (2012) toughened PLA with a commercially available ethylene acrylate copolymer impact modifier. PLA/impact modifier blends were partially miscible as confirmed by dynamic mechanical analysis. With increasing the impact modifier content, the stress-strain curves showed that the brittle behavior of PLA changed to ductile-failure. The blends showed some improvement in the elongation at break and notched impact strength, highlighting the toughening effects provided by the impact modifier again. In contrast, the yield stress and tensile modulus decreased with the increase in the impact modifier content (**Figure 12A**). Scanning electron microscopy micrographs revealed that the impact mechanisms among others involved shear-yielding or plastic deformation of the PLA matrix induced by interfacial de-bonding between the PLA and the impact modifier domains. In addition to shear-yielding of PLA, extensive deformation of the impact modifier domains was observed on the fractured surface, which accounts for the "partial" break of the blend after the impact test (**Figure 12B**).

PLA with polyethylene using PLLA-b-PE diblock copolymers as a compatibilizer (Anderson et al., 2003; Anderson and Hillmyer, 2004)

The addition of PLLA-b-PE block copolymers into the binary blend PLA/LLDPE resulted in improved interfacial adhesion and finer dispersion of LLDPE in PLA matrix. With the addition to the blend PLA/LLDPE (80/20, w/w) of 5 wt% of the block copolymer [with molecular weights for the PLA block above its entanglement molecular weight M_c, that is, PLLA-b-PE (30–30 w/w)], the impact strength was drastically increased to 460 J/m (**Figure 13**). This difference was attributed to the superior ability of the block copolymer from the long PLLA block to suppress the coalescence of dispersed phase. **Table 9** lists the impact strength properties as a function of the blend composition as well as some explanations of the occurring phenomena.

By increasing the amounts of PLLA-b-PE (30–30 wt/wt) block copolymer in the PLLA/LLDPE (80/20, wt/wt) blends, the size of dispersed LLDPE particles was gradually reduced. At 3 wt.% of block copolymer, the size of the dispersed LLDPE particles began leveling off at less than 1.0 μm, and the impact resistance drastically increased (**Figure 14** and **Table 10**).

Meng et al. successfully synthetized poly(butyl acrylate) (PBA) in order to melt-blend with PLA using a Haake Rheometer (Meng et al., 2012). Dynamic rheology, SEM and DSC results showed that PLA was partially miscible with PBA. The crystallinity of PLA increased with the content of PBA (<15 wt.%). By increasing PBA content, the tensile strength and modulus of the blend decreased slightly, while the elongation at break and toughness dramatically increased (**Table 11**). The failure mode changes from brittle to ductile fracture of the blend with PBA as well. SEM micrographs revealed that a de-bonding-initiated shear yielding mechanism is

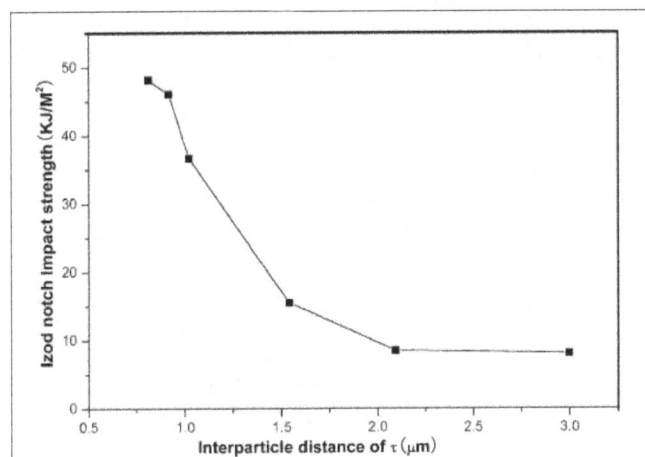

FIGURE 8 | The impact strength as a function of inter-particle distance of GMS. Copyright (2012) Wiley; used with permission from Ge et al. (2013).

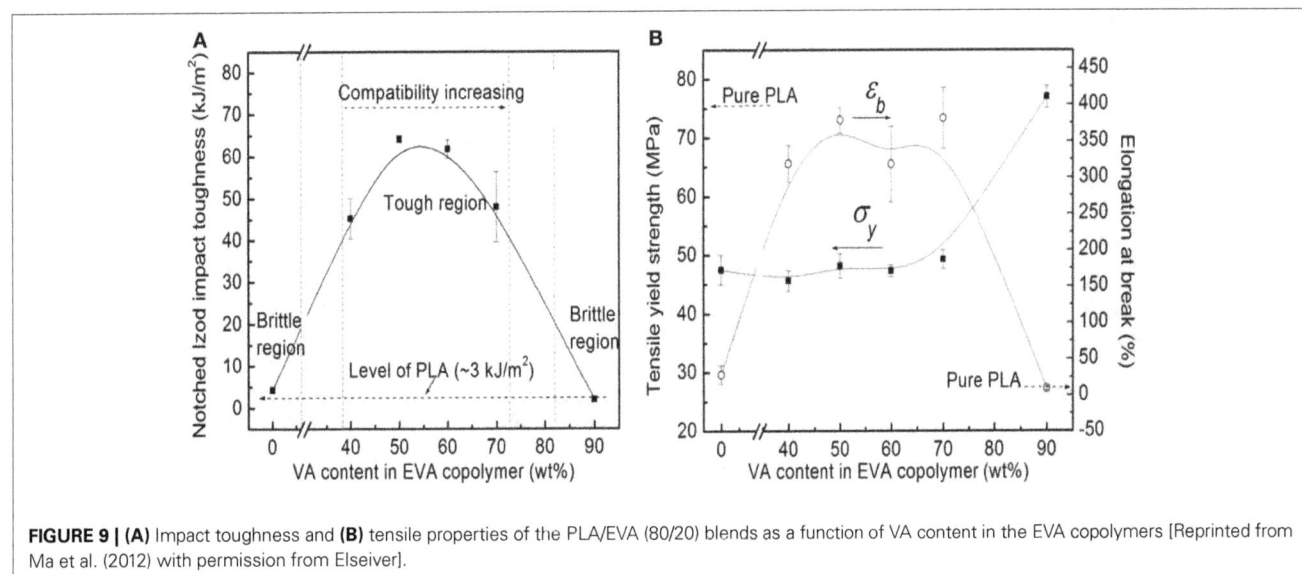

FIGURE 9 | (A) Impact toughness and **(B)** tensile properties of the PLA/EVA (80/20) blends as a function of VA content in the EVA copolymers [Reprinted from Ma et al. (2012) with permission from Elseiver].

involved in the toughening of the blend. Rheological investigation revealed that a phase segregation occurred at loading above 11 wt.% PBA. UV–vis light transmittance showed that PLA/PBA blends had a high transparency, but the transparency slightly decreased with the amount of PBA.

Commercially available impact modifiers for PLA. Recently, several polymeric impact modifiers have been specifically produced and commercialized in order to toughen brittle PLA (**Table 12**). These impact modifiers may be based on either linear thermoplastics/elastomers having a low glass transition temperature or crosslinked core-shell block copolymers, where the core is mainly a rubbery soft block encapsulated by a glassy and rigid shell that brings a good interfacial compatibilization with the matrix. In the optimal conditions (dispersion, compatibilization/adhesion, size and size distribution...), they dissipate the mechanical energy, retarding the initiation and propagation of micro-cracks through the polymer matrix.

Recently, Scaffaro et al. (2011) have compared toughening effects of OnCap™ BIO Impact T and Sukano® PLA im S550 on PLA. Both modifiers were immiscible with PLA, but Sukano® PLA im S550 displayed a more homogeneous dispersion in the PLA matrix. It was found that none of the impact modifiers

brought obvious increase in elongation to PLA. The maximum Izod impact strength of 141 J/m was achieved by adding 8 wt.% Sukano® PLA im S550, while the impact strength increased only to 124 J/m even with the addition of OnCap™ BIO Impact T. Murariu et al. (2008b) studied toughening effects of Biomax Strong® 100 on PLA and high-filled PLA/b-calcium sulphate anhydrite (AII) composites. Notched Izod impact strength of PLA containing 5 and 10 wt.% Biomax Strong® 100 increased from 2.6 kJ/m^2 of the neat PLA to 4.6 and 12.4 kJ/m^2, respectively. Elongation at break was more than 25% for the blend containing 10 wt.% of the impact modifier, while tensile strength and modulus of PLA gradually decreased with the addition of the impact modifier. Addition of 5 and 10 wt% of the impact modifier to the PLA/AII (70/30, wt/wt) composite also increased their impact strength to 4.5 and 5.7 kJ/m^2, respectively. Impact strength slightly decreased with further increase of the filler loading to 40 wt.%, but remained higher than that of both the unmodified composites and the neat PLA. On the other hand, for the PLA composites containing 40% of filler, tensile strength and elongation markedly decreased with the incorporation of the impact modifier. Zhu et al. (2009) studied the films of PLA blends containing either Biomax Strong® 100 or Sukano® PLA im S550 as a toughening agent. It was shown that the modulus decreased when increasing the concentration of Biomax Strong® 100 modifier, but was relatively independent of the concentration of Sukano® PLA im S550. The maximum elongation was of 255% in presence of 12 wt.% of BiomaxVR Strong 100 and of 240% in presence of 8 wt.% of Sukano® PLA im S550, while elongation at break of neat PLA was of about 90%. For a given composition, the latter impact modifier gave a clearer film than Biomax® Strong 100, but the clarity of films decreased as the concentration increased for both toughening agents. Afrifah and Matuana (2010) investigated the toughening mechanisms of PLA blended with an ethylene/acrylate copolymer (EAC) to show a mode of fracture through crazing or microcracking and debonding of impact modifier particles with the matrix. This resulted in brittle failure at low content. Higher impact modifier content than 10 wt% revealed fracture mechanisms including impact modifier debonding, fibrillation,

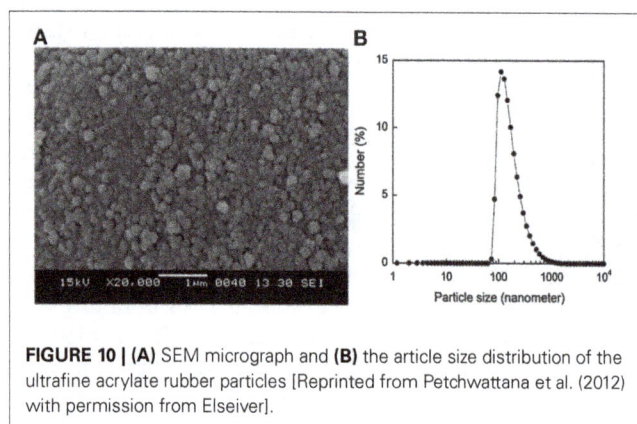

FIGURE 10 | (A) SEM micrograph and **(B)** the article size distribution of the ultrafine acrylate rubber particles [Reprinted from Petchwattana et al. (2012) with permission from Elseiver].

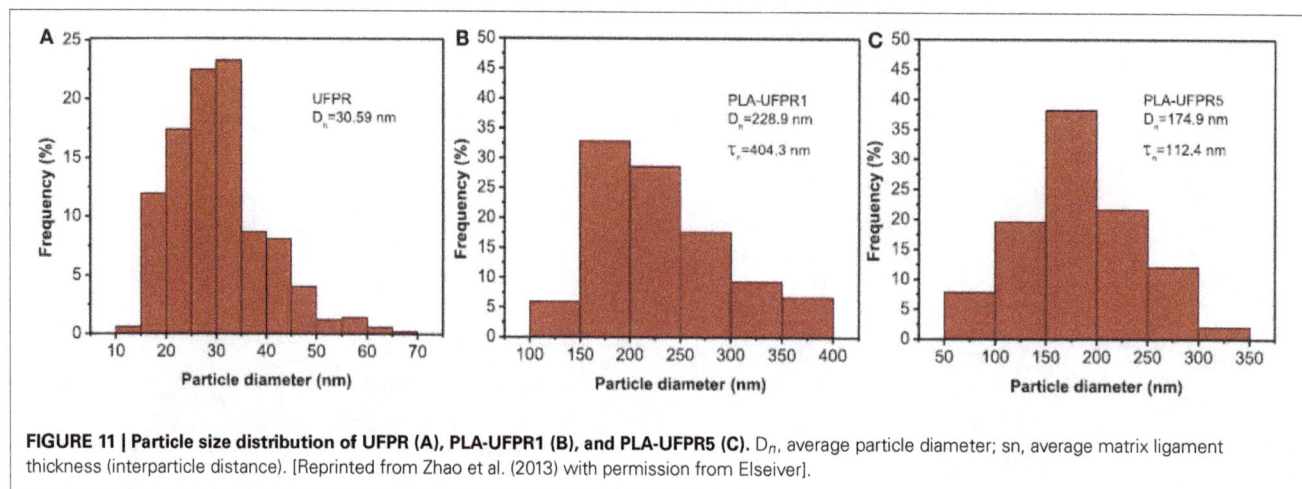

FIGURE 11 | Particle size distribution of UFPR (A), PLA-UFPR1 (B), and PLA-UFPR5 (C). D$_n$, average particle diameter; sn, average matrix ligament thickness (interparticle distance). [Reprinted from Zhao et al. (2013) with permission from Elseiver].

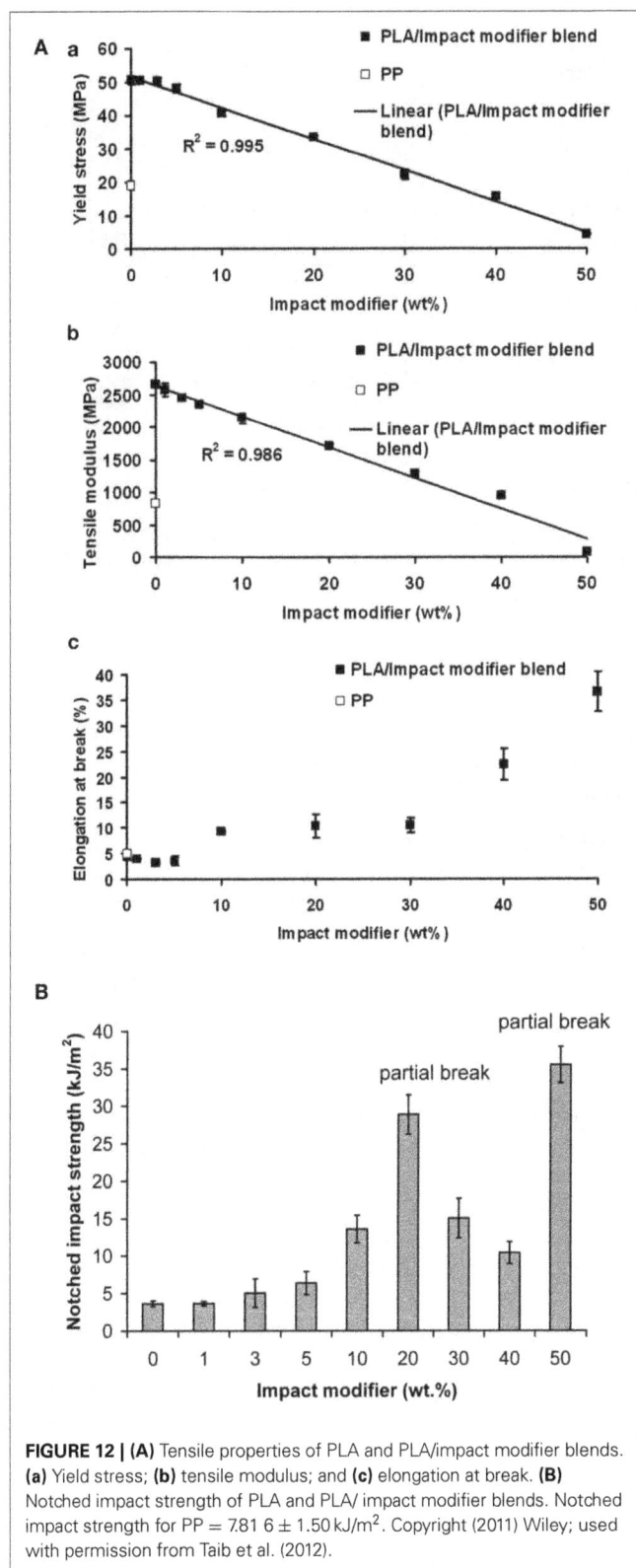

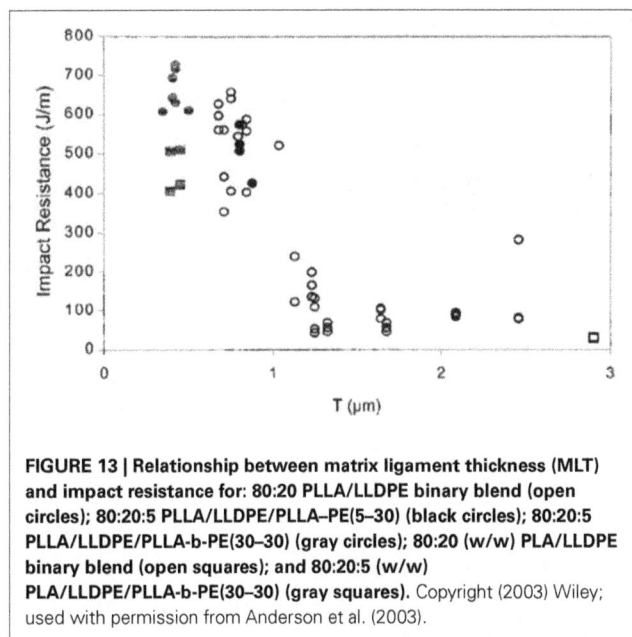

FIGURE 13 | Relationship between matrix ligament thickness (MLT) and impact resistance for: 80:20 PLLA/LLDPE binary blend (open circles); 80:20:5 PLLA/LLDPE/PLLA–PE(5–30) (black circles); 80:20:5 PLLA/LLDPE/PLLA-b-PE(30–30) (gray circles); 80:20 (w/w) PLA/LLDPE binary blend (open squares); and 80:20:5 (w/w) PLA/LLDPE/PLLA-b-PE(30–30) (gray squares). Copyright (2003) Wiley; used with permission from Anderson et al. (2003).

FIGURE 12 | (A) Tensile properties of PLA and PLA/impact modifier blends. (a) Yield stress; (b) tensile modulus; and (c) elongation at break. (B) Notched impact strength of PLA and PLA/ impact modifier blends. Notched impact strength for PP = 7.81 6 ± 1.50 kJ/m². Copyright (2011) Wiley; used with permission from Taib et al. (2012).

crack bridging, and matrix shear yielding, resulting in a ductile behavior. They also demonstrated that Biomax® Strong 100 yielded superior toughening on semi-crystalline PLA over amorphous PLA. With 40 wt.% of the toughening agent, the notched

Izod impact strength of the semi-crystalline PLA increased from 16.9 J/m for amorphous PLA to 248.4 J/m for semi-crystalline PLA. In addition, the presence of 15 wt.% Biomax® Strong 100 reduced the brittle-to-ductile transition temperature of PLA, as revealed by the notched Izod impact test data from the frozen specimens. Ito et al. (2010) investigated the fracture mechanism of neat PLA and PLA blends toughened with an acrylic core–shell modifier. The acrylic modifier was composed of a crosslinked alkyl acrylate rubber core and PMMA shell, and the particle size was in the range of 100–300 nm. Plane strain compression testing of PLA clearly showed strong softening after yielding. Because the stress for craze nucleation was close to that of yield stress, brittle fracture occurred for neat PLA. Addition of the acrylic modifier significantly lowered the yield stress and formed many microvoids. The release of strain constrained by microvoiding and the decrease of yield stress led to the relaxation of stress concentration, and therefore the toughness was improved moderately. **Table 9** summarizes the reported mechanical properties of some of highly toughened PLA blends prepared *via* melt-blending. From **Table 9**, it results that in order for a rubbery polymer to impart toughness to PLA or any other polymer, several criteria must be encountered as follows (Natureworks LLC, 2007):

– the rubber must be distributed as small domains (usually 0.1–1.0 μm) in the matrix polymer;
– the rubber must have a good interfacial adhesion to PLA;
– the glass transition temperature of the rubber must be at least 20°C lower that the test/use temperature;
– the molecular weight of the rubber must not be too low;
– the rubber should not be miscible, to a certain extent, with the polymer matrix;
– and the rubber must be thermally stable to PLA processing temperatures.

Table 9 | Results summary and explanation.

Blend system	Compatibilizer (PLLA-b-PE)	Impact strength (J/m)	Explanation
a-PLA	–	12	
a-PLA/LLDPE (80wt%/20wt%)	–	34	
a-PLA/LLDPE (80wt%/20wt%)	5 wt%[a]	36	The difference is attributed to superior ability of the PLLA-b-PE from the longer PLLA block to suppress the coalescence of the dispersed phase
a-PLA/LLDPE (80wt%/20wt%)	5 wt%[b]	460	
PLLA	–	20	
PLLA/LLDPE (80wt%/20wt%)	–	350	Adhesion test showed a superior interfacial adhesion when used semicrystalline PLLA instead of amorphous a-PLA
PLLA/LLDPE (80wt%/20wt%)	5 wt%[a]	510	The tacticity effects on either the Mc of PLA or the miscibility degree of PLA matrix with LLDPE phase accounted for the difference between the two binary blends
PLLA/LLDPE (80wt%/20wt%)	5 wt%[b]	660	

a-PLA, amorphous PLA; PLLA, semicrystalline PLA; LLDPE, linear low density polyethylene.

[a] The molecular weight of PLLA block in PLLA-b-PE is $5\,kg/mol < M_c = 9\,kg/mol$.

[b] The molecular weight of PLLA block in PLLA-b-PE is $30\,kg/mol > M_c = 9\,kg/mol$.

$M_c = 9\,kg/mol$ is the critical entanglement molecular weight of PLLA block in PLLA-b-PE.

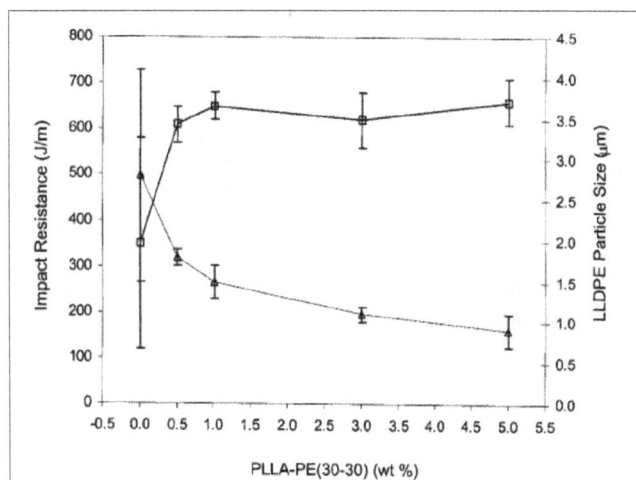

FIGURE 14 | Effect of the amount of PLLA–PE(30–30) block copolymer on the impact resistance (squares) and the LLDPE particle size (triangles) of 80: 20 PLLA/LLDPE blends. Copyright (2003) Wiley; used with permission from Anderson et al. (2003).

Table 10 | Particle size analysis and impact resistance of PLLA homopolymer and blends [Copyright (2003) Wiley; used with permission from Anderson et al. (2003)].

PLLA/LLDPE/ PLLA-PE block	PLLA-PE block	LLDPE particle size (μm)	Izod impact resistance (J/m)
100/0/0			20 ± 2
80/20/0		2.8 ± 1.3	350 ± 230
80/20/5	5–30	1.9 ± 0.2	510 ± 60
80/20/5	30–30	0.9 ± 0.2	660 ± 50

with rubbery polymers such as low modulus polyesters, linear elastomers, or cross-linked core-shell impact modifiers, which have been observed to impart the highest degree of toughening in PLA. These modifiers typically consist of a low T_g crosslinked rubbery core ($T_g < -10°C$) encapsulated by a glassy shell polymer ($T_g > 50°C$) that has good interfacial adhesion with the matrix polymer. When well-dispersed, these modifiers act as nanoscale or microscale rubbery domains that dissipate mechanical energy to retard or arrest crack initiation and propagation through the polymer. Some possible major drawbacks resulting from blending PLA with impact modifiers are the dispersion of the micro-domains in the PLA matrix and the transparency of the resulting material. The latter case depends to some extent on the dispersion of the micro-domains and their size. The addition of impact modifier to PLA often results in a substantial decrease in clarity of the toughened blend, although high clarity is required for many PLA applications. To retain the good clarity and transparency of PLA, for instance, small dispersed particles have to have a similar refractive index as PLA as well as the particle size has to be inferior to the visible light wavelength. The use of very small rubber particles with refractive indexes

These factors will allow the rubber to induce energy dissipation mechanisms in PLA, which retard crack-initiation and propagation and ultimately result in a material with improved toughness. PLA is similar to many polymers that can undergo plastic flow mechanisms, initiated by dispersed rubber domains. The increase in toughness comes from the transfer of the impact energy to plastic flow, either in the form of crazing or shear yielding mechanisms through a large volume fraction of polymer. In PLA, excellent toughness balance can be obtained with 15–25% of impact modifier, even if little improvement can be seen by the addition of 3–5% of impact modifier. Typically like most conventional thermoplastics, PLA can be toughened after blending

Table 11 | Mechanical properties of different PLA blends with non-biodegradable flexible/soft polymers.

Formulation (wt.%)	E_t (MPa)	σ_y (MPa)	ε_b(%)	NIIE (KJ/m^2)			
PLA/EVA50 (Ma et al., 2012)					Hardness (Shore D)	E_f(GPa)	σ_f (MPa)
100/0		75	9	3	86	3.7	105
95/5		68	310	2	85	3.3	90
90/10		61	390	5	84	2.9	75
85/15		54	430	32	82	2.7	70
80/20		45	340	64	80	2.4	65
70/30		37	400	83	76	1.9	50
PLA/EGA (Zhang et al., 2013c)							
100/0	1745 ± 39	60.0 ± 3.0	4.9 ± 0.3	3.0 ± 0.4			
90/10	1530 ± 35	44.3 ± 2.1	23.4 ± 3.6	3.9 ± 0.3			
80/20	1154 ± 42	33.8 ± 2.4	232.0 ± 26	59.8 ± 5.1			
70/30	945 ± 49	24.9 ± 1.3	126.0 ± 21	53.2 ± 8.4			
PLA/PEBA (Petchwattana et al., 2012)				NIIE (J/m)			
100/0	2750 ± 120	61.22 ± 1.42	3.46 ± 1.42	23.66 ± 1.33			
99.9/0.1	2660 ± 60	61.69 ± 1.99	3.53 ± 0.19	26.62 ± 1.87			
99.7/0.3	2680 ± 80	61.57 ± 1.76	5.01 ± 0.34	33.18 ± 2.01			
99.5/0.5	2550 ± 70	58.34 ± 0.94	8.94 ± 1.82	36.89 ± 2.43			
99.3/0.7	2310 ± 140	58.17 ± 1.83	15.1 ± 2.07	38.12 ± 1.95			
99/1	2350 ± 380	58.69 ± 0.91	19.8 ± 4.97	52.15 ± 2.57			
97/3	2050 ± 120	53.89 ± 0.84	53.7 ± 4.93	64.59 ± 3.46			
95/5	2150 ± 220	54.22 ± 0.97	124 ± 25.9	86.95 ± 4.65			
93/7	2040 ± 210	50.31 ± 0.93	167 ± 24.4	96.21 ± 4.99			
90/10	2000 ± 250	48.98 ± 1.79	198 ± 31.7	101.0 ± 5.63			
PLA/UFPR (Zhao et al., 2013)							
100/0	2062 ± 12	68.05 ± 1.42	6.08 ± 0.36	1.60 ± 0.21			
99.5/0.5	1922 ± 66	67.53 ± 1.99	106.60 ± 15.08	2.00 ± 0.15			
99/1	1896 ± 2	66.26 ± 1.76	219.93 ± 2.64	2.2 ± 0.23			
97/3	1768 ± 54	65.67 ± 0.94	231.45 ± 20.55	2.6 ± 0.37			
95/5	2029 ± 129	65.39 ± 1.83	215.63 ± 12.21	3.20 ± 0.19			
PLA/PBA (Meng et al., 2012)				Tensile toughness[a] (MJ/m^3)			
100/0	3510	68	4.52	2.13			
95/5	1540	51.77	31.52	3.7			
92/8	1490	44.79	74.62	17.0			
89/11	1440	40.82	173.98	41.74			
85/15	1330	41.01	174.52	47.02			

E_t, Tensile modulus; σ_y, Tensile yield stress; ε_b, Tensile elongation at break; NIIE, Notched Izod Impact Energy; E_f, Flexural modulus; σ_f, Flexural stress.
[a] Calculated as the area under stress-strain curve.

comparable to that of PLA can help maintaining transparency (refractive index in the range of from 1.430 to 1.485). This can be achieved if the added rubber is slightly compatible with PLA. Moreover, a poor compatibility and interfacial adhesion can also result in partially dispersed and large cluster-like domains responsible for the de-bonding of the rubber phase, void-formation, and a premature failure in a brittle mode. The main negative consequence resulting from the incorporation of toughening agents into PLA is the reduction of the material stiffness (elastic modulus). Accordingly, many researchers have investigated the incorporation of rigid fillers in order to compensate the loss of stiffness.

Generally, fillers or fibers are incorporated in PLA to either reduce the cost or modify the physical, rheological, or optical properties of the polymer. Starch is for instance an excellent example, which is available at less than $ 0.10/pound and which retains both renewability and biodegradability characteristics of PLA while enhancing some structural and mechanical properties at room or elevated temperature. Some additives (e.g., talc), can increase the nucleation rates and crystallinity, and

Table 12 | Injection moulded properties of PLA containing various commercial impact modifiers.

Manufacturer	Impact modifier (IM)	Nature	Characteristics of PLA/IM	Optimal load in PLA	IM loading (%)	Notched izod impact (J/m)	Tensile yield (MPa)	Elongation[a] (%)
Sukano	Sukano® PLA im S550 (Sukano Co., 2008; Scaffaro et al., 2011)	Linear elastomer	Transparent, compostable, cost effective, immiscible	4% (impact resistance improved by a factor of 10)	8	141 (87)		8 (3)
	Sukano® PLA im S555 (Co, 2010)	Linear elastomer	Transparent, compostable, cost effective					
PolyOne	OnCap™ BIO Impact T (http://www.polyone.com/en-us/docs/Documents/OnCap%20BIO%20Impact%20T.pdf. Accessed on 2008.; http://www.polyone.com/en-us/news/Press%20Release%20Attachments/Chinaplas%20OnCap%20BIO%20impact%20T.pdf. Accessed on April 19, 2010.; Scaffaro et al., 2011)	Linear elastomer	Transparent, tear resistant, immiscible		12.5	124 (87)		8 (3)
DuPont	Biomax® Strong 100 (DuPont Co. 2010).	Ethylene-acrylate copolymer	Non-food applications	<5% for good toughness and clarity	12			255
	Biomax® Strong 120	Ethylene-acrylate copolymer	Food packaging					
	Hytrel™3078	Aliphatic/aromatic copolyester	Very compatible	5–10% (High elongation at break and impact properties)	30	198 (27)	34 (62)	430 (10)
Rohm and Hass	Polaroid™ BPM-500	Acrylic-based	Transparent	3–5% (dart drop impact increased by four time with respect to neat PLA)				
	Polaroid™ KM 334				15	59 (27)	37 (62)	165 (10)
	Polaroid™ BTA 753				20	112 (27)	36 (62)	300 (10)
	Polaroid™ EXL 3691 A				20	64 (27)	43 (62)	280 (10)
	Polaroid™ EXL 2314				20	53 (27)	38 (62)	250 (10)

(Continued)

Table 12 | Continued

Manufacturer	Impact modifier (IM)	Nature	Characteristics of PLA/IM	Optimal load in PLA	IM loading (%)	Notched izod impact (J/m)	Tensile yield (MPa)	Elongation[a] (%)
Dow chemical company	Polaroid™ BPM-515	Acrylic-based	Transparent	1% (higher efficiency than Polaroid™ BPM-500)				
	Pellethane™2102-75A				30	769 (27)	42 (62)	410 (10)
Arkema	Biostrength™ 130	Acrylic core-shell structure	Transparent	2–6 wt.%				
	Biostrength™ 150	MMA-butadiene-styrene core-shell structure	Opaque	2–6 wt.%				
	Biostrength™ 280	Acrylic core-shell structure	Transparent	2–6 wt.%				
Procter and gamble metabolix	Nodax™	Aliphatic copolyester (PHA)		10–20 wt.% (rubber domains size 0.2–1.0 µm)				
Karton polymers LLC.	Karton™FG 1901X	Functionalized Elastomer			20	-	33 (62)	100 (10)
BASF	Ecoflex™	Aliphatic/aromatic copolyester	Very compatible	5–10% (High elongation at break and impact properties	30	128 (27)	43 (62)	100 (10)
Showa higher polymer	Bionelle™3001	Aliphatic polyester			15	48 (27)	59 (62)	230 (10)
Crompton corporation	Blendex™ 415				15	48 (27)	44 (62)	230 (10)
	Blendex™ 360	ABS resin (70% of butadiene)			20	107 (27)	47 (62)	280 (10)
	Blendex™ 338				20	518 (27)	43 (62)	281 (10)
Reference	PLA				0			10

[a] According to ASTM D-638 at 2.0 inch/min.

thereby improve heat resistance of PLA-based materials (Bopp and Whelan, 2003). Fillers and fibers can also increase the stiffness of PLA and, to some extent, enhance the toughness of PLA materials. This can be explained by the fact that the crystallization extent of PLA is enhanced on incorporation of fillers, and therefore yielding a ductile behavior for the resulting PLA-based materials (Urayama et al., 2003). In order to get the maximum benefits from the fibers or fillers, several factors must be considered. For instance, optimizing the extruder screw configuration, through-put rate, screw speed, temperature and other process parameters are necessary. The optimal particle size of the filler is generally in the range of 0.1–12 μm (Ikado et al., 1997). A good and uniform dispersion must be achieved as well. This is normally obtained by controlling the addition of the compatibilizers during processing (Kjeschke et al., 2001), which help the dispersion of filler/fiber as well as minimizes micro-defects in blends that can cause embrittlement. For instance, the affinity of organically modified clay in PLLA/PBS blends was found to be critical for the properties enhancement of resulting composites (Chen et al., 2005). When a commercially available nanoclays, i.e., 10 wt% Cloisite® 25A was used as compatibilizer into the PLLA/PBS (75/25, w/w) blend, the tensile modulus of blends increased from 1.08 GPa to 1.94 GPa, but the elongation at break was sacrificed from 71.8 to 3.6% at the same time, which was even lower than that of neat PLLA (6.9%). In contrast, using an epoxy-functionalized organoclay (TFC) at the same amounts that was able to react with the extremities (carboxylic/hydroxylic) of both polyesters, not only retained high tensile modulus, but

also increased elongation to 118%. Similar compatibilizing effect of TFC on the PLA/PBSA (75/25 w/w) blends were reported (Chen and Yoon, 2005). Odent et al. reported immiscible polymer blends made of PLA toughened with Biomax Strong® 100 in order to elaborate ultratough PLA-based materials mediated with nanoparticles (ref. Odent et al., 2013c). The co-addition of 10 wt% of Biomax Strong® 100 and 10 wt% of silica nanoparticles (CAB-O-SIL TS-530 from CABOT) provided an increase from $2.7 \, kJ/m^2$ to $30.2 \, kJ/m^2$, which is related to the formation of peculiar morphologies (from round-like nodules to elongated structure) mediated by the localization of nanoparticles at the interface PLA/impact modifier. Same improvement was also reached by replacing silica with organomodified layered aluminosilicate (clay), with value of 32.6, 37.6 and $21.9 \, kJ/m^2$ with only 1 wt% of Cloisite 20A, Cloisite 25A, and Cloisite 30B, respectively. Coupling agents are often used with glass fibers (Mochizuki and Suzuki, 2004) or coated fillers to enhance the interfacial adhesion of the additive to the matrix polymer, more particularly when polar additives are combined with non-polar polymers. Silane and titinate coupling agents with various structures, which depends on the polymer to be blended, are often embedded onto glass fibers and inorganic particulate fillers. These coupling agents can have beneficial effects on dispersion, toughness and rheology, and often allow higher levels of incorporation. However, the desired beneficial effects from the addition of fillers and fibers can have some negative consequences. High levels of fillers/fibers can significantly increase viscosity, cause shear heating and degradation (molecular weight loss and color formation), and

Table 13 | Mechanical properties of some fillers blended with PLA (http://www.natureworksllc.com/~/media/Technical_Resources/ Properties_Documents/PropertiesDocument_Fillers-and-Fibers_pdf.pdf).

PLA filled with	Filler loading (%)	Flex modulus[a] (MPa)	Dart impact at 23°C (KJ/m²)	IZOD impact[b] (notched) (J/m)	IZOD impact[b] (un-notched) (KJ/m²)
Specialty minerals MTAGD609 Talc	1.5	3940	123	43	331
	10	5002	112	27	272
	30	9248	69	27	176
Vicron 15–15 CaCO₃	1.5	3809	107	32	272
	10	4287	128	27	288
	30	5606	128	32	187
Specialty minerals Mica 5040	1.5	4009	123	32	256
	10	5366	139	27	198
	30	9874	85	32	123
Synthetic silicate	1.5	3855	144	32	304
	10	4345	117	27	214
	30	5761	96	21	112
Specialty minerals EMforce™ Bio	1.5	3876	128	32	203
	10	4457	134	32	171
	30	5687	1057	123	294
Unmodified NatureWorkx™ PLA 4032D		3651	160	37	235

[a] ASTM D 790.

[b] ASTM D 256-92.

affect the ability to fill thin walled parts. In the case of natural fibers, they contain high levels of moisture, requiring to apply extensive drying step before processing. It is worth noting that adding large amounts of natural fiber into the extruder requires using side-feeders for uniform extrusion operations. The batch-to-batch variation in natural fiber composition and quality can lead to consistency problems in the final blend as well (http://www.natureworksllc.com/~/media/Technical_Resources/ Properties_Documents/PropertiesDocument_Fillers-and-Fibers_ pdf.pdf).

Visual problems are also an issue with flow lines, poor colorability, and opacity. **Table 13** is listing the mechanical properties of some fillers blended with PLA.

Jiang et al. (Long et al., 2009) compared the effects of organo-modified montmorillonite (OMMT) and nanosized precipitated calcium carbonate (NPCC) on the mechanical properties of PLA/PBAT/nanofiller ternary composites. Mechanical testing demonstrated that the composites containing OMMT exhibited higher tensile strength and Young's modulus, but with lower elongation, as compared to those containing NPCC. When 25 wt.% PLA was replaced by maleic anhydride grafted PLA (PLA-g-MA), the elongation of the ternary composites was substantially increased, possibly as a result of the improved dispersion of the nanoparticles and enhanced interfacial adhesion from maleic anhydride moieties along the PLA backbone. Among these composites, PLA/10 wt.% PBAT/2.5 wt.% OMMT/25 wt.% PLA-g-MA demonstrated the best overall properties with 87% retention of tensile strength of pure PLA, slightly higher modulus and significantly improved elongation at break (16.5 times than that of neat PLA). Teamsinsungvon et al. (2010) also reinforced PLA/PBAT blends using microsized precipitated $CaCO_3$ and achieved similar toughening effects on PLA/PBAT (80/20, wt/wt) blends. The incorporation of talc particles significantly accelerated the crystallization process of the PLA matrix (Battegazzore et al., 2011). The presence of crystals improved the thermomechanical properties. Talc provides both reinforcing and toughening effects on PLA (Yu et al., 2012). The reinforcing effect of talc particles can be mainly ascribed to the good interfacial adhesion between PLA matrix and the orientation of talc layers during processing. Interfacial debonding of PLA/talc composite can induce massive crazing, meanwhile talc particles diffused in PLA matrix can prevent from the void coalescence and propagation of the crazes, which increase the toughness of PLA. Additionally, talc layers aligned along the tensile direction make its toughening effect on PLA more significant in tensile test because they

induce more advantageous shear yielding and prevent microcracks from propagation along fracture direction. Some aggregation of talc particles can appear in composites at higher talc content, which act as a stress concentration points or weak points and cause poor toughness of PLA/talc composites. Recently, NatureWorks® has succeeded to develop the Ingeo™ 3801X grade by co-adding impact modifier, crystal accelerant, reinforcing agent and nucleating agent into PLA. This PLA-based grade is a high heat and impact resistance material. **Table 14** shows its composition.

The bio-content of this material is of 70%. It was designed for non-food, opaque, semi-durable, and non-compostable applications. Due to its high crystallinity and rapid crystallization rate, the resulting blend has good thermal and dimensional stability. It is designed to be processed at fast cycle times. Its mechanical properties are summarized in **Table 15**.

Physical compounding of low or high molecular weight compounds offers a convenient approach to modifying biopolymers. Whether a good compatibility/affinity is present between both partners, the resulting blends exhibit good properties, being intermediate from that of each polymeric partner. However, only few biopolymer pairs are miscible or even compatible with each other. Therefore, chemical routes such as chemical modification or reactive compatibilization are required. Reactive blending has proven to be a simple, economically viable, and reliable technology for designing complex nanostructured polymeric blends. In this part, it will be pointed out, by reviewing the recent advances, that reactive extrusion technology represents the most promising approaches to tune the stiffness-toughness balance of bio-sourced polymeric materials. Reactive extrusion enables to manufacture biodegradable polymers through different routes of

Table 15 | Thermal and mechanical properties of the Ingeo™ 3801X grade (http://www.biocom.iastate.edu/workshop/2010workshop/ 2010workshop/presentations/dan_sawyer.pdf).

Mechanical properties	Value	ASTM standard
Tensile Modulus (MPa)	2980	D638
Tensile yield strength (MPa)	25.9	D638
Tensile elongation at break (%)	8.1	D638
Notched Izod impact (J/m)	144	D256
HDT B at 0.45 MPa (°C)	65	E2092
HDT at 0.114 MPa (°C)	140	E2092

Table 14 | Formulation of the Ingeo™ 3801X grade (http://www.biocom.iastate.edu/workshop/2010workshop/presentations/dan_sawyer.pdf).

Material	Commercial name	Supplier	Chemical	Formula weight fraction	25°C density [g/ml]
Matrix	Ingeo™ 3001D	NatureWorks LCC	PLA	0.711	1.24
Impact modifier	Biostrength® 150	Arkema Inc.	Proprietary core-shell copolymer	0.100	1.00
Crystal accelerant	Plasthall® DOA	The HallStar Company	Dioctyl adipate	0.090	0.98
Reinforcing agent	ULTRATALC® 609	Specialty Minerals Inc.	<0.9 μm particle 3MgO.4SiO$_2$.2H$_2$O	0.090	2.8
Nucleating agent	LAK-301	Takemoto Oil & Fat Co. LTD	Aromatic sulfonate derivative	0.009	1.00

reactive modification (polymerization, grafting, compatibilization, branching, functionalization,...) in a cost-effective polymer processing (Michaeli et al., 1993; Mani et al., 1999). Most of the researchers employ this technology for the reactive compatibilization of PLA with a rubbery modifier in order to impart toughness to PLA. In the review, the *in situ* reactive compatibilization is defined here as the melt-blending process in which two polymers containing mutually reactive functionalities react with each other at the interface to combine them during melt blending, generating *in situ* block or graft copolymer. The *in situ* generated block or graft copolymer will be then able to compatibilize the blend by reducing the interfacial tension and by improving the interfacial adhesion. This leads to a significantly improved dispersibility of the rubber into much smaller particles (Baker et al., 2001). Compatibility is defined in this context as the ability of the rubber modifier to finely disperse into the main PLA phase in order to form stable morphologies of fine rubber particle dispersions with reduced interfacial tension and improved adhesion to resist delamination. Early patent literature recognized the need for some types of grafting reactions or an associative interaction between the polymeric components of the blend to obtain sufficient compatibility for good impact modification (Kray and Bellet, 1968; Seddon et al., 1971; Murch, 1974; Mason and Tuller, 1983). Interfacial compatibilization and toughening is achieved through grafting copolymers generated *in situ* at the interface during the melt blending. There are four fundamental requirements to be addressed for the *in situ* reactive compatibilization in an extrusion device:

- Sufficient mixing to achieve the desired distribution and dispersion of one polymer into another polymer;
- The presence of reactive functionalities in each phase capable of reacting together across the interphase;
- Reaction to take place within the available residence time in the processing device;
- Formation of stable bonds during processing.

In the case of polymers with no reactive chemical function (such as polyolefins), peroxides are used to create free polymeric macroradicals in the blend. Like in some cases the compatibilization cannot be achieved through the direct reaction between the free polymeric macro-radicals, low molecular weight (macro)monomers, or a mixture of low molecular weight (macro)monomers have to be grafted on the free polymeric radicals in order to functionalize the polymeric chains. The role of these grafted chemical functions is to (1) stabilize the macroradicals, and thereby avoid any undesirable free radical side-reactions by localizing the free radical reactions at the polymers' interface, and (2) interact at the interface between the polymer components of the blend for compatibilization. PLA has been blended and reactive compatibilized with several biodegradable and non-biodegradable polymer modifiers that will be discussed here. We have to mention that other researchers have attempted using reactive extrusion technique to *in situ* synthetize the toughening agent for PLA.

Biodegradable hyperbranched polymers (HBP)

Hyperbranched polymer-based nanostructures (HBPs) have a globular molecular architecture with cavernous interiors, many peripheral functional end-groups, and low hydrodynamic volume and viscosity. They may have better miscibility with other polymers than the linear analogous (Seiler, 2002). Due to their inherently high surface area—volume ratio, structures engineered at the nano-meter length scales are increasingly played key-roles in the enhancement of the materials mechanical properties. In this regard, they have demonstrated a high potential to be used as impact modifiers for mechanical performances in a variety of industrial applications after a reactive process (Liu and Zhang, 2011). For instance, non-reactive melt-blending (physical compatibilization via H-bonding) of a hyperbranched biodegradable poly(ester amide) with PLA modestly enhance, yield strength. Moreover, the tensile failure mode changed from brittle to ductile fracture and led to a maximum tensile elongation at break of 50% compared to 3.7% for neat PLA using 20 wt.% HBP (Lin et al., 2007). This was explained by the fact that the dispersion of hyperbranched biodegradable poly(ester amide) was not fine enough

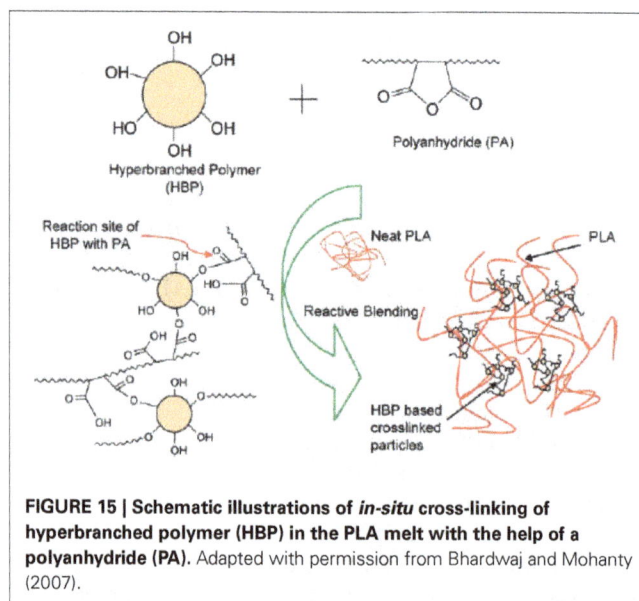

FIGURE 15 | Schematic illustrations of *in-situ* cross-linking of hyperbranched polymer (HBP) in the PLA melt with the help of a polyanhydride (PA). Adapted with permission from Bhardwaj and Mohanty (2007).

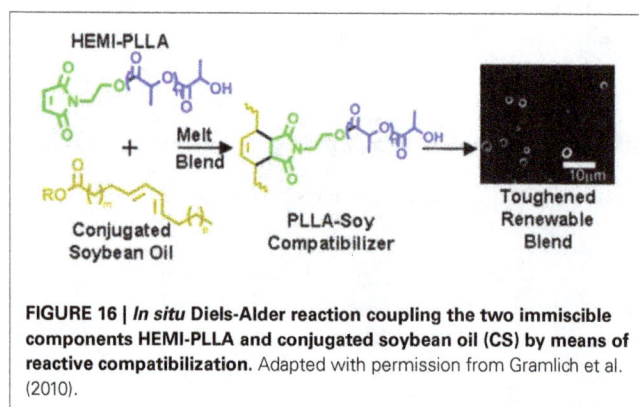

FIGURE 16 | *In situ* Diels-Alder reaction coupling the two immiscible components HEMI-PLLA and conjugated soybean oil (CS) by means of reactive compatibilization. Adapted with permission from Gramlich et al. (2010).

to get its maximum benefits. In this respect, Bhardwaj and Mohanty (2007) proposed and demonstrated a new industrially relevant methodology to develop PLA-based materials, having outstanding stiffness-toughness balance through *in situ* cross-linking reactions. They *in-situ* cross-linked a hydroxyl functional hyperbranched polymer (HBP) with a polyanhydride (PA) in the PLA matrix during melt-processing (**Figure 15**). Transmission electron microscopy (TEM) and atomic force microscopy (AFM) revealed the sea-island morphology of PLA-cross-linked HBP reactive blend. The domain size of cross-linked HBP particles in the PLA matrix was less than 100 nm. Compared to unmodified neat PLA, the PLA/HBP/PA (92/5.4/2.6, wt/wt/wt)

Table 16 | Physical properties of melt blends of CS with PLLA-49 and HEMI-PLLA-67 (Adapted with permission from Gramlich et al. (2010)].

Matrix polymer	W_{CS0} (%)[a]	W_{CS} (%)[b]	E (GPa)[c]	σ_b(MPa)[d]	ε_b(%)[e]	X_{HP} (%)[f]	X_{CS} (%)[g]	$d_{lm}(\mu m)$[h]	$\sigma_{lm}(\mu m)$[i]	MLT (μm)[j]
PLLA-49			2.4 ± 0.3	58 ± 3	5 ± 2					
HEMI-PLLA-67			3.0 ± 0.2	67 ± 9	4 ± 1					
PLLA-49	15	9	2.4 ± 0.3	28 ± 4	22 ± 7			1.81	1.8	3.1
HEMI-PLLA-67	15	7	2.0 ± 0.5	34 ± 2	50 ± 30	100	14	0.91	2.0	2.2
PLLA-49	5	4	2.1 ± 0.3	38 ± 1	30 ± 10			1.17	2.0	3.6
HEMI-PLLA-67	5	7	2.5 ± 0.2	37 ± 2	70 ± 30	98	44	0.70	2.1	2.0
PLLA-49/HEMI-PLLA-67[k]	5	6	2.3 ± 0.3	36 ± 3	20 ± 10	96	39	0.96	1.8	2.1
PLLA-49	2	2	2.6 ± 0.1	51 ± 1	5 ± 2			0.30	2.0	1.3
HEMI-PLLA-67	2	3	2.5 ± 0.2	54 ± 5	4 ± 2	66	70	0.35	1.3	0.5

[a] Weight fraction of CS added to melt mixer.

[b] Weight fraction of CS incorporated into blends, by 1HNMR spectroscopy.

[c] Elastic modulus.

[d] Stress at break.

[e] Elongation to break.

[f] Conversion of HEMI end-groups for blends with HEMI-PLLA-67.

[g] Conversion of E,E isomers of CS added to mixer.

[h] Log-mean average CS droplet diameter.

[i] Log-mean CS droplet size distribution parameter.

[j] Matrix ligament thickness.

[k] Matrix polymer was a 50/50 blend of PLLA-49 and HEMI-PLLA-67.

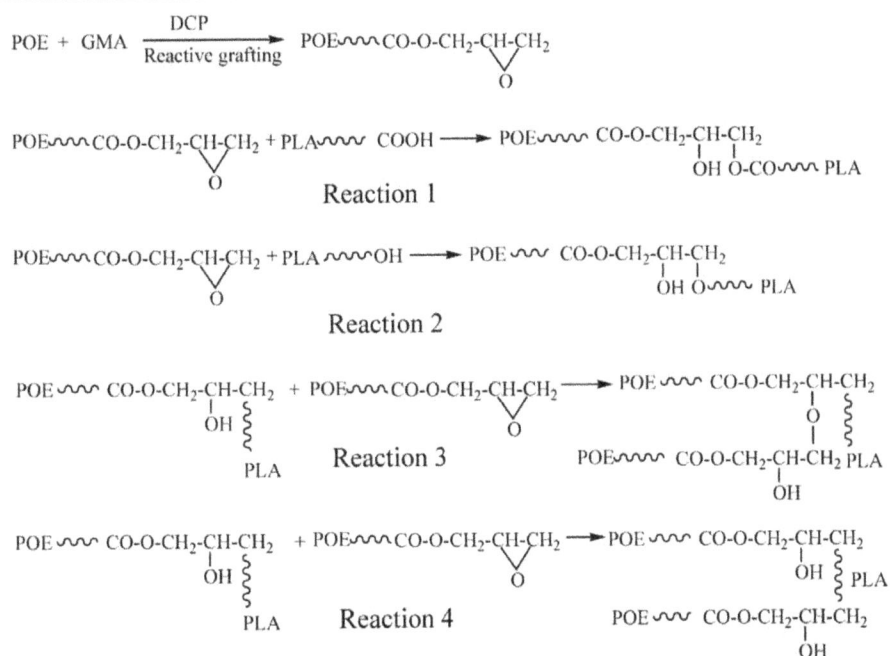

SCHEME 3 | The possible reactions in the melt reactive blends [Reprinted from Su et al. (2009) with permission from Elseiver].

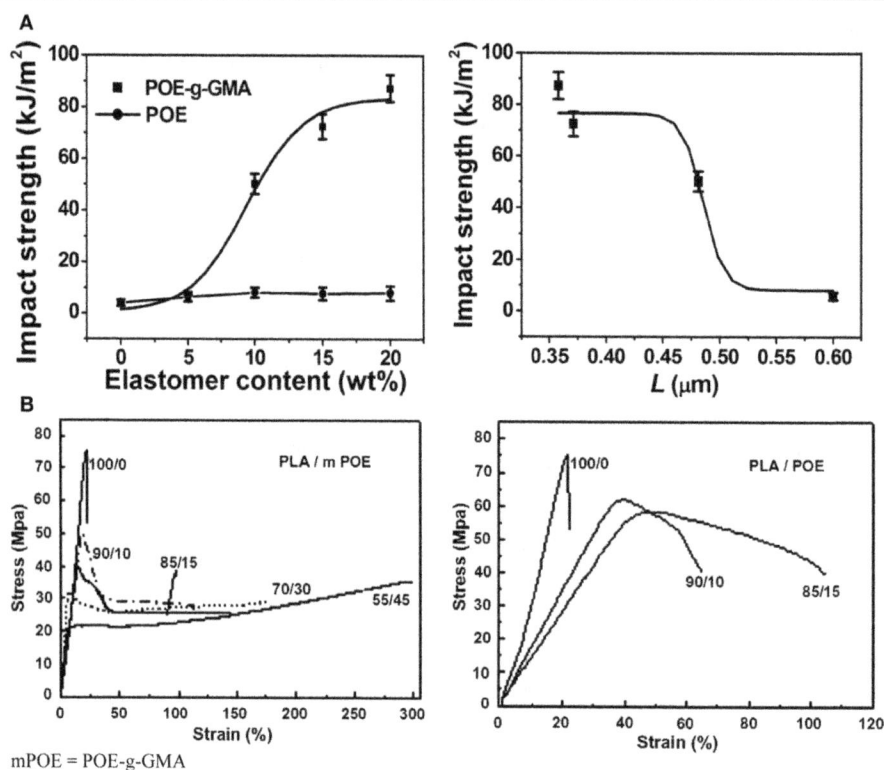

FIGURE 17 | (A) Impact strength as a function of elastomer contents (to the left), and interparticle distance (to the right) for PLA/POE-g-GMA blends. Copyright (2012) Wiley; used with permission from Feng et al. (2013). **(B)** Strain–stress curve of PLA/POE and PLA/ POE-g-GMA blends [Reprinted from Su et al. (2009) with permission from Elseiver].

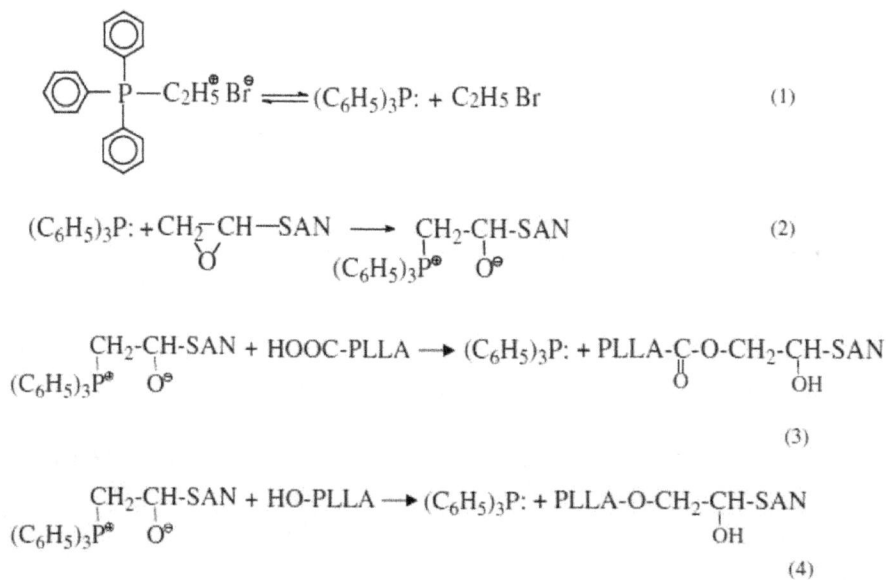

SCHEME 4 | Reaction of PLLA end groups with SAN-GMA under the catalyst of ETPB [Reprinted from Li and Shimizu (2009) with permission from Elseiver].

blend exhibited ~570% and ~847% improvement in the tensile toughness ($17.4\,MJ/m^3$ vs. $2.6\,MJ/m^3$ for neat PLA) and elongation at break (48.3% vs. 5.1% for neat PLA), respectively. However, tensile modulus and strength of the blend slightly decreased from 3.6 GPa (neat PLA) to 2.8 GPa and from 76.6 MPa (neat PLA) to 63.9 MPa, respectively. The authors ascribed the increase in the ductility of modified PLA to the stress-whitening and the multiple crazing initiated in the presence of cross-linked HBP particles. As revealed by rheological data, the formation of a networked interface was associated with enhanced compatibility of the PLA-cross-linked HBP blend as compared to the PLA/HBP blend.

The effects on mechanical properties of hydroxyl-terminated hyperbranched poly(ester amide) (HBP) and isocyanate-terminated prepolymer of butadiene (ITPB), alone and in combination, were investigated with the aim to make tough PLA (Nyambo et al., 2012). The glass transition temperature did not change from that of neat PLA. Interestingly, due to synergistic effects, impact strength and elongation at break of the PLA/HBP/ITPB ternary blend were improved by over 86 and 100%, respectively. Physical and chemical interactions between the hydroxyl-terminated HBP and the ITPB (**Scheme 2B**) may be responsible for the synergistic effect on the improvements in impact strength without scarifying the tensile modulus and strength. Scanning electron microscopy (SEM) images on impact fractures showed evidence of stretched and course surface,

which indicated a change in fracture behavior from brittle to ductile behavior after chemical modification. Accordingly, the impact strength of PLA can be easily enhanced using low additive loadings of 10 wt% *via* reactive extrusion with HBP and a suitable reactive compatibilizer such as ITPB. The modified PLA can address most issues related to neat brittle PLA, since it can exhibit a better stiffness–toughness balance and has the potential for use in durable commercial applications.

Soybean oil

PLLA/soybean oil binary blends containing unmodified soybean oil undergoes phase inversion at even low concentrations of soybean oil, leading to the release of the oil during blending. Therefore, the blends must be compatibilized (Chang et al., 2009). Ali et al. (2009) demonstrated that moderate improvements in the elongation at break of PLLA were gained by the addition of epoxidized soybean oil.

Robertson et al. (2010) explored how the polymerization and the optimization of soybean oil characteristics prior to blending improved its level of incorporation into PLLA and increased toughness compared to PLLA. They also demonstrated moderate improvements in the PLA/polysoybean oil blends regarding elongation at break and toughness of four and six times greater than those of unmodified PLLA, respectively. Gramlich et al. (2010) studied a more effective approach to toughen PLA consisting in the reactive compatibilization of conjugated soybean oil with PLLA. In a first step, bulk ring-opening polymerization via reactive extrusion (REx) of L-lactide using N-2-hydroxyethylmaleimide (HEMI) as a difunctional initiator and tin (II) 2-ethylhexanoate as a catalyst produced a high molecular weight reactive end-functionalized PLA (HEMI-PLLA). In a second step, REx of HEMI-PLLA and conjugated soybean oil (CS) was carried out through a Diels-Alder reaction in order to couple the two immiscible components *via* reactive compatibilization (**Figure 16**). Blends of HEMI-PLLA and 5 wt.% CS resulted in a greater than 17-fold increase in elongation to break compared to PLLA homopolymer and more than twice the elongation to break compared to a 5 wt.% CS blend with unreactive PLLA (**Table 16**). Analysis of the blend morphology indicated that the *in situ* formation of the compatibilizer at the HEMI-PLLA/CS interface decreased the CS droplet diameter to an optimal value ($0.7\,\mu m$) compared to unreactive binary blends, explaining the toughening PLLA with CS.

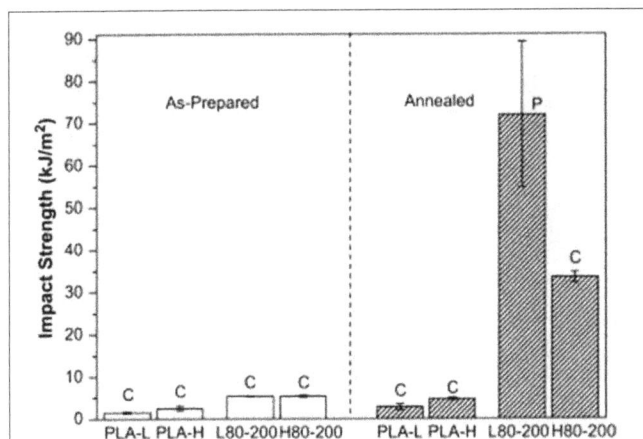

FIGURE 18 | Notched impact strength of PLAs and PLA/EGMA blends (C, complete break; P, partial break) [Oyama (2009) with permission from Elseiver].

Use of glycidyl methacrylate (GMA) and its copolymers

The grafting effect on mechanical properties of poly(ethylene octene) (POE) with PLA via glycidyl methacrylate (GMA) was

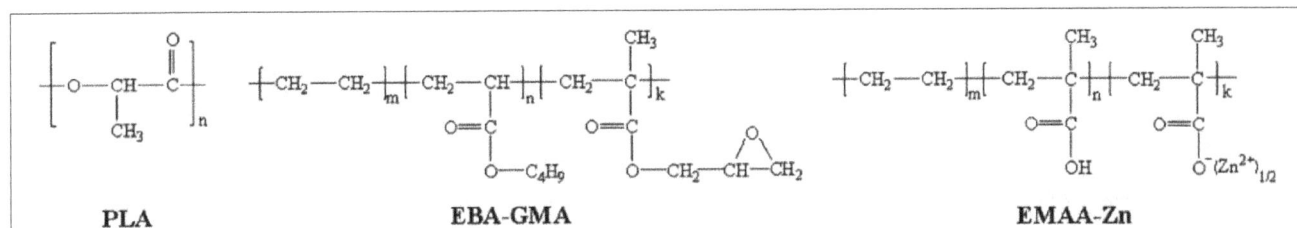

SCHEME 5 | **Chemical structure of the three polymers used in the study.** Adapted with permission from Liu et al. (2011).

investigated (Su et al., 2009; Feng et al., 2013). POE-*g*-GMA was used to prepare high impact modified PLA/POE-*g*-GMA reactive blends (**Scheme 3**). The presence of GMA moieties enhanced the blends compatibility due to the coupling reactions between the carboxyl and hydroxyl end-groups from PLA and the epoxy groups from POE-*g*-GMA (**Scheme 3**). Moreover, morphology analysis demonstrated better wetting of the dispersed phase by the PLA matrix and finer dispersed particles by reactive blending. Accordingly, the effective interfacial compatibilization promoted by the grafting reaction was mainly responsible for the significant improvement of PLA toughening (**Figures 17A,B**). Interestingly, the highest toughening effect was obtained at lower particle size and interparticle distance, which were submicronic (**Figure 17A**).

PLLA and acrylonitrile–butadiene–styrene copolymer (ABS) are thermodynamically immiscible and incompatible by simply melt blending them. Styrene acrylonitrile-glycidyl methacrylate copolymer (SAN-GMA) as a reactive compatibilizer and ethyltriphenyl phhosphonium bromide (ETPB) as a catalyst were thereby introduced during the reactive melt blending of PLA/ABS96 (Li and Shimizu, 2009). The epoxide group of SAN-GMA reacted with PLLA end-groups under the mixing conditions, and the addition of ETPB accelerated the reaction (**Scheme 4**). As a result, it was found that the size of the "salami-like" ABS domains in PLLA matrix significantly decreased and their dispersion improved by the addition of the reactive compatibilizer. A significant shift of glass transition temperatures for both PLLA and ABS indicated the improvement of the

compatibility between PLLA and ABS. As a result, the compatibilized PLLA/ABS blends exhibited a very nice stiffness-toughness balance, i.e., an improvement of the impact strength and the elongation at break with a slight reduction in the modulus. For instance, the addition of 5 phr of SAN-GMA to the PLLA/ABS (70/30 wt/wt) blend increased elongation at break from 3.1 to 20.5% and impact strength from 63.8 to 81.1 kJ/m^2. By further incorporating 0.02 phr ETPB, the elongation at break and impact strength of the blend increased to 23.8% and 123.9 kJ/m^2, respectively.

Low and high molecular weight PLA (L-PLA and H-PLA, respectively) were blended with 20% of poly(ethylene-*co*-glycidyl methacrylate) (EGMA) (Oyama, 2009). The resulting blend had a high elongation above 200% compared to 5% for neat PLA. The notched Charpy impact was only 2 times that of neat PLA. After annealing, the injection-moulded specimens of the L-PLA/EGMA (80/20 wt/wt) blend at 90°C for 2.5 h showed that the impact strength significantly increased to 72 kJ/m^2, about 50 times that of neat L-PLA. Moreover, the improvement in strength and modulus of the blend was accompanied by a significant decrease in elongation at break. With the higher molecular weight PLA (H-PLA) as matrix, such positive effect of annealing on impact strength appeared relatively less prominent (**Figure 18**). The author argued that the crystallization of the PLA matrix played a key-role in such significant improvement. It was demonstrated that the interfacial reaction (reactive compatibilization) between the polymeric components improved not

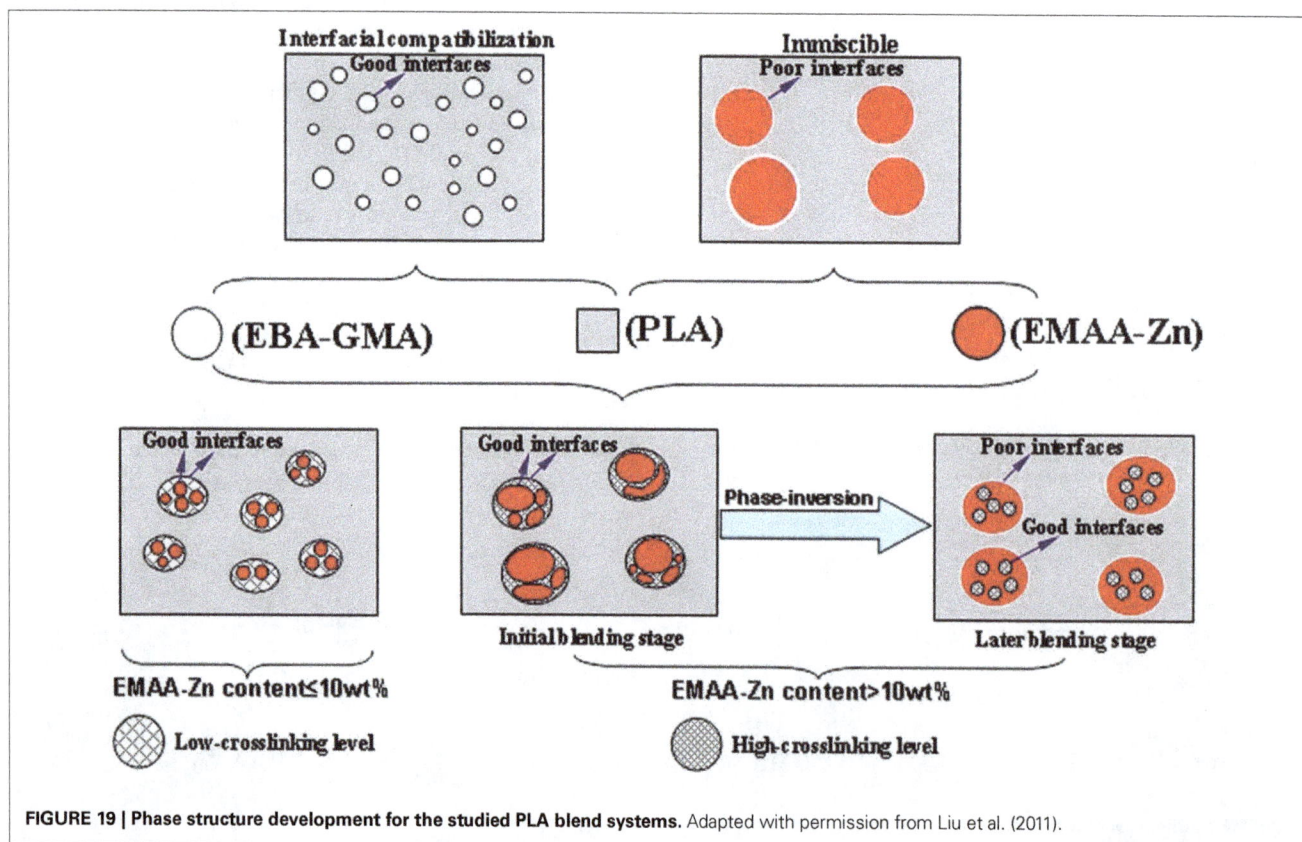

FIGURE 19 | Phase structure development for the studied PLA blend systems. Adapted with permission from Liu et al. (2011).

only the dispersion of the second component but also the bonding between the particles and matrix to expect combination of crazing and shear yielding, contributing to the formation of the super-touch PLA materials, superior to commercially available acrylonitrile-butadiene-styrene (ABS) resins. Furthermore, these improvements in mechanical properties were achieved without scarifying the heat resistance of the material. The material highlights again the importance of interface control in the preparation of multicomponent materials.

Liu et al. (2010, 2011) and Song et al. (2012) studied extensively the reactive ternary blends of PLA with ethylene/n-butyl acrylate/glycidyl methacrylate (EBA-GMA) terpolymer and a zinc ionomer of ethylene/methacrylic acid (EMAA-Zn) using a Leistritz ZSE 18 twin-screw extruder having a L/D ratio of 40. The three polymeric components are represented in **Scheme 5** and **Figure 19**.

The influence of the simultaneous dynamic vulcanization (crosslinking) and interfacial compatibilization and adhesion on mechanical and impact performance of the reactive PLA-based ternary blends was investigated. It was demonstrated that the EBA-GMA/EMMA-Zn ratio played a crucial role in determining the phase-morphology. Interestingly, the increase of the EMAA-Zn content gradually turned the phase of the latter from occluded sub-inclusions into a continuous phase within the "salami"-like micro-structure (domain-in-domain morphology) as revealed by TEM in the case of the ternary blends. It was reasonably proposed that when the EMAA-Zn content exceeded 10%, a phase inversion within the sub-structure of the dispersed phase domains could likely take place, which would account for the pronounced deterioration in interfacial wetting of the dispersed particles by the PLA matrix in these cases. The phase structure development for the studied PLA blend systems are schematized in **Figure 20**. The EMAA-Zn domains were finally occluded inside the EBA-GMA particles which were homogeneously dispersed in PLA.

Interestingly, it was demonstrated that at higher extrusion temperature (240 vs. 185°C), not only the carboxyl groups in the EMAA-Zn ionomer were able to trigger more cross-linking reactions *via* the epoxy groups in the EBA-GMA phase, but also more PLA macromolecules were grafted at the interface between PLA and the elastomer (**Figure 20**). The Zn ions further catalyzed the reactions. According to the SEM micrographs, this was confirmed by the better wetting of the dispersed phase by PLA matrix at higher blending temperature. Accordingly, effective interfacial compatibilization and adhesion were achieved at higher compounding temperature.

As a result, although increasing the extrusion temperature did not significantly influence the tensile properties (**Figure 21B**), both blending temperature and elastomer/ionomer ratio were

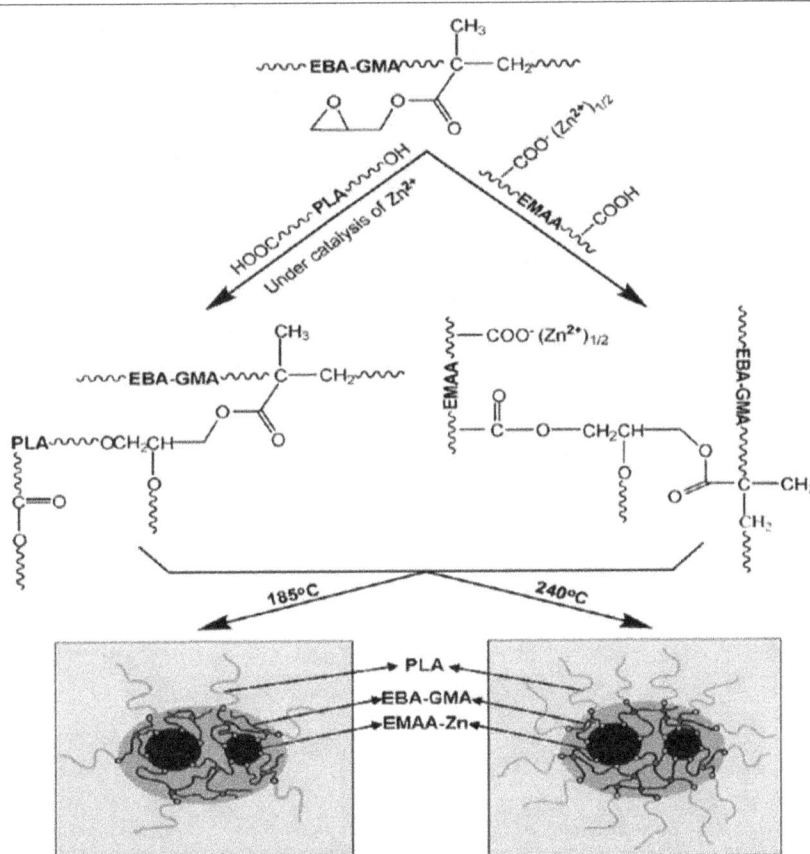

FIGURE 20 | Proposed reactions during the reactive blending process, together with schematic phase morphologies of the PLA/EBA-GMA/EMAA-Zn ternary blends extruded at 185 and 240°C respectively. Adapted with permission from Liu et al. (2010).

found to play keys-roles in achieving super-toughness (great improvement of impact strength and strain at break) of the PLA-based ternary reactive blends (**Figure 21A**). This can attributed to the effective interfacial compatibilization at higher temperature (240°C).

The correlation between the particle size and impact toughness had revealed that there existed an optimum submicronic range of particle sizes of the dispersed domains for PLA super-toughening in this ternary blend system (**Figure 22**). Preliminary analysis of micromechanical deformation suggested that the high impact toughness observed for some ternary blends was attributed to the low cavitation resistance of the dispersed particles coupled with suitable interfacial adhesion. It was found that debonding mainly occurred around the relatively large particles together with fibrillated crazes and no cavitation when blended with an ethylene/n-butyl acrylate/glycidyl methacrylate terpolymer (EBA-GMA). Addition of a zinc ionomer of ethylene/methacrylic acid copolymer (EMAA-Zn) within the PLA/EBA-GMA blend

gradually turned the morphology into a salami-like phase structure, which provides a low cavitation resistance coupled with suitable interfacial adhesion. Therefore, internal cavitation of the dispersed particles followed by the matrix shear yielding was predominant and resulted in the optimum impact strength. All of these examples regarding toughening mechanisms within PLA are not exhaustive but strengthens that toughness of PLA is a complex function which implies all of as-describe mechanisms (crazing, shear yielding, cavitation and debonding) and mode of fracture.

In a complementary study, Song et al. (2012) investigated the effect of the ionomer characteristics on reactions and properties of the PLA-based reactive ternary blends studied above (**Schemes 6A,B**). The ionomer was prepared by neutralizing the EMAA ionomer precursor with ZnO. It came out that the reactivity of the system and the interfacial compatibilization were drastically enhanced by increasing both the degree of neutralization (DN) of the ionomer and the methacrylic acid (MAA) content of ionomer precursor. As a result, the particle size and polydispersity of the dispersed phase reached the right optimum to greatly improve the impact toughness and tensile elongation at break of the material (**Figures 23A,B**).

Super-tough PLA alloy with greatly improved heat resistance

Hashima et al. (2010) toughened PLA by blending it with hydrogenated styrene-butadiene-styrene block copolymer (SEBS) with the aid of reactive compatibilizer, poly(ethylene-*co*-glycidyl methacrylate) (EGMA). The high temperature property (HDT) and thermal ageing resistance were improved by further incorporating a ductile polymer with a high glass transition temperature, that is, polycarbonate (PC). Based on TEM, differential scanning calorimetry (DSC), and dynamic mechanical analysis (DMA), the author explained that the origin of the outstanding toughness and ageing resistance of the 4 component alloy; e.g., PLA/PC/SEBS/EGMA 40/40/5/5 (wt.% ratio), seems

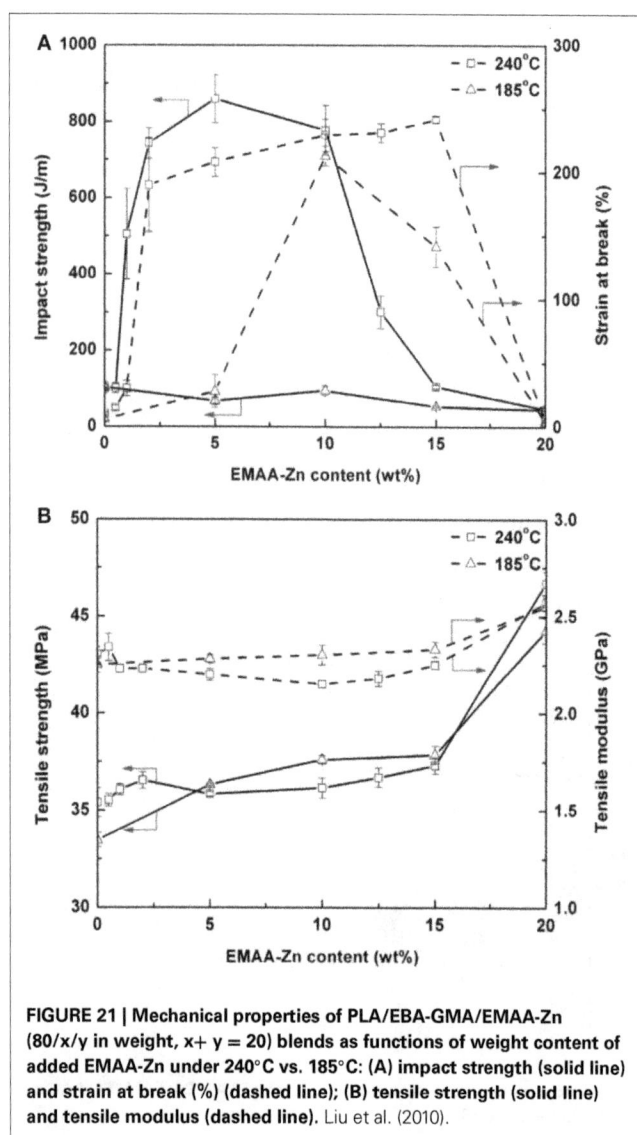

FIGURE 21 | Mechanical properties of PLA/EBA-GMA/EMAA-Zn (80/x/y in weight, x+ y = 20) blends as functions of weight content of added EMAA-Zn under 240°C vs. 185°C: (A) impact strength (solid line) and strain at break (%) (dashed line); (B) tensile strength (solid line) and tensile modulus (dashed line). Liu et al. (2010).

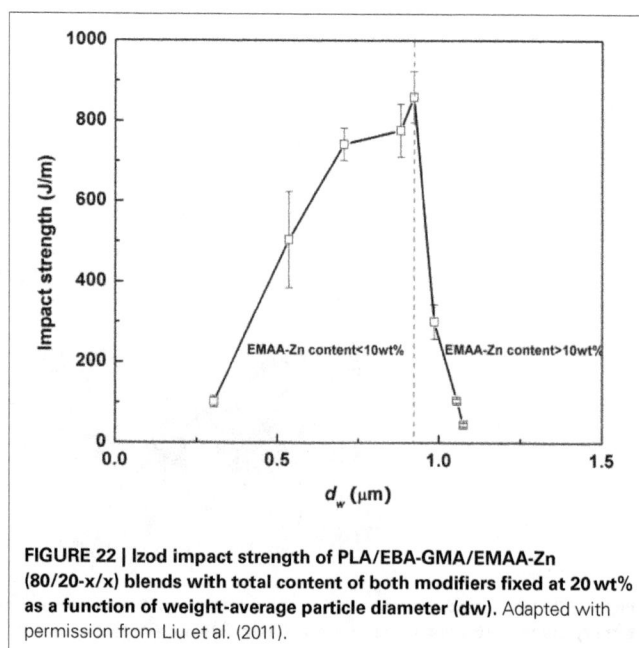

FIGURE 22 | Izod impact strength of PLA/EBA-GMA/EMAA-Zn (80/20-x/x) blends with total content of both modifiers fixed at 20 wt% as a function of weight-average particle diameter (dw). Adapted with permission from Liu et al. (2011).

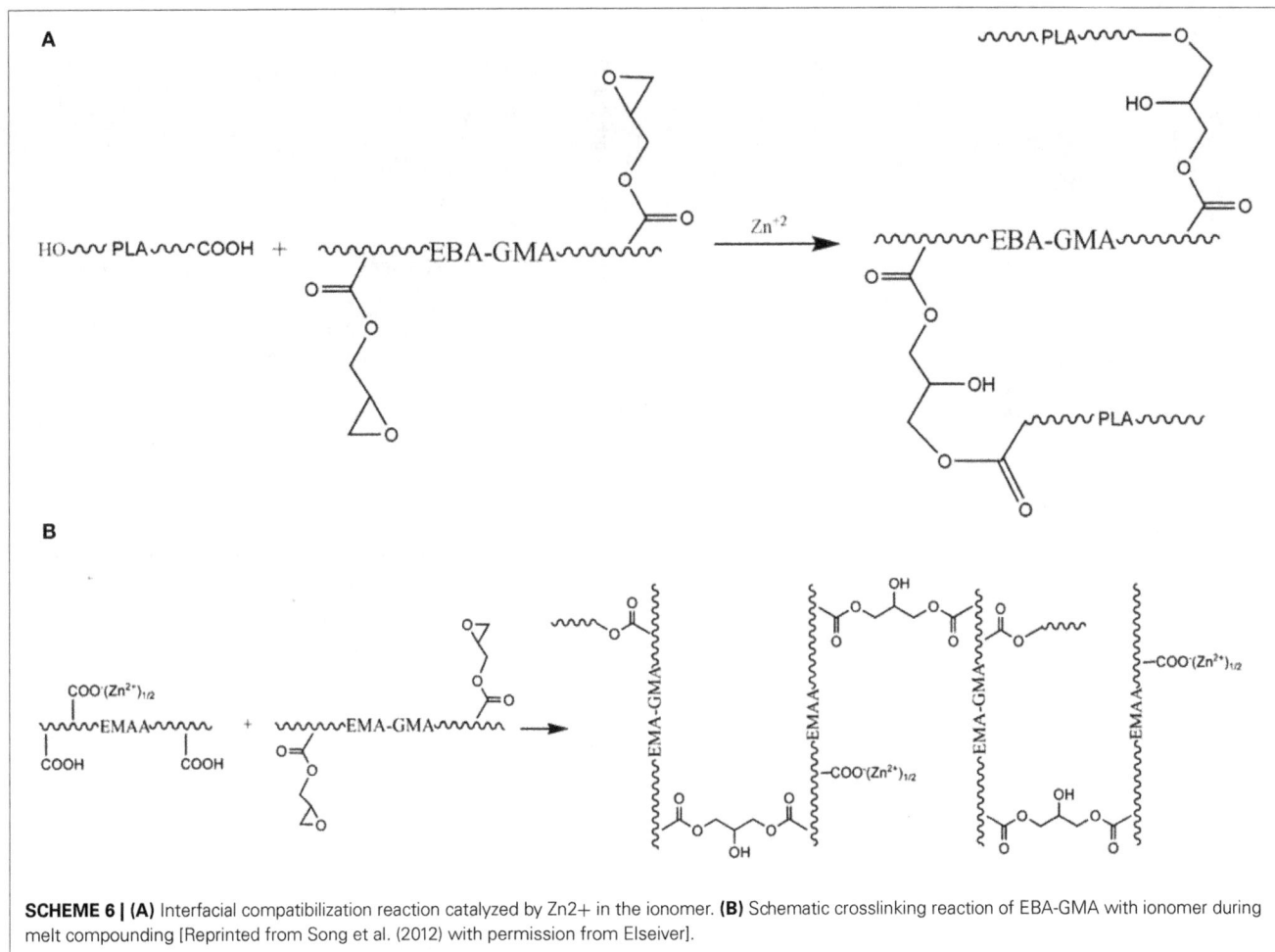

SCHEME 6 | (A) Interfacial compatibilization reaction catalyzed by Zn2+ in the ionomer. **(B)** Schematic crosslinking reaction of EBA-GMA with ionomer during melt compounding [Reprinted from Song et al. (2012) with permission from Elseiver].

to come from the negative pressure effect of SEBS that dilates the plastic matrix consisting of PLA and PC to enhance the local segment motions. The phenomenon is briefly summarized in **Figure 24**.

Jiang et al. (2012) blended PLA with various commercial rubber components, i.e., poly (ethylene-glycidyl methacrylate) (EGMA), maleic anhydride grafted poly(styrene-ethylene/butylene-styrene) triblock elastomer (m-SEBS), and poly(ethylene-co-octene) (EOR) and compared their toughening effect on PLA (**Figure 25**). It was observed that: (i) EGMA was highly compatible due to its reaction with PLA, (ii) m-SEBS was less compatible with PLA, and (iii) EOR was incompatible with PLA. SEM and TEM revealed that a fine 3-D co-continuous microlayer structure was formed in the injection-moulded PLA/EGMA blends. This led to polymer blends with high toughness and very low linear thermal expansion both in the flow direction and in the transverse direction. The microlayer thickness of rubber in PLA blends was found to play key-roles in reducing the linear thermal expansion and achieving high toughness of the blends. Therefore, PLA blends with the notched impact strength over 20 times higher (ca. $90\,kJ/m^2$) than that of the neat PLA (ca. $4\,kJ/m^2$) were obtained by reactive blending of PLA and EGMA at 40 wt.% of rubber loading. It should be highlighted that the PLA/EGMA blend having both high impact resistance and low thermal expansion coefficient is of great importance in applications.

CONCLUSION

In comparison with many other commodity thermoplastics, PLA presents many advantages, mainly its renewability, biodegradability, high stiffness and competitive cost production. The main problem for this biopolyester is its inherent brittleness due to a crazing deformation mechanism through which the polymer fails upon tensile and impact testing. Since many applications require high impact resistance and flexibility bio-based and/or biodegradable materials, several approaches aiming at toughening PLA has been investigated over the last decades. First of all, understanding the effect of the pristine microstructure modification on the mechanical performances was established. It has been demonstrated that de-aging and molecular orientation can improve the mechanical properties of PLA. However, such strategies require long and specific processes, which are not cost effective for an economical production of high performance PLA materials. Classically, compounding with softer polymers seems to be the best option for toughening PLA in

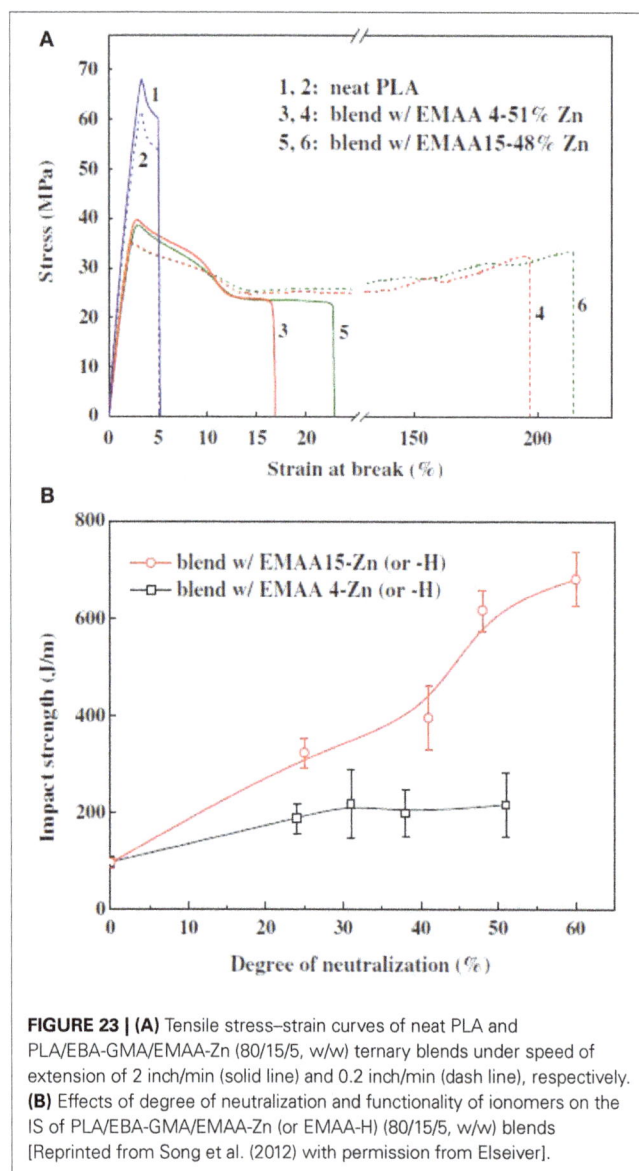

FIGURE 23 | (A) Tensile stress–strain curves of neat PLA and PLA/EBA-GMA/EMAA-Zn (80/15/5, w/w) ternary blends under speed of extension of 2 inch/min (solid line) and 0.2 inch/min (dash line), respectively. (B) Effects of degree of neutralization and functionality of ionomers on the IS of PLA/EBA-GMA/EMAA-Zn (or EMAA-H) (80/15/5, w/w) blends [Reprinted from Song et al. (2012) with permission from Elseiver].

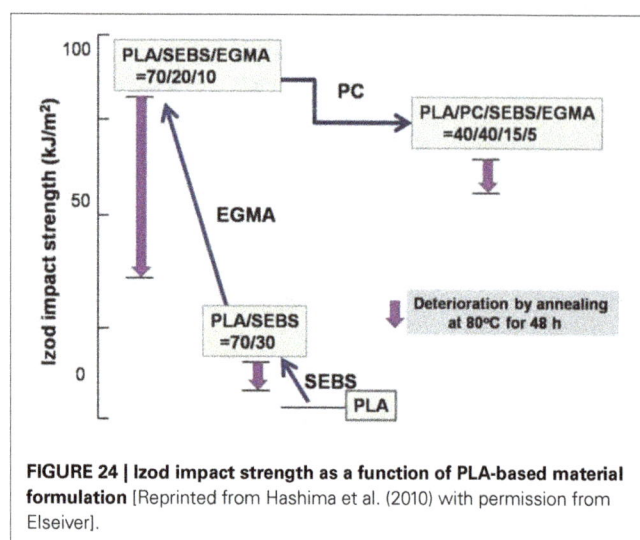

FIGURE 24 | Izod impact strength as a function of PLA-based material formulation [Reprinted from Hashima et al. (2010) with permission from Elseiver].

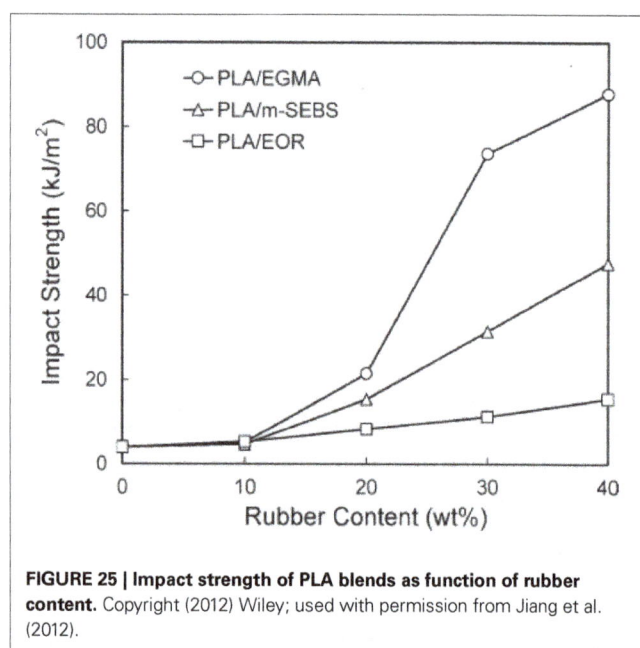

FIGURE 25 | Impact strength of PLA blends as function of rubber content. Copyright (2012) Wiley; used with permission from Jiang et al. (2012).

costless way. The toughening effects of PLA blends are complicated as many parameters are concerned including the high interfacial adhesion between the matrix and the toughener, the domain size of the dispersed phase that should be ideally between 0.1 and 1.0 μm to improve the blend compatibility. The most common compatibilization way consists on the incorporation of block copolymers. Recently, chemical compatibilization *via* reactive extrusion has proven to be a very promising technology and more effective in improving the toughness of PLA blends. In some cases, outstanding toughness was successfully achieved, but accompanied with a compromise of the biodegradability and the initial stiffness of PLA. Therefore, the challenge pursues to develop a fully bio-based and biodegradable PLA-based material with a balance of outstanding mechanical properties.

ACKNOWLEDGMENTS

Authors from CIRMAP are grateful to the "Région Wallonne" and European Community (FEDER, FSE) in the frame of "Pôle d'Excellence Materia Nova" for their financial support. CIRMAP thanks the "Belgian Federal Government Office Policy of Science (SSTC)" for general support in the frame of the PAI-6/27. Jean-Marie Raquez is "chercheur qualifié" by the F.R.S.-FNRS. Authors (from AMS and CIRMAP) thank the Fonds National de la Recherche-Luxembourg and the Centre de Recherche Public Henri Tudor for financial supports.

REFERENCES

Afrifah, K. A., and Matuana, L. M. (2010). Impact modification of polylactide with a biodegradable ethylene/acrylate copolymer. *Macromol. Mater. Eng.* 295, 802–811. doi: 10.1002/mame.201000107

Alcock, B., Cabrera, N. O., Barkoula, N. M., Loos, J., and Peijs, T. (2006). The mechanical properties of unidirectional all-polypropylene composites. *Compos. Part A Appl. Sci. Manuf.* 37, 716–726. doi: 10.1016/j.compositesa.2005.07.002

Ali, F., Chang, Y.-W., Kang, S. C., and Yoon, J. Y. (2009). Thermal, mechanical and rheological properties of poly (lactic acid)/epoxidized soybean oil blends. *Polym. Bull.* 62, 91–98. doi: 10.1007/s00289-008-1012-9

Anderson, K. S., and Hillmyer, M. A. (2004). The influence of block copolymer microstructure on the toughness of compatibilized polylactide/polyethylene blends. *Polymer* 45, 8809–8823. doi: 10.1016/j.polymer.2004.10.047

Anderson, K. S., Lim, S. H., and Hillmyer, M. A. (2003). Toughening of polylactide by melt blending with linear low-density polyethylene. *J. Appl. Polym. Sci.* 89, 3757–3768. doi: 10.1002/app.12462

Anderson, K. S., Schreck, K. M., and Hillmyer, M. A. (2008). Toughening polylactide. *Polymer Reviews* 48, 85–108. doi: 10.1080/15583720701834216

Argon, A. S., and Cohen, R. E. (1990). "Crazing and toughness of block copolymers and blends," in *Crazing in Polymers* Vol. 2, ed H. H. Kausch (Berlin Heidelberg: Springer), 301–351.

Babcock, L. M., Henton, D. E., and Tadesse, F. A. (2008). *Impact Modified Polylactide Resins*, U.S. Pat. WO 2008/051443 A1 patent application.

Baiardo, M., Frisoni, G., Scandola, M., Rimelen, M., Lips, D., Ruffieux, K., et al. (2003). Thermal and mechanical properties of plasticized poly(L-lactic acid). *J. Appl. Polym. Sci.* 90, 1731–1738. doi: 10.1002/app.12549

Baker, W. E.,, Scott, C. E., and Hu, G.-H. (2001). *Reactive Polymer Blending*, Cincinnati: Hanser Publishers; Munich Hanser Gardner Publications, Inc.

Barish, L. (1962). The study of cracking and fracturing of spherulitic isotactic polypropylene. *J. Appl. Polym. Sci.* 6, 617–623. doi: 10.1002/app.1962.070062403

Battegazzore, D., Bocchini, S., and Frache, A. (2011). Crystallization kinetics of poly(lactic acid)-talc composites. *eXPRESS Polym. Lett.* 5, 849–858. doi: 10.3144/expresspolymlett.2011.84

Bhardwaj, R., and Mohanty, A. K. (2007). Modification of Brittle Polylactide by Novel Hyperbranched Polymer-Based Nanostructures. *Biomacromolecules* 8, 2476–2484. doi: 10.1021/bm070367x

Bigg, D. M. (2005). Polylactide copolymers: effect of copolymer ratio and end capping on their properties. *Adv. Polym. Technol.* 24, 69–82. doi: 10.1002/adv.20032

Bitinis, N., Verdejo, R., Cassagnau, P., and Lopez-Manchado, M. A. (2011). Structure and properties of polylactide/natural rubber blends. *Mater. Chem. Phys.* 129, 823–831. doi: 10.1016/j.matchemphys.2011.05.016

Bopp, R., and Whelan, J. (2003). U.S. Pat. WO 200316015 A1 patent application.

Broz, M. E., Vanderhart, D. L., and Washburn, N. R. (2003). Structure and mechanical properties of poly(D,L-lactic acid)/poly(e-caprolactone) blends. *Biomaterials* 24, 4181–4190. doi: 10.1016/S0142-9612(03)00314-4

Bucknall, C. (1978). "Fracture and failure of multiphase polymers and polymer composites," in *Failure in Polymers*, ed E. H. Andrews (Berlin, Heidelberg: Springer), 121–148.

Bucknall, C. B. (2007). Quantitative approaches to particle cavitation, shear yielding, and crazing in rubber-toughened polymers. *J. Polym. Sci. B Polym. Phys.* 45, 1399–1409. doi: 10.1002/polb.21171

Bucknall, C. B., and Paul, D. R. (2009). Notched impact behaviour of polymer blends: Part 1: new model for particle size dependence. *Polymer* 50, 5539–5548. doi: 10.1016/j.polymer.2009.09.059

Carrasco, F., Pagès, P., Gámez-Pérez, J., Santana, O. O., and Maspoch, M. L. (2010). Processing of poly(lactic acid): characterization of chemical structure, thermal stability and mechanical properties. *Polym. Degrad. Stab.* 95, 116–125. doi: 10.1016/j.polymdegradstab.2009.11.045

Chang, K., Robertson, M. L., and Hillmyer, M. A. (2009). Phase inversion in polylactide/soybean oil blends compatibilized by poly(isoprene-b-lactide) block copolymers. *ACS Appl. Materials and Interfaces* 1, 2390–2399. doi: 10.1021/am900514v

Chen, G.-X., Kim, H.-S., Kim, E.-S., and Yoon, J.-S. (2005). Compatibilization-like effect of reactive organoclay on the poly(l-lactide)/poly(butylene succinate) blends. *Polymer* 46, 11829–11836. doi: 10.1016/j.polymer.2005.10.056

Chen, G.-X., and Yoon, J.-S. (2005). Morphology and thermal properties of poly(L-lactide)/poly(butylene succinate-co-butylene adipate) compounded with twice functionalized clay. *J. Polym. Sci. B Polym. Phys.* 43, 478–487. doi: 10.1002/polb.20345

Chen, X., McCarthy, S. P., and Gross, R. A. (1997). Synthesis and characterization of [L]-lactide–ethy- lene oxide multiblock copolymers. *Macromolecules* 30, 4295–4301. doi: 10.1021/ma970385e

Co, S. (2010). *Biogoes functional—Sukano biobased masterbacthes put performance into PLA [Online]*. Available: http://www.sukano.com/pdf/PC022010BIOMB.pdf (Accessed February, 2010).

Dompas, D., and Groeninckx, G. (1994). Toughening behaviour of rubber-modified thermoplastic polymers involving very small rubber particles: 1. *A criterion for internal rubber cavitation. Polymer* 35, 4743–4749. doi: 10.1016/0032-3861(94)90727-7

Dompas, D., Groeninckx, G., Isogawa, M., Hasegawa, T., and Kadokura, M. (1994). Toughening behaviour of rubber-modified thermoplastic polymers involving very small rubber particles: 2. *Rubber cavitation behaviour in poly(vinyl chloride)/methyl methacrylate-butadiene-styrene graft copolymer blends. Polymer* 35, 4750–4759. doi: 10.1016/0032-3861(94)90728-5

Donald, A. M., and Kramer, E. J. (1982). Craze initiation and growth in high-impact polystyrene. *J. Appl. Polym. Sci.* 27, 3729–3741. doi: 10.1002/app.1982.070271009

DuPont Co. (2010). Website. Product Data Sheet. Biomax Strong 100. Available online at: http://www2.dupont.com/Biomax/en_US/assets/ downloads/biomax_strong_100.pdf. Accessed on January 12, 2010.

Feng, F., and Ye, L. (2011). Morphologies and mechanical properties of polylactide/thermoplastic polyurethane elastomer blends. *J. Appl. Polym. Sci.* 119, 2778–2783. doi: 10.1002/app.32863

Feng, Y., Hu, Y., Yin, J., Zhao, G., and Jiang, W. (2013). High impact poly(lactic acid)/poly(ethylene octene) blends prepared by reactive blending. *Polym. Eng. Sci.* 53, 389–396. doi: 10.1002/pen.23265

Gamez-Perez, J. (2010). Fracture behaviour of quenched poly(lactic acid). *eXPRESS Polym. Lett.* 5, 82–91. doi: 10.3144/expresspolymlett.2011.9

Ge, H., Yang, F., Hao, Y., Wu, G., Zhang, H., and Dong, L. (2013). Thermal, mechanical, and rheological properties of plasticized poly(L-lactic acid). *J. Appl. Polym. Sci.* 127, 2832–2839. doi: 10.1002/app.37620

Ghosh, S., Viana, J. C., Reis, R. L., and Mano, J. F. (2008). Oriented morphology and enhanced mechanical properties of poly(l-lactic acid) from shear controlled orientation in injection moulding. *Mater. Sci. Eng. A* 490, 81–89. doi: 10.1016/j.msea.2008.01.003

Gramlich, W. M., Robertson, M. L., and Hillmyer, M. A. (2010). Reactive compatibilization of poly(l-lactide) and conjugated soybean oil. *Macromolecules* 43, 2313–2321. doi: 10.1021/ma902449x

Grijpma, D. W., Altpeter, H., Bevis, M. J., and Feijen, J. (2002). Improvement of the mechanical properties of poly(D,L-lactide) by orientation. *Polym. Int.* 51, 845–851. doi: 10.1002/pi.988

Gui, Z., Xu, Y., Cheng, S., Gao, Y., and Lu, C. (2013). Preparation and characterization of polylactide/poly(polyethylene glycol-co-citric acid) blends. *Polym. Bull.* 70, 325–342. doi: 10.1007/s00289-012-0810-2

Hammer, C. F., Koch, T. A., and Whitney, J. F. (1959). Fine structure of acetal resins and its effect on mechanical properties. *J. Appl. Polym. Sci.* 1, 169–178. doi: 10.1002/app.1959.070010207

Han, J.-J., and Huang, H.-X. (2011). Preparation and characterization of biodegradable polylactide/thermoplastic polyurethane elastomer blends. *J. Appl. Polym. Sci.* 120, 3217–3223. doi: 10.1002/app.33338

Han, L., Han, C., and Dong, L. (2013). Morphology and properties of the biosourced poly(lactic acid)/poly(ethylene oxide-b-amide-12) blends. *Polym. Compos.* 34, 122–130. doi: 10.1002/pc.22383

Harada, M., Iida, K., Okamoto, K., Hayashi, H., and Hirano, K. (2008). Reactive compatibilization of biodegradable poly(lactic acid)/poly(ε-caprolactone) blends with reactive processing agents. *Polym. Eng. Sci.* 48, 1359–1368. doi: 10.1002/pen.21088

Harada, M., Ohya, T., Iida, K., Hayashi, H., Hirano, K., and Fukuda, H. (2007). Increased impact strength of biodegradable poly(lactic acid)/poly(butylene succinate) blend composites by using isocyanate as a reactive processing agent. *J. Appl. Polym. Sci.* 106, 1813–1820. doi: 10.1002/app.26717

Hashima, K., Nishitsuji, S., and Inoue, T. (2010). Structure-properties of super-tough PLA alloy with excellent heat resistance. *Polymer* 51, 3934–3939. doi: 10.1016/j.polymer.2010.06.045

Hassouna, F., Raquez, J.-M., Addiego, F., Dubois, P., Toniazzo, V., and Ruch, D. (2011). New approach on the development of plasticized polylactide (PLA): Grafting of poly(ethylene glycol) (PEG) via reactive extrusion. *Eur. Polym. J.* 47, 2134–2144. doi: 10.1016/j.eurpolymj.2011.08.001

Hassouna, F., Raquez, J.-M., Addiego, F., Toniazzo, V., Dubois, P., and Ruch, D. (2012). New development on plasticized poly(lactide): Chemical grafting of citrate on PLA by reactive extrusion. *Eur. Polym. J.* 48, 404–415. doi: 10.1016/j.eurpolymj.2011.12.001

Hu, Y., Hu, Y. S., Topolkaraev, V., Hiltner, A., and Baer, E. (2003a). Aging of poly(lactide)/poly(ethylene glycol) blends. Part 2. Poly(lactide) with high stereoregularity. *Polymer* 44, 5711–5720. doi: 10.1016/S0032-3861(03)00615-3

Hu, Y., Rogunova, M., Topolkaraev, V., Hiltner, A., and Baer, E. (2003b). Aging of poly(lactide)/poly(ethylene glycol) blends. Part 1. Poly(lactide) with low stereoregularity. *Polymer* 44, 5701–5710.

Ikado, S., Kobayashi, N., Kurokit, T., Saruwatarim, M., Suzuki, K., and Wanibe, H. (1997). European Patent 776927 A1, assigned to Mitsui Toatsui Chem, Inc.

Ikeda, R. M. (1993). Shear yield and crazing stresses in selected glassy polymers. *J. Appl. Polym. Sci.* 47, 619–629. doi: 10.1002/app.1993.070470406

Ito, M., Abe, S., and Ishikawa, M. (2010). The fracture mechanism of polylactic acid resin and the improving mechanism of its toughness by addition of acrylic modifier. *J. Appl. Polym. Sci.* 115, 1454–1460. doi: 10.1002/app.31292

Jacobsen, S., and Fritz, H. G. (1999). Plasticizing polylactide – the effect of different plasticizers on the mechanical properties. *Polym. Eng. Sci.* 39, 1303–1310. doi: 10.1002/pen.11517

Jansen, B. J. P., Rastogi, S., Meijer, H. E. H., and Lemstra, P. J. (1999). rubber-modified glassy amorphous polymers prepared via chemically induced phase separation. 3. influence of the strain rate on the microscopic deformation mechanism. *Macromolecules* 32, 6283–6289. doi: 10.1021/ma981406n

Jiang, J., Su, L., Zhang, K., and Wu, G. (2012). Rubber-toughened PLA blends with low thermal expansion. *J. Appl. Polym. Sci.* 128, 3993–4000. doi: 10.1002/app.38642

Jiang, L., Wolcott, M. P., and Zhang, J. (2006). Study of biodegradable poly-lactide/poly(butylene adipate-co-terephthalate) blends. *Biomacromolecules* 7, 199–207.

Kambour, R. P. (1973). A review of crazing and fracture in thermoplastics. *J. Polym. Sci. Macromol. Rev.* 7, 1–154. doi: 10.1002/pol.1973.230070101

Kjeschke, K., Timmermann, R., and Voight, M. (2001). German Patent De 100279 A1, assigned to Bayer Ag.

Kmetty, Á., Bárány, T., and Karger-Kocsis, J. (2010). Self-reinforced polymeric materials: a review. *Prog. Polym. Sci.* 35, 1288–1310. doi: 10.1016/j.prog polymsci.2010.07.002

Kolstad, J. J. (1996). Crystallization kinetics of poly(L-lactide-co-meso-lactide). *J. Appl. Polym. Sci.* 62, 1079–1091.

Könczöl, L., Döll, W., and Michler, G. H. (1992). Study of the toughening mechanism of crazing in rubber modified thermoplastics. *Colloid Polym. Sci.* 270, 972–981. doi: 10.1007/BF00655966

Kowalczyk, M., and Piorkowska, E. (2012). Mechanisms of plastic deformation in biodegradable polylactide/poly(1,4-cis-isoprene) blends. *J. Appl. Polym. Sci.* 124, 4579–4589. doi: 10.1002/app.35489

Kramer, E. (1983). "Microscopic and molecular fundamentals of crazing," in *Crazing in Polymers*, ed H. H. Kausch (Berlin, Heidelberg: Springer), 1–56.

Kramer, E., and Berger, L. (1990). "Fundamental processes of craze growth and fracture," in *Crazing in Polymers*, Vol. 2, ed H. H. Kausch (Berlin, Heidelberg: Springer), 1–68.

Kray, R. J., and Bellet, R. J. (1968). U.S. Pat. 3,388,186 patent application.

Kulinski, Z., Piorkowska, E., Gadzinowska, K., and Stasiak, M. (2006). Plasticization of Poly(L-lactide) with Poly(propylene glycol). *Biomacromolecules* 7, 2128–2135. doi: 10.1021/bm060089m

Murch, L. E. (1974). U.S. Pat. 3,845,163 patent application.

Labrecque, L. V., Kumar, R. A., Davé, V., Gross, R. A., and Mccarthy, S. P. (1997). Citrate esters as plasticizers for poly(lactic acid). *J. Appl. Polym. Sci.* 66, 1507–1513.

Lapol, L. W. (2009). *Getting Started with LapolVR 108 Bioplasticizer* [Online]. Available online at: http://www.lapol.net/images/Lapol_Getting_Started_Guide_and_Essential_Documentation_100930.pdf

Lemmouchi, Y., Murariu, M., Santos, A. M. D., Amass, A. J., Schacht, E., and Dubois, P. (2009). Plasticization of poly(lactide) with blends of tributyl citrate and low molecular weight poly(d,l-lactide)-b-poly(ethylene glycol) copolymers. *Eur. Polym. J.* 45, 2839–2848. doi: 10.1016/j.eurpolymj.2009.07.006

Li, H., and Huneault, M. A. (2007). Effect of nucleation and plasticization on the crystallization of poly(lactic acid). *Polymer* 48, 6855–6866. doi: 10.1016/j.polymer.2007.09.020

Li, R., and Yao, D. (2008). Preparation of single poly(lactic acid) composites. *J. Appl. Polym. Sci.* 107, 2909–2916. doi: 10.1002/app.27406

Li, S. (1999). Hydrolytic degradation characteristics of aliphatic polyesters derived from lactic and glycolic acids. *J. Biomed. Mater. Res.* 48, 342–353.

Li, Y., and Shimizu, H. (2007). Toughening of polylactide by melt blending with a biodegradable poly(ether)urethane elastomer. *Macromol. Biosci.* 7, 921–928. doi: 10.1002/mabi.200700027

Li, Y., and Shimizu, H. (2009). Improvement in toughness of poly(l-lactide) (PLLA) through reactive blending with acrylonitrile–butadiene–styrene copolymer (ABS): morphology and properties. *Eur. Polym. J.* 45, 738–746. doi: 10.1016/j.eurpolymj.2008.12.010

Lim, L. T., Auras, R., and Rubino, M. (2008). Processing technologies for poly(lactic acid). *Prog. Polym. Sci.* 33, 820–852. doi: 10.1016/j.progpolymsci.2008.05.004

Lin, S., Guo, W., Chen, C., Ma, J., and Wang, B. (2012). Mechanical properties and morphology of biodegradable poly(lactic acid)/poly(butylene adipate-co-terephthalate) blends compatibilized by transesterification. *Mater. Des.* 36, 604–608. doi: 10.1016/j.matdes.2011.11.036

Lin, Y., Zhang, K.-Y., Dong, Z.-M., Dong, L.-S., and Li, Y.-S. (2007). Study of hydrogen-bonded blend of polylactide with biodegradable hyperbranched poly(ester amide). *Macromolecules* 40, 6257–6267. doi: 10.1021/ma070989a

Liu, H., Chen, F., Liu, B., Estep, G., and Zhang, J. (2010). Super tough-ened poly(lactic acid) ternary blends by simultaneous dynamic vulcaniza-tion and interfacial compatibilization. *Macromolecules* 43, 6058–6066. doi: 10.1021/ma101108g

Liu, H., Song, W., Chen, F., Guo, L., and Zhang, J. (2011). Interaction of microstructure and interfacial adhesion on impact performance of polylactide (PLA) ternary blends. *Macromolecules* 44, 1513–1522. doi: 10.1021/ma1026934

Liu, H., and Zhang, J. (2011). Research progress in toughening modifica-tion of poly(lactic acid). *J. Polym. Sci. B Polym. Phys.* 49, 1051–1083. doi: 10.1002/polb.22283

Ljungberg, N., Andersson, T., and Wesslén, B. (2003). Film extrusion and film weld-ability of poly(lactic acid) plasticized with triacetine and tributyl citrate. *J. Appl. Polym. Sci.* 88, 3239–3247. doi: 10.1002/app.12106

Ljungberg, N., Colombini, D., and Wesslén, B. (2005). Plasticization of poly(lactic acid) with oligomeric malonate esteramides: dynamic mechanical and thermal film properties. *J. Appl. Polym. Sci.* 96, 992–1002. doi: 10.1002/app.21163

Ljungberg, N., and Wesslén, B. (2002). The effects of plasticizers on the dynamic mechanical and thermal properties of poly(lactic acid). *J. Appl. Polym. Sci.* 86, 1227–1234. doi: 10.1002/app.11077

Ljungberg, N., and Wesslén, B. (2003). Tributyl citrate oligomers as plasticizers for poly (lactic acid): thermo-mechanical film properties and aging. *Polymer* 44, 7679–7688. doi: 10.1016/j.polymer.2003.09.055

Long, J., Bo, L., and Jinwen, Z. (2009). Properties of poly(lactic acid)/poly(butylene adipate- co -terephthalate)/nanoparticle ternary composites. *Ind. Eng. Chem. Res.* 48, 7594–7602. doi: 10.1021/ie900576f

López-Rodríguez, N., López-Arraiza, A., Meaurio, E., and Sarasua, J. R. (2006). Crystallization, morphology, and mechanical behaviour of polylactide/poly(ε-caprolactone) blends. *Polym. Eng. Sci.* 46, 1299–1308. doi: 10.1002/pen.20609

Lu, F., Cantwell, W. J., and Kausch, H. H. (1997). The role of cavitation and debond-ing in the toughening of core–shell rubber modified epoxy systems. *J. Mater. Sci.* 32, 3055–3059. doi: 10.1023/A:1018626012271

Ma, P., Hristova-Bogaerds, D. G., Goossens, J. G. P., Spoelstra, A. B., Zhang, Y., and Lemstra, P. J. (2012). Toughening of poly(lactic acid) by ethylene-co-vinyl acetate copolymer with different vinyl acetate contents. *Eur. Polym. J.* 48, 146–154. doi: 10.1016/j.eurpolymj.2011.10.015

Ma, P., Spoelstra, A. B., Schmit, P., and Lemstra, J. P. (2013). Toughening of poly(lactic acid) by poly(β-hydroxybutyrate-co- β-hydroxyvalerate) with high β-hydroxyvalerate content. *Eur. Polym. J.* 49, 1523–1531. doi: 10.1016/j.eurpolymj.2013.01.016

Mahajan, D. K., and Hartmaier, A. (2012). Mechanisms of crazing in glassy poly-mers revealed by molecular dynamics simulations. *Phys. Rev. E* 86, 021802. doi: 10.1103/PhysRevE.86.021802

Majola, A., Vainionpää, S., Rokkanen, P., Mikkola, H. M., and Törmälä, P. (1992). Absorbable self-reinforced polylactide (SR-PLA) composite rods for fracture fixation: strength and strength retention in the bone and subcuta-neous tissue of rabbits. *J. Mater. Sci. Mater. Med.* 3, 43–47. doi: 10.1007/BF00702943

Mäkelä, P., Pohjonen, T., Törmälä, P., Waris, T., and Ashammakhi, N. (2002). Strength retention properties of self-reinforced poly l-lactide (SR-PLLA) sutures compared with polyglyconate (MaxonR) and polydioxanone (PDS) sutures. An *in vitro* study. *Biomaterials* 23, 2587–2592. doi: 10.1016/S0142-9612(01)00396-9

Mani, R., Bhattacharya, M., and Tang, J. (1999). Functionalization of polyesters with maleic anhydride by reactive extrusion. *J. Polym. Sci. A Polym. Chem.* 37, 1693–1702.

Martin, O., and Avérous, L. (2001). Poly(lactic acid): plasticization and properties of biodegradable multiphase systems. *Polymer* 42, 6209–6219. doi: 10.1016/S0032-3861(01)00086-6

Martino, V. P., Jiménez, A., and Ruseckaite, R. A. (2009). Processing and characterization of poly(lactic acid) films plasticized with commercial adipates. *J. Appl. Polym. Sci.* 112, 2010–2018. doi: 10.1002/app.29784

Martino, V. P., Ruseckaite, R. A., and Jiménez, A. (2006). Thermal and mechanical characterization of plasticized poly (L-lactide-co-D,L-lactide) films for food packaging. *J. Therm. Anal. Calorim.* 86, 707–712. doi: 10.1007/s10973-006-7897-3

Mascia, L., and Xanthos, M. (1992). An overview of additives and modifiers for polymer blends: facts, deductions, and uncertainties. *Adv. Polym. Technol.* 11, 237–248. doi: 10.1002/adv.1992.060110402

Mason, C. D., and Tuller, H. W. (1983). *High impact nylon composition containing copolymer esters and ionic copolymers.* U.S. Pat. 4,404,325 A patent application. Available online at: http://www.google.nl/patents/US4404325.

Matabola, K. P., De Vries, A. R., Moolman, F. S., and Luyt, A. S. (2009). Single polymer composites: a review. *J. Mater. Sci.* 44, 6213–6222. doi: 10.1007/s10853-009-3792-1

McCarthy, S., and Song, X. (2002). Biodegradable plasticizers for polylactic acid. *J. Appl. Med. Polym.* 6, 64–69.

Meng, B., Deng, J., Liu, Q., Wu, Z., and Yang, W. (2012). Transparent and ductile poly(lactic acid)/poly(butyl acrylate) (PBA) blends: structure and properties. *Eur. Polym. J.* 48, 127–135. doi: 10.1016/j.eurpolymj.2011.10.009

Mercier, J. P., Aklonis, J. J., Litt, M., and Tobolsky, A. V. (1965). Viscoelastic behaviour of the polycarbonate of bisphenol A. *J. Appl. Polym. Sci.* 9, 447–459. doi: 10.1002/app.1965.070090206

Michaeli, W., Höcker, H., Berghaus, U., and Frings, W. (1993). Reactive extrusion of styrene polymers. *J. Appl. Polym. Sci.* 48, 871. doi: 10.1002/app.1993.070480512

Michler, G. H. (1989). Crazes in amorphous polymers I. Variety of the structure of crazes and classification of different types of crazes. *Colloid Polym. Sci.* 267, 377–388. doi: 10.1007/BF01410182

Mochizuki, T., and Suzuki, F. (2004). U.S. Patent Application 20040180990 A1, Assigned to Fuji Photo Co Ltd.

Murariu, M., Da Silva Ferreira, A., Alexandre, M., and Dubois, P. (2008a). Polylactide (PLA) designed with desired end-use properties: 1. PLA compositions with low molecular weight ester-like plasticizers and related performances. *Polym. Adv. Technol.* 19, 636–646. doi: 10.1002/pat.1131

Murariu, M., Ferreira, A. D. S., Duquesne, E., Bonnaud, L., and Dubois, P. (2008b). Polylactide (PLA) and highly filled PLA - calcium sulfate composites with improved impact properties. *Macromol. Symp.* 272, 1–12. doi: 10.1002/masy.200851201

Nair, L. S., and Laurencin, C. T. (2007). Biodegradable polymers as biomaterials. *Prog. Polym. Sci.* 32, 762–798. doi: 10.1016/j.progpolymsci.2007.05.017

Narisawa, I., and Yee, A. (2006). "Crazing and fracture of polymers," in *Materials Science and Technology* (Yonezaga; Ann Arbor: Wiley-VCH Verlag GmbH and Co. KGaA). doi: 10.1002/9783527603978.mst0146

Nascimento, L., Gamez-Perez, J., Santana, O. O., Velasco, J. I., Maspoch, M. L., and Franco-Urquiza, E. (2010). Effect of the Recycling and Annealing on the Mechanical and Fracture Properties of Poly(Lactic Acid). *J. Polym. Environ.* 18, 654–660. doi: 10.1007/s10924-010-0229-5

Natureworks® (2005). PLA Polymer 2002D – Data Sheet. Available online at: http://www.unicgroup.com/upfiles/file01170656495.pdf

Natureworks® (2006). PLA Polymer 4032D – Data Sheet. Available online at: http://www.natureworksllc.com/~/media/Technical_Resources/Technical_Data_Sheets/TechnicalDataSheet_4032D_films_pdf.pdf

Natureworks LLC (2007). Technology Focus Report: Toughened PLA. Availbale online at: http://www.natureworksllc.com/~/media/Technical_Resources/Properties_Documents/PropertiesDocument_Toughened-Ingeo_pdf.pdf

Noda, I., Green, P. R., Satkowski, M. M., and Schechtman, L. A. (2005). Preparation and properties of a novel class of polyhydroxyalkanoate copolymers. *Biomacromolecules* 6, 580–586. doi: 10.1021/bm049472m

Noda, I., Satkowski, M. M., Dowrey, A. E., and Marcott, C. (2004). Polymer Alloys of Nodax Copolymers and Poly(lactic acid). *Macromol. Biosci.* 4, 269–275. doi: 10.1002/mabi.200300093

Nyambo, C., Misra, M., and Mohanty, A. (2012). Toughening of brittle poly(lactide) with hyperbranched poly(ester-amide) and isocyanate-terminated prepolymer of polybutadiene. *J. Mater. Sci.* 47, 5158–5168. doi: 10.1007/s10853-012-6393-3

O'Connell, P. A., and McKenna, G. B. (2002). "Yield and crazing in polymers," in *Encyclopedia of Polymer Science and Technology* (Lubbock, Texas: Texas Tech University; John Wiley and Sons, Inc.). Available online at: http://onlinelibrary.wiley.com/book/10.1002/0471440264/homepage/Editors Contributors.html. doi: 10.1002/0471440264.pst463

Odent, J., Habibi, Y., Raquez, J.-M., and Dubois, P. (2013a). *Ultra-tough polylactide-based materials synergistically designed in the presence of rubbery ε-caprolactone-based copolyester and silica nanoparticles.* Composites Science and Technology. doi: 10.1016/j.compscitech.2013.05.003

Odent, J., Leclère, P., Raquez, J.-M., and Dubois, P. (2013b). Toughening of polylactide by tailoring phase-morphology with P[CL-co-LA] random copolyesters as biodegradable impact modifiers. *Eur. Polym. J.* 49, 914–922. doi: 10.1016/j.eurpolymj.2012.12.006

Odent, J., Habibi, Y., Raquez, J.-M., and Dubois, P. (2013c). "A new paradigm for the toughening of polylactide-based materials," in *Proceedings of the Polymer Processing Society 28th Annual Meeting ~ PPS-28.* Pattaya

Odent, J., Raquez, J.-M., Duquesne, E., and Dubois, P. (2012). Random aliphatic copolyesters as new biodegradable impact modifiers for polylactide materials. *Eur. Polym. J.* 48, 331–340. doi: 10.1016/j.eurpolymj.2011.11.002

Ohlberg, S. M., Roth, J., and Raff, R. A. V. (1959). Relationship between impact strength and spherulite growth in linear polyethylene. *J. Appl. Polym. Sci.* 1, 114–120. doi: 10.1002/app.1959.070010118

Oyama, H. T. (2009). Super-tough poly(lactic acid) materials: Reactive blending with ethylene copolymer. *Polymer* 50, 747–751. doi: 10.1016/j.polymer.2008.12.025

Park, S. D., Todo, M., Arakawa, K., and Koganemaru, M. (2006). Effect of crystallinity and loading-rate on mode I fracture behaviour of poly(lactic acid). *Polymer* 47, 1357–1363. doi: 10.1016/j.polymer.2005.12.046

Pearson Raymond, A. (2000). "Introduction to the Toughening of Polymers," in *Toughening of Plastics,* eds A. Pearson Raymond, H.-J. Sue, and A. F. Yee (Lehigh University; Texas A&M University; The University of Michigan: American Chemical Society), 1–12.

Pecorini, T. J., and Hertzberg, R. W. (1993). The fracture toughness and fatigue crack propagation behaviour of annealed PET. *Polymer* 34, 5053–5062. doi: 10.1016/0032-3861(93)90248-9

Perego, G., Cella, G. D., and Bastioli, C. (1996). Effect of molecular weight and crystallinity on poly(lactic acid) mechanical properties. *J. Appl. Polym. Sci.* 59, 37–43.

Perkins, W. G. (1999). Polymer toughness and impact resistance. *Polym. Eng. Sci.* 39, 2445–2460. doi: 10.1002/pen.11632

Petchwattana, N., Covavisaruch, S., and Euapanthasate, N. (2012). Utilization of ultrafine acrylate rubber particles as a toughening agent for poly(lactic acid). *Mater. Sci. Eng. A* 532, 64–70. doi: 10.1016/j.msea.2011.10.063

Pillin, I., Montrelay, N., and Grohens, Y. (2006). Thermo-mechanical characterization of plasticized PLA: Is the miscibility the only significant factor? *Polymer* 47, 4676–4682. doi: 10.1016/j.polymer.2006.04.013

Poirier, Y. (2002). Polyhydroxyalknoate synthesis in plants as a tool for biotechnology and basic studies of lipid metabolism. *Prog. Lipid Res.* 41, 131–155. doi: 10.1016/S0163-7827(01)00018-2

Rabetafika, H. N., Paquot, M., and Dubois, P. (2006). Les polymères issus du végétal matériaux à propriétés. *Biotechnol. Agron. Soc. Environ.* 10, 185–196.

Robertson, M. L., Chang, K., Gramlich, W. M., and Hillmyer, M. A. (2010). Toughening of polylactide with polymerized soybean oil. *Macromolecules* 43, 1807–1814. doi: 10.1021/ma9022795

Scaffaro, R., Morreale, M., Mirabella, F., and La Mantia, F. P. (2011). Preparation and Recycling of Plasticized PLA. *Macromol. Mater. Eng.* 296, 141–150. doi: 10.1002/mame.201000221

Schreck, K. M., and Hillmyer, M. A. (2007). Block copolymers and melt blends of polylactide with Nodax™ microbial polyesters: preparation and mechanical properties. *J. Biotechnol.* 132, 287–295. doi: 10.1016/j.jbiotec.2007.03.017

Seddon, J. D., Hepworth, S. J., and Priddle, J. E. (1971). United Kingdom 1,241,361 patent application.

Seelig, T., and Van Der Giessen, E. (2009). A cell model study of crazing and matrix plasticity in rubber-toughened glassy polymers. *Comput. Mater. Sci.* 45, 725–728. doi: 10.1016/j.commatsci.2008.05.024

Seiler, M. (2002). Dendritic polymers – interdisciplinary research and emerging applications from unique structural properties. *Chem. Eng. Technol.* 25, 237–253. doi: 10.1002/1521-4125(200203)25:3<237::AID-CEAT237>ù 3.0.CO;2-4

Semba, T., Kitagawa, K., Ishiaku, U. S., and Hamada, H. (2006). The effect of crosslinking on the mechanical properties of polylactic acid/polycaprolactone blends. *J. Appl. Polym. Sci.* 101, 1816–1825. doi: 10.1002/app.23589

Semba, T., Kitagawa, K., Ishiaku, U. S., Kotaki, M., and Hamada, H. (2007). Effect of compounding procedure on mechanical properties and dispersed phase morphology of poly(lactic acid)/polycaprolactone blends containing peroxide. *J. Appl. Polym. Sci.* 103, 1066–1074. doi: 10.1002/app.25311

Sierra, J., Noriega, M., Cardona, E., and Ospina, S. (2010). *Proceedings of Annual Technical Conference of the Society of Plastics Engineers (ANTEC 2010)* (Orlando, FL).

Sinclair, R. G. (1996). The case for polylactic acid as a commodity packaging plastic. *J. Macromol. Sci. A Pure Appl. Chem.* 33, 585–597. doi: 10.1080/10601329608010880

Song, W., Liu, H., Chen, F., and Zhang, J. (2012). Effects of ionomer characteristics on reactions and properties of poly(lactic acid) ternary blends prepared by reactive blending. *Polymer* 53, 2476–2484. doi: 10.1016/j.polymer.2012.03.050

Steinbüchel, A., and Valentin, H. E. (1995). Diversity of bacterial polyhydroxyalkanoic acids. *FEMS Microbiol. Lett.* 128, 219–228. doi: 10.1016/0378-1097(95)00125-O

Su, Z., Li, Q., Liu, Y., Hu, G.-H., and Wu, C. (2009). Compatibility and phase structure of binary blends of poly(lactic acid) and glycidyl methacrylate grafted poly(ethylene octane). *Eur. Polym. J.* 45, 2428–2433. doi: 10.1016/j.eurpolymj.2009.04.028

Sukano Co. (2008). Website. SUKANOVR announces unique transparent impact modifier for PLA. Available online at: http://www.sukano.com/downloads/newsletter/english/2008_02_PLA_IM_EN.pdf. Accessed on February 2008.

Taib, R. M., Ghaleb, Z. A., and Mohd Ishak, Z. A. (2012). Thermal, mechanical, and morphological properties of polylactic acid toughened with an impact modifier. *J. Appl. Polym. Sci.* 123, 2715–2725. doi: 10.1002/app.34884

Takayama, T., and Todo, M. (2006). Improvement of impact fracture properties of PLA/PCL polymer blend due to LTI addition. *J. Mater. Sci.* 41, 4989–4992. doi: 10.1007/s10853-006-0137-1

Takayama, T., Todo, M., Tsuji, H., and Arakawa, K. (2006). Effect of LTI content on impact fracture property of PLA/PCL/LTI polymer blends. *J. Mater. Sci.* 41, 6501–6504. doi: 10.1007/s10853-006-0611-9

Teamsinsungvon, A., Ruksakulpiwat, Y., and Jarukumjorn, K. (2010). *Mechanical and morphological properties of poly(lactic acid)/poly (butylene adipate-co-terephtalate)/calcium carbonate composite. 18th international conference on composite materials.* Avaliable online at: http://www.iccm-central.org/Proceedings/ICCM18proceedings/data/3.%20Poster%20Presentation/Aug23(Tuesday)/P2-18~35%20Green%20Composites/P2-27-IF1477.pdf

Törmälä, P. (1992). Biodegradable self-reinforced composite materials; Manufacturing structure and mechanical properties. *Clin. Mater.* 10, 29–34. doi: 10.1016/0267-6605(92)90081-4

Tormala, P., Rokkanen, P., Laiho, J., Tamminmaki, M., and Vainionpaa, S. (1988). Material for osteosynthesis devices. *US Patent Application 4,473,257*

Tsuji, H., Nakano, M., Hashimoto, M., Takashima, K., Katsura, S., and Mizuno, A. (2006). Electrospinning of Poly(lactic acid) Stereocomplex Nanofibers. *Biomacromolecules* 7, 3316–3320. doi: 10.1021/bm060786e

Urayama, H., Ma, C., and Kimura, Y. (2003). Mechanical and thermal properties of poly(L-lactide) incorporating various inorganic fillers with particle and whisker shapes. *Macromol. Mater. Eng.* 288, 562–568. doi: 10.1002/mame.200350004

Van Der Wal, A., and Gaymans, R. J. (1999). Polypropylene–rubber blends: 5. Deformation mechanism during fracture. *Polymer* 40, 6067–6075. doi: 10.1016/S0032-3861(99)00216-5

Vannaladsaysy, V., Todo, M., Takayama, T., Jaafar, M., Ahmad, Z., and Pasomsouk, K. (2009). Effects of lysine triisocyanate on the mode I fracture behaviour of polymer blend of poly (l-lactic acid) and poly (butylene succinate-co-l-lactate). *J. Mater. Sci.* 44, 3006–3009. doi: 10.1007/s10853-009-3428-5

Vilay, V., Mariatti, M., Ahmad, Z., Pasomsouk, K., and Todo, M. (2009). Characterization of the mechanical and thermal properties and morphological behaviour of biodegradable poly(L-lactide)/poly(ε-caprolactone) and poly(L-lactide)/poly(butylene succinate-co-L-lactate) polymeric blends. *J. Appl. Polym. Sci.* 114, 1784–1792. doi: 10.1002/app.30683

Vroman, I., and Tighzert, L. (2009). Biodegradable polymers. *Materials* 2, 307–344. doi: 10.3390/ma2020307

Wang, L., Ma, W., Gross, R. A., and McCarthy, S. P. (1998). Reactive compatibilization of biodegradable blends of poly(lactic acid) and poly(ε-caprolactone). *Polym. Degrad. Stab.* 59, 161–168. doi: 10.1016/S0141-3910(97)00196-1

Wang, R., Wang, S., Zhang, Y., Wan, C., and Ma, P. (2009). Toughening modification of PLLA/PBS blends via *in situ* compatibilization. *Polym. Eng. Sci.* 49, 26–33. doi: 10.1002/pen.21210

White, E. F. T. (1984). Fracture behaviour of polymers. Edited by A. J. Kinloch and R. J. Young, Applied Science Publishers, London and New York, 1983. Pp xxv + 496, Price £30.00. ISBN 0853341869. *Br. Polym. J.* 16, 114. doi: 10.1002/pi.4980160231f

Wright-Charlesworth, D. D., Miller, D. M., Miskioglu, I., and King, J. A. (2005). Nanoindentation of injection moulded PLA and self-reinforced composite PLA after *in vitro* conditioning for three months. *J. Biomed. Mater. Res. A* 74A, 388–396. doi: 10.1002/jbm.a.30353

Wu, S. (1990). Chain structure, phase morphology, and toughness relationships in polymers and blends. *Polym. Eng. Sci.* 30, 753–761. doi: 10.1002/pen.760301302

Xiao, H., Lu, W., and Yeh, J.-T. (2009). Effect of plasticizer on the crystallization behaviour of poly(lactic acid). *J. Appl. Polym. Sci.* 113, 112–121. doi: 10.1002/app.29955

Yu, F., Liu, T., Zhao, X., Yu, X., Lu, A., and Wang, J. (2012). Effects of Talc on the Mechanical and Thermal Properties of Polylactide. *J. Appl. Polym. Sci.* 125, E99–E109. doi: 10.1002/app.36260

Yu, L., Liu, H., Xie, F., Chen, L., and Li, X. (2008). Effect of annealing and orientation on microstructures and mechanical properties of polylactic acid. *Polym. Eng. Sci.* 48, 634–641. doi: 10.1002/pen.20970

Zhang, C., Wang, W., Huang, Y., Pan, Y., Jiang, L., Dan, Y., et al. (2013a). Thermal, mechanical and rheological properties of polylactide toughened by expoxidized natural rubber. *Mater. Des.* 45, 198–205. doi: 10.1016/j.matdes.2012.09.024

Zhang, H., Fang, J., Ge, H., Han, L., Wang, X., Hao, Y., et al. (2013b). Thermal, mechanical, and rheological properties of polylactide/poly(1,2-propylene glycol adipate). *Polym. Eng. Sci.* 53, 112–118. doi: 10.1002/pen.23238

Zhang, N., Wang, Q., Ren, J., and Wang, L. (2009a). Preparation and properties of biodegradable poly(lactic acid)/poly(butylene adipate-co-terephthalate) blend with glycidyl methacrylate as reactive processing agent. *J. Mater. Sci.* 44, 250–256. doi: 10.1007/s10853-008-3049-4

Zhang, W., Chen, L., and Zhang, Y. (2009b). Surprising shape-memory effect of polylactide resulted from toughening by polyamide elastomer. *Polymer* 50, 1311–1315. doi: 10.1016/j.polymer.2009.01.032

Zhang, X., Li, Y., Han, L., Han, C., Xu, K., Zhou, C., et al. (2013c). Improvement in toughness and crystallization of poly(L-lactic acid) by melt blending with ethylene/methyl acrylate/glycidyl methacrylate terpolymer. *Polym. Eng. Sci.* 53, 2498–2508. doi: 10.1002/app.23507

Zhao, P., Liu, W., Wu, Q., and Ren, J. (2010). Preparation, mechanical, and thermal properties of biodegradable polyesters/poly(lactic acid) blends. *J. Nanomater.* 2010. doi: 10.1155/2010/287082

Zhao, Q., Ding, Y., Yang, B., Ning, N., and Fu, Q. (2013). Highly efficient toughening effect of ultrafine full-vulcanized powdered rubber on poly(lactic acid)(PLA). *Polym. Test.* 32, 299–305. doi: 10.1016/j.polymertesting.2012.11.012

Zhu, S., Rasal, R., and Hirt, D. (2009). *Annual Technical Conference of the Society of Plastics Engineers (ANTEC 2009)*, (Chicago, Illinois), 1616–1620.

Zinn, M., Witholt, B., and Egli, T. (2001). Occurrence, synthesis and medical application of bacterial polyhydroxyalkanoate. *Adv. Drug Deliv. Rev.* 53, 5–21. doi: 10.1016/S0169-409X(01)00218-6

Conflict of Interest Statement: The authors declare that the research was conducted in the absence of any commercial or financial relationships that could be construed as a potential conflict of interest.

[1]H high resolution magic-angle coil spinning (HR-MACS) μNMR metabolic profiling of whole *Saccharomyces cervisiae* cells: a demonstrative study

Alan Wong[1]*, Céline Boutin[1] and Pedro M. Aguiar[2]

[1] *CEA Saclay, DSM, IRAMIS, UMR CEA/CNRS 3299 – NIMBE, Laboratoire Structure et Dynamique par Résonance Magnétique, Gif-sur-Yvette, France*
[2] *Department of Chemistry, University of York, Heslington, York, UK*

Edited by:
Martina Vermathen, University of Berne, Switzerland

Reviewed by:
Xue-Mei Li, Linyi University, China
Valeria Righi, University of Bologna, Italy

*Correspondence:
Alan Wong, CEA Saclay, DSM, IRAMIS, UMR CEA/CNRS 3299 – NIMBE, Laboratoire Structure et Dynamique par Résonance Magnétique, F-91191, Gif-sur-Yvette, France
e-mail: alan.wong@cea.fr

The low sensitivity and thus need for large sample volume is one of the major drawbacks of Nuclear Magnetic Resonance (NMR) spectroscopy. This is especially problematic for performing rich metabolic profiling of scarce samples such as whole cells or living organisms. This study evaluates a ^{1}H HR-MAS approach for metabolic profiling of small volumes (250 nl) of whole cells. We have applied an emerging micro-NMR technology, high-resolution magic-angle coil spinning (HR-MACS), to study whole *Saccharomyces cervisiae* cells. We find that high-resolution high-sensitivity spectra can be obtained with only 19 million cells and, as a demonstration of the metabolic profiling potential, we perform two independent metabolomics studies identifying the significant metabolites associated with osmotic stress and aging.

Keywords: micro-NMR, HR-MACS, Metabolic Profiling, *Saccharomyces cervisiae*, osmotic stress, cell growth

INTRODUCTION

[1]H nuclear magnetic resonance (NMR) spectroscopy has gained recognition as a key analytical technique for metabolic profiling of complex bio-systems (Dunn and Ellis, 2005). The non-destructive nature, simplicity of sample preparation and data acquisition provide advantages over other methods including mass spectrometry (MS) for high-precision investigation of metabolites in living specimens. Moreover, the observed signal intensities in a typical ^{1}H NMR spectrum provide a *direct* comparison of the metabolite contents in the samples without the need to construct calibration curves for every analyte, which is often the case for other analytical techniques. For these reasons, ^{1}H NMR spectroscopy is widely used in the study of metabolomes, offering a robust tool for rich-metabolic profiling (Reo, 2002). Many samples of interest are highly complex bio-mixtures (e.g., biopsies, whole living cells and organisms) and the heterogeneity in the magnetic susceptibility over the sample volume results in broadening of the observed NMR resonances; significantly reducing the ability to identify and quantify the metabolic content using traditional high-resolution NMR techniques. For such samples the application of [1]H-detected high-resolution magic-angle spinning (HR-MAS) NMR, has now emerged as a powerful analytical tool for the investigation of heterogenous samples such as intact biopsies (Lindon et al., 2009; Sitter et al., 2009; Beckonert et al., 2010) and whole living organisms (Blaise et al., 2007; Righi et al., 2010). The rapid spinning of the sample (ca. 2–6 kHz) about an axis at an angle of 54.74° (i.e., the magic-angle) with respect to the static magnetic field B_0 overcomes the broadening and yields well-resolved signals. ^{1}H HR-MAS NMR is considered a near universal technique for providing unbiased and high-precision fingerprints of abundant metabolites in heterogenous biosamples. Despite its utility, there are only a handful of studies on whole cells, mainly with robust eukaryotic cells such as marine unicellular microalgae cells (Chauton et al., 2003a,b) and bacterial cells (Himmelreich et al., 2003; Gudlavalleti et al., 2006; Palomino-Schätzlein et al., 2013; Righi et al., 2013), whose cell membranes provide additional protection and structural support significantly enhancing cell survival rates during the measurement. In a review by Li (2006), he has summarized a progress for *in vivo* studies of intact bacterial cell using multidimensional NMR experiments with HR-MAS. This paucity of examples for HR-MAS studies of whole cells, compared to tissues, is due in large part to the potential for cell lysis as a result of the large centrifugal forces that the cells are subject to under the rapid sample spinning. This is especially problematic for fragile animal and human cells, and can result in distortion of the intracellular metabolic composition. Thus, high-resolution NMR of cell extracts has been the preferred method for cell studies. However in this case, the cells must be subject to complicated chemical treatment protocols to extract specific metabolites; hydrophilic metabolites in aqueous extracts and lipophilic in organic extracts, and require larger sample quantities for the multiple NMR samples and experiments to obtain a rich profile of the metabolic response. Unlike with the analysis of extracts, it is unnecessary to chemically divide the hydrophilic and hydrophobic metabolites for HR-MAS, offering a direct analytical approach. Palomino-Schätzlein et al. have optimized a HR-MAS protocol for the study of whole cells using abundant *Saccharomyces cervisiae* cells and reported similar metabolic profiles to those obtained with high-resolution NMR of cell extracts (Palomino-Schätzlein et al., 2013).

NMR is an inherently insensitive technique, thus HR-MAS analysis often relies on large sample volumes for detection; typically about 100 million whole cells in a 30–50 μl volume for each of 3–5 replicate samples (for statistical analyses). In cases where sample size is limited (such as neuron cells), analysis of fewer cells—in a smaller volume—would ease the sample preparation and may improve the high-throughput efficiency (e.g., coupling with micro-fluidic devices for cell separation techniques). One promising approach for volume-, or mass-, limited bio-specimens is the uses of a high-resolution magic-angle coil spinning (HR-MACS) (Wong et al., 2012, 2013). HR-MACS, as with the original MACS experiment (Sakellariou et al., 2007), utilizes a secondary tuned circuit and a simple and robust rotor insert, designed to fit inside a standard MAS sample rotor to convert the MAS probe into a μMAS probe without any probe modification. HR-MACS can be readily coupled with any standard HR-MAS probe making it assessable to any laboratory with HR-MAS facilities. The use of a filling-factor optimized detector, susceptibility-optimized inserts and simultaneous spinning of the sample and detector have been demonstrated to yield high sensitivity and excellent spectral resolution (up to 2 ppb) allowing for high-precision metabolomic assessments of a sample volume less than 500 nl (Wong et al., 2013). Moreover, the reduced diameter of the sample also abates the centrifugal forces exerted on the cells under sample spinning diminishing the chances of cell lysis due to sample spinning.

The present study capitalizes on the new development of HR-MACS for high-sensitivity and high-resolution ^{1}H μNMR-based metabolomics study of whole cells. We build upon the ^{1}H HR-MAS study of whole *S. cervisiae* cells (Palomino-Schätzlein et al., 2013) to evaluate the utility of ^{1}H HR-MACS for profiling the metabolic composition of a small number of *S. cervisiae* cells (ca. 19 million cells in a 250 nl volume). We have performed two independent studies, monitoring the metabolic responses in *S. cervisiae* wild-type cells under osmotic stress and cell growth. There are a total of four cell groups submitted to the two studies: 24, 48, and 72-h cell growth and a 24-h cell growth subjected to a 60-min period of osmotic stress.

MATERIALS AND METHODS
SACCHAROMYCES CERVISIAE
Wild-type (WT) cells were grown in a glucose-rich medium YPD (1% yeast extract, 2% peptone, 2% glucose) at 30°C on a shaker (180 rpm). Each sample was inoculated with the same initial culture and with an initial OD_{600} of 0.1 and grown for 24, 48, and 72 h. For the osmotic cell stress study, 0.5 M NaCl (final concentration) was added to the 24-h culture cells for 60 min. Prior to the NMR sample preparation, *ca.* 300×10^6 cells were washed 3 times with ion-free distilled water. The cell pellet was re-suspended in 4 μl of D_2O for an immediate NMR acquisition. The use of high cell concentration was utilized to ensure significant numbers of cells were being transferred to the microsize capillary.

HR-MACS RESONATOR
A single sample-exchangeable HR-MACS resonator (**Figure 1**) was used in the study for data acquisition. The resonator was constructed by manually winding a 2 mm long (9-turn solenoid),

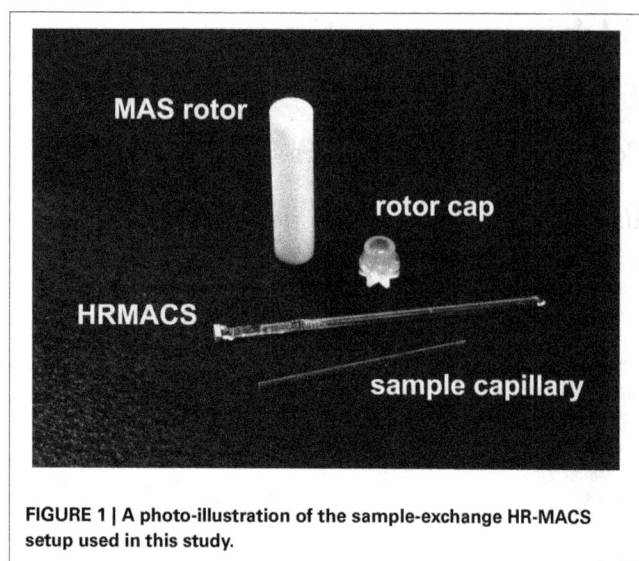

FIGURE 1 | A photo-illustration of the sample-exchange HR-MACS setup used in this study.

using 30-μm o.d. round copper wire around a 840/600-μm (outer/inner diameter) quartz capillary. The solenoid was then fixed in place with a thin layer of cyanoacrylate glue. A non-magnetic 2.2 pF capacitor (American Technical Ceramics, US) was soldered to the coil leads and affixed to the end of the capillary with cyanoacrylate glue. The resonator had a resonance frequency of 508 MHz with an unloaded coil quality factor, $Q = 26$. The resonator was secured in a Kel-F insert which fits tightly inside a Bruker ZrO_2 4-mm rotor allowing for an easy sampling with a 550/400-μm quartz capillary.

SAMPLE-PREPARATION
The sample preparation for HR-MACS was performed under a stereomicroscope. The *S. cervisiae* cells (ca. 300×10^6 cells in 4 μl of D_2O) were pipetted into a 550/400-μm capillary sealed at one end (with epoxy) using a micropipette equipped with a 20 μl GELoader® tip (Eppendorf, US). Very gentle centrifugation was applied for 2–5 s to remove air bubbles and ensure the cells displaced into the solenoid region. Within the coil detection region there are ca. 19×10^6 cells. The top of the sample capillary was then sealed with hot paraffin wax to prevent leakage during sample spinning, and inserted into the HR-MACS resonator already fitted inside the MAS rotor. The entire sample-preparation procedure was restricted to no more than 2 min to minimize the chances of sample tampering and degradation.

^{1}H HR-MACS NMR SPECTROSCOPY
^{1}H NMR experiments were performed on a widebore 11.7 T magnet equipped with a Bruker Avance II Spectrometer operating at 499.13 MHz with a standard Bruker 4-mm HX CP-MAS probe (with a HR-MACS resonator placed inside the probe). The MAS frequency was set using the manual settings of a standard Bruker MAS spin controller to 300 ± 1 Hz. The B_0 shimming was performed on a sucrose-D_2O sample under slow MAS and re-checked every five samples. The 90°-pulse for HR-MACS was 5.3 μs at 2 W of applied radio-frequency amplitude. An 8-step 2D-PASS sequence was used (Antzutkin et al., 1995), as

previously demonstrated with the addition of water-suppression (Wong et al., 2010, 2013), to acquire high-resolution sideband- and water-free spectra for all samples. ^{1}H PASS NMR spectra were acquired with 8 t_1 increments with each consisting of 96 co-added scans for 20 k data points over a spectral width of 20 ppm. A 1 s recycle delay was used resulting in a 26-min experiment time. The experiments were performed under sample temperature regulation at 10°C. ^{1}H chemical shifts were internally referenced to the alanine—CH$_3$—doublet at $\delta = 1.47$ ppm. For each group, ^{1}H NMR spectra of five replicate samples were recorded to ensure reproducibility and reliability of the spectral data for interpretation.

MULTIVARIATE DATA ANALYSIS

To reduce the complexity of the NMR data for the subsequent multivariate analysis, the spectra were reduced to 0.005 ppm-wide buckets over the spectral region between 0.87 and 9.00 ppm, with exclusion of the water region (4.7–5.1 ppm), and normalized by the total sum of intensities using MestReNova v8.1.4. Principal component analysis (PCA) and orthogonal partial least-square discriminant analysis (OPLS-DA) were applied to check the data homogeneity and identify the latent patterns and biomarkers using SIMCA-P 13 (Umetrics, Umea, Sweden). Variable data were centered prior to PCA and OPLS-DA.

RESULTS AND DISCUSSION

^{1}H HR-MACS NMR SPECTRA

The ^{1}H HR-MACS NMR spectra for all cell groups in this study (cell growth and osmotic study) are shown in **Figure 2A**. The overall spectral profiles and the resolution are consistent with the HR-MAS study using 30 μl volumes and ca. 75 million cells (Palomino-Schätzlein et al., 2013). Despite using a CP-MAS rather than a HR-MAS probe with field-locking capabilities, we nonetheless obtained excellent quality as previously shown (Wong et al., 2013), but better performance is expected with a HR-MAS probe. The use of the PASS experiment under slow MAS (300 Hz), which consists a train of 180-degree refocussing pulses within one rotor period (3.3 ms), provides a T$_2$-filter similar to the T$_2$-edited CPMG experiments for suppressing signals from large molecules that may mask the metabolite signals and distort the baselines. All cell groups were found to have high glucose ($\delta = 3.5$–4.0 ppm) contents, which is attributed to the use of a glucose-rich growth medium (YPD) for the cultivation of cells. The observable fine J-splittings in many resonances are measureable from the spectra. For example, the distinct doublets of valine at 0.98 and 1.04 ppm ($J = 7.3$ Hz), and the triplet of α-glucose at 3.45 ppm ($J = 9.6$ Hz) are apparent in all HR-MACS spectra. The excellent spectral resolution allows extraction of this vital second parameter—in addition to chemical shift—permitting greater precision in peak assignments. A total of 22 metabolites have been identified from the two studies and are summarized in Table S1 of the supporting information (SI). The capability of such rich-profiling from a sub-microliter sample volume (250 nl) is owed to the fact that HR-MACS offers a 4.8-fold sensitivity enhancement in signal-to-noise (SNR) per-unit-mass compared to the coupled HR-MAS probe. This enhancement factor has been calculated based on B_1^{HRMACS}/B_1^{HRMAS} at a given radio frequency input power (Hoult, 2000), where the B_1 field can be

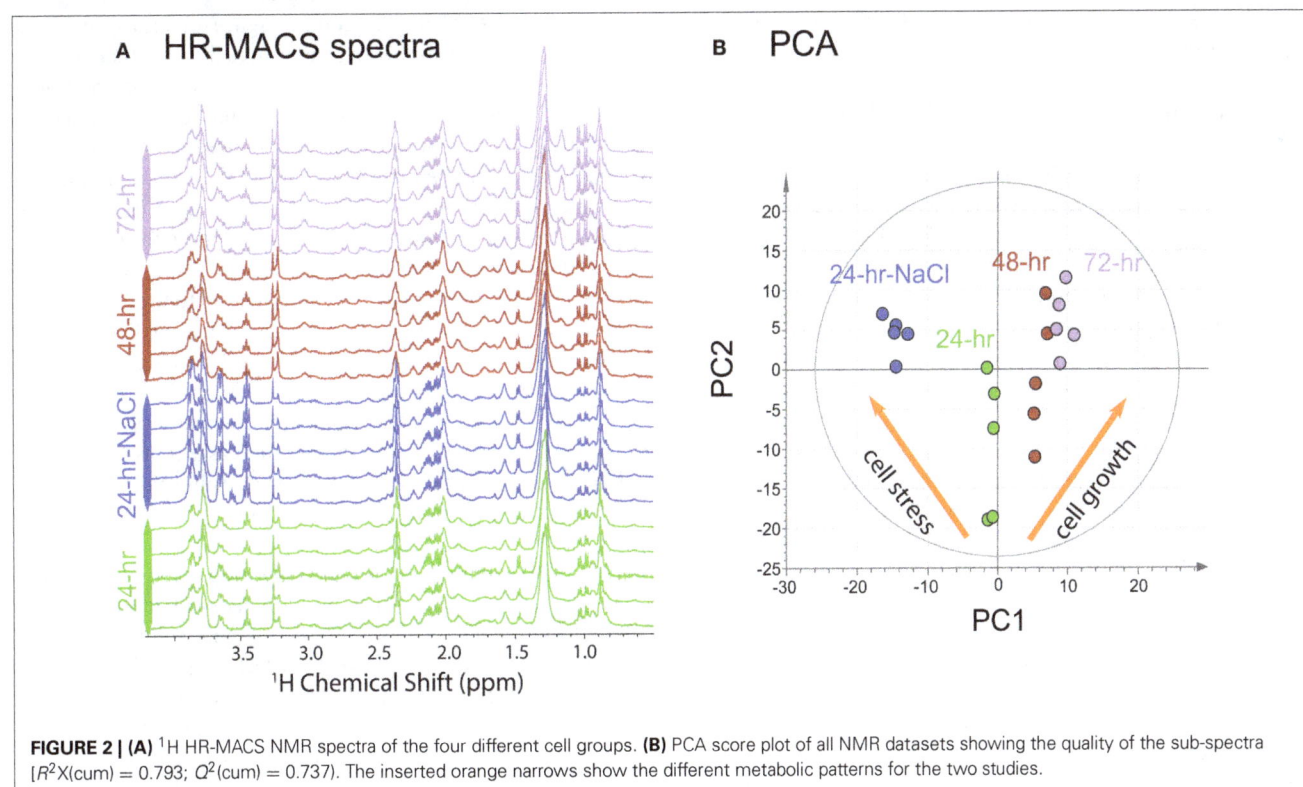

FIGURE 2 | (A) ^{1}H HR-MACS NMR spectra of the four different cell groups. **(B)** PCA score plot of all NMR datasets showing the quality of the sub-spectra [R^2X(cum) = 0.793; Q^2(cum) = 0.737). The inserted orange narrows show the different metabolic patterns for the two studies.

determined from a standard nutation experiment. We found that with an estimated 19 million cells in 250 nl, we obtained an average SNR (using the 2.5–2.0 ppm spectral region) of 120 ± 10 for a 26-min experimental time. Such high SNR should allow for further reductions in sample size, or in signal-averaging for samples which are prone to rapid decay. Although, there are multiple factors that contribute to the SNR using the inductively-coupled HR-MACS resonator such as coil volume, sampling spin and voltage noise (more detailed descriptions can be found in SI), using our measured SNR and a value of 10:1 (SNR) as the minimum required for metabolite analysis, these results would indicate that with this HR-MACS spectra should be obtainable from ca. 2 million cells under the same experimental conditions.

Visual inspection of the spectra in **Figure 2A** allows discrimination of the spectral differences among the different cell groups. For example, higher intensities are found in the region between 3.5–4.0 ppm (glycerol and α-glucose) in the 24-h-NaCl group (cells under osmotic stress); larger signals at 1.0 ppm (valine) and at 3.2 ppm (phosphorylcholine and glycerophosphocholine) are also observed in the 72-h cells. Upon a more careful inspection, additional metabolites (choline, creatine, tyrosine, and phenylalanine) are identified in the aging cells as compared to stressed cells (see Table S1). An unsupervised PCA score plot is shown in **Figure 2B**. It reveals partitioning into four discrete clusters corresponding to the different cell groups. Interestingly, the plot also reveals a clear distinction between the two independent studies, with the corresponding cells responding in an opposite fashion along the first principal component (PC1). The PCA loading plot (Figure S1b), provides further evidence for metabolite differences between the two studies. As the cells age both phosphorylcholine and glycerophosphocholine increase; whereas, the metabolites glycerol, glucose and glutamine show the largest response to the osmatic shock induced by the additional NaCl.

METABOLIC PROFILING OF *S. CERVISIAE* WT CELLS UNDER OSMOTIC STRESS

Figure 3A shows a HR-MACS spectral comparison between the non-stressed and stressed cell groups. Both are average spectra, constructed using 5 replicate samples for each group. The difference spectrum, ΔS, consists only of the metabolites which vary under a 60-min osmotic shock; positive signals indicate an increase, and negative signals a decrease in a given metabolite upon osmotic shock. Substantial increases are found for both glucose and glycerol, and a decrease in a few amino acids (lysine, arginine, and valine) is observed. These changes have been quantified (Table S2 of SI) and are shown graphically in the spectral integral analysis of **Figure 3B**. For a more accurate metabolic analysis of the latent patterns, the acquired spectra were subjected to a supervised statistical analysis using OPLS-DA. The score plot (**Figure 3C**) displays a clear discrimination between the two groups. Interestingly, it shows a greater metabolic variation (i.e., along the orthogonal component t_{orth}) among the non-stressed cells agreeing with the observed scatter pattern in the unsupervised PCA (**Figure 2B**). The corresponding loadings plot (**Figure 3D**) illustrates the metabolites associated with the separation in the score plot. As with the ΔS spectrum for **Figure 3A**, the positive signals are associated with the metabolites in higher concentration for the stressed cells and the negative peaks represent higher concentration in the non-stressed cells. The color scale provides a measure of the correlation of that spectral region's (i.e., metabolites') variation to the multivariate discrimination. Similar to the spectral comparison in **Figures 3A,B**, significant productions of glucose and glycerol metabolites are evident in the stressed cells (in agreement with the PCA loading plot Figure S2). This is inline with studies which have found that glycerol plays an osmoregulatory role in yeast cells (Tamas et al., 1999; Costenoble et al., 2000; Dihazi et al., 2004), and is produced when the yeast cells are subject to osmotic shock. The OPLS-DA analysis also reveals the latent metabolic changes; decreases in select amino acids (lysine, aspartate, arginine, and valine) may suggest an altered glucose metabolism that involves pyruvate production. Unfortunately, the chemical shift signature (a singlet at 2.36 ppm) for pyruvate is not readily identifiable in the spectra. The metabolic results obtained from ¹H HR-MACS are coherent with the previous ¹H HR-MAS study (Palomino-Schätzlein et al., 2013) using a large sample volume with greater number of cells, validating the used of HR-MACS.

METABOLIC PROFILING OF *S. CERVISIAE* WT CELLS AT THREE DIFFERENT STAGES OF CELL GROWTH

We also monitored the metabolic profiles at three different stages of cell growth: 24-, 48- and 72-h. Typically, the *S. cervisiae* cells are in the exponential growth phase between 24- and 48-h interval, and reach a steady stage at 48-h. At 72-h, the cells may begin to breakdown the nutrients. **Figure 4A** shows the average spectral comparison ($n = 5$) between the different stages, with a differential ΔS spectrum of 24- and 72-h cell groups; the positive signals indicate an increase, and negative peaks a decrease of a given metabolite in the 72-h cell groups. **Figure 4A** clearly exhibits a metabolic spectral profile that is different from that for the stressed cells, suggesting that the metabolic variations are different in both studies. Increases in the metabolite content of ethanol, lipid, lactate, alanine, arginine, creatine, phosphorylcholine, and glycerophosphocholine are accompanied by decreases mainly in glutamate and glutamine. These changes are also found in the OPLS-DA analysis (**Figure 4B**), but with additional quantification of their contribution to the variation. The production of ethanol and depletion of glucose are attributed to the fact that the *S. cervisiae* cells are cultivated in a glucose-rich medium, resulting in initiation of the fermentation process (utilizing glucose and producing ethanol) and repression of respiration. The biosynthesis of the phosphocholine derivatives, the major phospholipid component of eukaryotic cell membranes, as well as choline are derived from its synthesis and catabolism (phosphatidylcholine metabolism) contributing to cell growth (Howe and McMaster, 2001).

It should be emphasized that good spectral quality data is vital for reliable metabolic identification and assessment, and acquisition of high quality data from small sample volumes is not a trivial task. It is often hindered by a lack of sensitivity and/or resolution, due to poor sample filling factor, or magnetic susceptibility gradients. The success of acquiring excellent spectral quality with HR-MACS is because it offers a near optimal filling factor and minimizes susceptibility-induced broadening by

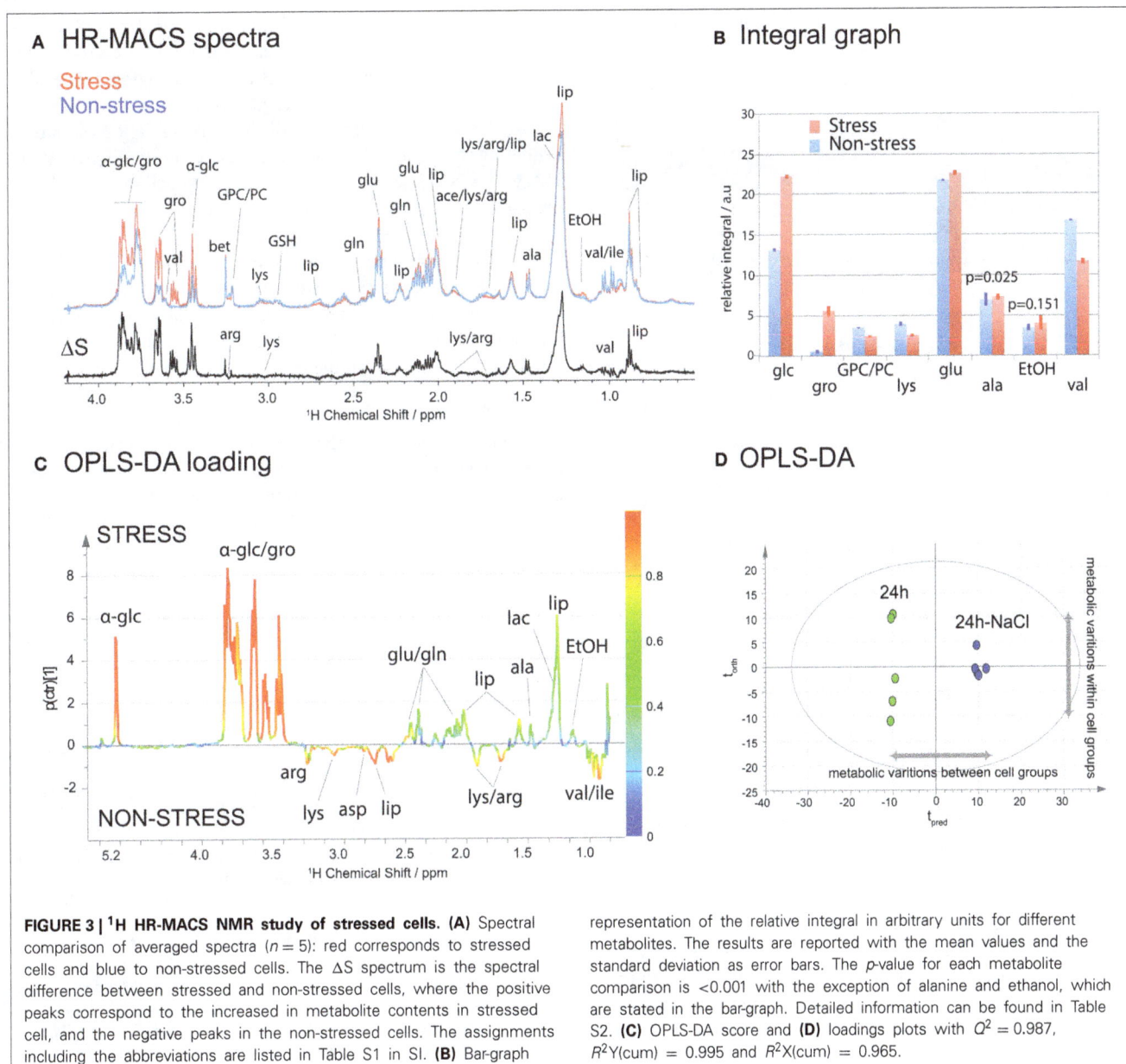

FIGURE 3 | ¹H HR-MACS NMR study of stressed cells. (A) Spectral comparison of averaged spectra (n = 5): red corresponds to stressed cells and blue to non-stressed cells. The ΔS spectrum is the spectral difference between stressed and non-stressed cells, where the positive peaks correspond to the increased in metabolite contents in stressed cell, and the negative peaks in the non-stressed cells. The assignments including the abbreviations are listed in Table S1 in SI. **(B)** Bar-graph representation of the relative integral in arbitrary units for different metabolites. The results are reported with the mean values and the standard deviation as error bars. The p-value for each metabolite comparison is <0.001 with the exception of alanine and ethanol, which are stated in the bar-graph. Detailed information can be found in Table S2. **(C)** OPLS-DA score and **(D)** loadings plots with $Q^2 = 0.987$, $R^2Y(cum) = 0.995$ and $R^2X(cum) = 0.965$.

ensemble magic-angle spinning of the resonator with the sample. To our knowledge, there are no other analytical techniques capable of offering rich-metabolic analyses with such a small sample volume (250 nl) without damaging or interfering with the sample anatomy, including MS.

We would like to briefly discuss the benefits and drawbacks of HR-MACS for small-volume analysis of cells. HR-MACS offers a simple and cost-effective approach to high-quality μNMR analysis (materials cost for one coil < €20). The combination of slow sample spinning (300 Hz) and small sample diameter (i.e., 400 μm i.d. HR-MACS) minimize the centrifugal forces exerted upon the cells under rotation, thus making it amenable to analysis of more fragile cells. Another advantage is reduction in numbers of required cells; a 10-fold reduction in total cell quantity from that typically used (i.e., 20 samples of 100×10^6 cells) should be

readily achievable. ¹H-detected multiple-resonance NMR experiments are also compatible with the HR-MACS approach (Aguiar et al., 2011), permitting the use of HSQC (and related) experiments for in-cell NMR spectroscopic investigations of the bioactivities inside the living cells (Li, 2006; Maldonado et al., 2011). The ability to analyze nano-scale volumes of cells opens new opportunities for coupling with cell-sorting microfluidic devices for a potent NMR cell screening. Microfluidic ¹H NMR has recently emerged as a micro-NMR diagnostic device by differentiating the spin relaxation mechanisms between healthy and unhealthy cells; however, it yields minimal spectral information due to the poor resolution (Haun et al., 2011).

Although the cost is low, the manufacture of the small and delicate HR-MACS resonator can be challenging. Replacing the manual fabrication of HR-MACS with an automated system may

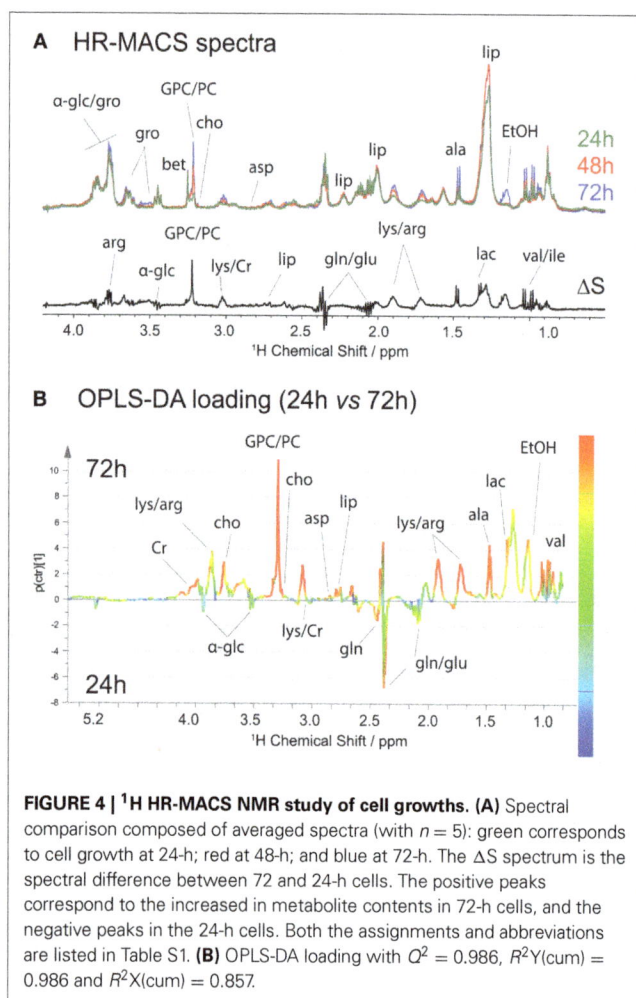

FIGURE 4 | ^{1}H HR-MACS NMR study of cell growths. (A) Spectral comparison composed of averaged spectra (with $n = 5$): green corresponds to cell growth at 24-h; red at 48-h; and blue at 72-h. The ΔS spectrum is the spectral difference between 72 and 24-h cells. The positive peaks correspond to the increased in metabolite contents in 72-h cells, and the negative peaks in the 24-h cells. Both the assignments and abbreviations are listed in Table S1. **(B)** OPLS-DA loading with $Q^2 = 0.986$, R^2Y(cum) = 0.986 and R^2X(cum) = 0.857.

Significant increases in phosphatidylcholine and glycerophosphocholine are found in relation to cell growth and the production of ethanol via the fermentation process in the aging cells is also evident. The metabolic results found in the stressed cells using HR-MACS are in agreement with the previous HR-MAS study, validating the use of HR-MACS for small cell quantity study. ^{1}H HR-MACS NMR spectroscopy opens new possibilities for high-precision investigation of small-volume sample, and may widen the scope of yeast metabolome and other bacterial cells, and the limited number of neuron cells in organisms. The ability to acquire ^{1}H NMR spectra of such small volumes puts HR-MACS closer to the scales utilized for microfluidic-based cell sorting and manipulation techniques, facilitating their coupling for a potent micro-scale cell NMR sample screening pipeline.

ACKNOWLEDGMENTS
We would like to thank the French National Research Agency for financial support under ANR33HRMACSZ; and Mr. Angelo Guiga (Saclay, France) for fabricating the Kel-F insert.

make this technique more widespread (Malba et al., 2003; Badilita et al., 2012). The very-low absolute metabolite contents—despite the fact that HR-MACS enhances the detection sensitivity—hinder their detection, without resorting to greater signal averaging and thus, longer experiment times. Coupling with other sensitivity enhancement techniques (i.e., nuclear polarization) would shorten the experimental times. Another option to further improve the sensitivity is to build a standalone μHR-MAS probe (without HR-MACS) for eliminating the any noise contribution from the HR-MACS resonator itself; however, all above options would involve complex instrumentation developments.

CONCLUDING REMARKS
In this short report, we have demonstrated, for the first time, the ^{1}H NMR metabolic profiling of a small quantity of whole cells on a nanolitre volume. The ^{1}H NMR spectra acquired with HR-MACS are of excellent-quality and exhibit high-reproducibility facilitating a rich-metabolic profiling. The two cellular studies demonstrated here illustrate the ability of ^{1}H HR-MACS coupled with statistical methods to provide high precision metabolic analyses of S. cervisiae cells in different conditions. Under an osmotic shock, the bio-production of glucose and glycerol metabolites and depleting glutamate and glutamine in S. cervisiae cells is evident.

REFERENCES
Aguiar, P. M., Jacquinot, J.-F., and Sakellariou, D. (2011). A convenient, high-sensitivity approach to multiple-resonance NMR at nanolitre volumes with inductively-coupled micro-coils. Chem. Commun. 47, 2119–2121. doi: 10.1039/c0cc04607h

Antzutkin, O. N., Shekar, S. C., and Levitt, M. H. (1995). Two-dimensional side-band separation in magic-angle-spinning NMR. J. Magn. Reson. A 115, 7–19. doi: 10.1006/jmra.1995.1142

Badilita, V., Fassbender, B., Kratt, K., Wong, A., Bonhomme, C., Sakellariou, D., et al. (2012). Microfabricated inserts for magic angle coil spinning (MACS) wireless NMR spectroscopy. PLoS ONE 7:e42848. doi: 10.1371/journal.pone.0042848

Beckonert, O., Coen, M., Keun, H. C., Wanga Y., Ebbels, T. M. D., Holmes. E., et al. (2010). High-resolution magic-angle-spinning NMR spectroscopy for metabolic profiling of intact tissues. Nat. Protoc. 5, 1019–1032. doi: 10.1038/nprot.2010.45

Blaise, B. J., Giacomotto, J., Elena, B., Dumas, M.-E., Toulhoat, P., Ségalat, L., et al. (2007). Metabotyping of Caenorhabditis elegans reveals latent phenotypes. Proc. Natl. Acad. Sci. U.S.A. 104, 19808–19812. doi: 10.1073/pnas.0707393104

Chauton, M. S., Optun, O. I., Bathen, T. F., Volent, Z., Gribbestand, I. S., and Johnsen, G. (2003a). HR MAS ^{1}H NMR spectroscopy analysis of marine microalgal cells. Mar. Echo. Prog. Ser. 256, 57–62. doi: 10.3354/meps256057

Chauton, M. S., Størseth, T. R., and Johnsen, G. (2003b). High-resolution magic angle spinning 1H NMR analysis of whole cells of Thalassiosira pseudonana (Bacillariophyceae): broad range analysis of metabolic composition and nutritional value. J. Appl. Phycol. 15, 533–542. doi: 10.1023/B:JAPH.0000004355.11837.1d

Costenoble, R., Valadi, H., Gustafsson, L., Niklasson, C., and Franzen, C. J. (2000). Microaerobic glycerol formation in Saccharomyces cerevisiae. Yeast 16, 1483–1495. doi: 10.1002/1097-0061(200012)16:16<1483::AID-YEA642>3.0.CO;2-K

Dihazi, H., Kessler, R., and Eschrich, K. (2004). High osmolarity glycerol (HOG) pathway-induced phosphorylation and activation of 6- phosphofructo-2-kinase are essential for glycerol accumulation and yeast cell proliferation under hyperosmotic stress. J. Biol. Chem. 279, 23961–23968. doi: 10.1074/jbc.M312974200

Dunn, W. B., and Ellis, D. I. (2005). Metabolomics: current analytical platforms and methodologies. Trends Anal. Chem. 24, 285–294. doi: 10.1016/j.trac.2004.11.021

Gudlavalleti, S. K., Szymanski, C. M., Jarrell, H. C., and Stephens, D. S. (2006). In vivo determination of Neisseria meningitidis serogroup A capsular polysaccharide by whole cell high-resolution magic angle spinning NMR spectroscopy. *Carbohydr. Res.* 341, 557–562. doi: 10.1016/j.carres.2005.11.036

Haun. J. B., Castro, C. M., Wang, R., Peterson, V. M., Marinelli, B. S., Lee, H., et al. (2011). Micro-NMR for rapid molecular analysis of human tumor samples. *Sci. Transl. Med.* 3, 71ra16. doi: 10.1126/scitranslmed.3002048

Himmelreich, U., Somorjai, R. L., Dolenko, B., Lee, O. C., Daniel, H.-M., Murray, R., et al. (2003). Rapid identification of Candida species by using nuclear magnetic resonance spectroscopy and a statistical classification strategy. *Appl. Environ. Microbiol.* 69, 4566–4574. doi: 10.1128/AEM.69.8.4566-4574.2003

Hoult, D. I. (2000). The principle of reciprocity in signal strength calculations - a mathematical guide. *Concept Magn. Reson.* 12, 173–187. doi: 10.1002/1099-0534(2000)12:4<173::AID-CMR1>3.0.CO;2-Q

Howe, A. G., and McMaster, C. R. (2001). Regulation of vesicle trafficking, transcription, and meiosis: lessons learned from yeast regarding the disparate biologies of phosphatidylcholine. *Biochim. Biophys. Acta.* 1534, 65–77. doi: 10.1016/S1388-1981(01)00181-0

Li, W. (2006). Multidimensional HRMAS NMR: a platform for *in vivo* studies using intact bacterial cells. *Analyst* 131, 777–781. doi: 10.1039/b605110c

Lindon, J. C., Beckonert, O. P., Holmes, E., and Nicholson, J. K. (2009). High-resolution magic angle spinning NMR spectroscopy: application to biomedical studies. *Prog. Nucl. Magn. Reson.* 55, 79–100. doi: 10.1016/j.pnmrs.2008.11.004

Malba, V., Maxwell, R., Evans, L. B., Bernhardt, A. E., Cosman, M., and Yan, K. (2003). Laser-lathe lithography - a novel method for manufacturing nuclear magnetic resonance microcoils. *Biomed. Microdevices* 5, 21–27. doi: 10.1023/A:1024407231584

Maldonado, A. Y., Burz, D. S., and Shekhtman, A. (2011). In-cell NMR spectroscopy. *Prog. Nucl. Magn. Reson. Spectrosc.* 59, 197–212. doi: 10.1016/j.pnmrs.2010.11.002

Palomino-Schätzlein, M., Molina-Navarro, M. M., Tormos-Pérez, M., Rodríguez-Navarro, S., and Pineda-Lucena, A. (2013). Optimised protocols for the metabolic profiling of S. çerevisiae by ^{1}H-NMR and HR-MAS spectroscopy. *Anal. Bioanal. Chem.* 405, 8431–8441. doi: 10.1007/s00216-013-7271-9

Reo, N. V. (2002). NMR-based metabolomics. *Drug Chem. Toxicol.* 25, 375–382. doi: 10.1081/DCT-120014789

Righi, V., Apidianakis, Y., Mintzopoulos, D., Astrakas, L., Rahme, L. G., and Tzika, A. A. (2010). *In vivo* high-resolution magic angle spinning magnetic resonance spectroscopy of Drosophila melanogaster at 14.1 T shows trauma in aging and in innate immune-deficiency is linked to reduced insulin signaling. *Int. J. Mol. Med.* 26, 175–184. doi: 10.3892/ijmm_00000450

Righi, V., Constantinou, C., Kearwani, M., Rahme, L. G., and Tzika, A. A. (2013). Live-cell high resolution magic angle spinning magnetic resonance spectroscopy

for *in vivo* analysis of *Pseudomonas aeruginosa* metabolomics. *Biomed. Rep.* 1, 707–712. doi: 10.3892/br.2013.148

Sakellariou, D., Le Goff, G., and Jacquinot, J.-F. (2007). High-resolution, high-sensitivity NMR of nanolitre anisotropic samples by coil spinning. *Nature* 447, 694–698. doi: 10.1038/nature05897

Sitter, B., Bateh, T. F., Tessem, M.-B., and Gribbestad, I. S. (2009). High-resolution magic angle spinning (HR MAS) MR spectroscopy in metabolic characterization of human cancer. *Prog. Nucl. Magn. Reson.* 54, 239–254. doi: 10.1016/j.pnmrs.2008.10.001

Tamas, M. J., Luyten, K., Sutherland, F. C., Hernandez, A., Albertyn, J., Valadi, H., et al. (1999). Fps1p controls the accumulation and release of the compatible solute glycerol in yeast osmoregulation. *Mol. Microbiol.* 31, 1087–1104. doi: 10.1046/j.1365-2958.1999.01248.x

Wong, A., Aguiar, P. M., and Sakellariou, D. (2010). Slow magic-angle coil spinning: a high-sensitivity and high-resolution NMR strategy for microscopic biological specimens. *Magn. Reson. Med.* 63, 269–274. doi: 10.1002/mrm.22231

Wong, A., Jiménez, B., Li, X., Holmes, E., Nicholson, J. K., Lindon, J. C., et al. (2012). Evaluation of high resolution magic-angle coil spinning NMR spectroscopy for metabolic profiling of nanoliter tissue biopsies. *Anal. Chem.* 84, 3843–3848. doi: 10.1021/ac300153k

Wong, A., Li, X., and Sakellariou, D. (2013). Refined magic-angle coil spinning resonator for nanoliter NMR spectroscopy: enhanced spectral resolution. *Anal. Chem.* 85, 2021–2026. doi: 10.1021/ac400188b

Conflict of Interest Statement: The authors declare that the research was conducted in the absence of any commercial or financial relationships that could be construed as a potential conflict of interest.

3

TiO$_2$@C core-shell nanoparticles formed by polymeric nano-encapsulation

Mitra Vasei, Paramita Das, Hayet Cherfouth, Benoît Marsan and Jerome P. Claverie*

Department of Chemistry, NanoQAM, Québec Center for Functional Materials, Université du Québec à Montréal, Montreal, QC, Canada

Edited by:
Muthukaruppan Alagar, Anna University, India

Reviewed by:
Ahmad A. Mousa, Balqa Applied University, Jordan
Marino Lavorgna, Italy's National Reserach Council, Italy

*Correspondence:
Jerome P. Claverie, NanoQAM, Québec Center for Functional Materials, Department of Chemistry, Université du Québec à Montréal, Succ Centre-Ville, CP8888, Montreal, QC, Canada
e-mail: claverie.jerome@uqam.ca

TiO$_2$ semiconducting nanoparticles are known to be photocatalysts of moderate activity due to their high band-gap and high rate of electron-hole recombination. The formation of a shell of carbon around the core of TiO$_2$, i.e., the formation of TiO$_2$@C nanoparticles, is believed to partly alleviate these problems. It is usually achieved by a hydrothermal treatment in a presence of a sugar derivative. We present here a novel method for the formation of highly uniform C shell around TiO$_2$ nanoparticles. For this purpose, TiO$_2$ nanoparticles were dispersed in water using an oligomeric dispersant prepared by Reversible Addition-Fragmentation chain Transfer (RAFT) polymerization. Then the nanoparticles were engaged into an emulsion polymerization of acrylonitrile, resulting in the formation of a shell of polyacrylonitrile (PAN) around each TiO$_2$ nanoparticles. Upon pyrolysis, the PAN was transformed into carbon, resulting in the formation of TiO$_2$@C nanoparticles. The structure of the resulting particles was elucidated by X-Ray diffraction, FTIR, UV-VIS and Raman spectroscopy as well as TEM microscopy. Preliminary results about the use of the TiO$_2$@C particles as photocatalysts for the splitting of water are presented. They indicate that the presence of the C shell is responsible for a significant enhancement of the photocurrent.

Keywords: RAFT polymerization, encapsulation, TiO$_2$, photocatalysis, carbon, polyacrylonitrile

INTRODUCTION

Since the landmark report on the photoelectrochemical splitting of water by TiO$_2$ (Fujishima and Honda, 1972) TiO$_2$ nanoparticles have attracted much attention (Linsebigler et al., 1995) and found applications in dye-sensitized photovoltaic devices (O'Regan and Grätzel, 1991; Grätzel, 2009) for the photodegradation of organic pollutants (Wold, 1993; Hoffmann et al., 1995) and for the production of hydrogen (Ni et al., 2007). TiO$_2$ is a choice material for these usages because of its high stability, low cost and low toxicity. However, due to its high band gap energy (3.2 eV), TiO$_2$ (anatase) is only activated by UV-light, which merely constitutes 5% of solar spectrum. The extension of the TiO$_2$ active window to the visible portion of sunlight spectrum is the object of considerable scrutiny. For example, TiO$_2$ can be doped with nitrogen (Burda et al., 2003), carbon (Park et al., 2006), boron (Chen et al., 2006), or sulfur (Tang and Li, 2008) in order to achieve this goal. However, the dopant tends to distort the TiO$_2$ lattice, resulting in an increase in the rate of electron/hole recombination. As an alternative, TiO$_2$- C composites have been investigated (Inagaki et al., 2005; Shanmugam et al., 2006; Zhang et al., 2010; Zhao et al., 2010; Jang et al., 2012; Olurode et al., 2012; Zheng et al., 2013), because the presence of carbon is beneficial to transport the charges. In these materials, it is believed that electron hole pairs are photogenerated in the TiO$_2$, but rapidly recombine due to the presence of defects in the crystalline structure. With its zero-band gap, the carbon layer acts as a sink for the electrons. Thus electron-hole recombination is attenuated due

to a better charge separation, and the electrons are channeled from the semi-conductor to the device electrodes, thus permitting the flow of the electrical current. Furthermore, due to the coupling of π states of the graphitic material with the conduction band of TiO$_2$, the activity window is extended to the visible spectrum.

The deposition of a carbon layer on TiO$_2$ nanoparticles is usually achieved via a hydrothermal process, whereby the particles are coated with an organic precursor which is then thermally decomposed to yield a carbon layer. For example, glucose (Zhang et al., 2011; Zheng et al., 2013), dextrose (Olurode et al., 2012), carboxy methyl cellulose (Inagaki et al., 2005) and polyvinyl alcohol (Tsumura et al., 2002) have been used in the past as precursors to coat TiO$_2$ nanoparticles. However, this process does not usually allow control over the carbon layer thickness and uniformity, resulting in products which are opaque to light. Thus, the advantages bestowed by the presence of a carbon layer (larger surface area, easier charge transport, reduced recombination rate) can be offset by the reduction of UV intensity actually reaching the TiO$_2$ surface.

In this work, we propose a novel templating process to prepare TiO$_2$@C nanoparticles, that is to say core-shell particles where the core is TiO$_2$ and the shell graphitic carbon. For this purpose, we have exploited the ability of polyacrylonitrile, PAN, to convert cleanly into graphite when heated under controlled atmospheric conditions. Thus, highly uniform TiO$_2$@PAN nanoparticles were prepared, using an *in-situ* nanoencapsulation method that we recently devised (Das et al., 2011; Das and Claverie, 2012; Zhong et al., 2012) to encapsulate quantum dots and carbon nanotubes (**Figure 1**). This method is based on the use of a so-called controlled polymerization technique, Reversible Addition-Fragmentation chain Transfer (RAFT) polymerization. After carbonization, the TiO_2@C nanoparticles were obtained with a highly uniform shell of carbon. Advantageously, the thickness of the shell could be tailored by simply changing the amount of PAN. To our knowledge, the use of RAFT polymerization technique for the encapsulation is the only method which allows forming 1 to 1 core shell nanoparticles, whereby each TiO_2 nanoparticle is surrounded by a continuous and size-tunable shell of PAN, and no free PAN or TiO_2 nanoparticles are observed. This paper concentrates on the fabrication and characterization of these novel hybrid nanoparticles, and their conversion into TiO_2@C, upon pyrolysis. Preliminary results on the photocatalytic activity of TiO_2@C particles for the water splitting reaction are also presented.

MATERIALS AND METHODS
MATERIALS
All chemicals were purchased from Sigma-Aldrich and used without further purification unless specifically mentioned. Indium tin oxide (ITO) glass was graciously offered by the Pilkington company. Just prior use, acrylonitrile (AN) was passed over a column of basic alumina in order to remove the polymerization inhibitor. The random copolymer PABA, containing an average of five butyl acrylate and 10 acrylic acid units was prepared according to literature (Nguyen et al., 2008; Ali et al., 2009). Outmost care should be used to manipulate acrylonitrile which is highly toxic. For this work, three different kinds of TiO_2 were used. Rutile nanoparticles of average diameter 20 nm (specific surface 30 m^2/g), anatase nanoparticles of average diameter 20 nm (specific surface 200–220 m^2/g) and P25 nanoparticles from Evonik of

average size 25 nm (80:20 mixture of anatase and rutile, specific surface 30–65 m^2/g).

PREPARATION OF THE TIO₂@C NANOPARTICLES AND ELECTRODES
Preparation of TiO₂@PAN
Concentrated NaOH (19.4 mol/L) was added dropwise to a suspension of PABA (12.5 g/L) in water until pH increased to 6, point at which the PABA was fully dissolved (slight heating was necessary). Dry TiO_2 nanoparticles (2g) were added to 20 mL of PABA solution and stirred for half an hour. Then, the suspension was sonicated with a Fisher scientific sonic dismembrator model 500 for 15 min at 30% amplitude (400 W). During the sonication, the suspension was maintained at room temperature using a cooling bath and was also magnetically stirred, and care was taken in order the magnetic stir bar not to touch the sonicating probe. A colloidaly stable dispersion of TiO_2 in water was thus obtained. To this dispersion, 50 mg of sodium dodecyl sulfate (SDS) and AN (amount as listed in **Table 1**) were added, and was degased by bubbling nitrogen for 20 min. The emulsion was then stirred and heated at 70°C under a continuous blanket of nitrogen, and 20 mL of a degassed solution of 4,4 azobis (4 cyanovaleric acid), ABVA, ($c = 8.9$ mmol/L) was added over 4 h (5 ml/h rate) under nitrogen atmosphere. Then, the reaction was stirred for an additional

Table 1 | Composition of encapsulated TiO_2 nanoparticles.

Sample	TiO₂	TiO₂ (g)	PABA (g)	AN (g)	ABVA (g)	SDS (g)
R1	Rutile	2.0	0.25	2.0	0.05	0.05
P1	P25	2.0	0.25	2.0	0.05	0.05
A1	Anatase	2.0	0.25	2.0	0.05	0.05
A2	Anatase	2.0	0.25	4.0	0.05	0.05
A3	Anatase	2.0	0.25	1.0	0.05	0.05

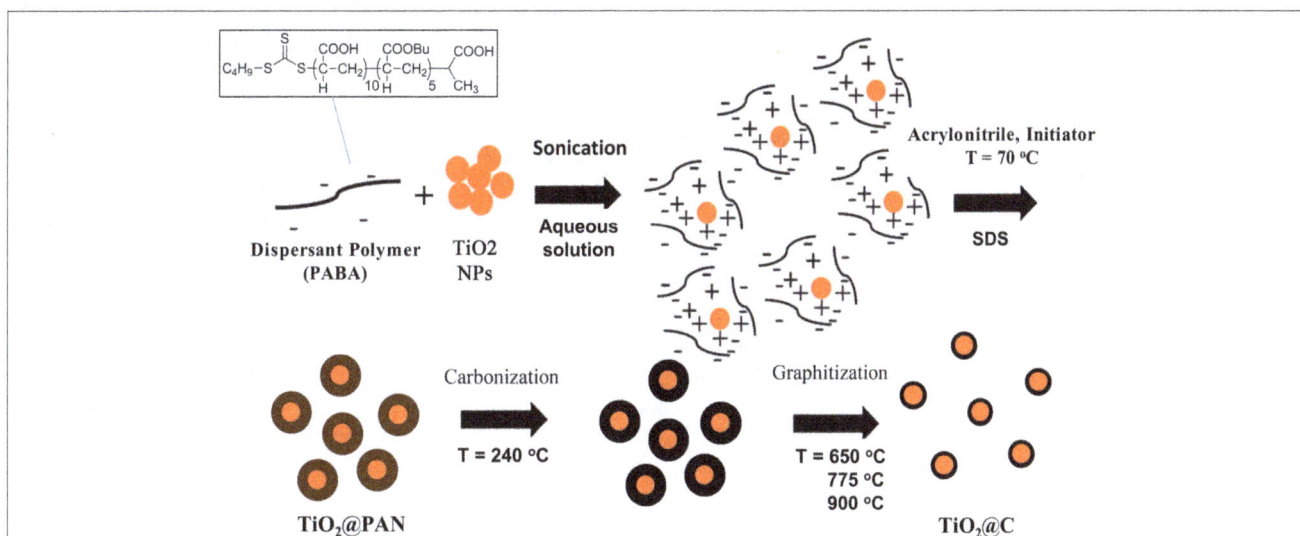

FIGURE 1 | Process for the formation of TiO_2@C nanoparticles. The first step is the dispersion of agglomerated TiO_2 nanoparticles in water with a small amount of PABA dispersant. Electrostatic interaction occurs between positively charged TiO_2 surface and negatively charged PABA. The second step is an emulsion polymerization process resulting in the formation of a layer of polyacrylonitrile. The third or fourth steps are respectively carbonization and graphitization steps, whereby PAN is converted in graphite.

2 h at 70°C in order to complete the polymerization, yielding a white colloidaly stable dispersion of TiO_2@PAN nanoparticles.

Preparation of TiO_2@C

In a Petri dish, 1 mL of the TiO_2@PAN dispersion was dried in air, to yield a white powder which was transferred to a ceramic crucible. In an oven, the powder was heated under air to 240°C at a rate of 11°C/min. This carbonization step promotes the intramolecular cyclization of PAN, to yield a conjugated ladder structure which acts as a precursor for graphite-like materials. During this carbonization step, release of hydrocyanic (HCN) acid and organic nitriles occurs (Usami et al., 1990), and suitable measures were taken to trap them and ensure safety of the experimentator. Once the temperature of 240°C reached, the oven was switched to nitrogen, and heated at a rate of 5°C/min. Three final temperatures were tested: 650, 775, and 900°C (**Figure 1**).

Preparation of TiO_2@C electrode

To analyze the electrochemical behavior of the TiO_2@C nanoparticles, the powder samples were applied on ITO coated glass plates. However, it was soon discovered that due to high surface tension of ITO, the TiO_2@C powder did not adhere to the ITO surface when immersed in water. Thus, in order to render ITO surface more hydrophobic and to obtain a uniform and continuous coating of TiO_2@C, the ITO surface was first treated with n-octyltrichlorosilane in a n-heptane: tetrachloromethane (7:3 v:v) solution. The detailed procedure for this silanization treatment can be found in (Choi et al., 2001). In 150 μL of N-methyl-pyrrolidone (NMP), polyvinylidene difluoride (PVDF, 0.3% (w/v) in NMP) and TiO_2@C powder (4.85 mg) were added. The resulting suspension was sonicated for 5 min, using the same conditions as described in Preparation of TiO_2@PAN. The resulting dispersion was then dripped on silanized ITO and spread to form a uniform film, and the resulting film was placed in a vacuum-oven at 40°C for 48 h.

CHARACTERIZATIONS
Fourier-Transform InfraRed (FTIR) Spectroscopy

Dry samples were analyzed on a Nicolet 6700 Spectrometer equipped with Smart ATR accessory (ThermoSci).

Raman spectroscopy

In order to assess the nature of the carbon layer, TiO_2@C nanoparticles were characterized by Raman spectroscopy, using a Renishaw RM3000 RAMN microscope equipped with 514.5 nm excitation laser and a CCD detector. The RAMAN signals were acquired in a backscattering geometry with a 50× magnification.

Transmission Electron Microscopy (TEM)

TEM analysis was performed on carbon coated copper grids (mesh 200) using a JEOL JEM-2100F microscope operating at 200 kV acceleration voltage. Samples were diluted with nanopure water in order to reach an approximate concentration of 0.05 wt% in water. For TiO_2@C nanoparticles, the dispersion was sonicated after dilution in order to prevent the formation of aggregates (this step was not necessary for TiO_2@PAN nanoparticles as they were stabilized by surfactant). A 10 μL drop

was deposited on each grid, and left to dry in air for 12 h. A large number of TEM pictures are presented as supplementary material.

X-Ray diffraction (XRD)

All samples were analyzed by XRD in order to assess possible structural changes to TiO_2 crystalline structure, using a Panalytical X'Pert diffractometer with Cu K_α source at 1.5405 Å, operating with a maximum voltage of 50 kV and current of 40 mA, and by varying 2θ between 20 and 80°.

ELECTROCHEMICAL CHARACTERIZATION
Cyclic voltammetry (CV)

CV experiments were performed at room temperature in a one-compartment 125 mL glass cell using a three-electrode configuration, a TiO_2@C electrode of surface 0.25 cm^2 as working electrode, a counter electrode of platinum and a reference electrode of calomel. The electrolyte was an aqueous solution of Na_2SO_4 (0.5 mol/L) acidified with sulfuric acid to pH = 2.5. Voltage was swept from 0.5 to 1.4 V at scan rate varied between 25 and 200 mV/s. Several CVs can be found in supplementary material section.

Electrochemical Impedance Spectroscopy (EIS)

EIS experiments were performed at room temperature in a custom-made one compartment 125 mL glass cell fitted with a large opening on its top in order light not to be blocked. A three electrode configuration was used, using, a TiO_2@C electrode of surface 1 cm^2 as work electrode, a counter electrode of platinum and a reference electrode of Ag/AgCl (0.242 V vs. ENH). The electrolyte was an aqueous solution of Na_2SO_4 (0.5 mol/L) acidified with sulfuric acid to pH = 2.5. The electrolyte was purged with argon before the experiment. The signals were generated with a Solartron 2255B with an AC perturbation signal of 10 mV, and frequencies varied between 100 mHz and 1 MHz. The data points were recorded and analyzed with the help of the Zplot/ZView software. The EIS experiments were performed in darkness, under solar illumination provided by a solar simulator SS500W from Sciencetech or under UV illumination provided by a 500 W high-pressure Xenon lamp. The distance from the mirror to the electrode was 20 cm. The nominal light intensity was 80 mW/cm^2, and the recorded intensity at the electrode surface was 3 mW/cm^2 for the UV lamp and 3.7 mW/cm^2 for the solar simulator.

Potentiostatic measurements

Photocurrents were measured in experiments were performed at room temperature in a custom-made one compartment 125 mL glass cell fitted with a large opening on its top in order light not to be blocked. A three electrode configuration was used, using, a TiO_2@C electrode of surface 1 cm^2 as work electrode, a counter electrode of platinum and a reference electrode of Ag/AgCl. The electrolyte was an aqueous solution of Na_2SO_4 (0.5 mol/L) acidified with sulfuric acid to pH = 2.5. The electrolyte was purged with argon before the experiment. The voltage was maintained at 1.5 V, and the current was recorded while alternating 1 min cycles of light and darkness. Two different light sources were used: solar illumination provided by a solar simulator SS500W from Sciencetech or UV illumination provided by a 500 W high-pressure Xenon lamp. The distance from the conducting mirror to the electrode was 20 cm. The nominal light intensity in both cases was 80 mW/cm^2, and the recorded intensity at the electrode surface was 3 mW/cm^2 for the UV lamp and 3.7 mW/cm^2 for the solar simulator. Several potentiostatic curves can be found in Supplementary Material section.

RESULTS

PREPARATION OF TIO$_2$@PAN NANOPARTICLES

The first step for the preparation of the core-shell nanoparticles consists in the preparation of a colloidaly stable dispersion of TiO$_2$ nanoparticles in water, using a polymeric dispersant prepared by RAFT polymerization. In this work we used the RAFT dispersant PABA which is a random copolymer of acrylic acids (10 units) and butyl acrylate (5 units), which was first reported by Ferguson et al. (2005) and used at several occasions either by the Hawkett group (Nguyen et al., 2008, 2013; Ali et al., 2009) or by our group (Das et al., 2011; Das and Claverie, 2012; Zhong et al., 2012). Other polymeric dispersants were also assessed, such as random copolymers of styrene and acrylic acid (Zhong et al., 2012) or homopolymers of acrylic acid (Daigle and Claverie, 2008), but they were found to be less efficient during the subsequent encapsulation, and they were not further investigated. Dispersions prepared with 2.0 g of TiO$_2$ (rutile, anatase or P25) and 0.25 g of PABA in 20 mL of water were found to be colloidaly stable over several days, as shown by the absence of visual aggregates. In past studies (Daigle and Claverie, 2008; Zhong et al., 2012), we have shown that PABA is adsorbed at the surface of the nanoparticle owing to a combination of hydrophobic interactions between the butyl acrylate and the surface and electrostatic interactions between negatively charged acrylic acids and positively charged surface patches.

The TiO$_2$ dispersion in water was then engaged in an emulsion polymerization of AN. If the TiO$_2$ nanoparticles were stabilized by a conventional surfactant, the emulsion polymerization process would lead to the formation of a PAN latex, that is to say separated PAN nanoparticles and TiO$_2$ nanoparticles. However, in our case the dispersant has been prepared by RAFT polymerization which is a controlled radical polymerization process. Therefore, when AN is consumed, a block copolymer is formed whereby the first block is PABA and the second is a growing chain of PAN. Thus, PAN is covalently anchored to the PABA dispersant with itself is located at the TiO$_2$ surface. As a large excess of AN is used compared to PABA (**Table 1**), the TiO$_2$ is effectively engulfed within PAN and core-shell TiO$_2$@PAN nanoparticles are thus formed. In **Figure 2**, the nanoparticles are characterized by TEM, demonstrating that neither free TiO$_2$ particles nor free PAN particles are present in the sample. The flaky aspect of the polymer is due to the high glass transition temperature, Tg, of PAN (95°C) as well as its native crystallinity (Hobson and Windle, 1993) which prevents the polymer chains from adopting a spherical shape. Interestingly, the sample is colloidaly stable after the emulsion polymerization process, and it is devoid of aggregates. Thus, the use of the RAFT polymerization process is an efficient and simple method to form TiO$_2$@PAN nanoparticles. It should be mentioned that no attempt was made to check whether the polymerization of AN was actually well-controlled (for example by isolating the polymer and by checking its polydispersity), as the subsequent step consists in its pyrolysis.

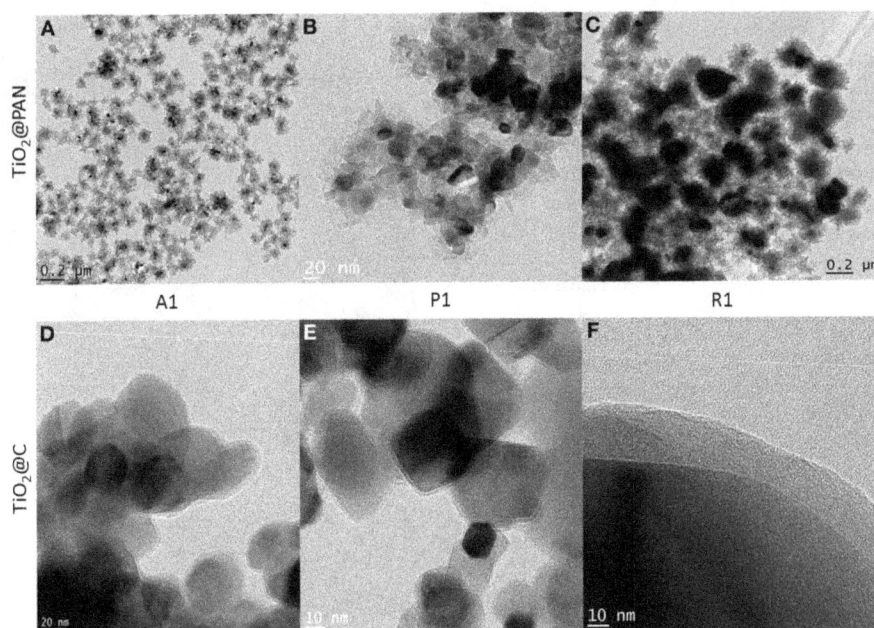

FIGURE 2 | TEM pictures of TiO$_2$@PAN nanoparticles (A–C) and TiO$_2$@C nanoparticles (D–F). Anatase: (A,D), P25: (B,E), rutile (C,F). The corresponding sample composition (A1, P1 or R1) is indicated in **Table 1** and the carbon layer thickness in **Table 2**. In (D,E), several TiO$_2$@C nanoparticles are superimposed (as shown by darker contrast).

PREPARATION AND CHARACTERIZATION OF TIO$_2$@C NANOPARTICLES

The formation of graphitic carbon upon pyrolysis of organic molecules, and most notably PAN, is by now a well-established technique (Zussman et al., 2005; Ismail and Li, 2008; Kong et al., 2010; Thomassin et al., 2010). The layer of carbon is obtained by a two-step process (**Figure 1**). First a carbonization or stabilization process, performed in air by heating the TiO$_2$@PAN nanoparticles up to 240°C, leads to the formation of conjugated ladder-like structures. Then, this carbonaceous material is heated under nitrogen to three different temperatures 650, 775, and 900°C, in order to investigate the effect of the final graphitization temperature on the structure of the hybrid nanoparticle.

The formation of a carbon shell was evident in TEM pictures (**Figures 2D–F**), where TiO$_2$ and C were identified by electron diffraction (the carbon layer did not diffract) and by energy dispersive X-ray (EDX) analysis (Supplementary material). The carbon shell is also remarkably uniform and continuous: no large carbon aggregates or carbon nanoparticles could be observed by TEM. The thickness of the carbon shell was measured on TEM pictures at more than 50 different locations in order to obtain a statistical sampling (**Table 2**). For the sake of comparison, TiO$_2$@C nanoparticles were prepared using the conventional hydrothermal process, whereby the TiO$_2$ nanoparticles are first impregnated with dextrose, and the sugar is then decomposed under heat and pressure (**Figure 3**).

Evidence for the clean conversion of PAN into carbon is brought by FTIR spectroscopy. The band near 700 cm^{-1} corresponds to the Ti-O stretching vibration, and is present in TiO$_2$, TiO$_2$@PAN and TiO$_2$@C samples (**Figure 4A**). However, the band at 2243 cm^{-1}, characteristic of CN stretching, is only present in TiO$_2$@PAN and not in TiO$_2$ and TiO$_2$@C, indicating that the nitrile functionality completely disappears during the pyrolysis process.

Importantly, the pyrolysis process did not affect the structure of TiO$_2$ provided the temperature was kept below 775°C. It is well known that at higher temperatures, anatase is converted in rutile, and indeed, the sample which was pyrolyzed up to 900°C contained a significant amount of rutile, as shown by XRD analysis (**Figure 4B**). By TEM, spherical anatase particles are observed for samples pyrolyzed at 650 and 775°C, but samples pyrolyzed at 900°C exhibit a rod-like structure characteristic of rutile (see supplementary material). The anatase to rutile transition is known to occur at around 600°C for pure TiO$_2$, but for carbon covered TiO$_2$ nanoparticles this temperature is shifted to higher temperatures due to the shielding effect of the graphite coating (Inagaki et al., 2005). Excepted for the transition to rutile at 900°C, the shape and size of the TiO$_2$@C nanoparticles is conserved during the graphitization process. Thus, for sample A2, the particle size as measured by TEM is respectively 17, 19, and 25 nm for the sample graphitized at 650, 775, and 900°C. The nanoparticle size is also in good agreement with the crystallite size measured by XRD using the Debye-Scherrer equation, which was respectively 14.7, 16.0, and 17.0 nm for the sample graphitized at 650, 775, and 900°C.

Raman spectroscopy was used to investigate the type of carbon allotropes formed through thermal treatment (**Figure 5**). The peaks D and G at around 1346 and 1590 cm^{-1}, respectively correspond to sp^2 carbon atoms in a disordered environment and to carbons in extended p conjugated graphite-like arrangements (Ferrari and Robertson, 2000; Ferrari, 2007) The ratio of the intensity of D to G peaks, called graphitization degree, is an indication of the amount of graphite carbon formed at the surface of the nanoparticle (lower values indicate greater amounts). The degree of graphitization (**Table 2**) decreases with graphitization temperature. As the temperature is raised, the thermal energy of the system becomes sufficient to convert structural defects

Table 2 | Degree of graphitization (D/G) obtained from RAMAN spectroscopy, band gap energy measured by Tauc's plot (Tauc et al., 1966), thickness and standard deviation *SD* of the C shell as measured by TEM microscopy.

	Degree of graphitization			Band gap (eV)	C Thickness (nm)	
	650°C	775°C	900°C		Average (nm)	SD (%)
R1	1.21	nd	nd	2.9	15.8	31
rutile@C	nd	nd	nd	nd	3.2	83
P1	1.28	nd	nd	2.7	1.4	24
P25@C	nd	nd	nd	nd	3.1	79
A1	1.31	0.96	0.93	2.6	8.9	26
A2	1.33	1.09	0.83	2.2	12.4	31
A3	1.30	0.85	0.86	2.5	5.1	21
anatase@C	nd	nd	nd	nd	2.2	146

Bandgap and C thickness were determined for samples graphitized at 650°C. The samples rutile@C, P25@C and anatase@C are control experiments performed via hydrothermal treatment of dextrose (Olurode et al., 2012).

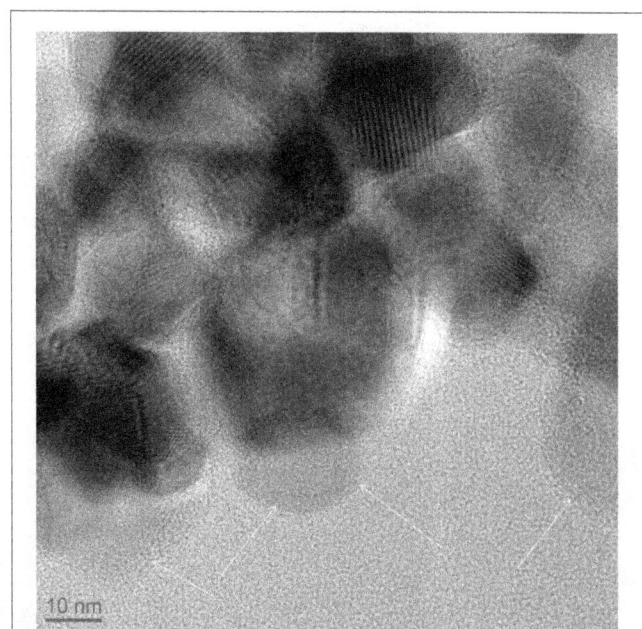

FIGURE 3 | TiO$_2$@C (TiO$_2$ = anatase) nanoparticles prepared by a conventional hydrothermal synthesis, following the procedure of (Olurode et al., 2012). The arrows outline carbon particles.

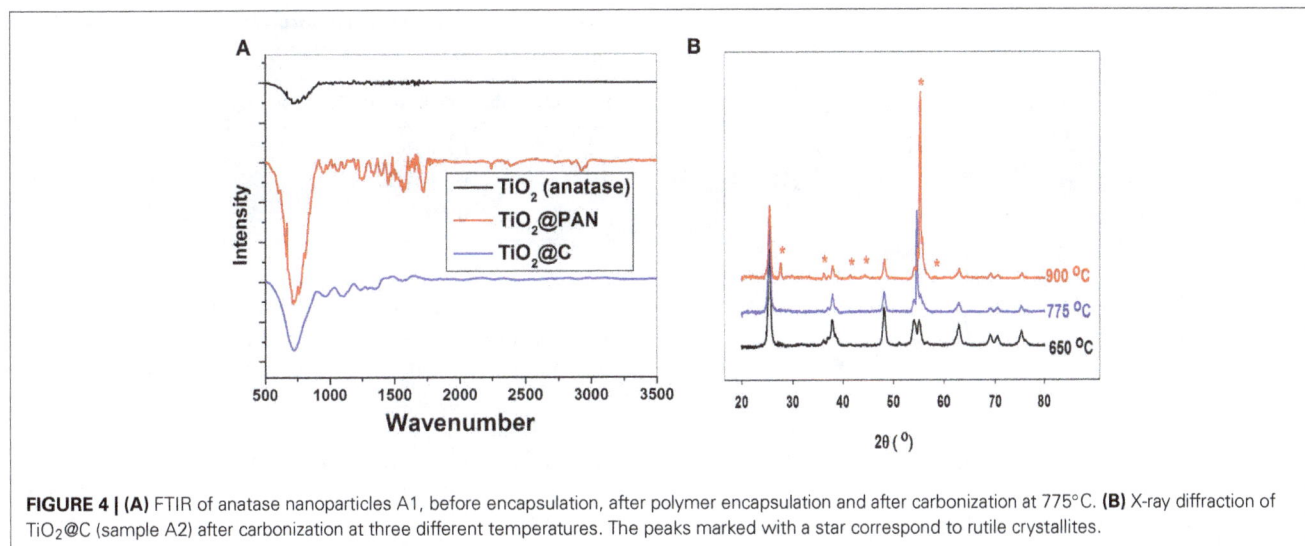

FIGURE 4 | (A) FTIR of anatase nanoparticles A1, before encapsulation, after polymer encapsulation and after carbonization at 775°C. (B) X-ray diffraction of TiO$_2$@C (sample A2) after carbonization at three different temperatures. The peaks marked with a star correspond to rutile crystallites.

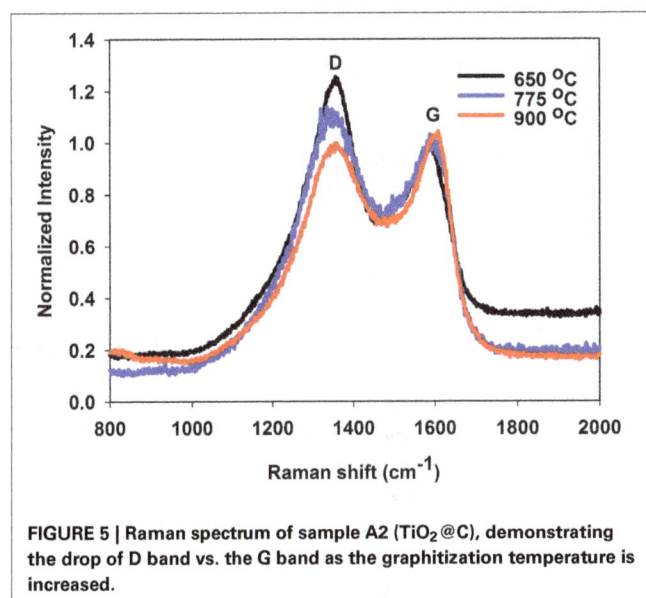

FIGURE 5 | Raman spectrum of sample A2 (TiO$_2$@C), demonstrating the drop of D band vs. the G band as the graphitization temperature is increased.

into graphitic carbons. Interestingly, the graphitization degree obtained through our PAN thermal decomposition process at 775°C (0.85–0.96) is slightly lower than the one obtained via conventional hydrothermal treatment process at 800°C (1.00–1.18) (Zhang et al., 2008), an effect that we attribute to the preorganization of the extended carbon structure in the ladder -like carbonized PAN. The use of a PAN as template for the formation of the carbon layer is also advantageous as it allows tailoring the thickness of the carbon shell (**Table 2**).

PHOTOCATALYTIC ACTIVITY OF THE TIO$_2$@C NANOPARTICLES

In the previous section, we have demonstrated that the encapsulation process is applicable for rutile, anatase and P25 (a composite of anatase and rutile). However, despite its greater band gap energy (3.2 vs. 3.0 eV) anatase is a better photocatalyst than rutile (Luttrell et al., 2014), and therefore we concentrated

on anatase samples for the rest of this work. All further experiments were performed with samples graphitized at 650°C. The band gap energy of each sample, E$_g$, was determined using Tauc's equation (Tauc et al., 1966; Zhang et al., 2011):

$$(\alpha h\nu)^{0.5} = K\left(h\nu - E_g\right) \qquad (1)$$

where α is the absorption coefficient, K is a frequency independent constant, h is Planck's constant and ν is the frequency. The data were presented under the form of Tauc's plot (supplementary information), whereby $(\alpha h\nu)^{0.5}$ was plotted vs. hν, and the band gap energy was found at the intersect between the steep part of the curve and the x axis. The band gap energy of the TiO$_2$@C samples (2.6–2.7 eV) is lower than the one of pure anatase (**Table 2**), indicating that the ability to absorb visible light is enhanced, and therefore TiO$_2$@C are promising photocatalysts.

Electrochemical characterization was performed by preparing films of the TiO$_2$@C nanoparticles deposited on transparent ITO glass. The preparation of the film was found to be of the crucial to insure reproducibility and stability of the electrical signals. The TiO$_2$@C nanoparticles were dispersed in NMP containing a small amount of PVDF as binder. The sample was then spin-coated on ITO and dried. When immersed in water, the film was found to turn hazy and to blister, with a gradual loss of adhesion. Thus, ITO was first silanized prior spin-coating in order to ensure long term adhesion of the TiO$_2$@C film. After silanization, no blistering and no hazing occurred, and the electrochemical measurements became reproducible.

Nyquist plots (**Figure 6**) clearly demonstrate the role played by the carbon layer. For TiO$_2$, the impedance is large and it is not significantly affected by the presence of UV light. In stark contrast, the impedance of the TiO$_2$@C sample is much lower, indicating that the conducting carbon layer facilitates the transport of the charge carriers through the system. Furthermore, in the presence of UV light, the impedance is further decreased, indicating that the carbon layer is also facilitating the separation of photogenerated carriers.

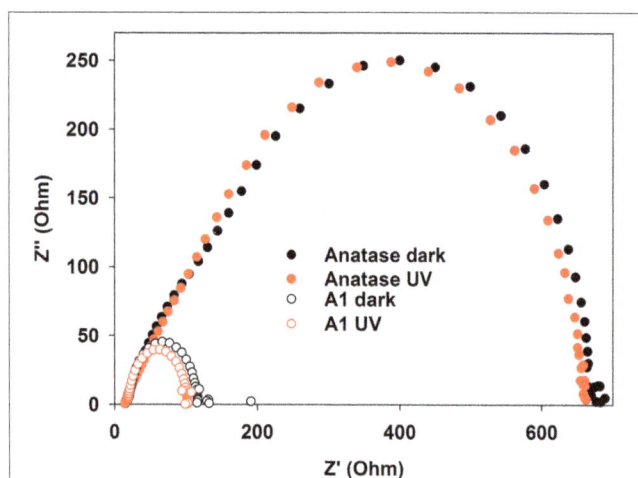

FIGURE 6 | Imaginary vs. real impedance (Nyquist graph) for anatase and TiO$_2$@C (sample A1 graphitized at 650°C). UV (Xe lamp) illumination of nominal intensity 80 mW/cm^2, measured intensity at surface: 3 mW/cm^2.

Table 3 | Comparison of induced photocurrent under UV and solar illumination (samples pyrolyzed at 650°C).

Sample	UV induced current (mA)	Solar light induced current (μA)
Anatase	0.039	2.0
A1	0.130	1.4
A2	0.244	5.8

Cyclic voltammetry was used to assess the nature of the electrochemical events taking place in this system (several CV are shown in supplementary material section). The zero current potential, V_{oc}, was found to be equal to 0.27 V. At higher imposed potentials, positive currents corresponding to oxidation of water are observed, and at an imposed potential 1.5 V, the oxidation becomes dominant. Thus, potentiostatic experiments were performed at an imposed potential of 1.5 V with alternated periods of darkness and light, in order assess the enhancement generated by the photocatalyzed generation of electron-holes pairs. Light was either produced by a UV source or by a solar simulator. In all cases, the presence of a carbon layer resulted in an enhancement of the photocurrent, enhancement which could be as high as 6 times greater. The response to UV is considerably greater than the response to solar illumination, as the band gap of anatase (3.2 eV) is more closely matched with the UV spectrum (**Table 3**).

DISCUSSION

Several salient features of the encapsulation method presented here are worthy of being emphasized. First, the method is versatile, as it works equally well on anatase, rutile and P25 nanoparticles. Our method does only rely on a physical adsorption mechanism of a hydrophobic oligomer (PABA) in water, and it does not require covalent bonding between TiO$_2$ surface sites and the polymer. As a result, this method is efficient irrespective of the nature of the TiO$_2$ facet which is exposed at the surface. Furthermore, the absence of covalent bonding is noteworthy, as covalent bonds may significantly alter the electronic and optical properties of the TiO$_2$, for example via the formation of a heterostructure. Importantly, the encapsulation method is not only applicable to TiO$_2$ particles, but also to a variety of other inorganic species, as we (Das et al., 2011; Das and Claverie, 2012; Zhong et al., 2012) and others (Nguyen et al., 2008, 2013; Ali et al., 2009) have demonstrated in the past. Although we and others have worked so far in water, it should also be mentioned that we recently

adapted this encapsulation technique for the encapsulation of nanotubes in organic solvents (Nguendia et al., 2014). Thus, this versatile nanoencapsulation method could be used to generate carbon shells on a great number of particles, and could become an interesting alternative to the hydrothermal treatment.

Another important feature of this nanoencapsulation technique is that due to the preorganization of the polymer around the polymer prior carbonization, the resulting carbon shell is highly uniform. The C shell thickness was measured on over 50 different locations in TEM pictures and was found to be relatively constant (the relative standard deviation is less than 35%, **Table 2**). In stark contrast, the usual hydrothermal synthesis leads to significantly more disperse shells, with standard deviations of 146% (resp 83, 79%) in the case of anatase (resp rutile, P25) nanoparticles (see supplementary information for relevant TEM pictures). Furthermore, conventional hydrothermal synthesis leads to shells which are often very thin. Although the average shell thickness is around 2–3 nm, most of the coating is in fact around 1 nm thick, but a few spots are much thicker (carbon deposits), resulting in an average thickness which is greater than 1 nm. Our nanoencapsulation method ensures that carbon is uniformly distributed, and therefore carbon deposits are absent. As a result, the thickness of the carbon layer can be very simply adjusted by solely adjusting the amount of AN vs. TiO$_2$ in the emulsion polymerization recipe, as demonstrated for samples A1, A2, and A3. Preliminary results on the photocatalytic activity presented here indicate that carbon layer thickness indeed influences the amount of photocurrent (**Table 3**), and therefore the catalytic activity of the TiO$_2$ particles. The results presented here establish a proof-of-concept that polymeric nanoencapsulation is a viable method to consistently construct uniform carbon layers with tunable thickness, and we intend to apply this method in the future to derive structure-activity relationships, relationships which are crucially lacking currently in literature, as, to our knowledge, no other method allows to efficiently control the thickness of the carbon layer.

ACKNOWLEDGMENTS

We thank A. Al-Shbool for the preparation of TiO$_2$@C by hydrothermal treatment, and Mr. J. P. Masse of the CM^2 facility (Ecole Polytechnique) for TEM pictures. The authors thank National Science and Engineering Research Council of Canada as well as Fonds de la Recherche du Québec Nature et Technologies (FRQNT) for financial support as well as NanoQuébec for infrastructure support. Mitra Vasei is grateful for a tuition exemption scholarship from UQAM and Paramita Das for a FRQNT scholarship.

REFERENCES

Ali, S. I., Heuts, J. P. A., Hawkett, B. S., and van Herk, A. M. (2009). Polymer encapsulated gibbsite nanoparticles: efficient preparation of anisotropic composite latex particles by RAFT-based starved feed emulsion polymerization. Langmuir 25, 10523–10533. doi: 10.1021/la9012697

Burda, C., Lou, Y., Chen, X., Samia, A. C. S., Stout, J., and Gole, J. L. (2003). Enhanced nitrogen doping in TiO_2 nanoparticles. Nano Lett. 3, 1049–1051. doi: 10.1021/nl034332o

Chen, D., Yang, D., Wang, Q., and Jiang, Z. (2006). Effects of boron doping on photocatalytic activity and microstructure of titanium dioxide nanoparticles. Ind. Eng. Chem. Res. 45, 4110–4116. doi: 10.1021/ie0600902

Choi, B., Rhee, J., and Lee, H. H. (2001). Tailoring of self-assembled monolayer for polymer light-emitting diodes. Appl. Phys. Lett. 79, 2109. doi: 10.1063/1.1398327

Daigle, J.-C., and Claverie, J. P. (2008). A simple method for forming hybrid core-shell nanoparticles suspended in water. J. Nanomater. 2008, 1–9. doi: 10.1155/2008/609184

Das, P., and Claverie, J. P. (2012). Synthesis of single-core and multiple-core core-shell nanoparticles by RAFT emulsion polymerization: lead sulfide-copolymer nanocomposites. J. Polym. Sci. A Polym. Chem. 50, 2802–2808. doi: 10.1002/pola.26070

Das, P., Zhong, W., and Claverie, J. P. (2011). Copolymer nanosphere encapsulated CdS quantum dots prepared by RAFT copolymerization: synthesis, characterization and mechanism of formation. Colloid Polym. Sci. 289, 1519–1533. doi: 10.1007/s00396-011-2466-0

Ferguson, C. J., Hughes, R. J., Nguyen, D., Pham, B. T. T., Gilbert, R. G., Serelis, A. K., et al. (2005). Ab Initio emulsion polymerization by RAFT-controlled self-sssembly. Macromolecules 38, 2191–2204. doi: 10.1021/ma048787r

Ferrari, A., and Robertson, J. (2000). Interpretation of Raman spectra of disordered and amorphous carbon. Phys. Rev. B 61, 14095–14107. doi: 10.1103/PhysRevB.61.14095

Ferrari, A. C. (2007). Raman spectroscopy of graphene and graphite: disorder, electron–phonon coupling, doping and nonadiabatic effects. Solid State Commun. 143, 47–57. doi: 10.1016/j.ssc.2007.03.052

Fujishima, A., and Honda, K. (1972). Electrochemical photolysis of water at a semiconductor electrode. Nature 238, 37–38. doi: 10.1038/238037a0

Grätzel, M. (2009). Recent advances in sensitized mesoscopic solar cells. Acc. Chem. Res. 42, 1788–1798. doi: 10.1021/ar900141y

Hobson, R. J., and Windle, A. H. (1993). Crystalline structure of atactic polyacrylonitrile. Macromolecules 26, 6903–6907. doi: 10.1021/ma00077a030

Hoffmann, M. R., Martin, S. T., Choi, W., and Bahnemann, D. W. (1995). Environmental applications of semiconductor photocatalysis. Chem. Rev. 95, 69–96. doi: 10.1021/cr00033a004

Inagaki, M., Kojin, F., Tryba, B., and Toyoda, M. (2005). Carbon-coated anatase: the role of the carbon layer for photocatalytic performance. Carbon 43, 1652–1659. doi: 10.1016/j.carbon.2005.01.043

Ismail, A. F., and Li, K. (2008). "Inorganic membranes: synthesis, characterization and applications," in Membrane Science and Technology, Vol. 13, eds R. Mallada and M. Menéndez (Amsterdam: Elsevier), 81–119.

Jang, Y. H., Xin, X., Byun, M., Jang, Y. J., Lin, Z., and Kim, D. H. (2012). An unconventional route to high-efficiency dye-sensitized solar cells via embedding graphitic thin films into TiO2 nanoparticle photoanode. Nano Lett. 12, 479–485. doi: 10.1021/nl203901m

Kong, J., Tan, H. R., Tan, S. Y., Li, F., Wong, S. Y., Li, X., et al. (2010). A generic approach for preparing core-shell carbon-metal oxide nanofibers: morphological evolution and its mechanism. Chem. Commun. (Camb.) 46, 8773–8775. doi: 10.1039/c0cc03006f

Linsebigler, A. L., Lu, G., and Yates, J. T. (1995). Photocatalysis on TiO2 surfaces: principles, mechanisms, and selected results. Chem. Rev. 95, 735–758. doi: 10.1021/cr00035a013

Luttrell, T., Halpegamage, S., Tao, J., Kramer, A., Sutter, E., and Batzill, M. (2014). Why is anatase a better photocatalyst than rutile?–Model studies on epitaxial TiO2 films. Sci. Rep. 4, 4043. doi: 10.1038/srep04043

Nguendia, J. Z., Zhong, W., Fleury, A., De Grandpré, G., Soldera, A., Sabat, R. G., et al. (2014). Supramolecular complexes of multivalent cholesterol-containing polymers to solubilize carbon nanotubes in apolar organic solvents. Chem. Asian J. 9, 1356–1364. doi: 10.1002/asia.201301687

Nguyen, D., Such, C. H., and Hawkett, B. S. (2013). Polymer coating of carboxylic acid functionalized multiwalled carbon nanotubes via reversible addition-fragmentation chain transfer mediated emulsion polymerization. J. Polym. Sci. A Polym. Chem. 51, 250–257. doi: 10.1002/pola.26389

Nguyen, D., Zondanos, H. S., Farrugia, J. M., Serelis, A. K., Such, C. H., and Hawkett, B. S. (2008). Pigment encapsulation by emulsion polymerization using macro-RAFT copolymers. Langmuir 24, 2140–2150. doi: 10.1021/la7027466

Ni, M., Leung, M. K. H., Leung, D. Y. C., and Sumathy, K. (2007). A review and recent developments in photocatalytic water-splitting using TiO2 for hydrogen production. Renew. Sustain. Energy Rev. 11, 401–425. doi: 10.1016/j.rser.2005.01.009

O'Regan, B., and Grätzel, M. (1991). A low-cost, high-efficiency solar cell based on dye-sensitized colloidal TiO2 films. Nature 353, 737–740.

Olurode, K., Neelgund, G. M., Oki, A., and Luo, Z. (2012). A facile hydrothermal approach for construction of carbon coating on TiO2 nanoparticles. Spectrochim. Acta. A. Mol. Biomol. Spectrosc. 89, 333–336. doi: 10.1016/j.saa.2011.12.025

Park, J. H., Kim, S., and Bard, A. J. (2006). Novel carbon-doped TiO2 nanotube arrays with high aspect ratios for efficient solar water splitting. Nano Lett. 6, 24–28. doi: 10.1021/nl051807y

Shanmugam, S., Gabashvili, A., Jacob, D. S., Yu, J. C., and Gedanken, A. (2006). Synthesis and characterization of TiO2 @C core-shell composite nanoparticles and evaluation of their photocatalytic activities. Chem. Mater. 18, 2275–2282. doi: 10.1021/cm052790n

Tang, X., and Li, D. (2008). Sulfur-doped highly ordered TiO2 nanotubular arrays with visible light response. J. Phys. Chem. C 112, 5405–5409. doi: 10.1021/jp710468a

Tauc, J., Grigorovici, R., and Vancu, A. (1966). Optical properties and electronic structure of amorphous germanium. Phys. Status Solidi 15, 627–637. doi: 10.1002/pssb.19660150224

Thomassin, J.-M., Debuigne, A., Jérôme, C., and Detrembleur, C. (2010). Design of mesoporous carbon fibers from a poly(acrylonitrile) based block copolymer by a simple templating compression moulding process. Polymer (Guildf). 51, 2965–2971. doi: 10.1016/j.polymer.2010.05.010

Tsumura, T., Kojitani, N., Izumi, I., Iwashita, N., Toyoda, M., and Inagaki, M. (2002). Carbon coating of anatase-type TiO2 and photoactivity. J. Mater. Chem. 12, 1391–1396. doi: 10.1039/b201942f

Usami, T., Itoh, T., Ohtani, H., and Tsuge, S. (1990). Structural study of polyacrylonitrile fibers during oxidative thermal degradation by pyrolysis-gas chromatography, solid-state carbon-13 NMR, and fourier-transform infrared spectroscopy. Macromolecules 23, 2460–2465. doi: 10.1021/ma00211a009

Wold, A. (1993). Photocatalytic properties of titanium dioxide (TiO2). Chem. Mater. 5, 280–283. doi: 10.1021/cm00027a008

Zhang, D., Yang, X., Zhu, J., Zhang, Y., Zhang, P., and Li, G. (2011). Graphite-like carbon deposited anatase TiO2 single crystals as efficient visible-light photocatalysts. J. Sol-Gel Sci. Technol. 58, 594–601. doi: 10.1007/s10971-011-2433-8

Zhang, H., Lv, X., Li, Y., Wang, Y., and Li, J. (2010). P25-graphene composite as a high performance photocatalyst. ACS Nano 4, 380–386. doi: 10.1021/nn901221k

Zhang, L.-W., Fu, H.-B., and Zhu, Y.-F. (2008). Efficient TiO2 photocatalysts from Ssurface hybridization of TiO2 particles with graphite-like carbon. Adv. Funct. Mater. 18, 2180–2189. doi: 10.1002/adfm.200701478

Zhao, L., Chen, X., Wang, X., Zhang, Y., Wei, W., Sun, Y., et al. (2010). One-step solvothermal synthesis of a carbon@TiO2 dyade structure effectively promoting visible-light photocatalysis. Adv. Mater. 22, 3317–3321. doi: 10.1002/adma.201000660

Zheng, H., Zhai, T., Yu, M., Xie, S., Liang, C., Zhao, W., et al. (2013). TiO2@C core-shell nanowires for high-performance and flexible solid-state supercapacitors. J. Mater. Chem. C 1, 225. doi: 10.1039/C2TC00047D

Zhong, W., Zeuna, J. N., and Claverie, J. P. (2012). A versatile encapsulation method of noncovalently modified carbon nanotubes by RAFT polymerization. *J. Polym. Sci. A Polym. Chem.* 50, 4403–4407. doi: 10.1002/pola. 26252

Zussman, E., Chen, X., Ding, W., Calabri, L., Dikin, D. A., Quintana, J. P., et al. (2005). Mechanical and structural characterization of electrospun PAN-derived carbon nanofibers. *Carbon* 43, 2175–2185. doi: 10.1016/j.carbon.2005.03.031

Conflict of Interest Statement: The authors declare that the research was conducted in the absence of any commercial or financial relationships that could be construed as a potential conflict of interest.

4

Hydrogen-bonds structure in poly(2-hydroxyethyl methacrylate) studied by temperature-dependent infrared spectroscopy

*Shigeaki Morita**

Department of Engineering Science, Osaka Electro-Communication University, Neyagawa, Japan

Edited by:
Young Mee Jung, Kangwon National University, South Korea

Reviewed by:
Veronica Ambrogi, Dipartimento di Ingegneria Chimica dei Materiali e Della Produzione Industriale, Italy
Marianna Pannico, Institute of Chemistry and Technology of Polymers, Italy

*Correspondence:
Shigeaki Morita, Department of Engineering Science, Osaka Electro-Communication University, 18-8 Hatsucho, Neyagawa 572-8530, Japan
e-mail: smorita@isc.osakac.ac.jp

Hydrogen-bonds structure in poly(2-hydroxyethyl methacrylate) (PHEMA) were investigated by means of temperature-dependent infrared (IR) spectroscopy. Spectral variations involved with the OH$\cdots$OH and C=O$\cdots$HO types of hydrogen-bonds were found around the glass transition temperature of 80°C. Hydrogen-bonds among the hydroxyl groups gradually dissociate with increasing temperature. In contrast, discontinuous variation in the carbonyl bands was observed around the glass transition temperature. An association of the C=O$\cdots$HO type of hydrogen-bond with increasing temperature above the glass transition temperature was revealed. These were concluded from the present study that hydrogen-bonds among the hydroxyl groups in each side chain terminal suppress the main chain mobility in the polymer matrix below the glass transition temperature, while the dissociation of the OH$\cdots$OH type of hydrogen-bonds induces the association of the C=O$\cdots$HO type of hydrogen-bond. As a result, the mobility of the main chain is induced by the change in hydrogen-bonds structure at the glass transition temperature.

Keywords: PHEMA, hydrogen-bond, glass transition, infrared spectroscopy

INTRODUCTION

Poly(2-hydroxyethyl methacrylate) (PHEMA) contains one carbonyl (C=O) and one hydroxyl (OH) groups on each side chain (Montheard et al., 1992). The OH group acts as both proton donor and proton acceptor, while the C=O group as only proton acceptor (Jeffrey, 1997; Marechal, 2007). Thus, both OH$\cdots$OH and C=O$\cdots$HO types of hydrogen-bonds are acceptable in PHEMA. Not only dimer structure (OH$\cdots$OH) but also aggregates structure ($\cdots$OH$\cdots$OH$\cdots$OH$\cdots$) have been found in many systems including liquid alcohols (Kristiansson, 1999; Ohno et al., 2008) and solid polymers (Morita et al., 2008, 2009). Such the hydrogen-bonds structure in polymers plays important roles for their macromolecular functions in artificial polymers (Brunsveld et al., 2001) as well as biopolymers (Watanabe et al., 2006, 2007). Our recent study revealed that 47.3% of the OH group on the PHEMA side chain terminal are engaged in the OH$\cdots$O=C type of hydrogen-bond, while the remaining 53.7% contributes to the OH$\cdots$OH type of hydrogen-bond at ambient temperature (Morita et al., 2009).

In the present study, change in hydrogen-bonds structure in PHEMA in the vicinity of glass transition temperature was explored by means of temperature-dependent infrared (IR) spectroscopy (Perova et al., 1997; Morita et al., 2008; Morita and Kitagawa, 2010). Although PHEMA is water insoluble, large amounts of water is sorbed into a PHEMA matrix with an equilibrium water content of ca. 40 wt% (Tanaka et al., 2002). A dry PHEMA solid is brittle, since its glass transition temperature is higher than ambient temperature. On the other hand, a PHEMA hydrogel, i.e., a water sorbed PHEMA, becomes soft

material, because its glass transition temperature is reduced to be lower than ambient temperature (Roorda et al., 1988). These imply macromolecular properties in PHEMA are characterized by non-covalent interactions of hydrogen-bonds among the polymer chains as well as the hydrated water molecules.

MATERIALS AND METHODS

An atactic PHEMA with a viscosity-averaged molecular weight of ca. 3.0×10^5 was purchased from Aldrich and used without further purification. A glass transition temperature of the PHEMA sample evaluated by differential scanning calorimetry (DSC) was 80°C, which was performed using a PerkinElmer Pyris 6 at a heating rate of 10°C min^{-1}. An evidence of crystalline phase in the solid was not detected by DSC, demonstrating that the PHEMA sample used in the present study is amorphous. A film sample was prepared on a calcium fluoride substrate by solvent casting from a methanol solution. A thickness of the film sample was controlled as all the IR signals in an absorbance unit become less than 1. The film sample was enough dried at an ambient temperature before the measurement. Temperature-dependent IR spectra of the PHEMA film were collected over a temperature range of 25–150°C with an increment of 1°C using a Fourier transform IR spectrometer (Varian, FTS-3000) equipped with a deuterated triglycine sulfate detector at a nitrogen atmosphere. A total of 128 scans were co-added to obtain each spectrum.

RESULTS

Figure 1 shows temperature-dependent IR spectra of the PHEMA film. No IR signals arising from the residual solvent of methanol

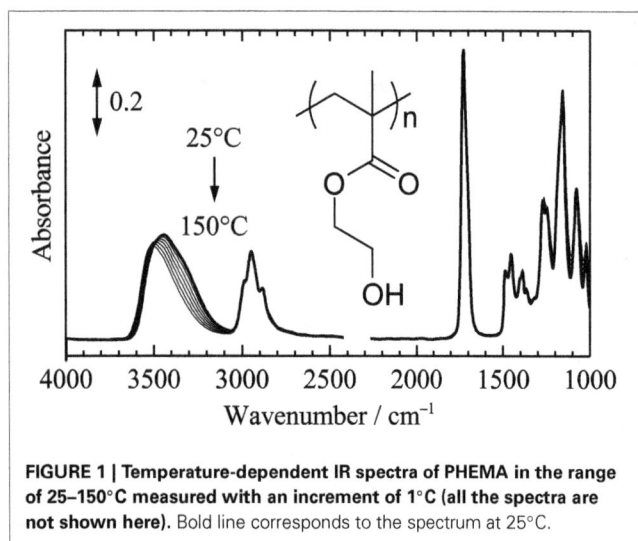

FIGURE 1 | Temperature-dependent IR spectra of PHEMA in the range of 25–150°C measured with an increment of 1°C (all the spectra are not shown here). Bold line corresponds to the spectrum at 25°C.

Table 1 | Assignments of selected IR absorption bands in PHEMA.

Wavenumber/cm^{-1}	Assignments
3666	Free OH (trans)
3624	Free OH (gauche)
3534	Hydrogen-bonded OH with C=O (OH···O=C), dimer OH (OH···OH)
3434	First overtone of C=O stretching
3320	Aggregates OH (··· OH···OH···OH ···)
1730	Free C=O
1703	Hydrogen-bonded C=O with OH (C=O···HO)

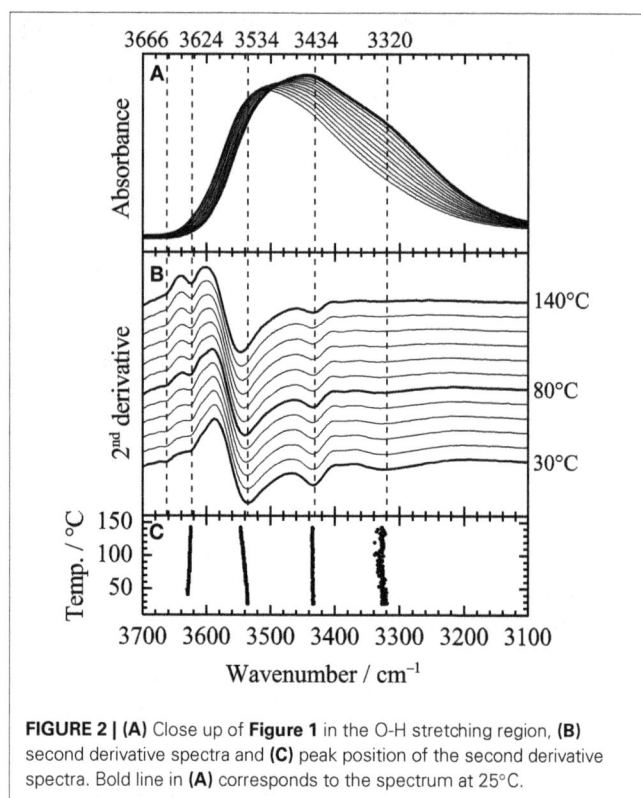

FIGURE 2 | (A) Close up of **Figure 1** in the O-H stretching region, **(B)** second derivative spectra and **(C)** peak position of the second derivative spectra. Bold line in **(A)** corresponds to the spectrum at 25°C.

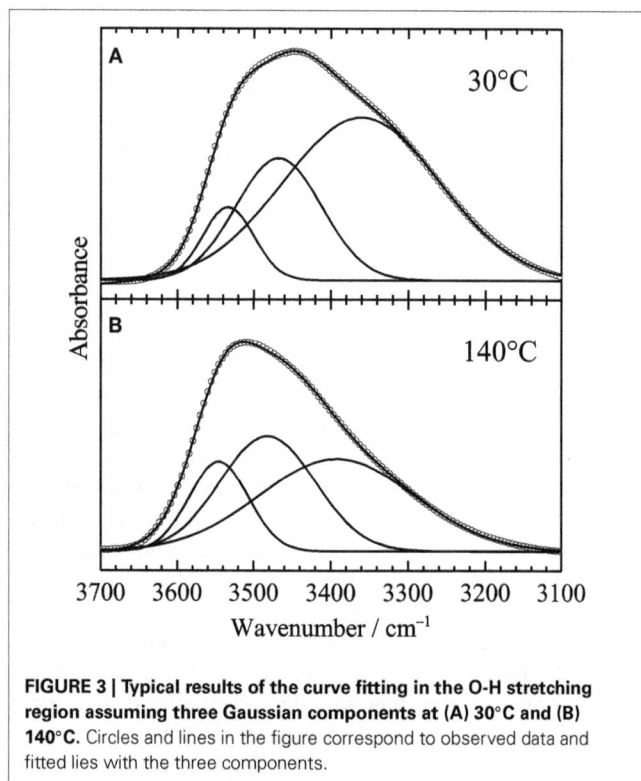

FIGURE 3 | Typical results of the curve fitting in the O-H stretching region assuming three Gaussian components at (A) 30°C and (B) 140°C. Circles and lines in the figure correspond to observed data and fitted lies with the three components.

and hydrated water from ambient air were detected, representing that a sufficiently dry film was obtained at the nitrogen atmosphere. This represents O-H stretching band in the spectra is arising from only PHEMA. Assignments of IR absorption bands in the spectrum of PHEMA have been reported previously (Ferreira et al., 2000; Morita et al., 2009). The assignments are summarized in **Table 1**. A large spectral shape variation in the O-H stretching region around 3700–3100 cm^{-1} was observed, while only weak variations were detected in the other spectral regions. Detailed spectral variations in the O-H stretching and the C=O stretching regions are discussed below.

O-H STRETCHING REGION

Figure 2A shows the temperature-dependent IR spectra of PHEMA in the O-H stretching region (close up of **Figure 1**). Second derivative spectra calculated from the obtained spectra shown in **Figure 2A** and peak positions of the second derivative spectra plotted as a function of temperature are also depicted in **Figures 2B,C**, respectively. At least five contributions around 3666, 3624, 3534, 3434, and 3320 cm^{-1} are identified in the O-H stretching region. Our recent study using model compounds of methanol, methyl acetate and 2-hydroxyethyl methacrylate monomer revealed their assignments as summarized in **Table 1** (Morita et al., 2009). Although very weak signals in the obtained spectra, the bands arising from free OH, i.e., OH group not donating hydrogen-bond, are clearly identified at 3666 and 3624 cm^{-1} in the second derivative spectra above the glass transition temperature of 80°C.

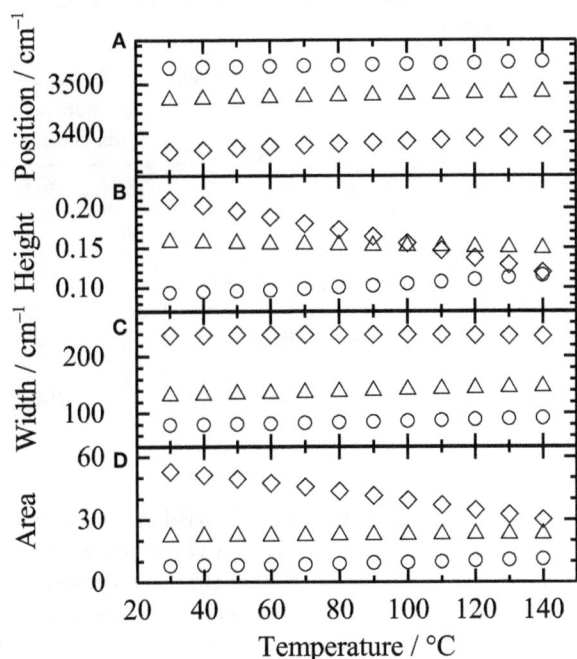

FIGURE 4 | Fitting parameters of (A) peak position, (B) peak height, (C) peak width and (D) area intensity in the O-H stretching region plotted as a function of temperature. Symbols of circle, triangle and diamond correspond to the bands around 3534, 3434, 3320 cm^{-1}, respectively.

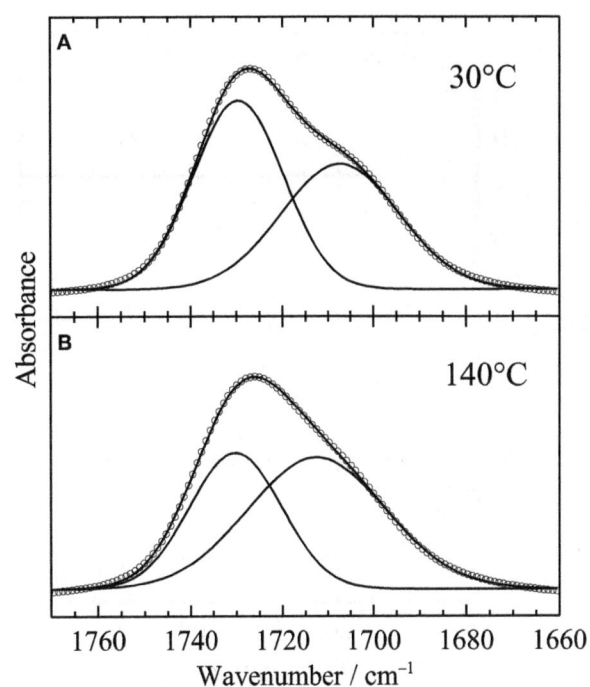

FIGURE 6 | Typical results of the curve fitting in the C=O stretching region assuming two Gaussian components at (A) 30°C and (B) 140°C. Circles and lines in the figure correspond to observed data and fitted lies with the two components.

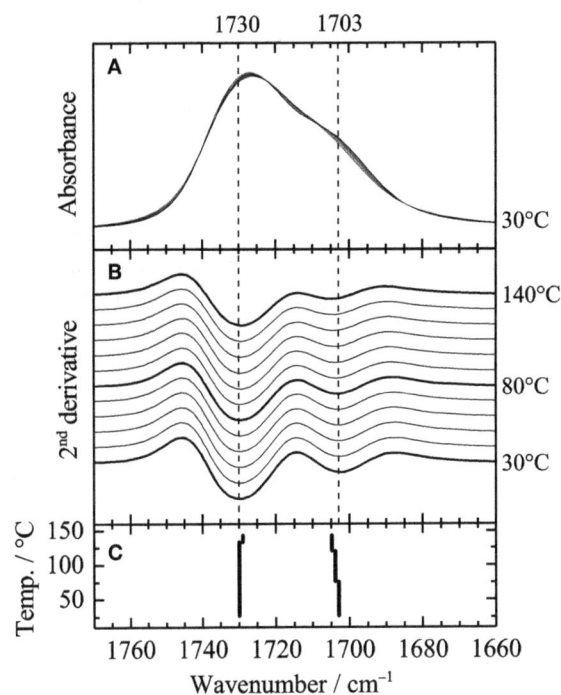

FIGURE 5 | (A) Close up of Figure 1 in the C=O stretching region, (B) second derivative spectra and (C) peak position of the second derivative spectra.

In order to clarify the spectral variations in the O-H stretching region, the spectral shapes in the region were fitted using the following Gaussian functions:

$$A(v) = \sum_i h_i \exp\left(-4\ln2 \frac{(v - v_i)^2}{w_i^2} \right)$$

were, h_i, v_i and w_i are peak height, peak position and peak width, respectively. **Figure 3** shows the fitting results at (**A**) 30 and (**B**) 140°C, respectively. All the spectra were well fitted assuming three Gaussian contributions, since the bands at 3666 and 3624 are very weak. **Figure 4** depicts the fitting parameters for the O-H stretching region plotted as a function of temperature. Variations of the peak position and the peak width are relatively small, while those of peak height or area intensity, which is calculated as

$$A_i = \int A_i(v)\, dv$$

are intense. The intensity variation of the band at 3434 cm^{-1}, which is assigned to the first overtone of the C=O stretching, is not clearly identified. In contrast, the intensity of the band at 3320 cm^{-1} gradually decreases with increasing temperature, while that at 3534 cm^{-1} gradually increases. It should be noted that the intensity variations at 3320 and 3534 cm^{-1} are not discontinuous with temperature at the glass transition temperature of 80°C.

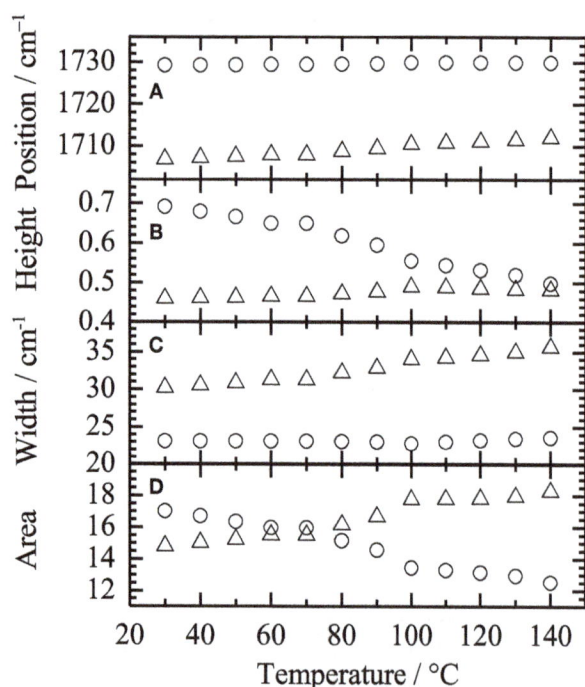

FIGURE 7 | Fitting parameters of (A) peak position, (B) peak height, (C) peak width and (D) area intensity in the C=O stretching region plotted as a function of temperature. Symbols of circle and triangle correspond to the bands around 1730 and 1703 cm^{-1}, respectively.

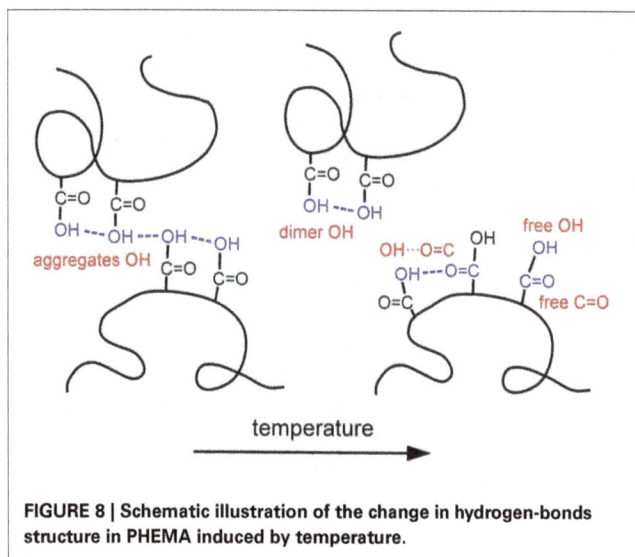

FIGURE 8 | Schematic illustration of the change in hydrogen-bonds structure in PHEMA induced by temperature.

C=O STRETCHING REGION

Figure 5A shows the temperature-dependent IR spectra of PHEMA in the C=O stretching region (close up of **Figure 1**). Second derivative spectra and peak positions of the second derivative spectra are also plotted in **Figures 5B,C**, respectively. Two contributions around 1730 and 1703 cm^{-1} are identified in the C=O stretching region. Assignments of the two bands given in our previous study (Morita et al., 2009) are summarized in

Table 2 | Typical glass transition temperature for poly(acrylate)s (Morita et al., 2004).

	Side chain structure	Typical glass transition temperature/°C
Poly(2-methoxyethyl acrylate) (PMEA)	-COO(CH$_2$)$_2$OCH$_3$	−50
Poly(ethyl acrylate) (PEA)	-COOCH$_2$CH$_3$	−24
Poly(n-butyl methacrylate) (PBMA)	-CH$_3$, -COO(CH$_2$)$_3$CH$_3$	20
Poly(2-hydroxyethyl methacrylate) (PHEMA)	-COO(CH$_2$)$_2$OH	80
Poly(methyl methacrylate) (PMMA)	-CH$_3$, -COOCH$_3$	105

Table 1. In order to clarify the spectral variations, the shapes in the region were fitted by two Gaussian components. **Figure 6** shows the fitting results at (**A**) 30 and (**B**) 140°C, respectively. All the spectra were well fitted assuming two Gaussian contributions. **Figure 7** depicts the fitting parameters for the C=O stretching region plotted as function of temperature. The peak position of the band around 1730 scarcely changed with temperature, whereas that around 1703 cm^{-1} shifts toward to the higher wavenumber as similar to the peak position evaluated by the second derivatives as shown in **Figure 5C**. The peak width of the band at 1730 also scarcely changed, while that at 1703 cm^{-1} becomes broad with increasing temperature. It is of particular to note that the peak height or the area intensity of the band at 1703 cm^{-1} assigned to hydrogen-bonded C=O discontinuously increases above the glass transition temperature of 80°C, whereas that at 1730 cm^{-1} assigned to free C=O discontinuously decreases at the temperature.

DISCUSSION

An evidence of gradual dissociation of the OH$\cdots$OH type of hydrogen-bonds with increasing temperature was found in the O-H stretching region. In contrast, it was found in the C=O stretching region that association of the C=O$\cdots$HO type of hydrogen-bond occurs discontinuously above the glass transition temperature. The second derivative spectra in the O-H stretching region revealed that the free OH appears above the glass transition temperature.

A schematic illustration of the change in hydrogen-bonds structure in PHEMA induced by temperature speculated from the spectral variations is described in **Figure 8**. At ambient temperature, 53.7% of the OH groups in the side chain terminal are associated with each other via the$\cdots$OH$\cdots$OH$\cdots$OH$\cdots$type of hydrogen-bonds. As a result, mobility of the main chain is expected to be suppressed by the non-covalent interactions among the side chains. However, the aggregates OH are easily dissociated by increasing temperature. At the glass transition temperature, which relates to the mobility of polymer main chain, the OH groups which dissociated with the other OH groups are associated with the C=O groups with the OH$\cdots$O=C type of hydrogen-bond. It is likely that the mobility of the main chain is induced by the change in the hydrogen-bonds structure at the glass transition temperature.

Glass transition temperature for analogous poly(acrylate)s are generally lower than that for PHEMA as summarized in **Table 2** (Morita et al., 2004). Only glass transition temperature for PMMA is higher than that for PHEMA because of non-bulky side chain, which enhances the main chain interaction. Glass transition temperature for PHEMA hydrogels is reduced by increasing a content of water in the matrix (Roorda et al., 1988). These also support the conclusion that the mobility of the PHEMA main chain is induced by the dissociation of the OH$\cdots$OH type of hydrogen-bonds among the side chains, since water molecules also hydrated to both OH and C=O groups in the PHEMA side chain via the OH$\cdots$OH and C=O$\cdots$HO types of hydrogen-bonds (Tsuruta, 2010). Miwa et al. found the evidences of strong OH$\cdots$OH and C=O$\cdots$HO types of hydrogen-bonds between the PHEMA side chain and water molecule using NMR spectroscopy (Miwa et al., 2010). In the case of PHEMA hydrogels, it is likely that the hydrogen-bonds among the PHEMA side chains are partially inhibited by water molecules. As a result, glass transition temperature for PHEMA is reduced by the content of water.

REFERENCES

Brunsveld, L., Folmer, B., Meijer, E., and Sijbesma, R. (2001). Supramolecular polymers. *Chem. Rev.* 101, 4071–4098. doi: 10.1021/cr990125q

Ferreira, L., Vidal, M., and Gil, M. (2000). Evaluation of poly(2-hydroxyethyl methacrylate) gels as drug delivery systems at different pH values. *Int. J. Pharm.* 194, 169–180. doi: 10.1016/S0378-5173(99)00375-0

Jeffrey, G. A. (1997). *An Introduction to Hydrogen Bonding*. New York, NY: Oxford University Press.

Kristiansson, O. (1999). Investigation of the OH stretching vibration of CD3 OH in CCl4. *J. Mol. Struct.* 477, 105–111. doi: 10.1016/S0022-2860(98)00591-2

Marechal, Y. (2007). *The Hydrogen Bond and the Water Molecule: The Physics and Chemistry of Water, Aqueous and Bio-Media*. Amsterdam: Elsevier.

Miwa, Y., Ishida, H., Tanaka, M., and Mochizuki, A. (2010). 2H-NMR and 13C-NMR study of the hydration behavior of poly(2-methoxyethyl acrylate), poly(2-hydroxyethyl methacrylate) and poly(tetrahydrofurfuryl acrylate) in relation to their blood compatibility as biomaterials. *J. Biomater. Sci. –Polym. Ed.* 21, 1911–1924. doi: 10.1163/092050610X489682

Montheard, J.-P., Chatzopoulos, M., and Chappard, D. (1992). 2-Hydroxyethyl methacrylate (HEMA): chemical properties and applications in biomedical fields. *J. Macromol. Sci. Part C Polym. Rev.* 32, 1–34. doi: 10.1080/15321799208018377

Morita, S., and Kitagawa, K. (2010). Temperature-dependent structure changes in Nafion ionomer studied by PCMW2D IR correlation spectroscopy. *J. Mol. Struct.* 974, 56–59. doi: 10.1016/j.molstruc.2009.12.040

Morita, S., Kitagawa, K., Noda, I., and Ozaki, Y. (2008). Perturbation-correlation moving-window 2D correlation analysis of temperature-dependent infrared spectra of a poly(vinyl alcohol) film. *J. Mol. Struct.* 883, 181–186. doi: 10.1016/j.molstruc.2007.12.004

Morita, S., Kitagawa, K., and Ozaki, Y. (2009). Hydrogen-bond structures in poly(2-hydroxyethyl methacrylate): infrared spectroscopy and quantum chemical calculations with model compounds. *Vib. Spectrosc.* 51, 28–33. doi: 10.1016/j.vibspec.2008.09.008

Morita, S., Ye, S., Li, G. F., and Osawa, M. (2004). Effect of glass transition temperature (Tg) on the absorption of bisphenol A in poly(acrylate)s thin films. *Vib. Spectrosc.* 35, 15–19. doi: 10.1016/j.vibspec.2003.11.020

Ohno, K., Shimoaka, T., Akai, N., and Katsumoto, Y. (2008). Relationship between the broad OH stretching band of methanol and hydrogen-bonding patterns in the liquid phase. *J. Phys. Chem. A* 112, 7342–7348. doi: 10.1021/jp800995m

Perova, T. S., Vij, J. K., and Xu, H. (1997). Fourier transform infrared study of poly(2-hydroxyethyl methacrylate) PHEMA. *Colloid Polym. Sci.* 275, 323–332. doi: 10.1007/s003960050089

Roorda, W. E., Bouwstra, J. A., De Vries, M. A., and Junginger, H. E. (1988). Thermal behavior of poly hydroxy ethyl methacrylate (pHEMA) hydrogels. *Pharm. Res.* 5, 722–725. doi: 10.1023/A:1015912028859

Tanaka, M., Mochizuki, A., Ishii, N., Motomura, T., and Hatakeyama, T. (2002). Study of blood compatibility with poly(2-methoxyethyl acrylate). Relationship between water structure and platelet compatibility in poly(2-methoxyethylacrylate-co-2-hydroxyethylmethacrylate). *Biomacromolecules* 3, 36–41. doi: 10.1021/bm010072y

Tsuruta, T. (2010). On the role of water molecules in the interface between biological systems and polymers. *J. Biomater. Sci. Polym. Ed.* 21, 1831–1848. doi: 10.1163/092050610X488269

Watanabe, A., Morita, S., and Ozaki, Y. (2006). Study on temperature-dependent changes in hydrogen bonds in cellulose I beta by infrared spectroscopy with perturbation-correlation moving-window two-dimensional correlation spectroscopy. *Biomacromolecules* 7, 3164–3170. doi: 10.1021/bm0603591

Watanabe, A., Morita, S., and Ozaki, Y. (2007). Temperature-dependent changes in hydrogen bonds in cellulose I alpha studied by infrared spectroscopy in combination with perturbation-correlation moving-window two-dimensional correlation spectroscopy: comparison with cellulose I beta. *Biomacromolecules* 8, 2969–2975. doi: 10.1021/bm700678u

Conflict of Interest Statement: The author declares that the research was conducted in the absence of any commercial or financial relationships that could be construed as a potential conflict of interest.

5

The surging role of Chromogranin A in cardiovascular homeostasis

Bruno Tota*, Tommaso Angelone and Maria C. Cerra

Department of Biology, Ecology and Earth Sciences, University of Calabria, Arcavacata di Rende (CS), Italy

Edited by:
Bulent Mutus, University of Windsor, Canada

Reviewed by:
Ruchi Chaube, Case Western Reserve University, USA
Claudia Penna, University of Torino, Italy

*Correspondence:
Bruno Tota, Department of Biology, Ecology and Earth Sciences, University of Calabria, 87036 Arcavacata di Rende (CS), Italy
e-mail: bruno.tota@unical.it

Together with Chromogranin B and Secretogranins, Chromogranin A (CGA) is stored in secretory (chromaffin) granules of the diffuse neuroendocrine system and released with noradrenalin and adrenalin. Co-stored within the granule together with neuropeptideY, cardiac natriuretic peptide hormones, several prohormones and their proteolytic enzymes, CGA is a multifunctional protein and a major marker of the sympatho-adrenal neuroendocrine activity. Due to its partial processing to several biologically active peptides, CGA appears an important pro-hormone implicated in relevant modulatory actions on endocrine, cardiovascular, metabolic, and immune systems through both direct and indirect sympatho-adrenergic interactions. As a part of this scenario, we here illustrate the emerging role exerted by the full-length CGA and its three derived fragments, i.e., Vasostatin 1, catestatin and serpinin, in the control of circulatory homeostasis with particular emphasis on their cardio-vascular actions under both physiological and physio-pathological conditions. The Vasostatin 1- and catestatin-induced cardiodepressive influences are achieved through anti-beta-adrenergic-NO-cGMP signaling, while serpinin acts like beta1-adrenergic agonist through AD-cAMP-independent NO signaling. On the whole, these actions contribute to widen our knowledge regarding the sympatho-chromaffin control of the cardiovascular system and its highly integrated "whip-brake" networks.

Keywords: Chromogranin A, Vasostatin 1, Catestatin, Serpinin, adreno-sympathetic control, cardioprotection, vasoactive peptides, endothelial signaling

INTRODUCTION

The granins are a structurally and functionally related family of proteins including Chromogranin A (CGA), Chromogranin B (CGB) and Secretogranins (SG) II-VII which are stored in secretory (chromaffin) granules of the diffuse neuroendocrine system and released together with noradrenalin and adrenalin (Winkler and Fischer-Colbrie, 1992; Montero-Hadjadje et al., 2008; Bartolomucci et al., 2011). Detailed information on chromosomal positions, genomic structure, cDNA, and proteins encoded by their three paralogous genes, that likely arose by gene duplication within the vertebrate lineage, has been reported by Mahata et al. (1996), Zhang et al. (2011) and Bartolomucci et al. (2011). The adrenal medulla and the adrenergic terminals are a main source of granins which are therefore been employed as markers of the sympatho-adrenal neuroendocrine (SAN) activity. Granins within the vesicle stabilize the core osmotically by binding catecholamines (CAs) and ATP, and are also co-stored with neuropeptide Y (NPY), the cardiac natriuretic peptide hormones (NPs), several prohormones and their proteolytic enzymes (Videen et al., 1992). Their partial processing to biologically active fragments together with milieu acidification contribute to the maturation of the granule. The mechanisms of granin sorting into regulated secretory pathway granules, including the CGA domains that are required for directing it into the secretory granules, as well as the CGA N- and C-terminus that may have

targeting information, have been summarized by Bartolomucci et al. (2011). Among granins, the soluble acidic 439- residue long CGA is the most abundant protein of the granule, accounting for almost 50% of its soluble protein content, and is the most studied member, being considered the major indicator and multifunctional effector of the SAN tone (Cryer et al., 1991; O'Connor et al., 2008). Its wide spectrum of biological and physio-pathological activities ranges from intracellular to organ and system levels.

A fundamental intracellular function of CGA is its granulogenic role in the formation of dense core secretory vesicles in (neuro) endocrine cells (Kim et al., 2001; Kim and Loh, 2006). In fact, CGA depletion by gene targeting causes reduction of adrenal chromaffin granules in number, size, and electron density, as well as disruption of transmitter secretion from the regulated pathway (Mahapatra et al., 2005). Together with other granins, CGA binds to the inner layer of the vesicle membrane affecting the release of calcium from secretory granules to the cytosolic exocytotic machinery through the inositol 1,4,5-trisphosphate receptor/Ca^{2+} channel (Yoo et al., 2002). CGA has a long evolutionary history and CGA-like proteins have been detected in mammals, birds, amphibians, fish (including CGA mRNA in zebrafish) and arthropods (Xie et al., 2008). In addition to its expression in the neuroendocrine cells, as detailed below, CGA has also been detected in other cell types, including the myocardiocytes of various vertebrate species, e.g., amphibians

(Krylova, 2007), rodents (Steiner et al., 1990; Biswas et al., 2010; Pasqua et al., 2013) and humans, particularly in patients affected by cardiomyopathy and heart failure (HF: Pieroni et al., 2007). In normal conditions, upon stimulation of regulated secretion, CGA is exocytotically released in the extracellular space and then in the circulation, thus being able to exert systemic and/or organ and tissue modulatory effects (Helle et al., 2007; Bartolomucci et al., 2011; Angelone et al., 2012a and references therein). These actions are mainly related to the prohormone/cytokin ability of CGA to undergo a finely regulated clevage. That is, following stimulus- and differential cell-type-specific or tissue-specific proteolytic processing at dibasic sites, CGA generates within secretory vesicles several peptides in aggregate which can then exert relevant modulatory actions on endocrine, cardiovascular, metabolic, and immune systems through both direct and indirect SAN interactions (Metz-Boutigue et al., 1993; Parmer et al., 2000; Taupenot et al., 2002). The CGA biologically active fragments include the amino terminal cardio and vasoactive vasostatin 1 (VS-1), the antimicrobial chromacin, the dysglicemic peptide pancreastatin (PST), the parathormon release modulator parastatin, the catecholamine release inhibitor catestatin (CST), and the recently discovered serpinin (see **Figure 1**).

With reference to the cardiovascular focus of this review, we will illustrate the actions of both circulating CGA and its three fragments, VS-1, CST and serpinin (Tota et al., 2010, 2012), which exert a broad spectrum of regulatory influences on the cardio-circulatory system. Although nothing is known on how these activities and the underpinning proteolytic events are spatio-temporally coordinated, it is conceivable that both the systemic actions of CGA and those induced by the CGA-derived peptides at organ/tissue (heart and vessels) level may be synergically implicated in circulatory homeostasis, coordinating and counteracting SAN overactivity under normal and perturbed conditions. Namely, these substances can operate as "integral" controller components, bringing the controlled variable back to set "point" at any steady-state disturbance, according to the Koeslag et al.'s (1999) concept of the "zero steady-state error" homeostasis achieved by pairs of counter-regulatory hormones (originally applied for CST and PST). Before discussing this issue, for the non-expert reader we will very briefly summarize the physio-pathological implications of heightened SAN activity.

PHYSIO-PATHOLOGICAL ASPECTS OF SAN OVERACTIVATION

There is a wide spectrum of SAN-induced actions on the circulation, ranging from blood coagulation, platelet adhesiveness, smooth muscle cell hyperplasia, arterial wall tone, hyperlipidemia, denervated myocardium etc. to immunologic responses in relation to cardiovascular changes. Therefore, it is not surprising that SAN overstimulation (the "adrenergic storm") may act on a number of different targets, directly impinging the heart and the vasculature. The corresponding physio-pathological changes have long been known, ranging from the experimental necrotic damage induced by CAs in the rodent heart (Samuels, 2007) to the extensive clinical evidence showing that the initial heart response to prolonged SAN overactivity leads to compensatory remodeling, cardiac hypertrophy and, if the stress will overwhelm the system, HF (Chien et al., 1991 and references therein). If left uncontrolled, the SAN overactivation may result to be more deleterious than the actual stress placed on the heart. In human HF, chronic heightened adrenergic activation, mainly *via* CAs signaling, has adverse prognostic significance, accelerating

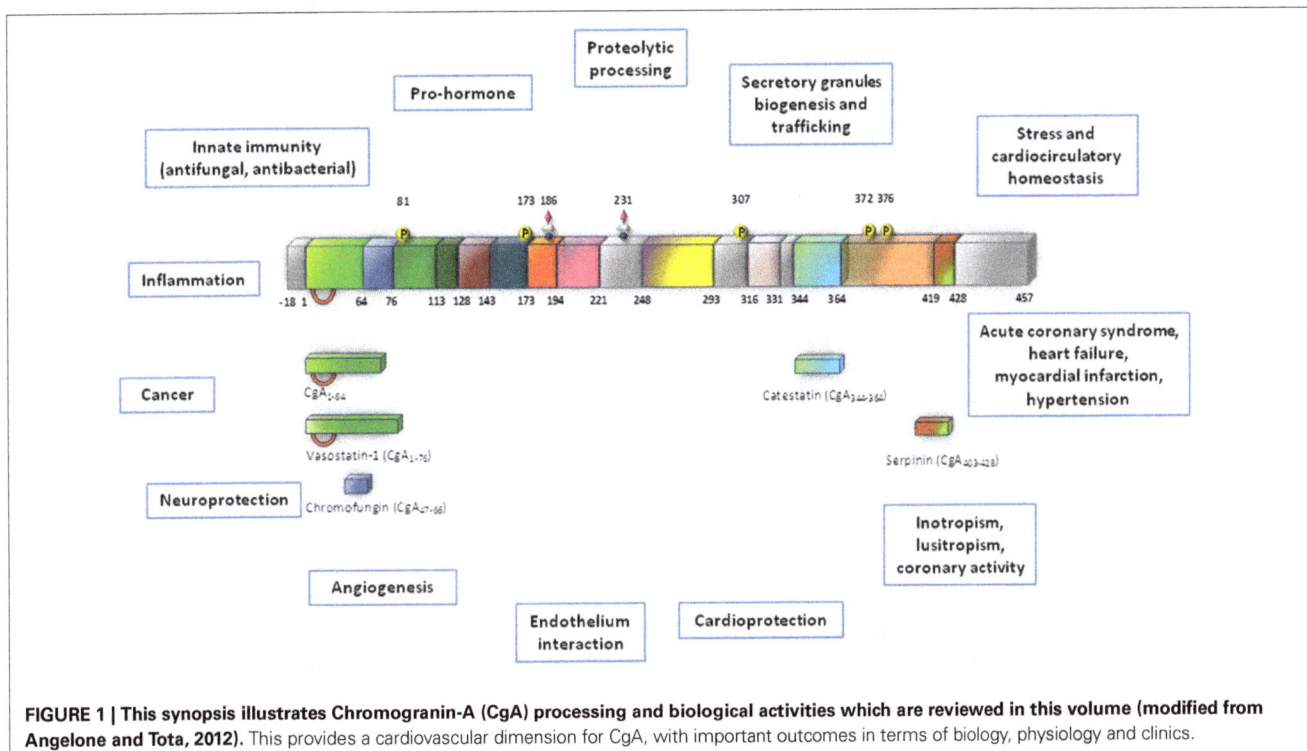

FIGURE 1 | This synopsis illustrates Chromogranin-A (CgA) processing and biological activities which are reviewed in this volume (modified from Angelone and Tota, 2012). This provides a cardiovascular dimension for CgA, with important outcomes in terms of biology, physiology and clinics.

the pathological processes (Cohn and Yellin, 1984). Apart from being of clinical relevance, these cardiovascular studies have provided the rationale for anti-adrenergic drug therapy, including the beta-adrenergic-blockers, still amongst the most used drugs. The growing evidence regarding the emerging cardiovascular role of CGA and CGA-derived VS-1, CST and serpinin may represent the next breakthrough in this field (see **Figure 2**).

THE CARDIO-CIRCULATORY PROFILE OF FULL-LENGTH CGA

Circulating CGA (normal values: 0.5–2 nM: Helle et al., 2007; Crippa et al., 2013) increases under conditions of stress-elicited SAN over-activation and physio-pathological conditions, e.g., chronic inflammation, neuroendocrine tumors, acute coronary syndromes and chronic HF. Therefore, CGA plasma levels have been used as prognostic indicators in these conditions (Helle et al., 2007; Angelone et al., 2012a; D'amico et al., 2014). It is important to note that pertinent information on plasma levels of CGA and its derived peptides can be best obtained by serological studies that in addition to processing-independent radioimmunoassays include also region-specific processing-dependent assays. In fact, only the latter can allow to analyse the plasma levels of the various granin-derived fragments which exhibit relevant and sometime opposite biological functions and prognostic significance (Crippa et al., 2013; Goetze et al., 2014). The CGA *in vivo* long half-life (~18 min) and its relatively elevated circulating concentrations (also under normal conditions), reduce the possibility of false measurements and facilitate blood collection, pre-analytic handling and final determinations (O'Connor et al., 1989).

Plasma CGA concentrations are increased up to 10–20 nM (500–1000 ng/ml) in patients with essential hypertension (Takiyyuddin et al., 1995), myocardial infarction (Omland et al., 2003), acute destabilized HF (Dieplinger et al., 2009), acute coronary syndromes (Jansson et al., 2009), chronic HF (Ceconi et al., 2002) and decompensated hypertrophic cardimyopathy (Pieroni et al., 2007). As firstly documented by Ceconi et al. (2002), circulating CGA levels significantly parallel the severity of the dysfunction, representing an independent predictor for mortality. Accordingly, from a clinical point of view, CGA is now emerging as a potentially new diagnostic and prognostic cardiovascular biomarker independent from conventional markers.

Studies in twins indicated that basal plasma CGA concentration is highly heritable (Takiyyuddin et al., 1995). Compared with age-matched normotensive counterparts, patients with essential hypertension show increased plasma CGA and an increased release of stored CGA in response to adrenal medullary stimulation by insulin-evoked hypoglycemia (Takiyyuddin et al., 1995). Among others, these observations confirm the correlation between CGA and SAN activity.

A relevant circulatory function of CGA is related to the regulation of endothelial barrier (Ferrero et al., 2004) and tumor-induced vascular remodeling (Veschini et al., 2011). Both CGA and VS-1 are potent inhibitors of the proangiogenic Vascular Endothelial Growth Factor (VEGF), as well as the thrombin-induced endothelial cell permeability (Ferrero et al., 2004) and inhibit the TNF-elicited changes on endothelial cells, i.e., gap formation, disassembly of vascular endothelial-cadherin

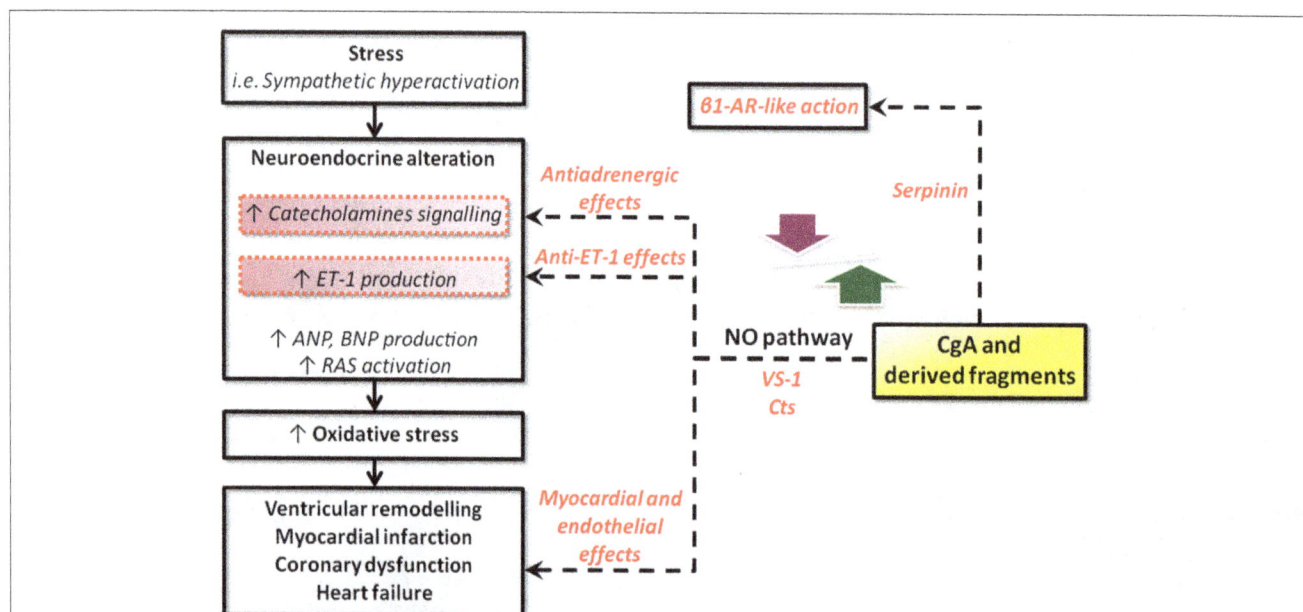

FIGURE 2 | Schematic representation of the possible sites for intervention of CgA and its derived peptides in heart failure. CgA and its fragments could operate at two, non-exclusive, levels: systemic and local. At the systemic level, CgA may work together with other factors (catecholamines, ANGII, cytochines, chemochines, etc) in the stress response, as in the case of the neuroendocrine scenario activated in CHF. At the local (heart) level, systemic and/or intracardiac physical and chemical stimuli could trigger CgA processing to generate cardioactive peptides, i.e., VSs, CST, serpinin (modified from Angelone et al., 2012a).

adherence junctions and vascular leakage (Ferrero et al., 2004; Dondossola et al., 2011). Systemic administration of CGA (1 µg) to lymphoma-bearing mice potently reduces the TNF-elicited penetration of a synthetic dye (patent blue) in tumor tissues (Dondossola et al., 2011), confirming previous observations that CGA can affect host/tumor interactions (Colombo et al., 2002).

Recently, Crippa et al. (2013) have reported in healthy subjects the presence of biologically relevant plasma levels of full-length CGA, CGA 1–76 (antiangiogenic) and fragments lacking the C terminal region (proangiogenic) and have demonstrated that blood coagulation triggers a thrombin-dependent almost complete conversion of circulating CGA into fragments lacking the C- terminal region. This uncovers a novel role of CGA as an angiogenesis activator which, under conditions of perturbed angiogenesis (wound healing, cancer, etc.), can contribute to circulatory and vascular homeostasis through the opposite angiogenic effects of its fragments (possibly VS-1 and CST) generated by tightly spatio-temporally regulated proteolysis.

A similar modulatory strategy exerted by CGA has been suggested at the heart level on the basis of morphological, biochemical and physio-pharmacological evidences briefly summarized below.

THE INTRACARDIAC LOCALIZATION OF CGA

Using immunohistochemical technique, Steiner et al. (1990) demonstrated in the myoendocrine granules of the rat heart the co-localization of CGA and ANP. Their immunoblotting finding suggested a more extensive myocardial CGA processing compared to that of the adrenal medulla. As remarked by Miserez et al. (1992), an additional source of cardiac CGA and/or CGA-derived fragments may result from the nerve termini innervating the heart. Weiergräber et al. (2000) showed that CGA was also present in rat Purkinje conduction fibers, in both rat atrium and ventricle, as well as in H9c2 rat cardiomyocytes. More recently, Biswas et al. (2010) confirmed the presence of CGA, as well as that of CGB and SG in the secretory granules of the mouse myocardium.

Importantly from a physio-pathological point of view, Pieroni et al. (2007) provided immunohistochemical evidence of CGA-positive intracellular staining in the human myocardium. They used confocal microscopy to show that in ventricular cardiomyocytes of dilated and hypertrophic human hearts CGA is colocalized with Brain Natriuretic Peptide (BNP). RT-PCR corroborated this finding documenting the myocardial presence of CGA-mRNA, while ELISA assays with four different monoclonal antibodies allowed to measure more than 0.5 µg of CGA per gram of left ventricular myocardial tissue. Assuming a constant myocardial release of CGA and considering that the plasma half-life of CGA is 18.4 min Corti et al. (1996), Pieroni et al. (2007) postulated a significant cardiac contribution to the increased circulating CGA levels reported in their patients. The possible correlation of CGA with the NPs system could contribute to regulate cardiovascular function through short- and long-term depressing influences, inducing tonic vasodilation, hypotension and cardioprotection against adreno-sympathetic hyperactivation. Moreover, because of the strong association between plasma CGA/NPs concentrations and the degree of hemodynamic dysfunction in HF, both hormones have been used as promising prognostic indicators of the severity of HF (see Dieplinger et al., 2009 and references therein). Furthermore, the significant correlation between CGA levels and left ventricle end diastolic pressure may implicate for myocardial CGA the operation of stretch-elicited release and transcriptional up-regulation mechanisms similar to that reported for BNP (Tota et al., 2010 and references therein).

In the rat heart, intracardiac presence and processing of CGA were biochemically demonstrated by Glattard et al. (2006). By submitting the RPHPLC purified CGA-immunoreactive fractions from cardiac extracts to western blot and MS analysis (TOF/TOF technique), the authors characterized four endogenous N terminal CgA-derived peptides, i.e., CGA4–113, CGA1–124, CGA1–135 and CGA1–199, containing the VSs sequence. The intact CGA was also detected among these and other C-terminal truncated fragments. It is important to note the cell-specific feature of this proteolytic CGA fragmentation, as contrasted by the rat adrenal gland in which almost no intact CGA is found; on the other hand, pancreatic beta cells produce betagranin corresponding s to the N-terminal portion of CGA (Hutton et al., 1987). The comparison of these and other observations seems to indicate that in the heart the maturation process can be incomplete and specific. It is of particular interest the finding, more recently confirmed by Pasqua et al. (2013), that the cardioactive motif (the VS-1 sequence or a portion of it) is present among the low-molecular-mass fragments identified. Therefore, the possibility exists that, under normal or stressfull conditions the heart responds to a specific physical (e.g., stretch) or chemical (e.g., CAs) stimulus activating proteolytic CGA processing with subsequent increase in lower-molecular-mass cardioactive fragments. This supports our working hypothesis (Pasqua et al., 2013, and references therein) that CGA, in addition to its endocrine and systemic role, can exert a direct autocrine/paracrine modulation on the heart.

PHYSIO-PHARMACOLOGICAL EVIDENCE OF DIRECT CGA CARDIOACTIVITY

Using isolated and Langendorff perfused hearts of normotensive and spontaneously hypertensive rats (SHR), Pasqua et al. (2013) have shown for the first time that the exogenous full-length CGA (1–4 nM) directly affects myocardial contractility and coronary vasomotion (dilation) by Akt/NOS/NO/cGMP/PKG pathway. At the same time, the rat heart in response to hemodynamic and chemical (β-adrenergic and ET-1) excitatory stimuli generates CGA fragments, including the cardioactive VS-1. Therefore, this evidence, which conceptually interlocks the systemic (endocrine) and intracardiac (paracrine/autocrine) actions of full-length CGA and its derived cardioactive peptides, provides a rationale for future investigations on the putative multilevel switches of the SAN/CGA axis that may operate under normal and physio-pathological conditions. In this perspective, here below we will briefly illustrate the cardiovascular profiles of VS-1, CST and serpinin, that have been documented in rodent heart preparations. For space economy, we will not discuss the cardiac effects that VS-1 and CST exert on cold vertebrate hearts (frog and eel) which confirm the peptide profiles observed in the rodent heart (see for references, Tota et al., 2010). Our aim is to highlight the intriguing characteristics of a protein like CGA which has revealed itself to function both as a pleiotropic hormone and a cytokine.

CARDIAC ACTIONS OF VS-1

Numerous studies conducted by our group have demonstrated that the N terminal human recombinant (hr) CGA-derived fragments VS-1 (hrCGA1–78) and VS-2 (hrCGA 1–113), as well as the corresponding fragments from bovine CGA and rat CGA (rCGA1-64), exert relevant cardiodepressive and anti-adrenergic influence on several vertebrate (rat, frog and eel) heart preparations (reviewed by Tota et al., 2010). Both VS-1 and VS-2 concentration-dependently inhibit myocardial contractility (negative inotropism) and relaxation (negative lusitropism), thus directly modulating the mechanical performance of intact isolated and perfused hearts beating under basal, i.e., non-stimulated, and adrenergically stimulated, conditions (Tota et al., 2010). Since the VS-1-elicited cardiotropism resulted more relevant than that of VS-2, here we will only focus on VS-1.

The isolated Langendorff-perfused rat heart was used to analyse in detail the VSs influence on myocardial contractility (Cerra et al., 2006). In particular, hrCgA 1–78, containing the VS-1 (CgA 1–76) sequence, at all concentrations tested (11–165 nM) under basal conditions dose-dependently depressed both contractile activity and cardiac work, as indicated by the decrease of left ventricular pressure (LVP) and rate pressure product (RPP: HRx LVP), respectively. No effect on coronary vasomotion was detected, as indicated by the unchanged coronary pressure (CP). Noteworthy, as shown by Pieroni et al. (2007) on the isolated Langendorff rat heart, hrVS-1 is also a negative lusitropic agent since concentration-dependently it decreases myocardial relaxation, i.e., it reduces the maximal rate of the left ventricular pressure decline of LVP $[-(LVdP/dt)max]$, the half time relaxation (HTR), and T/−t ratio obtained by $+(LVdP/dt)max/-(LVdP/dt)max$. The peptide counteracted the β-adrenergic (Isoproterenol, ISO)-induced positive inotropic and lusitropic effects through a non-competitive mechanism. Taken together, these peptide-induced actions point to VS-1 as a relevant modulator of rat heart mechanical performance under basal and adrenergically stimulated conditions. However, these findings beg the question as to whether or not the peptide-elicited cardiac effects are species-specific. To clarify this issue, using rat cardiac preparations (the isolated Langendorff perfused heart and the papillary ventricular preparation), Cerra et al. (2008) analyzed the cardiac actions of the native (rat) CGA 1-64 (rCGA1-64), i.e., a highly conserved cleavage N-terminal site that reproduces the native rat sequence, corresponding to human N-terminal CGA-derived VS-1 (Metz-Boutigue et al., 1993; Helle et al., 2007). Of note, the concentrations used were the same of its precursor, CGA, in the human serum (i.e., normal levels: 0.5–4 nM, neuroendocrine tumors and last stages of chronic HF: >10 nM) (Helle et al., 2007). The study not only confirmed that, the rCgA1-64 fragment from 33 to 165 nM elicited a significant negative inotropic and lusitropic effects without modifying HR, but also demonstrated that the peptide was a coronary vasodilator, as indicated by a significant reduction of CP. Accordingly, this vasoactivity, for the first time detected on an intact whole coronary bed, further corroborated the "vasostatin" (ad litteram)

profile of this fragment, as previously reported on segments of bovine coronary resistance arteries, intrathoracic artery and saphenous vein exposed to the hrCgA 1–78 active domain 1–40 (Brekke et al., 2002). The absence of coronary activity of the human recombinant hrCgA1-78, previously evidenced by Cerra et al. (2006), suggests that distinct species-specific sensitivities of the vascular tissues toward VS peptides may account for these different responses. This, however, does not appear to be the case of the myocardium, since the comparison of the contractile effects of rCGA 1-64 and hrCGA1-78 highlights their substantial similarity. This suggests that the rat heart may be a useful model for medically-oriented studies regarding the potential of VS−1as a therapeutic agent. The treatment with rCGA65-76 did not modify cardiac performance at all concentrations tested, except for a small chronotropic effect from 11 to 110 nM, and did not modify the β-adrenergic (ISO)-induced intrinsic activity. An important characteristic of rCGA 1-64 is its ability to counteract ISO (1 μM)- and endothelin 1 (ET-1)-elicited positive contractility, as well as the potent ET-1-induced coronary constriction. Experiments on isolated papillary muscles, i.e., an experimental model in which contractility is analyzed independently from HR and coronary flow, confirmed that the peptide depresses basal and ISO-elicited contractility, without affecting calcium transients on isolated ventricular cells.

The analysis of the percentage of variations of LVP, which provides the EC50 values in the presence of either increasing concentrations of ISO alone or of ISO plus rCGA 1-64 (11, 33, and 65 nM), showed that rCGA1-64 elicits its anti-β-adrenergic action through a functional non-competitive antagonism, confirming the results obtained with the human VS-1 (Cerra et al., 2006).

Furthermore, to have an insight on structure-function relationship, three modified peptides were tested on both rat heart and papillary muscles, showing that the di-sulfide bridge was necessary for the cardiac activity.

In the heart, similarly to the hrVS-1 signal-transduction (Cerra et al., 2006), the rCGA1-64 signals through a Gi/o protein-PI3K-NO-cGMP-PKG-dependent mechanism. In particular, as clarified by the results obtained on the isolated papillary muscle, the action mechanism appears to implicate a calcium-independent/PI3K-dependent NO release by endothelial cells. In fact, rCGA1-64, had no effect on intracellular calcium concentration in isolated ventricular cells, but elicited NO release from cultured bovine aortic endothelial cells (BAE-1) through a calcium-independent mechanism. The eventual involvement of the endocardial endothelium in this mechanism remains to be evaluated. The evidence that the rCGA1-64-induced coronary dilation is also abolished by inhibitors of the NO-cGMP-PKG pathway suggests the likely endothelial release of vasodilator autacoids such as NO.

The control of the NO/NOS system on myocardial contractility and its common tonic depressive influence have been extensively documented. In the ventricular myocardium of the rat NOS-generated NO depresses contractility through sGC-PKG mechanism which decreases L-type Ca^{2+} current (Abi-Gerges et al., 2001) and troponin I phosphorylation (Hove-Madsen et al., 1996). Calcium-independent eNOS activation has been shown to take place after stimulation of endothelium with insulin, insulin-like growth factor-1 (IGF-1) and estrogens (Hartell et al., 2005) through the possible involvement of Akt dependent NOS phosphorylation (Shaul et al., 2002). Conceivably, our experiments on papillary muscles and on BAE-1 cells treated with the PI3K inhibitor wortmannin strongly suggest that the rCGA1-64-elicited NO synthesis depends on PI3K activation. Maniatis et al. (2006) have proposed a calcium-independent mechanism of eNOS activation involving caveolae-mediated endocytosis elicited by the albumin-binding protein gp60 and activation of downstream Src, Akt and PI3K pathways. It has been hypothesized that VSs may interact with caveolar domain (see for references, Tota et al., 2007), and that endothelial cells internalize CGA1-78 (Ferrero et al., 2004). Therefore, it is possible that a similar mechanism may explain VS-1 dependent NOS activation in BAE-1 cells, an issue that needs further research. On the whole, these results strongly support our hypothesis (Tota et al., 2008, 2010) that in the rat heart VS-1, and particularly the homologous rCGA1-64 fragment may act as an autocrine/paracrine modulator of myocardial and coronary performance, functioning as homeostatic stabilizer against heightened SAN tone.

CARDIAC ACTIONS OF CST

CST was initially identified as the most potent endogenous antagonist of nicotinic-cholinergic receptor (nAChR), exerting antagonistic inhibition of nicotine-evoked CAs secretion in a non-competitive way (Mahata et al., 1997). However, it has revealed itself to act as a multifunctional peptide with different action mechanisms. The *in vitro* and *in vivo* vasoactive and relevant anti-hypertensive properties of CST have been extensively reviewed (e.g., Mahata et al., 2010) and will not be considered here.

The cardiotropic actions of wild-type CST (WT-CST, human CGA352-372) and its naturally occurring variants (G364S-CST and P370L-CST) were demonstrated for the first time by Angelone et al. (2008) who used the isolated and Langendorff perfused rat heart to evaluate the peptide-induced cardiac effects independently from the minute-to-minute control exerted by SAN over cardiac output (CO) and vascular tone. WT-CST (from 11 to 200 nM) dose-dependently decreased left ventricular pressure (LVP, index of contractility), rate pressure product (index of cardiac work) and both positive and negative LVdP/dt (index of maximal rate of left ventricular contraction and relaxation, respectively), while increased CP. While G364S-CST was ineffective on basal mechanical performance, P370L-CST elicited only a negative inotropism. Human CST variants counteracted the positive β-adrenergic (ISO)-induced inotropic and lusitropic effects with different rank order of potency (for the ISO-induced positive inotropism: WT-CST>G364S-CST>P370L-CST; for the ISO-induced positive lusitropism: G364S-CST>WT-CST>P370L-CST (Angelone et al., 2008). In both the isolated Langendorff rat heart (Angelone et al., 2012b) and in rat papillary muscle (Bassino et al., 2011), CST before eliciting these major prolonged myocardial actions, induced an early transient positive inotropic effect that disappeared after 5 min from administration. This effect was mediated by H1 histamine receptors, since it was abolished by H1 blockade, consistent with previous evidence that in the rat the activation of myocardial H1 receptors mediates the histamine-dependent positive inotropism (Matsuda

et al., 2004). The relevant tonic CST-induced negative inotropism and lusitropism are attained through pertussis toxin-sensitive (PTX), receptor-independent activation *via* heterotrimeric G proteins and Gαi/o subunits. Likewise, Gi/o proteins activation could limit the Gs-mediated positive contractile effects. The peptide signaling involves both beta2-AR and beta3-AR, with a higher affinity for the first one (as evidenced by low IC50 values) but not beta1-AR, being unaffected by cholinergic receptor inhibition, respectively (Angelone et al., 2008). It is known that beta1-AR, coupled to Gs proteins, is responsible for positive inotropism and lusitropism, while beta2-AR, mainly coupled to Gi/o proteins, is responsible for the opposite effects on contractility and relaxation (Xiao et al., 1999). Moreover, cardiac beta2-AR/Gi stimulation can activate PI3K with consequent negative inotropism (Yano et al., 2007). Interestingly, an important component of the CST signal-transduction is represented by the PI3K/Akt/eNOS/NO/cGMP-dependent pathway, as shown by the PI3K blockade which abolishes the peptide-induced inotropism and lusitropism (Angelone et al., 2012b). The CST signaling also requires an endothelium-derived bioactive NO mechanism, since it depends from the functional integrity of the endothelium. To further analyse the signal-transduction mechanism of CST and its variants, Bassino et al. (2011) measured contractility and Ca^{2+} transients respectively on papillary muscles and isolated cardiomyocytes in basal conditions and after beta-adrenergic stimulation, evaluating on BAE-1 NO production and eNOS phosphorylation ($P^{Ser1179}$eNOS). Their data show that CST dose-dependently (5–50 nM) reduces the effect of beta-adrenergic stimulation which, rather than resulting from a direct myocardial action of the peptide, depends from a Ca^{2+}-independent/PI3K-dependent NO release from endocardial endothelial cells. Consistent with this, CST induces in BAE-1 cells a Wortmannin-sensitive, Ca^{2+}-independent increase of NO production and $P^{Ser1179}$eNOS. The variant P370L-CST, but not G364S-CST, exerted an anti-adrenergic effect and an increased NO release comparable to that elicited by WT-CST. As mentioned above (*Cardiac actions of VS-1*), such calcium-independent, caveolae-mediated endocytosis mechanism for activation of Akt-PI3K-eNOS pathway, also proposed for insulin, insulin-like growth factor-1, and estrogens, has been suggested for VS-1 (Ramella et al., 2010 and references therein). In agreement with the NO-sGC-cGMP-dependent signaling, in the peptide-treated hearts cGMP was significantly increased and at least two of its targets appeared implicated in the CST action mechanism. One is PKG, known to depress myocardial contractility by reducing both L-type Ca^{2+} current and troponin C affinity for calcium (Angelone et al., 2012b). The other target is phosphodiesterases type 2 (PDE2) (Angelone et al., 2012b). Its selective inhibition by EHNA abrogates the CST-induced inotropic and lusitropic effects both under basal and stimulated (ISO) conditions, indicating a relevant PDE2 involvement. Conceptually important in the context of temporal SAN/CGA interactions, Angelone et al. (2012a) also demonstrated the involvement of both phospholamban (PLN) and beta-arrestin S-nitrosylation. beta-arrestin is implicated in the desensitization and internalization of G-protein-coupled receptors (including beta1-AR), while its S-nitrosylation elicits and accelerates beta1-AR desensitization (Ozawa et al., 2008). PLN controls sarcoplasmic reticulum Ca-ATPase (SERCA2a) by a phosphorylation/dephosphorylation mechanism so that while dephosphorylated PLN inhibits SERCA2a-dependent SR Ca^{2+} sequestration (Reddy et al., 1999), phosphorylated PLN at Ser16 by PKA relieves its inhibition on SERCA2a (Schmidt et al., 2001). Thus, PLN S-nitrosylation appears a selective target of the CST-induced NO release able to regulate SR Ca^{2+} fluxes and Ca^{2+} availability for inotropy and lusitropy. These data highlight a temporally tuned and more sophisticated CST-promoted anti-beta-adrenergic cardiac modulation than previously perceived, which consists in rapid (PDE2 and PLN) and medium-term (beta-arrestin-mediated beta1-AR desensitization) regulatory switches. Accordingly, a PDE2-dependent short-term signal switches off β adrenergic activity, while a medium-term signal promotes more prolonged adrenergic counteraction through beta-arrestin-mediated beta1-AR desensitization. In the rat heart the coronary profile of CST appears non-univocal. CST under basal conditions dose-dependently increases CP (maximum response at 200 nM) and abolishes the ISO-dependent vasodilation; in contrast, it potently vasodilates the ET-1 preconstricted coronaries (Angelone et al., 2008), confirming the vasodilation promoted by exogenous CST in human subjects (O'Connor et al., 2002; Fung et al., 2010). Since it is known that in mammals the stimulation of cardiac ETB subtype receptors promote an endothelium-dependent negative inotropism (Brás-Silva and Leite-Moreira, 2005), CST signal was tested in presence of the selective ETB antagonist BQ788. The consequent abrogation of the ET-1- induced effects demonstrated the relevant ET-1/CST cross talk, suggesting at the same time that the CST signal at plasma membrane level significantly converges on the sympatho-inhibitory NO pathway. In conclusion, this growing evidence indicates that CST, in addition to its antihypertensive profile, directly modulates myocardial and coronary functions under both basal and stimulated (β-adrenergic and ET-1) conditions. The peptide-induced negative contractility/relaxation influence, as well as its coronary vasomotion might be viewed as relevant components of a homeostatic counteraction against heightened SAN over-activation, e.g., prolonged stress, HF, hypertensive cardiomyopathy, namely, conditions characterized by a potentially harmful spill-over of CAs, ET-1 and RAS agonists. Conceivably, these cardiac properties of the peptide, together with its antihypertensive and vasoactive profile, suggest that CST can function as a novel autocrine-paracrine modulator cooperating with full-length CGA and its derived VS-1 in the multilevel processes required for cardio-circulatory homeostasis (see **Figure 3A**).

CARDIOPROTECTIVE INFLUENCE OF VS-1 AND CST

Reperfusion is indispensable for salvaging viable myocardium, following an acute heart infarction. Both ischemic preconditioning (PreC) and post-conditioning (PostC) have become standard procedures for evaluating the cardioprotective ability of an agent when applied respectively before or after an infarcting ischemia can enhance heart function recovery, limiting infarct size (Hausenloy, 2009, 2012; Penna et al., 2009). Various peptides can elicit cardioprotection, triggering both pharmacological pre- and post-conditioning signaling pathways. In

rodents, these include the *Reperfusion Injury Salvage Kinase* (RISK) and the *Survivor Activating Factor Enhancement* (SAFE) pathways (Sivaraman et al., 2007; Hausenloy et al., 2011). The actions of Survival kinases, such as phosphatidylinositol-3-kinase (PI3K)/protein kinase B (Akt/PKB) and protein kinase C (PKC), may converge on downstream mitochondrial targets to open ATP-sensitive potassium (mitoKATP) channels, thereby affecting cellular survival through a decrease of necrosis and apoptosis (Zatta et al., 2006; Boengler, 2011). Cardioprotection involves endothelial and endothelial derived NO, as well as adrenergic components (Bell and Yellon, 2003; Pagliaro et al., 2003; Cappello et al., 2007). The acknowledged anti-adrenergic and endothelial PI3K-induced NO signaling of both VS-1 and CST have addressed several studies to verify their eventual cardioprotective profiles. The comparison of the cardioprotective effects of VS-1 and CST in ischemic conditioning highlights a remarkable similarity and subtle differences. VS-1 appears to act as a pre-conditioning inducer while CST acts as a post-conditioning agent (reviewed by Penna et al., 2012). Human recombinant VS-1 protects against the extension of myocardial infarction converging on PKC through two different pathways, one mediated by adenosine A1 receptors and the other mediated by NO release (Cappello et al., 2007). On the other hand, the CST-induced protection reduces infarct size and improves post-ischemic cardiac function *via* PI3K/Akt, PKCs and mito-KATP channel activation, which may implicate a ROS signaling (Perrelli et al., 2013). CST is also able to exert direct myocardial protection *via* an endothelium-independent mechanism, as shown by Penna et al. (2010) in isolated adult cardiomyocytes exposed to simulated I/R, where the peptide has been shown to induce a partial mitochondrial depolarization (Perrelli et al., 2013) (see **Figure 3B**). Very recently, it has been also reported that CST protects in the post-ischemic SHR heart by increasing the expression of anti-apoptotic and pro-angiogenic factors, supporting its potential therapeutic role, even in the presence of comorbidities, such as hypertension and cardiac hypertrophy (Penna et al., 2014a, June Accepted). These data are of interest also taking into account the angiogenic properties of CST reported by Theurl et al. (2010). A CST-induced S-nitrosylation of calcium channels in the post-ischemic phase has been recently reported by Penna et al. (2014a). Of note, this posttranslational modification of a L-type calcium channel subunit has already been described in preconditioning cardioprotection by Murphy et al. (2014) and in non-ischemic hearts treated with CST by Angelone et al. (2012b). The CST-Post-elicited S-nitrosylation of calcium channel may be functionally important, since the oxidative/nitrosative signaling is known to play a major role in cardioprotection against ischemia/reperfusion injury in both preconditioning and post-conditioning (Pagliaro et al., 2011; Tullio et al., 2013; Penna et al., 2014b).

CARDIAC ACTIONS OF SERPININ

The recent story of serpinin began with Kim and Loh (2006) examining the processing of the C-terminal domain of CGA in mouse pituitary cell line, (AtT-20). They showed that a CGA fragment, but not intact CGA, secreted in an activity-dependent manner, increased granule biogenesis by up-regulating protease nexin-1 (PN-1), a serine protease inhibitor protein, which then

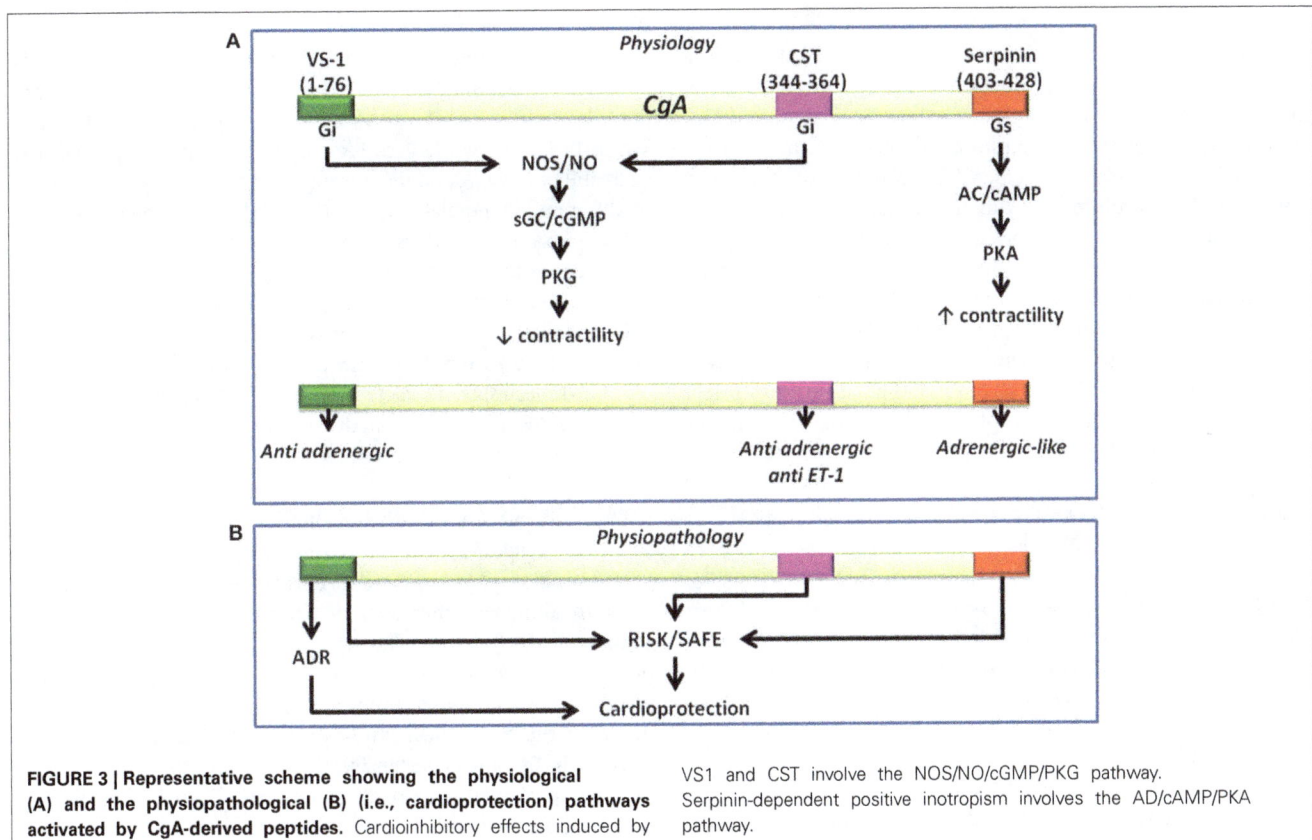

FIGURE 3 | Representative scheme showing the physiological (A) and the physiopathological (B) (i.e., cardioprotection) pathways activated by CgA-derived peptides. Cardioinhibitory effects induced by VS1 and CST involve the NOS/NO/cGMP/PKG pathway. Serpinin-dependent positive inotropism involves the AD/cAMP/PKA pathway.

stabilizes granule proteins to enhance their levels in the Golgi complex. Successively, Koshimizu et al. (2010, 2011) demonstrated that, in AtT-20 pituitary cells, both mouse CGA435–460, a synthetic 26 amino acid residue peptide named serpinin, and endogenous-serpinin related peptides were able to induce PN-1 mRNA up-regulation. In particular, they identified in AtT-20 cell-conditioned medium a 23-mer serpinin-like fragment, pyroglutaminated (pGlu-23Leu) serpinin, which was present in the highest amount compared to the other serpinin-related peptides serpinin (Ala26Leu), and serpinin-Arg-Arg-Gly (Ala29Gly), (Koshimizu et al., 2011). The recognition that the serpinin region of CGA is highly conserved in mammals suggests that peptides derived from this domain may have relevant physiological functions.

pGlu-serpinin upregulates PN-1 mRNA expression in AtT-20 cells via a cAMP-protein kinase and exerts an antiapoptotic effect on these cells and on cultured CNS neurons exposed to oxidative stress. Since PN-1 expression is upregulated by CGA in AtT-20 cells (Kim and Loh, 2006) and PN-1 is a potent inhibitor of plasmin which causes apoptosis in chronically injured neurons (Ho-Tin-Noé et al., 2009), it is possible that pGlu-serpinin can be implicated in plasmin inhibition.

Soon after, Tota et al. (2012) demonstrated the involvement of serpinin in heart function. Using HPLC and ELISA methods, they detected in the rat heart serpinin peptides, Ala29Gly and pGlu-serpinin being the predominant fragments. This means that the rodent heart is able to process the C terminal domain of CGA as it does with its N terminal domain. Moreover, using

the Langendorff perfused rat heart to evaluate the hemodynamic responses, Tota et al. (2012) showed that serpinin and pGlu-serpinin dose-dependently (11–165 nM) increase inotropism and lusitropism within the first 5 min after administration. pGlu-serpinin appears more potent than serpinin, its action starting from 1 nM. These effects were corroborated by the results obtained on the isolated rat papillary muscle preparation which allows to measure contractility in terms of tension development and muscle length. It was found that while pGlu-serpinin induces positive inotropism, Ala29Gly is unaffected. Both pGlu-serpinin and serpinin act through a beta1-Adrenergic Receptor/Adenylate Cyclase/cAMP/PKA pathway (Tota et al., 2012), a finding of particular interest in view of the opposite beta-depressive profile of the two CGA-derived peptides, VS-1 and CST. The beta-adrenergic-like agonist profile of serpinin is further confirmed by the results showing that pGlu-serpinin increases intracardiac cAMP levels. pGlu-serpinin and serpinin at nanomolar range act as beta 1-adrenergic-like agonists, mimicking the intracardiac sympathetic neurotransmitters and/or circulating CAs. Like these agents, they increase cellular levels of cAMP, thereby remarkably affecting myocardial mechanical performance. Namely, they increase the rate and extent of tension development during systole (positive inotropy), hence augmenting stroke volume; at the same time, they accelerate myocardial relaxation (positive lusitropy), hence shortening the overall duration of diastole. As demonstrated by the experiments on isolated papillary muscle, the serpinin action is independent from any possible alteration in HR and coronary flow rate, as well as from norepinephrine release from sympathetic nerve terminals (as demonstrated by tyramine treatment). Studies are needed to clarify the earliest events underpinning the transduction of the serpinin signal. Since no receptor or direct binding partner has been so far identified for serpinin or pGlu-serpinin, on the basis of the reported functional antagonism, Tota et al. (2012) have proposed that serpinin and pGlu-serpinin may function as allosteric modulators of the beta-adrenergic receptor independently from the ligand binding site, thereby triggering the well-known beta-adrenergic-induced cascade. Therefore, serpinin peptides resemble the other CGA-derived peptides in such apparent lack of classical receptors. According to a mechanism common for cAMP elevating agonists (Koshimizu et al., 2010), serpinin and pGlu-serpinin might bind to a G protein-coupled receptor (GPCR) enhancing the cardiac cAMP levels, with consequent hemodynamic effects that are indeed abolished by selective inhibition of AD and PKA. The major downstream targets of the AD-cAMP signaling are PKAs which phosphorylate a number of proteins, including SERCA and its associated modulatory protein, PLN, both crucial regulators of myocardial inotropy and lusitropy (Katz, 1990). SERCA-dependent Ca^{++} uptake within SR promotes cation removal during diastole, thus affecting relaxation and the subsequent contraction (Satoh et al., 2011). Noteworthy, blockade of SERCA activity by thapsigargin, abolishes the pGlu-serpinin-elicited inotropy and lusitropy. It is well known that Ca^{++} sensitivity for SERCA is enhanced by PKA-induced phosphorylation of PLN, a downstream target of the beta-adrenergic-PKA cascade which plays a determinant role in the adreno-sympathetic modulation of myocardial contractility/relaxation (Mattiazzi et al., 2007). It is thus physiologically relevant that pGlu-serpinin induces PNL phosphorylation at Ser16, residue (see **Figure 3A**).

PUTATIVE CARDIOPROTECTIVE INFLUENCE

The pGlu-serpinin-PKA cascade also appears to induce phosphorylation of ERK1/2 and GSK3beta known to mediate PGE2 and EP4 signaling in neonatal ventricular myocytes (Tota et al., 2012 and references therein). These proteins are components of the protective RISK (*reperfusion injury signaling kinase*) pathway implicated in myocardial protection against ischemia-reperfusion injury in rodents (Hausenloy et al., 2011) and, possibly, in the pGlu-serpinin-induced anti-apoptotic effects against reactive oxygen species (ROS) reported in cultured cerebral neurons by Koshimizu et al. (2011). This observation has prompted preliminary experiments aimed to test the cardioprotective influence of serpinin against I/R injury in the rodent heart. Very recently, Tota et al. (2012) reported that, given in pre- and post-conditioning, pGlu-serpinin reduces infarct size and preserves the hemodynamic function of both isolated normotensive and SHR hearts, being more protective in the latter. Moreover, in both normotensive and hypertensive hearts, pGlu-serpinin while inducing a mild cardioprotection in pre-conditioning, exerts streaking cardioprotection in post-conditioning. All these effects appear to involve the protective RISK pathway (Tota et al., 2012) (see **Figure 3B**).

In conclusion, the recent recognition that the C terminal domain of CGA can generate in the heart serpinin fragments that enhance myocardial contractility/relaxation through beta1-adrenergic/AD/cAMP signaling adds new evidence to the

sympatho-chromaffin profile of CGA. It is likely that in response to perturbed conditions, a tissue-specific and spatio-temporally concerted processing of CGA can produce counter-regulatory peptides able to reset cardio-circulatory homeostasis through "whip-brake" networks (see **Figure 4**). It might be expected that such ability of CGA and its derived VS-1, CST and serpinin could be differentially regulated in relation to short- or long-term activation of SAN, but this issue is still a closed book. In this context, the evidence that pGlu-serpinin acts on beta1-AR in a functional manner, i.e., independent of a direct classical interaction with AR active site, may be of putative therapeutical interest, being an alternative to direct adreno-receptor stimulation when this can be undesirable. In fact, it is known that prolonged, long-term beta-AR stimulation can directly induce maladaptive cardiac hypertrophy, eventually, leading to HF.

CONCLUSIONS AND PERSPECTIVES

Studies in the past 15 years have convincingly revealed the cardiovascular activities of CGA and its derived peptides VS-1, CST and serpinin, along with their striking adreno-sympathetic regulatory influences (see **Table 1**). This knowledge, which has widened the prohormone/cytokine profile of CGA, has been paralleled and integrated by a growing number of clinical studies that have documented the biomedical implications of CGA and its peptides (especially the antihypertensive CST) in various cardiovascular diseases, particularly in relation to their diagnostic and prognostic value. It is well known that conditions characterized by perturbed cardio-circulatory homeostasis, especially in the case of severe heart diseases (i.e., myocardial infarction, acute coronary syndromes, ventricular hypertrophy), activate complex neurohumoral networks (SAN and NPs, to mention two), finely integrated at both local and systemic levels, that tend to re-establish the constancy of the internal milieu (Angelone et al., 2012a and references therein). As illustrated in this review, the available evidence strongly supports the view that CGA and its cardiotropic

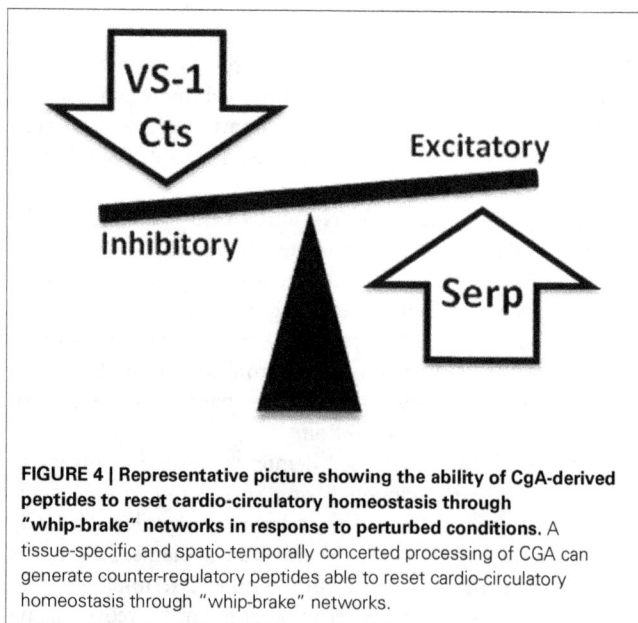

FIGURE 4 | Representative picture showing the ability of CgA-derived peptides to reset cardio-circulatory homeostasis through "whip-brake" networks in response to perturbed conditions. A tissue-specific and spatio-temporally concerted processing of CGA can generate counter-regulatory peptides able to reset cardio-circulatory homeostasis through "whip-brake" networks.

Table 1 | Synopsis of the cardiac effects of CgA and its derived peptides.

Peptide	Tissue	Contractility/ Relaxation	Heart rate	Vasoactivity	Doses	Adrenergic Stimulation	References
CgA	*Ex vivo rat* heart	Reduction	No changes	Vasodilation	1 pM ÷ 4 nM	-	Pasqua et al., 2013
VS1	*Ex vivo rat* heart	Reduction	No changes	No changes	11 ÷ 165nM	Non-competitive Antagonism	Cerra et al., 2008
rCGA1-64	*Ex vivo rat* heart Rat Papillary muscles	Reduction	No changes	Vasodilation	33 ÷ 165 nM 10 ÷ 100 nM	Non-competitive Antagonism	Cerra et al., 2008
CST	*Ex vivo rat* heart	Reduction	No changes	Vasodilation	11 ÷ 200 nM	Non-competitive Antagonism	Angelone et al., 2008, 2012a
Serpinin	*Ex vivo rat* heart Rat Papillary muscles	Increase	No changes	No changes	11 ÷ 165 nM 1 ÷ 33 nM	Beta-adrenergic like inotropim	Tota et al., 2012

fragments represent an important component of these neuroendocrine networks, challenging cross-disciplinary contributions in this direction. However, the excitement that accompanies this new knowledge is paired with a deeper perception of the many questions that need to be clarified. We neither know to which extent the extensive *ex vivo* evidence can be extrapolated to the *in vivo* situation, nor if the peptide-induced cardiac actions are beneficial for the diseased heart or they may concur to aggravate its pathology. For example, in acute coronary syndromes the vasodilatory, negative inotropic and lusitropic properties, and the anti-adrenergic effect elicited by the two CGA fragment VS-1 and CST, might be either detrimental, thus contributing to HF development, or compensatory, at least at the beginning of the pathology (Kubota, 1997). Similarly, in the context of circulatory homeostasis, there is a need to deepen the potential beneficial role of the CGA/NO axis in relation to the CGA-induced "anti-inflammatory" ability and protection of the endothelial barrier against TNF-alpha-induced vascular permeability (Ferrero et al., 2002). At the same time, an open window remains concerning the putative value of CGA and its fragments as cardiovascular diagnostic/prognostic biomarkers. In this regard, caution is especially needed because of two main reasons. The first is to understand whether these peptides provide incremental information with respect to conventional biomarkers. The second is to prevent errors in evaluating CGA and its fragments in biological samples by developing appropriate and sensitive methods for their detection and measurements, excluding interference induced by food intake and drug therapies (i.e., proton pump inhibitors, H2 antagonists, and glucocorticoids (Giusti et al., 2004).

The following years will tell us how can our existing view of CGA biology be modified in face of the new discoveries and how are we to deal with the cross-disciplinary challenges provided by the upcoming research.

REFERENCES

Abi-Gerges, N., Fischmeister, R., and Méry, P. F. (2001). G protein-mediated inhibitory effect of a nitric oxide donor on the L-type Ca2+ current in rat ventricular myocytes. *J. Physiol.* 1, 117–130. doi: 10.1111/j.1469-7793.2001.0117j.x

Angelone, T., Mazza, R., and Cerra, M. C. (2012a). Chromogranin-A: a multifaceted cardiovascular role in health and disease. *Curr. Med. Chem.* 24, 4042–4050. doi: 10.2174/092986712802430009

Angelone, T., Quintieri, A. M., Brar, B. K., Limchaiyawat, P. T., Tota, B., Mahata, S. K., et al. (2008). The antihypertensive Chromogranin A peptide catestatin acts as a novel endocrine/ paracrine modulator of cardiac inotropism and lusitropism. *Endocrinology* 10, 4780–4793. doi: 10.1210/en.2008-0318

Angelone, T., Quintieri, A. M., Pasqua, T., Gentile, S., Tota, B., Mahata, S. K., et al. (2012b). Phosphodiesterase type-2 and NO-dependent S-nitrosylation mediate the cardioinhibition of the antihypertensive catestatin. *Am. J. Physiol. Heart Circ. Physiol.* 302, H431–H442. doi: 10.1152/ajpheart.00491.2011

Angelone, T., and Tota, B. (2012). Editorial: Chromogranin A at the cross-roads of health and disease. *Curr. Med. Chem.* 24, 4039–4041. doi: 10.2174/092986712802430027

Bartolomucci, A., Possenti, R., Mahata, S. K., Fischer-Colbrie, R., Loh, Y. P., and Salton, S. R. (2011). The extended granin family: structure, function, and biomedical implications. *Endocr. Rev.* 6, 755–797. doi: 10.1210/er.2010-0027

Bassino, E., Fornero, S., Gallo, M. P., Ramella, R., Mahata, S. K., Tota, B., et al. (2011). A novel catestatin-induced antiadrenergic mechanism triggered by the endothelial PI3K-eNOS pathway in the myocardium. *Cardiovasc. Res.* 4, 617–624. doi: 10.1093/cvr/cvr129

Bell, R. M., and Yellon, D. M. (2003). Bradykinin limits infarction when administered as an adjunct to reperfusion in mouse heart: the role of PI3K, Akt and eNOS. *J. Mol. Cell Cardiol.* 2, 185–193. doi: 10.1016/S0022-2828(02)00310-3

Biswas, N., Curello, E., O'Connor, D. T., and Mahata, S. K. (2010). Chromogranin/secretogranin proteins in murine heart: myocardial production of Chromogranin A fragment catestatin (Chga(364-384)). *Cell Tissue Res.* 3, 353–361. doi: 10.1007/s00441-010-1059-4

Boengler, K. (2011). Ischemia/reperfusion injury: the benefit of having STAT3 in the heart. *J. Mol. Cell Cardiol.* 4, 587–588. doi: 10.1016/j.yjmcc.2011.01.009

Brás-Silva, C., and Leite-Moreira, A. F. (2005). Modulation of the myocardial effects of selective ETB receptor stimulation and its implications for heart failure. *Rev. Port. Cardiol.* 9, 1125–1133.

Brekke, J. F., Osol, G. J., and Helle, K. B. (2002). N-terminal chromogranin-derived peptides as dilators of bovine coronary resistance arteries. *Regul. Pept.* 2, 93–100. doi: 10.1016/S0167-0115(02)00004-6

Cappello, S., Angelone, T., Tota, B., Pagliaro, P., Penna, C., Rastaldo, R., et al. (2007). Human recombinant Chromogranin A-derived vasostatin-1 mimics preconditioning via an adenosine/nitric oxide signaling mechanism. *Am. J. Physiol. Heart Circ. Physiol.* 1, H719–H727. doi: 10.1152/ajpheart.01352.2006

Ceconi, C., Ferrari, R., Bachetti, T., Opasich, C., Volterrani, M., Colombo, B., et al. (2002). Chromogranin A in heart failure; a novel neurohumoral factor and a predictor for mortality. *Eur. Heart J.* 12, 967–974. doi: 10.1053/euhj.2001.2977

Cerra, M. C., De Iuri, L., Angelone, T., Corti, A., and Tota, B. (2006). Recombinant N-terminal fragments of chromogranin-A modulate cardiac function of the Langendorff-perfused rat heart. *Basic Res. Cardiol.* 1, 43–52. doi: 10.1007/s00395-005-0547-2

Cerra, M. C., Gallo, M. P., Angelone, T., Quintieri, A. M., Pulerà E., Filice, E., et al. (2008). The homologous rat Chromogranin A1-64 (rCGA1-64) modulates myocardial and coronary function in rat heart to counteract adrenergic stimulation indirectly via endothelium-derived nitric oxide. *FASEB J.* 11, 3992–4004. doi: 10.1096/fj.08-110239

Chien, K. R., Knowlton, K. U., Zhu, H., and Chien, S. (1991). Regulation of cardiac gene expression during myocardial growth and hypertrophy: molecular studies of an adaptive physiologic response. *FASEB J.* 5, 3037–3046.

Cohn, J. N., and Yellin, A. M. (1984). Learned precise cardiovascular control through graded central sympathetic stimulation. *J. Hypertens. Suppl.* 2, S77–S79.

Colombo, B., Curnis, F., Foglieni, C., Monno, A., Arrigoni, G., and Corti, A. (2002). Chromogranin A expression in neoplastic cells affects tumor growth and morphogenesis in mouse models. *Cancer Res.* 3, 941–946.

Corti, A., Gasparri, A., Chen, F. X., Pelagi, M., Brandazza, A., Sidoli, A., et al. (1996). Characterisation of circulating Chromogranin A in human cancer patients. *Br. J. Cancer.* 8, 924–932. doi: 10.1038/bjc.1996.183

Crippa, L., Bianco, M., Colombo, B., Gasparri, A. M., Ferrero, E., Loh, Y. P., et al. (2013). A new Chromogranin A-dependent angiogenic switch activated by thrombin. *Blood* 2, 392–402. doi: 10.1182/blood-2012-05-430314

Cryer, P. E., Wortsman, J., Shah, S. D., Nowak, R. M., and Deftos, L. J. (1991). Plasma Chromogranin A as a marker of sympathochromaffin activity in humans. *Am. J. Physiol.* 260 (Pt 21), E243–E246.

D'amico, M. A., Ghinassi, B., Izzicupo, P., Manzoli, L., and Di Baldassarre, A. (2014). Biological function and clinical relevance of Chromogranin A and derived peptides. *Endocr. Connect.* 2, 45–54. doi: 10.1530/EC-14-0027

Dieplinger, B., Gegenhuber, A., Haltmayer, M., and Mueller, T. (2009). Evaluation of novel biomarkers for the diagnosis of acute destabilised heart failure in patients with shortness of breath. *Heart* 18, 1508–1513. doi: 10.1136/hrt.2009.170696

Dondossola, E., Gasparri, A. M., Colombo, B., Sacchi, A., Curnis, F., and Corti, A. (2011). Chromogranin A restricts drug penetration and limits the ability of NGR-TNF to enhance chemotherapeutic efficacy. *Cancer Res.* 17, 5881–5890. doi: 10.1158/0008-5472.CAN-11-1273

Ferrero, E., Magni, E., Curnis, F., Villa, A., Ferrero, M. E., and Corti, A. (2002). Regulation of endothelial cell shape and barrier function by Chromogranin A. *Ann. N.Y. Acad. Sci.* 971, 355–358. doi: 10.1111/j.1749-6632.2002.tb04495.x

Ferrero, E., Scabini, S., Magni, E., Foglieni, C., Belloni, D., Colombo, B., et al. (2004). Chromogranin A protects vessels against tumor necrosis factor alpha-induced vascular leakage. *FASEB J.* 3, 554–556. doi: 10.1096/fj.03-0922fje

Fung, M. M., Salem, R. M., Mehtani, P., Thomas, B., Lu, C. F., Perez, B., et al. (2010). Direct vasoactive effects of the Chromogranin A (CHGA) peptide catestatin in humans *in vivo*. *Clin. Exp. Hypertens.* 5, 278–287. doi: 10.3109/10641960903265246

Giusti, M., Sidoti, M., Augeri, C., Rabitti, C., and Minuto, F. (2004). Effect of short-term treatment with low dosages of the proton-pump inhibitor omeprazole on serum Chromogranin A levels in man. *Eur. J. Endocrinol.* 3, 299–303. doi: 10.1530/eje.0.1500299

Glattard, E., Angelone, T., Strub, J. M., Corti, A., Aunis, D., Tota, B., et al. (2006). Characterization of natural vasostatin-containing peptides in rat heart. *FEBS J.* 14, 3311–3321. doi: 10.1111/j.1742-4658.2006.05334.x

Goetze, J. P., Alehagen, U., Flyvbjerg, A., and Rehfeld, J. F. (2014). Chromogranin A as a biomarker in cardiovascular disease. *Biomark Med.* 1, 133–140. doi: 10.2217/bmm.13.102

Hartell, N. A., Archer, H. E., and Bailey, C. J. (2005). Insulin-stimulated endothelial nitric oxide release is calcium independent and mediated via protein kinase B. *Biochem. Pharmacol.* 5, 781–790. doi: 10.1016/j.bcp.2004.11.022

Hausenloy, D. J. (2009). Signalling pathways in ischaemic postconditioning. *Thromb Haemost* 4, 626–634. doi: 10.1160/TH08-11-0734

Hausenloy, D. J. (2012). Conditioning the heart to prevent myocardial reperfusion injury during PPCI. *Eur. Heart. J. Acute Cardiovasc. Care* 1, 13–32. doi: 10.1177/2048872612438805

Hausenloy, D. J., Lecour, S., and Yellon, D. M. (2011). Reperfusion injury salvage kinase and survivor activating factor enhancement prosurvival signaling pathways in ischemic postconditioning: two sides of the same coin. *Antioxid Redox Signal* 5, 893–907. doi: 10.1089/ars.2010.3360

Helle, K. B., Corti, A., Metz-Boutigue, M. H., and Tota, B. (2007). The endocrine role for Chromogranin A: a prohormone for peptides with regulatory properties. *Cell. Mol. Life Sci.* 22, 2863–2886. doi: 10.1007/s00018-007-7254-0

Ho-Tin-Noé, B., Enslen, H., Doeuvre, L., Corsi, J. M., Lijnen, H. R., and Anglés-Cano, E. (2009). Role of plasminogen activation in neuronal organization and survival. *Mol.Cell. Neurosci.* 4, 288–295. doi: 10.1016/j.mcn.2009.08.001

Hove-Madsen, L., Méry, P. F., Jurevicius, J., Skeberdis, A. V., and Fischmeister, R. (1996). Regulation of myocardial calcium channels by cyclic AMP metabolism. *Basic Res. Cardiol.* 2, 1–8. doi: 10.1007/BF00795355

Hutton, J. C., Davidson, H. W., Grimaldi, K. A., and Peshavaria, M. (1987). Biosynthesis of betagranin in pancreatic beta-cells. Identification of a Chromogranin A-like precursor and its parallel processing with proinsulin. *Biochem. J.* 2, 449–456.

Jansson, A. M., Røsjø, H., Omland, T., Karlsson, T., Hartford, M., Flyvbjerg, A., et al. (2009). Prognostic value of circulating Chromogranin A levels in acute coronary syndromes. *Eur. Heart J.* 1, 25–32. doi: 10.1093/eurheartj/ehn513

Katz, A. M. (1990). Inotropic and lusitropic abnormalities in heart failure. *Eur. Heart J.* 11 (Suppl. A), 27–31. doi: 10.1093/eurheartj/11.suppl_A.27

Kim, T., and Loh, Y. P. (2006). Protease nexin-1 promotes secretory granule biogenesis by preventing granule protein degradation. *Mol. Biol. Cell.* 2, 789–798. doi: 10.1091/mbc.E05-08-0755

Kim, T., Tao-Cheng, J. H., Eiden, L. E., and Loh, Y. P. (2001). Chromogranin A, an "on/off" switch controlling dense-core secretory granule biogenesis. *Cell.* 4, 499–509. doi: 10.1016/S0092-8674(01)00459-7

Koeslag, J. H., Saunders, P. T., and Wessels, J. A. (1999). The chromogranins and the counter-regulatory hormones: do they make homeostatic sense? *J. Physiol.* 3, 643–649. doi: 10.1111/j.1469-7793.1999.0643s.x

Koshimizu, H., Cawley, N. X., Yergy, A. L., and Loh, Y. P. (2011). Role of pGlu-serpinin, a novel Chromogranin A-derived peptide in inhibition of cell death. *J. Mol. Neurosci.* 2, 294–303. doi: 10.1007/s12031-011-9521-7

Koshimizu, H., Kim, T., Cawley, N. X., and Loh, Y. P. (2010). Reprint of: Chromogranin A: a new proposal for trafficking, processing and induction of granule biogenesis. *Regul. Pept.* 1, 95–101. doi: 10.1016/j.regpep.2010.09.006

Krylova, M. I. (2007). Chromogranin A: immunocytochemical localization in secretory granules of frog atrial cardiomyocytes. *Tsitologiia* 7, 538–543.

Kubota, M. (1997). Vasopressin: basic and clinical aspects. *Nihon Shinkei Seishin Yakurigaku Zasshi* 3, 113–121.

Mahapatra, N. R., O'Connor, D. T., Vaingankar, S. M., Hikim, A. P., Mahata, M., Ray, S., et al. (2005). Hypertension from targeted ablation of Chromogranin A can be rescued by the human ortholog. *J. Clin. Invest.* 7, 1942–1952. doi: 10.1172/JCI24354

Mahata, S. K., Kozak, C. A., Szpirer, J., Szpirer, C., Modi, W. S., Gerdes, H. H., et al. (1996). Dispersion of chromogranin/secretogranin secretory protein family loci in mammalian genomes. *Genomics* 1, 135–139. doi: 10.1006/geno.1996.0171

Mahata, S. K., Mahata, M., Fung, M. M., and O'Connor, D. T. (2010). Reprint of: catestatin: a multifunctional peptide from Chromogranin A. *Regul. Pept.* 1, 52–62. doi: 10.1016/j.regpep.2010.09.007

Mahata, S. K., O'Connor, D. T., Mahata, M., Yoo, S. H., Taupenot, L., Wu, H., et al. (1997). Novel autocrine feedback control of catecholamine release. A discrete Chromogranin A fragment is a noncompetitive nicotinic cholinergic antagonist. *J. Clin. Invest.* 6, 1623–1633. doi: 10.1172/JCI119686

Maniatis, N. A., Brovkovych, V., Allen, S. E., John, T. A., Shajahan, A. N., Tiruppathi, C., et al. (2006). Novel mechanism of endothelial nitric oxide synthase activation mediated by caveolae internalization in endothelial cells. *Circ. Res.* 8, 870–877. doi: 10.1161/01.RES.0000245187.08026.47

Matsuda, N., Jesmin, S., Takahashi, Y., Hatta, E., Kobayashi, M., Matsuyama, K., et al. (2004). Histamine H1 and H2 receptor gene and protein levels are differentially expressed in the hearts of rodents and humans. *J. Pharmacol. Exp. Ther.* 2, 786–795. doi: 10.1124/jpet.103.063065

Mattiazzi, A., Vittone, L., and Mundiña-Weilenmann, C. (2007). C.Ca2_/calmodulin-dependent protein kinase: a key component in the contractile recovery from acidosis. *Cardiovasc. Res.* 73, 648–656. doi: 10.1016/j.cardiores.2006.12.002

Metz-Boutigue, M. H., Garcia-Sablone, P., Hogue-Angeletti, R., and Aunis, D. (1993). Intracellular and extracellular processing of Chromogranin A. Determination of cleavage sites. *Eur. J. Biochem.* 1, 247–257. doi: 10.1111/j.1432-1033.1993.tb18240.x

Miserez, B., Annaert, W., Dillen, L., Aunis, D., and De Potter, W. (1992). Chromogranin A processing in sympathetic neurons and release of Chromogranin A fragments from sheep spleen. *FEBS Lett.* 2, 122–124. doi: 10.1016/0014-5793(92)80956-H

Montero-Hadjadje, M., Vaingankar, S., Elias, S., Tostivint, H., Mahata, S. K., and Anouar, Y. (2008). Chromogranins A and B and secretogranin II: evolutionary and functional aspects. *Acta Physiol. (Oxf).* 2, 309–324. doi: 10.1111/j.1748-1716.2007.01806.x

Murphy, E., Kohr, M., Menazza, S., Nguyen, T., Evangelista, A., Sun, J., et al. (2014). Signaling by S-nitrosylation in the heart. *Mol. Cell. Cardiol.* 73, 18–25. doi: 10.1016/j.yjmcc.2014.01.003

O'Connor, D. T., Kailasam, M. T., Kennedy, B. P., Ziegler, M. G., Yanaihara, N., and Parmer, R. J. (2002). Early decline in the catecholamine release-inhibitory peptide catestatin in humans at genetic risk of hypertension. *J. Hypertens.* 7, 1335–1345. doi: 10.1097/00004872-200207000-00020

O'Connor, D. T., Pandlan, M. R., Carlton, E., Cervenka, J. H., and Hslao, R. J. (1989). Rapid radioimmunoassay of circulating Chromogranin A: *in vitro* stability, exploration of the neuroendocrine character of neoplasia, and assessment of the effects of organ failure. *Clin. Chem.* 35, 1631–1637.

O'Connor, D. T., Zhu, G., Rao, F., Taupenot, L., Fung, M. M., Das, M., et al. (2008). Heritability and genome-wide linkage in US and australian twins identify novel genomic regions controlling Chromogranin A: implications for secretion and blood pressure. *Circulation* 3, 247–257. doi: 10.1161/CIRCULATIONAHA.107.709105

Omland, T., Dickstein, K., and Syversen, U. (2003). Association between plasma Chromogranin A concentration and long-term mortality after myocardial infarction. *Am. J. Med.* 1, 25–30. doi: 10.1016/S0002-9343(02)01425-0

Ozawa, K., Whalen, E. J., Nelson, C. D., Mu, Y., Hess, D. T., Lefkowitz, R. J., et al. (2008). S-nitrosylation of beta-arrestin regulates beta-adrenergic receptor trafficking. *Mol. Cell.* 3, 395–405. doi: 10.1016/j.molcel.2008.05.024

Pagliaro, P., Chiribiri, A., Mancardi, D., Rastaldo, R., Gattullo, D., and Losano, G. (2003). Coronary endothelial dysfunction after ischemia and reperfusion and its prevention by ischemic preconditioning. *Ital Heart J.* 6, 383–394.

Pagliaro, P., Moro, F., Tullio, F., Perrelli, M. G., and Penna, C. (2011). Cardioprotective pathways during reperfusion: focus on redox signaling and other modalities of cell signaling. *Antioxid Redox Signal* 14, 833–850. doi: 10.1089/ars.2010.3245

Parmer, R. J., Mahata, M., Gong, Y., Mahata, S. K., Jiang, Q., O'Connor, D. T., et al. (2000). Processing of Chromogranin A by plasmin provides a novel mechanism for regulating catecholamine secretion. *J. Clin. Invest.* 7, 907–915. doi: 10.1172/JCI7394

Pasqua, T., Corti, A., Gentile, S., Pochini, L., Bianco, M., Metz-Boutigue, M. H., et al. (2013). Full-length human chromogranin-A cardioactivity: myocardial, coronary, and stimulus-induced processing evidence in normotensive and hypertensive male rat hearts. *Endocrinology* 9, 3353–3365. doi: 10.1210/en.2012-2210

Penna, C., Alloatti, G., Gallo, M. P., Cerra, M. C., Levi, R., Tullio, F., et al. (2010). Catestatin improves post-ischemic left ventricular function and decreases ischemia/reperfusion injury in heart. *Cell Mol. Neurobiol.* 8, 1171–1179. doi: 10.1007/s10571-010-9598-5

Penna, C., Angotti, C., and Pagliaro, P. (2014b). Protein S-nitrosylation in preconditioning and postconditioning. *Exp. Biol. Med.* 239, 647–662. doi: 10.1177/1535370214522935

Penna, C., Mancardi, D., Rastaldo, R., and Pagliaro, P. (2009). Cardioprotection: a radical view Free radicals in pre and postconditioning. *Biochim. Biophys. Acta.* 7, 781–793. doi: 10.1016/j.bbabio.2009.02.008

Penna, C., Pasqua, T., Amelio, D., Perrelli, M. G., Angotti, C., Tullio, F., et al. (2014a). Catestatin increases the expression of anti-apoptotic and pro-angiogenetic factors in the post-ischemic hypertrophied heart of SHR. *PLoS ONE* 9:e102536. doi: 10.1371/Journal.pone.0102536

Penna, C., Tullio, F., Perrelli, M. G., Mancardi, D., and Pagliaro, P. (2012). Cardioprotection against ischemia/reperfusion injury and Chromogranin A-derived peptides. *Curr. Med. Chem.* 24, 4074–4085. doi: 10.2174/0929867128 02429966

Perrelli, M. G., Tullio, F., Angotti, C., Cerra, M. C., Angelone, T., Tota, B., et al. (2013). P.Catestatin reduces myocardial ischaemia/reperfusion injury: involvement of PI3K/Akt, PKCs, mitochondrial KATP channels and ROS signalling. *P. Pflugers Arch.* 7, 1031–1040. doi: 10.1007/s00424-013-1217-0

Pieroni, M., Corti, A., Tota, B., Curnis, F., Angelone, T., Colombo, B., et al. (2007). Myocardial production of Chromogranin A in human heart: a new regulatory peptide of cardiac function. *Eur. Heart J.* 9, 1117–1127. doi: 10.1093/eur-heartj/ehm022

Ramella, R., Boero, O., Alloatti, G., Angelone, T., Levi, R., and Gallo, M. P. (2010). Vasostatin 1 activates eNOS in endothelial cells through a proteoglycan-dependent mechanism. *J. Cell Biochem.* 110, 70–79. doi: 10.1002/jcb.22510

Reddy, A. G., Mishra, S. K., Prakash, V. R., Tandan, S. K., Tripathi, H. C., and Varshney, V. P. (1999). Evaluation of the activity of receptor-operated Ca2$^+$

channels in rat portal vein in induced hyperthyroidism. *Indian J. Physiol. Pharmacol.* 3, 389–392.

Samuels, M. A. (2007). The brain-heart connection. *Circulation* 1, 77–84. doi: 10.1161/CIRCULATIONAHA.106.678995

Satoh, K., Matsu-Ura, T., Enomoto, M., Nakamura, H., Michikawa, T., and Mikoshiba, K. (2011). Highly cooperative dependence of sarco/endoplasmic reticulum calcium ATPase(SERCA) 2a pump activity on cytosolic calcium in living cells. *J. Biol. Chem.* 286, 20591–20599. doi: 10.1074/jbc.M110. 204685

Schmidt, A. G., Edes, I., and Kranias, E. G. (2001). Phospholamban: a promising therapeutic target in heart failure? *Cardiovasc Drugs Ther.* 15, 387–396. doi: 10.1023/A:1013381204658

Shaul, P. W., Afshar, S., Gibson, L. L., Sherman, T. S., Kerecman, J. D., Grubb, P. H., et al. (2002). Developmental changes in nitric oxide synthase isoform expression and nitric oxide production in fetal baboon lung. *Am. J. Physiol. Lung. Cell Mol. Physiol.* 6, L1192–L1199.

Sivaraman, V., Mudalagiri, N. R., Di Salvo, C., Kolvekar, S., Hayward, M., Yap, J., et al. (2007). Postconditioning protects human atrial muscle through the activation of the RISK pathway. *Basic Res. Cardiol.* 5, 453–459. doi: 10.1007/s00395-007-0664-1

Steiner, H. J., Weiler, R., Ludescher, C., Schmid, K. W., and Winkler, H. (1990). Chromogranins A and B are co-localized with atrial natriuretic peptides in secretory granules of rat heart. *J. Histochem. Cytochem.* 6, 845–850. doi: 10.1177/38.6.2139887

Takiyyuddin, M. A., Parmer, R. J., Kailasam, M. T., Cervenka, J. H., Kennedy, B., Ziegler, M. G., et al. (1995). Chromogranin A in human hypertension. Influence of heredity. *Hypertension* 1, 213–220. doi: 10.1161/01.HYP. 26.1.213

Taupenot, L., Harper, K. L., and O'Connor, D. T. (2002). The Chromogranin Secretogranin Family. *N. Engl. J. Med.* 348, 1134–1149. doi: 10.1056/NEJMra021405

Theurl, M., Schgoer, W., Albrecht, K., Jeschke, J., Egger, M., Beer, A. G., et al. (2010). The neuropeptide catestatin acts as a novel angiogenic cytokine via a basic fibroblast growth factor-dependent mechanism. *Circ Res.* 11, 1326–1335. doi: 10.1161/CIRCRESAHA.110.219493

Tota, B., Angelone, T., Mazza, R., and Cerra, M. C. (2008). The Chromogranin A-derived vasostatins: new players in the endocrine heart. *Curr. Med. Chem.* 14, 1444–1451. doi: 10.2174/092986708784567662

Tota, B., Cerra, M. C., and Gattuso, A. (2010). Catecholamines, cardiac natriuretic peptides and Chromogranin A: evolution and physiopathology of a "whip-brake" system of the endocrine heart. *J. Exp. Biol.* 18, 3081–3103. doi: 10.1242/jeb.027391

Tota, B., Gentile, S., Pasqua, T., Bassino, E., Koshimizu, H., Cawley, N. X., et al. (2012). The novel Chromogranin A-derived serpinin and pyroglutaminated serpinin peptides are positive cardiac β-adrenergic-like inotropes. *FASEB J.* 7, 2888–2898. doi: 10.1096/fj.11-201111

Tota, B., Quintieri, A. M., Di Felice, V., and Cerra, M. C. (2007). New biological aspects of Chromogranin A-derived peptides: focus on vasostatins. *Comp. Biochem. Physiol. A Mol. Integr. Physiol.* 1, 11–18. doi: 10.1016/j.cbpa.2006.05.016

Tullio, F., Angotti, C., Perrelli, M. G., Penna, C., and Pagliaro, P. (2013). Redox balance and cardioprotection. *Basic Res. Cardiol.* 108:392. doi: 10.1007/s00395-013-0392-7

Veschini, L., Crippa, L., Dondossola, E., Doglioni, C., Corti, A., and Ferrero, E. (2011). The vasostatin-1 fragment of Chromogranin A preserves a quiescent phenotype in hypoxia-driven endothelial cells and regulates tumor neovascularization. *FASEB J.* 11, 3906–3914. doi: 10.1096/fj.11-182410

Videen, J. S., Mezger, M. S., Chang, Y. M., and O'Connor, D. T. (1992). Calcium and catecholamine interactions with adrenal chromogranins. Comparison of driving forces in binding and aggregation. *J. Biol. Chem.* 5, 3066–3073.

Weiergräber, M., Pereverzev, A., Vajna, R., Henry, M., Schramm, M., Nastainczyk, W., et al. (2000). Immunodetection of alpha1E voltage-gated Ca(2$^+$) channel in chromogranin-positive muscle cells of rat heart, and in distal tubules of human kidney. *J. Histochem. Cytochem.* 6, 807–819. doi: 10.1177/0022155400048 00609

Winkler, H., and Fischer-Colbrie, R. (1992). The chromogranins A and B: the first 25 years and future perspectives. *Neuroscience* 3, 497–528. doi: 10.1016/0306-4522(92)90222-N

Xiao, R. P., Cheng, H., Zhou, Y. Y., Kuschel, M., and Lakatta, E. G. (1999). Recent advances in cardiac beta(2)-adrenergic signal transduction. *Circ. Res.* 11, 1092–1100. doi: 10.1161/01.RES.85.11.1092

Xie, J., Wang, W. Q., Liu, T. X., Deng, M., and Ning, G. (2008). Spatio-temporal expression of Chromogranin A during zebrafish embryogenesis. *J. Endocrinol.* 3, 451–458. doi: 10.1677/JOE-08-0221

Yano, N., Ianus, V., Zhao, T. C., Tseng, A., Padbury, J. F., and Tseng, Y. T. (2007). A novel signaling pathway for beta-adrenergic receptor-mediated activation of phosphoinositide 3-kinase in H9c2 cardiomyocytes. *Am. J. Physiol. Heart Circ. Physiol.* 1, H385–H393. doi: 10.1152/ajpheart.01318.2006

Yoo, S. H., So, S. H., Huh, Y. H., and Park, H. Y. (2002). Inositol 1,4,5-trisphosphate receptor/Ca(2+) channel modulatory role of chromogranins A and B. *Ann. N.Y. Acad. Sci.* 971, 300–310. doi: 10.1111/j.1749-6632.2002.tb04484.x

Zatta, A. J., Kin, H., Lee, G., Wang, N., Jiang, R., Lust, R., et al. (2006). Infarct-sparing effect of myocardial postconditioning is dependent on protein kinase C signalling. *Cardiovasc Res.* 2, 315–324. doi: 10.1016/j.cardiores.2005.11.030

Zhang, K., Chen, Y., Wen, G., Mahata, M., Rao, F., Fung, M. M., et al. (2011). Catecholamine storage vesicles: role of core protein genetic polymorphisms in hypertension. *Curr. Hypertens Rep.* 1, 36–45. doi: 10.1007/s11906-010-0170-y

Conflict of Interest Statement: The authors declare that the research was conducted in the absence of any commercial or financial relationships that could be construed as a potential conflict of interest.

6

Membrane transporters for the special amino acid glutamine: structure/function relationships and relevance to human health

Lorena Pochini†, Mariafrancesca Scalise†, Michele Galluccio and Cesare Indiveri*

Department DiBEST (Biologia, Ecologia, Scienze della Terra) Unit of Biochemistry and Molecular Biotechnology, University of Calabria, Arcavacata di Rende, Italy

Edited by:
Cecila Giulivi, University of
California, Davis, USA

Reviewed by:
Imogen R. Coe, Ryerson University,
Canada
Laurent Counillon, University of
Nice-Sophia Antipolis, France

*Correspondence:
Cesare Indiveri, Department DiBEST
(Biologia, Ecologia, Scienze della
Terra) Unit of Biochemistry and
Molecular Biotechnology, University
of Calabria, via Bucci 4C,
87036 Arcavacata di Rende, Italy
e-mail: cesare.indiveri@unical.it

†These authors have contributed
equally to this work.

Glutamine together with glucose is essential for body's homeostasis. It is the most abundant amino acid and is involved in many biosynthetic, regulatory and energy production processes. Several membrane transporters which differ in transport modes, ensure glutamine homeostasis by coordinating its absorption, reabsorption and delivery to tissues. These transporters belong to different protein families, are redundant and ubiquitous. Their classification, originally based on functional properties, has recently been associated with the SLC nomenclature. Function of glutamine transporters is studied in cells over-expressing the transporters or, more recently in proteoliposomes harboring the proteins extracted from animal tissues or over-expressed in microorganisms. The role of the glutamine transporters is linked to their transport modes and coupling with Na^+ and H^+. Most transporters share specificity for other neutral or cationic amino acids. Na^+-dependent co-transporters efficiently accumulate glutamine while antiporters regulate the pools of glutamine and other amino acids. The most acknowledged glutamine transporters belong to the SLC1, 6, 7, and 38 families. The members involved in the homeostasis are the co-transporters B0AT1 and the SNAT members 1, 2, 3, 5, and 7; the antiporters ASCT2, LAT1 and 2. The last two are associated to the ancillary CD98 protein. Some information on regulation of the glutamine transporters exist, which, however, need to be deepened. No information at all is available on structures, besides some homology models obtained using similar bacterial transporters as templates. Some models of rat and human glutamine transporters highlight very similar structures between the orthologs. Moreover the presence of glycosylation and/or phosphorylation sites located at the extracellular or intracellular faces has been predicted. ASCT2 and LAT1 are over-expressed in several cancers, thus representing potential targets for pharmacological intervention.

Keywords: glutamine, amino acids, nutrients, membrane, transporters, cancer, homology models

INTRODUCTION

The pivotal role of glutamine in cell metabolism of mammals is well recognized. Besides glucose, glutamine represents a primary nutrient for maintenance of body's homeostasis (Bode, 2001; Newsholme et al., 2003; McGivan and Bungard, 2007). Therefore glutamine plasmatic concentration must be regulated and kept constant. Glutamine is the most abundant amino acid: the intracellular concentration ranges from 2 to 20 mM, while the extracellular one ranges from 0.2 to 0.8 mM (Bode, 2001; Newsholme et al., 2003). It is involved in many cell processes, the most important of which are depicted in **Figure 1**. Glutamine is precursor for proteins, amino sugar, purines and pyrimidines synthesis in different tissues; it is essential for acid-base buffering in kidney, being the most important donor of NH_3 excreted in urine (Busque and Wagner, 2009; Verrey et al., 2009). Interestingly, in kidney and liver the carbon skeleton of glutamine enters the TCA cycle participating to gluconeogenesis. The amount of glucose produced through this pathway accounts for 25% of the circulating glucose which increases in case of diabetes (Stumvoll et al., 1999; Curi et al., 2005; Daye and Wellen, 2012). Glutamine represents fuel also for intestine, where it plays the additional role in maintaining gut integrity. Several reports show that glutamine

Abbreviations: TCA, tricarboxylic acid cycle; MeAIB, methylaminoisobutyrate; BCH, 2-aminobicyclo(2,2,1)heptane-2-carboxylic acid; SLC, Solute Carrier transporters; mTOR, mammalian target of rapamycin; FXR, farnesoid X receptor; RXR, 9-cis-retinoic acid receptor; IR-1, inverted repeat 1; EGF, epidermal growth factor; PKC, protein kinase C; MEK, tyrosine/threonine kinase; IGF, insulin-like growth factor; PI3K, phosphoinositide 3-kinase; SGK, serum/glucocorticoid regulated kinase; PKB, protein kinase B; MCT, lactate membrane transporter; ACE2, angiotensin-converting enzyme 2; APN, aminopeptidase N/CD13; CATs, cationic amino acid transporters; LATs, light subunits of amino acid transporters; HATs, heterodimeric amino acid transporters; TMS, trans-membrane domain; IPTG, isopropyl β-D-1-thiogalactopyranoside; DHT, dihydrotestosterone; ERK1/2, mitogen-activated protein kinase; APC, amino acid polyamine-organo cation family; EAAT1, excitatory amino acid transporter SLC1A3; EAAT2, excitatory amino acid transporter SLC1A2; PKA, protein kinase A; GLS1, glutaminase 1; OCTN, organic cation transporter novel; SERT, serotonine transporter; RMSD, Root Mean Square Deviation.

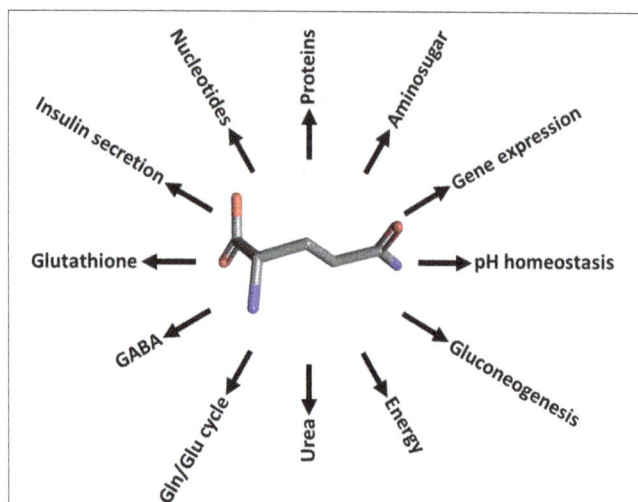

FIGURE 1 | The glutamine roles in cell pathways. Schematic representation of the cell processes involving glutamine. Proteins, protein synthesis; Aminosugar, aminosugar synthesis; Nucleotides, purine and pyrimidine synthesis; pH homeostasis, mainteinance of acid-base balance; Gluconeogenesis, precursor synthesis; Energy, providing carbon atoms for TCA; Urea, release of NH_3 in liver for urea synthesis; Gln/Glu cycle and GABA, neurotransmission regulation; Glutathione, GSH synthesis and redox balance regulation; Insulin secretion, glucose concentration regulation; Gene expression, gene expression regulation.

administration helps recovery of intestinal integrity from pathological conditions (Ziegler et al., 2000). In liver glutamine provides nitrogen atoms for the urea cycle, a metabolic pathway compartmentalized between cytosol and mitochondria (Indiveri et al., 1994). In brain, besides a role as scavenger of NH_3, glutamine participates to the glutamine/glutamate cycle, in which glutamine is synthesized from glutamate reabsorbed from synaptic cleft (Broer and Brookes, 2001; Bak et al., 2006; Conti and Melone, 2006). Glutamine is involved in the maintenance of redox potential balance in terms of GSH/GSSG ratio. GSH synthesis needs glycine, cysteine and glutamate, the latter deriving from glutamine through glutaminase action (Shanware et al., 2011; Daye and Wellen, 2012); moreover, glutamine entering the TCA cycle, gives rise to reducing equivalents, NADPH, used to increase GSH/GSSG ratio (Curi et al., 2005). Glutamine is linked to insulin secretion by pancreatic β–cells where it stimulates glucose oxidation with consequent increase of ATP/ADP ratio. This event induces membrane depolarization and results in increase of the intracellular Ca^{2+} level and, thus, insulin release (Newsholme et al., 2003). Lastly, an important role for glutamine is the regulation of several genes involved in signal transduction, metabolism, cell proliferation, cell defense and repair (Curi et al., 2005) and refs herein) (Figure 1). Taken together, these information highlight that glutamine has both acute and chronic effects on cell metabolism and functions. Therefore, glutamine must be considered "essential" even though it can be synthesized endogenously. Indeed, patients with critical illness (HIV infections, muscle disorders and side effects of chemotherapy) take advantage of glutamine supplementation. On the contrary, some concerns on glutamine-rich diet in healthy individual have been

recently raised. Thus, further studies are needed to clarify these points (Holecek, 2013). The significance of glutamine for cell metabolism is proven by its role also in pathological conditions: cancer cells, in fact, have a special need for amino acids, such as glutamine, arginine and leucine which are those mainly required as reported in Section "Glutamine transporters in human pathology." From the depicted scenario it emerges that the processes of glutamine absorption from diet, renal reabsorption and delivery to different tissues are essential and must be well coordinated. Therefore, different transport systems should work in a concerted manner to allow this special amino acid performing all its functions.

GLUTAMINE TRANSPORTERS: FROM FUNCTIONAL TO MOLECULAR CLASSIFICATION

Glutamine transport processes are guaranteed by a number of membrane transporters which share specificity for glutamine but show differences in transport modes. The pleiotropic role of glutamine may be the reason why glutamine transporters, belonging to several protein families, are redundant and ubiquitous (Figure 2). These proteins are, in general, specific for several neutral amino acids, therefore their classification, as well as their functional identification, were not unequivocal. The classification of these transporters was originally based on functional properties, such as substrate specificity, ion and pH dependence, kinetics and regulatory properties. This original classification led to cluster in "systems." System A and System L were the first to be so-defined indicating "alanine-preferring" and "leucine-preferring," respectively (Oxender and Christensen, 1963). The systems included one or even more transporters, still not identified as specific proteins, due to obvious difficulties to discriminate a single function in entire cells.

Over the years, a broad classification was adopted which differentiated between Na^+-dependent and Na^+-independent systems (Bode, 2001). Several transporters sharing the specificity for glutamine were described on the basis of specificity toward other amino acids, inhibitors and modes of transport such as System N which exhibits narrow specificity to glutamine, histidine and asparagine (Schioth et al., 2013), System ASC specific for alanine, serine and cysteine and System B0 with broad specificity to neutral amino acids. Among the Na^+-dependent transporters, the best described are the transporters belonging to Systems A and N, the ASCT2 transporter which belongs to System ASC, B0AT1 belonging to system B0.

Among the Na^+-independent transporters, LAT1 and LAT2 which belong to system L, are the best characterized. Some other less known transporters have been described whose relationship with glutamine homeostasis is, however, not unequivocal. These minor transporters will be dealt with to some extent in the sections below. A further feature used to distinguish different transporters was the high (such as systems N) or low (such as ASCT2) tolerance toward the substitution of Na^+ by Li^+. However, in 2003 Mackenzie et al suggested that Li^+ tolerance is not per se a useful criterion to classify amino acid transporters, since in some instance it depends on membrane potential and on the parameters used in experiments (Mackenzie et al., 2003). Moreover, transporters can be characterized by the different specificity for

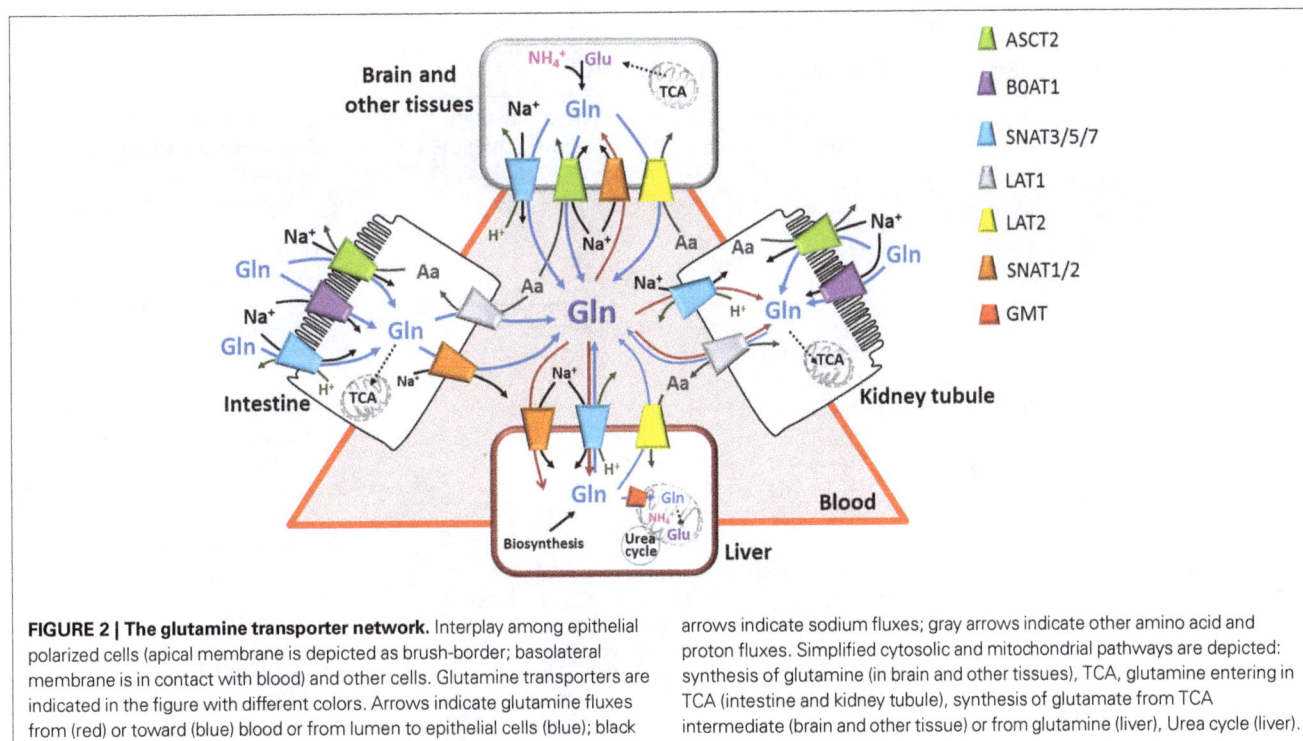

FIGURE 2 | The glutamine transporter network. Interplay among epithelial polarized cells (apical membrane is depicted as brush-border; basolateral membrane is in contact with blood) and other cells. Glutamine transporters are indicated in the figure with different colors. Arrows indicate glutamine fluxes from (red) or toward (blue) blood or from lumen to epithelial cells (blue); black arrows indicate sodium fluxes; gray arrows indicate other amino acid and proton fluxes. Simplified cytosolic and mitochondrial pathways are depicted: synthesis of glutamine (in brain and other tissues), TCA, glutamine entering in TCA (intestine and kidney tubule), synthesis of glutamate from TCA intermediate (brain and other tissue) or from glutamine (liver), Urea cycle (liver).

amino acids and sensitivity to inhibitors; for example, the Systems A and L could be functionally identified by response to the amino acid analog MeAIB and BCH, respectively. Even though inhibitors allow to follow a transport reaction of a specific protein excluding that of other transporters, in the case of amino acid transporters, this criterion was not unequivocal due to the redundancy and the broad specificity of the glutamine transporters in cell.

Upon the identification of genes coding for amino acid transporters and the progress in wide genome sequencing, the classification of the glutamine transporters should be revised taking into account the sequence similarity and the presence of motifs corresponding to specific protein families. The SLC nomenclature may be helpful for a more structured clustering of the transporters. Several reviews already refer to SLC based classification for amino acid transporters (Mackenzie and Erickson, 2004; Hagglund et al., 2011; Bodoy et al., 2013; El-Gebali et al., 2013; Fotiadis et al., 2013; Kanai et al., 2013; Pramod et al., 2013; Rask-Andersen et al., 2013).

Referring to SLC, the classification of glutamine transporters is a little more complicated, since in some instances they are homologous to transporters for amino acids with different chemical properties. For these reasons, due to some discrepancies among similarity in function and similarity in amino acid sequences, the functional classification does not match completely the sequence similarity classification. SLC was previously used for the glutamine transporters by McGivan and Bungard together with the different names associated to the specific transporters (McGivan and Bungard, 2007). **Table 1** resumes the two classifications for the glutamine transporters. The different modes of transport displayed by these transporters determine their roles in the different tissues and in polarized membranes (**Figure 2**). Thus, Na$^+$-dependent symporters perform the function of efficiently

absorbing glutamine in epithelia using the electrochemical gradient; this allows net accumulation of the amino acid against transmembrane gradient. Antiporters perform the function of regulating the balance between pools of glutamine and other amino acids or protons in specific tissues such as the neuronal subtypes; these transporters allow also glutamine efflux from specific tissues. Interaction of some of the glutamine transporters with accessory proteins has been described. In the case of the glutamine transporters LAT1 and LAT2, it seems that the accessory proteins are part of the transport competent complex or at least are important for stabilizing the protein into the plasma membrane (Costa et al., 2013; Rosell et al., 2014). We expect increasing evidences of interaction of the glutamine transporters with other proteins for regulative purposes. Regulation of the glutamine transporters is, indeed, essential for accomplishing its homeostasis, also in response to variations of intracellular and extracellular glutamine levels. However, relatively few information is available on this topic, so far.

Function of the glutamine transporters has been initially studied mainly in cell systems over-expressing the transporters. Significant advancements to the knowledge of the glutamine transporters derived from studies in artificial systems which give the important advantage to exclude interferences ascribed to other transporters and/or cell enzymes. In these systems, native transporters extracted from model animals were firstly reconstituted (Oppedisano et al., 2004; Oppedisano and Indiveri, 2008). The most promising approach is currently represented by the heterologous expression of the human isoforms of transporters in microorganisms. So far, three glutamine transporters have been successfully over- expressed in *E. coli* or in the yeast *P. pastoris*, interestingly, all at the same time (Costa et al., 2013; Galluccio et al., 2013; Pingitore et al., 2013), establishing the basic strategies

Table 1 | The basic characteristics of the glutamine transporters.

Family	Member	System	Aliases	Mechanism	Variant	Regulator/ mode
SLC1	A5	ASC	ASCT2, ATB0	Na^+-glutamine/ neutral amino acids antiport	Isoform1: NP_005619.1 Isoform 2: NP_001138616.1 Isoform 3: NP_001138617.1	Glutamine/protein expression EGF/trafficking and activity Insulin and IGF/activity mTOR/protein expression Leptin/trafficking and gene expression pRb/protein expression Aldosterone/protein expression
SLC6	A14	$B^{(0,+)}$	$ATB^{0,+}$	$2Na^+$-$1Cl^-$-glutamine co-transport (electrogenic)	NP_009162.1	EGF and GH/expression
	A19	B or B^0	B^0AT1	Na^+- glutamine co-transport, (electrogenic)	NP_001003841.1	collectrin (kidney), ACE2/trafficking APN (intestine)/activity and trafficking Leptin/trafficking and gene expression potassium/activity JAK2 (Janus kinase-2)/trafficking PKB-Akt and SGK/trafficking
	A15		B^0AT2	Na^+-glutamine co-transport, (electrogenic)	Isoform 1: NP_877499.1 Isoform 2: NP_060527.2 Isoform 3: NP_001139807.1	
SLC7	A5	L	LAT1	glutamine/ large neutral amino acids antiport	NP_003477.4 XP_006721350.1* XP_006721349.1*	4F2hc/trafficking c-Myc/protein expression EAA/protein expression Glucose/up regulation Aldosterone /protein expression insulin/increases mRNA abundance
	A8		LAT2	glutamine/ small neutral amino acids antiport	Isoform 1: NP_036376.2 Isoform 2: NP_877392.1 Isoform 3: NP_001253965.1 Isoform 4: NP_001253966.1	4F2hc/trafficking Aldosterone/protein expression mTORC1/trafficking DHT/protein expression
	A6		y+LAT2	Na^+ -glutamine /cationic amino acids antiport	NP_001070253.1	4F2hc/ trafficking
SLC38	A1	A	SNAT1, ATA1, SAT1, NAT2	Na^+-glutamine cotransport (electrogenic)	Isoform 1: NP_001265317.1 Isoform 2: NP_001265319.1	
	A2		SNAT2, ATA2, SAT2, SA1	Na^+-glutamine cotransport (electrogenic)	Isoform 1: NP_061849.2 Isoform 2: XP_005269040.1*Isoform 3: BAG57253.1*	DHT/activity glucagon/expression
	A3	N	SNAT3, SN1, NAT	Na^+-glutamine/H^+antiport (electroneutral)	NP_006832.1	Insulin/trafficking PKC/trafficking Manganese/degradation
	A5		SNAT5, SN2	Na^+-glutamine/H^+ antiport (electroneutral)	Isoform 1: NP_277053.2 Isoform 2: XP_005272752.1*	c-myc/expression
	A7		SNAT7	Na^+-glutamine/H^+antiport (electroneutral)	Isoform 1: NP_060701.1 Isoform 2: XP_006721292.1*	

The SLC and System classifications are reported; Mechanism of transport is described; variants reported in GeneBank are listed; known regulators and their effect are indicated.

*Predicted protein.

(**Figure 3**) for further advancement of the knowledge on the glutamine transporter function and structure.

In the following paragraphs an update is given on the glutamine transporter knowledge concerning functional, regulatory and structural advancements and relationships with human health, respect to the last excellent comprehensive reviews on the glutamine transporters (Bode, 2001; McGivan and Bungard, 2007).

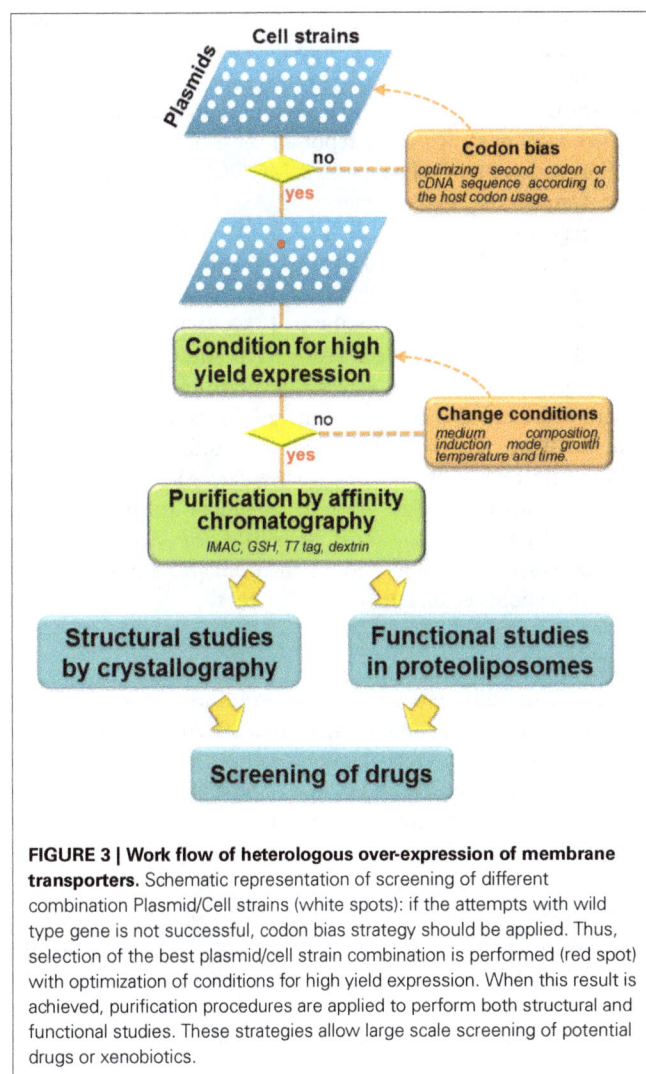

FIGURE 3 | Work flow of heterologous over-expression of membrane transporters. Schematic representation of screening of different combination Plasmid/Cell strains (white spots): if the attempts with wild type gene is not successful, codon bias strategy should be applied. Thus, selection of the best plasmid/cell strain combination is performed (red spot) with optimization of conditions for high yield expression. When this result is achieved, purification procedures are applied to perform both structural and functional studies. These strategies allow large scale screening of potential drugs or xenobiotics.

SLC1A5: ASCT2
GENE AND TISSUE LOCALIZATION

SLC1A5, known as ASCT2, is a glutamine transporter belonging to the *SLC*1 family which encompasses five high affinity glutamate transporters (*SLC1A1, SLC1A2, SLC1A3, SLC1A6, SLC1A7*) and two neutral amino acid transporters (*SLC1A4* and *SLC1A5*). In humans, the five glutamate transporters possess 44–55% amino acid sequence identity with each other, whereas the two neutral amino acid transporters exhibit 57% identity. The human isoform of *SLC1A5* gene, isolated in 1996 from human placenta (Kekuda et al., 1996), is annotated on the chromosome 19q13.3. Three different transcripts exist in GeneBank, deriving from different translation start (Kanai et al., 2013); up to now the functional and kinetic characterizations have been conducted only on the first variant (NM_005628) constituted by 2873 nucleotides and 8 exons (**Table 1**). This transcript encodes a peptide of 541 amino acids. The dbSNP database reports more than 400 SNPs both in coding and non-coding region. Only the variant *SLC1A5-P17A* (rs3027956) is associated with breast cancer (Savas et al., 2006). *SLC1A5* gene is expressed in several tissues: kidney, intestine, brain, lung, skeletal muscle, placenta and pancreas (Kekuda et al., 1996; Utsunomiya-Tate et al., 1996; Broer and Brookes, 2001; Deitmer et al., 2003; Gliddon et al., 2009; Indiveri et al., 2014).

FUNCTION

The acronym ASCT2 stands for AlaSerCys Transporter 2, even though the preferred substrate is glutamine. Indeed, the importance of ASCT2 for human health is linked to the ability to mediate delivery of this amino acid. The transport mechanism of this transporter has been widely studied in different experimental models: cell systems (Torres-Zamorano et al., 1998) as well as proteoliposomes reconstituted with the rat transporter extracted from kidney (Oppedisano et al., 2004, 2007; Pingitore et al., 2013). More recently, the human protein was over-expressed in *P. pastoris* and reconstituted in fully active form in proteoliposomes (Pingitore et al., 2013). This gave a further advancement to the knowledge of the human transporter. All the experimental approaches revealed similar basic properties of the transporter which works as a strictly Na^+-dependent obligatory antiporter of neutral amino acids (**Table 1**). Na^+ cannot be substituted by Li^+. Experiments conducted with radioactive substrates as well as competition data confirmed that also alanine, threonine, serine, leucine, valine, asparagine, methionine, isoleucine, tryptophan, histidine, and phenylalanine are substrates of this transporter. Glutamate, lysine, arginine, MeAIB and BCH are not transported. By using proteoliposomes, important novelties on the human ASCT2 substrate specificity have been revealed. In particular, human ASCT2 showed an asymmetric specificity for amino acids. Glutamine, serine, asparagine, and threonine are bi-directionally transported while alanine, valine and methionine can be only inwardly transported. The functional asymmetry was also confirmed by kinetic analysis: different Km values were measured on the external and internal sides of proteoliposomes, 0.097 and 1.8 mM, respectively (Pingitore et al., 2013). The Na^+_{ex}:amino $acid_{ex}$ stoichiometry of transport is 1:1. The electrical aspect of the transport reaction is, however, not completely clarified: in fact, the inwardly-directed transport of Na^+ should evoke electrical currents, leading to an electrogenic transport (Kekuda et al., 1996, 1997; Torres-Zamorano et al., 1998). Some groups, on the contrary, described an electroneutral transport (Utsunomiya-Tate et al., 1996; Broer et al., 1999). Interestingly, anion conductance activity has been described by electrophysiological measurements (Grewer and Grabsch, 2004). Very recently, using computational and experimental approaches a transport mechanism in which more than one Na^+ ion is involved in amino acid translocation has been suggested. The study showed that even if the amino acid exchange occurs in an electroneutral mode, the transport cycle is indeed electrogenic due to the movement of Na^+ ions across membrane. However, the used experimental model did not allow assay at the intracellular side (Zander et al., 2013). Thus, to gain further insights in the electrical properties of this transporter, a more appropriate model is needed, in which the intracellular environment might be better controlled; this would be useful also for defining the overall catalytic mechanism which is still underneath. The physiological role of hASCT2 has been extensively investigated in different tissue. ASCT2 in brain contributes to the glutamine-glutamate cycle (**Figure 1**) mediating the efflux of glutamine from astrocytes to recover the glutamate released in synaptic cleft (Broer et al., 1999). The same cycle occurs in placenta where glutamine enters the fetal liver to synthesize glutamate, which is then used for fetus metabolism (Torres-Zamorano et al., 1998). An evolutionary role has also been proposed for the human ASCT2 isoform: a group of retroviruses used this transporter as a receptor to infect human cells, giving rise to a co-evolution phenomenon (Marin et al., 2003). Moreover, among them, the human endogenous retroviruses (HERVs) have been detected as transcripts and proteins in the central nervous system. Associations among these retroviruses with syndromes including multiple sclerosis (MS) and several psychiatric disorders have been described. Binding of HERV proteins to ASCT-1 or -2 receptors might reduce the amino acid intake (Antony et al., 2011). One of the most important signaling function reported for ASCT2 relies on the link with the mTOR pathway. mTOR is an highly conserved Ser/Thr kinase that forms two complexes (mTORC1 and mTORC2) by interacting with several other proteins. mTOR pathway integrates signals from five major routes: growth factors, stress, energy status, oxygen and amino acids. In particular, leucine and arginine should be involved in activating mTOR. However, the link with the amino acid signals is still underneath. In this scenario, glutamine taken up by ASCT2 stimulates leucine uptake by a parallel leucine/glutamine antiport catalyzed by LAT1 (Nicklin et al., 2009).

REGULATORY ASPECTS

Several data on regulation of ASCT2 expression is reported both in physiological and pathological conditions, even though the molecular mechanisms of regulation are mostly unknown (**Table 1**). One of the first report dealing with this issue showed that glutamine itself is able to regulate ASCT2 expression in human hepatoma cells (Bungard and McGivan, 2004). In the mentioned study, a region of 907 bp flanking the 5′ of ASCT2 gene has been cloned in the luciferase reporter system. In this system, glutamine strongly induces the promoter activity possibly via FXR/RXR dimer formation. This transcription factor complex binds to an IR-1 repeat of 24 bp in the promoter region of ASCT2. This has been further demonstrated by the site-directed mutagenesis of the IR-1 region. Furthermore, it has been postulated that the expression of FXR, which is linked to glucose availability, may positively regulate glutamine (Bungard and McGivan, 2005). Soon after, other groups proposed that glutamine transport through ASCT2 is stimulated by EGF signaling pathway (Palmada et al., 2005; Avissar et al., 2008). This should involve activation of PKC and protein kinase MEK. The uptake of amino acid is also regulated by insulin and IGF, which activate a cascade involving PI3K and the downstream targets SGK and PKB. The regulation of ASCT2 is due to increase of its abundance in plasma membrane even though ASCT2 does not harbor a canonical consensus sequence for these kinases (Palmada et al., 2005). This is in line with the observation that the transporter trafficking is a PI3K-dependent phenomenon and that the transporter stabilization in membrane is due to the GTPase Rho (Avissar et al., 2008). The link of ASCT2 with mTOR pathway has been suggested in 2007 by Fuchs et al.; the study showed that silencing ASCT2 in hepatoma cells causes reduction of mTOR activity,

leading to apoptosis (Fuchs et al., 2007). Rapamycin decreases ASCT2 expression, indicating a reciprocal effect between this transporter and mTOR activity. The phenomenon underlying this mechanism is still underneath (Fuchs et al., 2007). Interestingly, in 2005 a study by Xu et al, reported the isolation of a supercomplex constituted by LAT1/CD98, MCT/CD147 (lactate membrane transporter) heterodimers and ASCT2, which would be under regulation of AMPK and mTOR (Xu and Hemler, 2005). This would fit with the need of simultaneously regulating such transporters, which are fundamental for energetic metabolism, particularly in cancers (Ganapathy et al., 2009). Down-regulation of ASCT2 expression has been reported in intestine by the multifunction hormone leptin, which is involved in modulating the activity of several transporters of energy rich molecules (Ducroc et al., 2010). Lastly, the tumor suppressor pRb via E2F-3 transcription factor has been shown to regulate ASCT2 expression as well as other proteins involved in glutamine metabolism. Interestingly, in cancer the constitutive pRb degradation leads to continuous activation of E2F-3 and, then, enhanced expression of ASCT2 (Reynolds et al., 2014).

STRUCTURAL ASPECTS

In spite of the different physiological roles in mammals, *SLC1* family members share common structural features. The tridimensional structure of the archeal homolog of the *SLC1* family, Gltph from *P. horikoshii*, was solved by X-ray crystallography (Yernool et al., 2004). The structure of Gltph has a trimeric architecture forming an aqueous cavity. In 2007 the same group solved the structure in presence of substrate and of a competitive inhibitor (Boudker et al., 2007), opening the possibility of docking analysis with substrates and inhibitors. Each protomer is an independent transport unit and is formed by eight transmembrane domains and two hairpin loops (HP1 and HP2) which are thought to be responsible of substrate translocation, being the mobile part of the transporter (Boudker et al., 2007; Reyes et al., 2009).

The first homology model of rat ASCT2 has been constructed in 2010 (Oppedisano et al., 2010), followed by others which confirmed the putative structure of the glutamine transporter (Albers et al., 2012; Zander et al., 2013). More recently the human ortholog has been modeled (Pingitore et al., 2013). Homology models of human and rat ASCT2 are shown in **Figure 4**. Despite the 79% sequence identity between the two orthologous proteins, a very low identity degree (14%) has been found in local stretches (aa 200–229) as shown by the model (**Figure 4A**). The localization of this region is predicted in the vicinity of the substrate binding site (Oppedisano et al., 2010). Two putative glycosylation sites are reported in the homology model (N163, N212 for human and N164, N215 for rat) which are conserved and exposed toward the extracellular face, according with the proposed orientation of the protein in membrane (**Figure 4B**).

Another relevant difference between the two orthologs is the number of cysteines. 8 Cys residues are conserved in both the proteins, while the rat isoform contains additional 8 Cys residues. Two of these residues (C207 and C210) form, in the rat ASCT2, a CXXC metal binding motif missing in the human isoform. Function/structure relationships studied in proteoliposomes highlighted the importance of Cys residues in

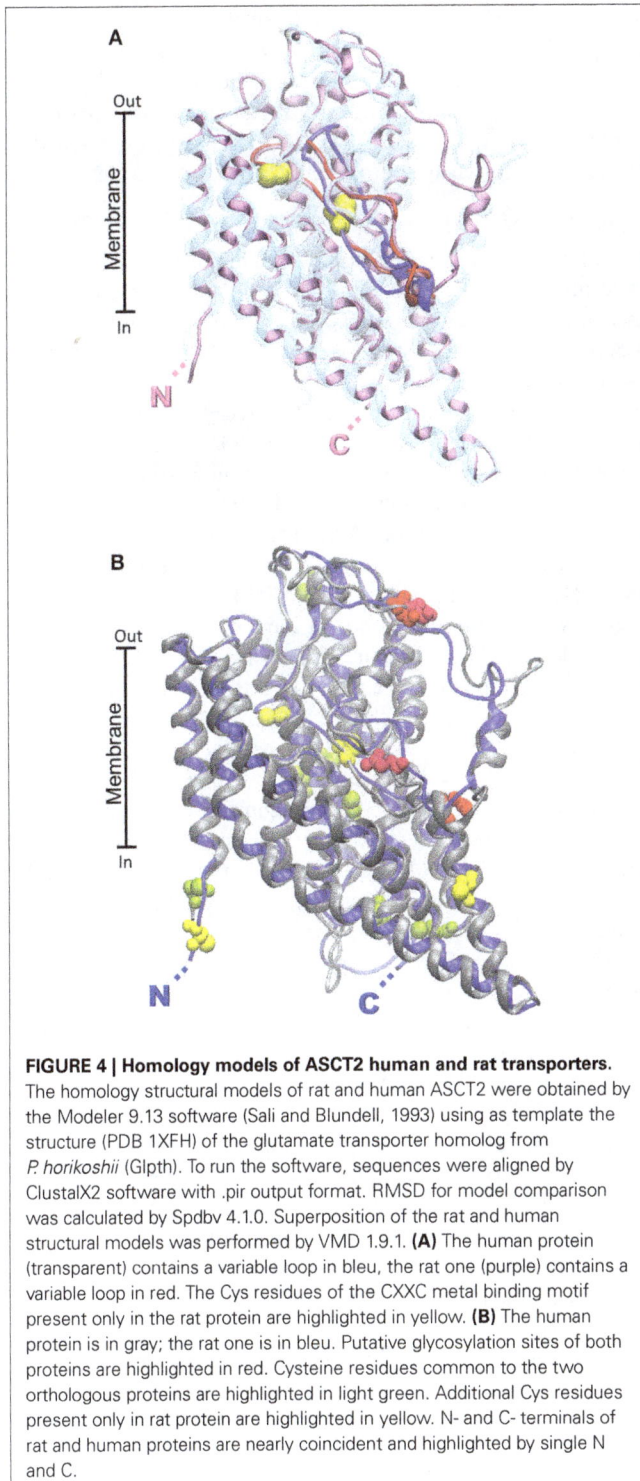

FIGURE 4 | Homology models of ASCT2 human and rat transporters.
The homology structural models of rat and human ASCT2 were obtained by
the Modeler 9.13 software (Sali and Blundell, 1993) using as template the
structure (PDB 1XFH) of the glutamate transporter homolog from
P. horikoshii (Glpth). To run the software, sequences were aligned by
ClustalX2 software with .pir output format. RMSD for model comparison
was calculated by Spdbv 4.1.0. Superposition of the rat and human
structural models was performed by VMD 1.9.1. **(A)** The human protein
(transparent) contains a variable loop in bleu, the rat one (purple) contains a
variable loop in red. The Cys residues of the CXXC metal binding motif
present only in the rat protein are highlighted in yellow. **(B)** The human
protein is in gray; the rat one is in bleu. Putative glycosylation sites of both
proteins are highlighted in red. Cysteine residues common to the two
orthologous proteins are highlighted in light green. Additional Cys residues
present only in rat protein are highlighted in yellow. N- and C- terminals of
rat and human proteins are nearly coincident and highlighted by single N
and C.

substrate translocation which is potently inhibited by specific
Cys reagents. Interestingly, ASCT2 harbors the two hairpins,
HP1 and HP2, which are important for substrate translocation
(Pingitore et al., 2013). This finding suggests that the transloca-
tion mechanism may be similar to that of Glpth, even though
ASCT2 works by an antiport mode of transport. Therefore,

additional constraints are needed for coupling substrate transport
in opposed directions.

SLC6A19: B0AT1
GENE AND TISSUE LOCALIZATION

SLC6A19 belongs to the *SLC6* also called Na^+-dependent neurotransmitter transporter family in which only one out of 21
members is a well characterized glutamine transporter. Few other
members of the family recognize glutamine but with lower affinity respect to other substrates (Verrey et al., 2005; Broer, 2006;
Pramod et al., 2013).

B0AT1 was identified and cloned from mouse cDNA library;
it shows all properties of the system previously defined as B0
(Broer et al., 2004). The transporter localization was found in
proximal tubules and intestinal microvillus. B0AT1 was then recognized as one of the major transporters of neutral amino acids in
these epithelia (Bohmer et al., 2005; Camargo et al., 2005; Broer
et al., 2011) (**Figure 2**). Also in humans, B0AT1 has been localized in kidney and intestine. Lower, but significant expression has
also been found in skin, pancreas, prostate, stomach, liver (Kleta
et al., 2004; Seow et al., 2004). The genomic localization of the
mouse B0AT1 is on chromosome 13 in cytoband C1, syntenic to
human chromosome 5p15. The human gene is constituted by 12
exons. Two transcripts are reported in the data bases, but only
one (5174 bp) codes for a protein, while the other one (3337 bp)
undergoes nonsense mediated decay (**Table 1**).

FUNCTION

The function of mouse and human B0AT1 was initially studied in
oocytes. The transporter recognizes, besides glutamine, leucine,
cysteine, valine, isoleucine, methionine, phenylalanine, alanine,
serine, and asparagine as main substrates. Threonine, glycine,
proline, histidine, tyrosine, tryptophan and BCH showed also
some affinity for the transporter even though lower than glutamine. While arginine, lysine, aspartate, glutamate and MeAIB
are not substrate. Half saturation constants of the mouse transporter for glutamine, measured by different authors, range from
0.5 mM to more than 3 mM (Broer et al., 2004; Bohmer et al.,
2005). Km of the human transporter was reported to be 0.25 mM
(Souba et al., 1992). However, the Km on the internal side of
the transporter could not be measured in oocytes. Transport catalyzed by B0AT1 is well recognized to be electrogenic. An Hill
coefficient of 1.5 of the Na^+ dependence suggests that either
two Na^+ are co-transported with amino acids or that 1 Na^+ is
co-transported and a seconds second ion binds to an allosteric
site (Broer et al., 2004). Further studies demonstrated a 1:1 stoichiometry (Bohmer et al., 2005; Camargo et al., 2005). The
replacement of Na^+ with Li^+ is not tolerated. It was also suggested that the transporter may provide pathways for Na^+, H^+,
and/or Cl^- that are not thermodynamically coupled to the amino
acid transport indicating not saturable channel-like, as opposed
to carrier-mediated activity (Camargo et al., 2005). This phenomenon is similar to that found for the mitochondrial carriers
but has not a physiological explanation (Tonazzi and Indiveri,
2003) and refs herein). Regarding the mechanism of transport,
either a random model was proposed (Bohmer et al., 2005) or an
ordered mechanism (Camargo et al., 2005) in which Na^+ would

follow substrate binding. The discrepancies among the different studies are caused by differences in the experimental systems, by the confounding presence of multiple transport systems and by the complexity of the entire cell model (Camargo et al., 2005; McGivan and Bungard, 2007; Oppedisano and Indiveri, 2008). Significant clarification and advancement of the knowledge on B0AT1 were provided by the more recent studies of the protein extracted from rat kidney and inserted in proteoliposomes. In this experimental model, the transporter was inserted right side out respect to the cell orientation, thus giving the possibility to gain information without the interferences but ascribable to the cell context. Besides confirming some basic properties, such as substrate, ion specificity and inhibitor sensitivity, the artificial system gave information on the electrical nature of transport. Electrogenicity originates from Na^+ transport coupled to glutamine, while Cl^- has no influence on the electrogenic behavior. Regulation by K^+ was also found (see below). The half-saturation constant for glutamine in proteoliposomes corresponded to a Km measured in oocytes (Broer et al., 2004), i.e., 0.55 mM. It was clarified that the Na^+-glutamine co-transport displays a 1:1 stoichiometry and the transport mechanism is random simultaneous. Interestingly, in proteoliposomes the intracellular Km could be measured. Its value is 2.0 mM indicating an asymmetric nature of the transporter (Oppedisano and Indiveri, 2008; Oppedisano et al., 2011). This data correlates well with the intracellular and extracellular glutamine concentrations. Indeed the concentration inside (about 2 mM) is much higher than that outside the cell (about 0.5 mM) (Cynober, 2002). These data allows to speculate that B0AT1 might, under some condition, catalyze also efflux of glutamine (Oppedisano et al., 2011).

It can be concluded that the absorption of glutamine and other neutral amino acids from intestine and reabsorption in kidney are the primary functions of this transporter (**Figure 2**).

REGULATORY ASPECTS

Study on regulation of B0AT1 is at an advanced stage (**Table 1**). The carboxypeptidase ACE2 and collectrin (non-peptidase homolog of ACE2) have been described as key regulators of intestinal and renal amino acid uptake (Fairweather et al., 2012). Interestingly, collectrin was shown to specifically increase B0AT1 activity probably by enhancing its surface expression (Danilczyk et al., 2006), not the intrinsic transport function. Proteoliposome experiments demonstrated that collectrin has no essential role in the catalysis of glutamine transport by B0AT1 (Oppedisano et al., 2011). B0AT1 forms complexes with APN that increases the affinity of the transporter for substrate up to 2.5-fold besides increasing its surface expression (Fairweather et al., 2012). Leptin modulates intestinal glutamine absorption through reduction of B0AT1 trafficking to the plasma membrane (Ducroc et al., 2010).

Allosteric regulation by K^+ has been described. K^+ exerts a biphasic modulation of the transporter interacting at the intracellular site. Up to 50 mM internal (intracellular) K^+, transport was stimulated while at higher K^+ level it was inhibited. Since the intracellular K^+ level depends on ATP concentration, it can be assessed that the nucleotide indirectly modulates the B0AT1 transporter activity. High ATP level, causing increase in K^+, would impair B0AT1 activity; lower ATP level will stimulate

B0AT1 (Oppedisano and Indiveri, 2008). This regulation is opposed to the regulation by ATP found for the rat ASCT2.

The Janus Kinase 2 enhances the protein abundance in the cell membrane (Bhavsar et al., 2011). PKB/Akt up-regulates *SLC6A19* activity, which may foster amino acid uptake into PKB/Akt-expressing epithelial and tumor cells (Bogatikov et al., 2012).

STRUCTURAL ASPECTS

According to protein similarity, B0AT1 displays LeuT fold. The main properties influencing the substrate affinity are: the presence of the charged amino and carboxyl groups, L-configuration (not D-) of the α-carbon, net neutral charge, size and, to a lesser extent, hydrophobicity of the side-chains (Yamashita et al., 2005). Uncharged amino acids are favored due to the predicted interaction with a hydrophobic pocket. B0AT1 has similar substrate specificity of LeuT with the exception of tryptophan, that is weakly transported only by B0AT1 (Camargo et al., 2005; Yamashita et al., 2005; Broer, 2006; O'mara et al., 2006). Another difference is the co-transport of 2 Na^+ in the case of LeuT, while only 1 Na^+ is co-transported by B0AT1. The role of the seconds second Na^+-binding site in B0AT1 remains unclear. The homology model highlighted a structural asymmetry which correlates with the function and explains the structural basis of the loss of function of some mutants (Broer, 2008, 2009).

More recently, the dopamine transporter from *Drosophila melanogaster* (PDB 4M48) has been crystallized and resolved at 2.95 Å (Penmatsa et al., 2013). The alignment of the B0AT1 with the dopamine transporter has higher identity (33%) respect to the LeuT transporter (18%), thus a more accurate modeling could be obtained. The superposition of the rat and human structural models (0.43 Å RMSD) shows that the two orthologs are very similar (**Figure 5**), as expected from the 87% identity between the human and rat protein and differently from ASCT2 (**Figure 4A**). This similarity correlates well to the similar functional properties of the two orthologs.

Database analysis (ExPASy PROSITE, http://www.expasy.ch/prosite) of regulatory sites within the SLC6A19 amino acid sequence predicted five N-glycosylation sites. Three (N158, N182, N258) are reported in the homology model (**Figure 5**). The others are not present since the region 332–377 containing these residues cannot be modeled. According to the predicted orientation, the glycosylation sites protrude toward the extracellular side. The SLC6A19 sequence contains also a conserved intracellular consensus site (S100) for phosphorylation (**Figure 5**) which is known to regulate a variety of channels and transporters (Bohmer et al., 2010). Moreover, **Figure 5** highlights the presence of two conserved metal binding motifs in the middle of the transporter, the CXXC (containing C200/C203) and the CXXXC motifs (containing C45/C49). The first is located close to the membrane face, far from the hypothetical substrate binding site; the seconds second is in the core of the protein, closer to the substrate binding pocket. The metal binding sites correlate well with the interaction of heavy metals with the transporter, as described below (Oppedisano et al., 2011).

SLC6A14: ATB0,+

SLC6A14 is the molecular counterpart of $ATB^{0,+}$ belonging to the system B^{0+}. The $hATB^{0,+}$ was the first cloned amino

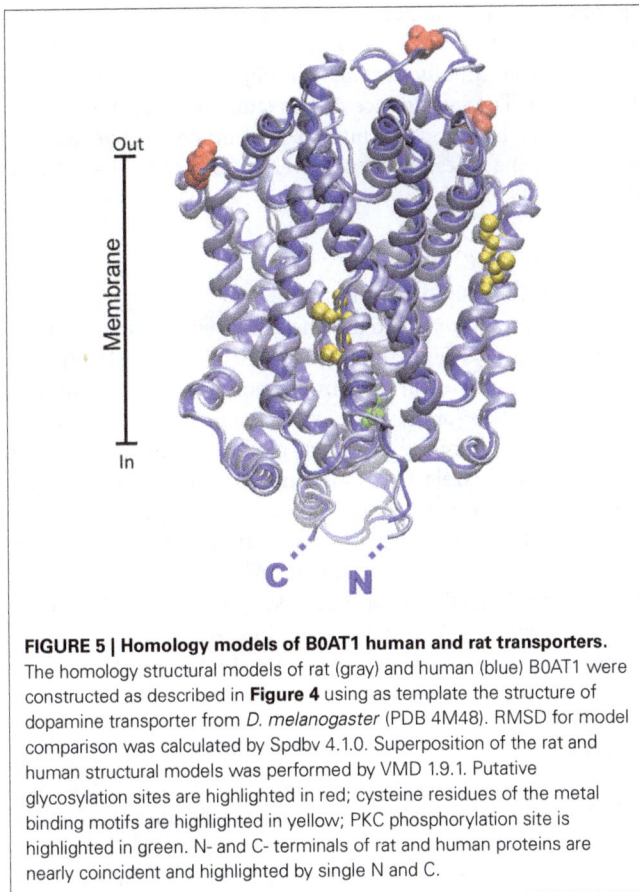

FIGURE 5 | Homology models of B0AT1 human and rat transporters.
The homology structural models of rat (gray) and human (blue) B0AT1 were constructed as described in **Figure 4** using as template the structure of dopamine transporter from *D. melanogaster* (PDB 4M48). RMSD for model comparison was calculated by Spdbv 4.1.0. Superposition of the rat and human structural models was performed by VMD 1.9.1. Putative glycosylation sites are highlighted in red; cysteine residues of the metal binding motifs are highlighted in yellow; PKC phosphorylation site is highlighted in green. N- and C- terminals of rat and human proteins are nearly coincident and highlighted by single N and C.

acid transporter of system B^{0+} (Sloan and Mager, 1999). It is expressed in colon, lung (Sloan et al., 2003), eye (Ganapathy and Ganapathy, 2005) and mammary gland (Nakanishi et al., 2001) (**Table 1**). This is a broad specificity transporter which recognizes glutamine with relatively low affinity (Bode, 2001) Km 0.6 mM (Sloan and Mager, 1999) and other neutral as well as cationic amino acids but also carnitine and carnitine derivatives. Therefore, it can be considered a minor glutamine transporter. Transport catalyzed by $ATB^{0,+}$ is dependent on Na^+ and Cl^- transmembrane gradients and is sensitive to membrane potential. 2 Na^+ and 1 Cl^- are involved in the transport (Nakanishi et al., 2001). As found for ASCT2, $ATB0^{.+}$ may be up-regulated in cancer to meet the increasing demand of arginine (Gupta et al., 2005, 2006), representing a potential target of cancer therapy (Karunakaran et al., 2011). In this sense the selective blocker methyl-DL-tryptophan (α-MT) or the BCH acid might represent an important chemical scaffold for drug design (Sloan and Mager, 1999).

SLC6A15: B0AT2

In 2006 it was demonstrated that the mouse v7-3 gene encodes a transporter for neutral amino acids named B0AT2, due to the high similarity with B0AT1 (Broer, 2006) (**Table 1**). The results indicated that B0AT2 mediates Na^+-dependent uptake of leucine, isoleucine, valine, proline, methionine and with a very low affinity glutamine. The transport is electrogenic with a stoichiometry of 1 Na^+:1 amino acid. Besides kidney, it is highly expressed in cerebellum and brain. This localization indicates that

B0AT2 might be mainly involved in providing neurotransmitter precursors to neurons. Nitrogen from leucine, in fact, is transferred to oxoglutarate to form glutamate; isoleucine, methionine and valine can be metabolized to succynil-CoA for glutamate biosynthesis. Thus, B0AT2 has marginal if any role in glutamine homeostasis.

GLUTAMINE TRANSPORTERS OF THE SLC7 FAMILY

Important glutamine transporters belong to the *SLC7* family which accounts for 13 members divided in two subgroups: the Cationic Amino acid Transporters (CATs) and the Light subunits of Amino acid Transporters (LATs) of the Heterodimeric Amino acid Transporters. Some general information will be given here to highlight the peculiarity of the glutamine transporters belonging to this family. HATs and CATs originate from a common ancestor with 12 trans-membrane domains (TMD) which duplicated the last two TMD originating a 14 TMD structure (Verrey et al., 2004; Hansen et al., 2011; Fotiadis et al., 2013). Only the HATs group includes glutamine transporters. They mediate obligatory antiport of a broad spectrum of substrates (Fotiadis et al., 2013). HATs emerged in metazoan and represent one of the few example of transporters composed by two different subunits. The Light subunit of HATs, LATs, are known as glycoprotein-associated amino acid transporters since they associate with heavy subunit belonging to the SLC3 family via a conserved disulphide bridge (Broer and Brookes, 2001; Wagner et al., 2001; Palacin and Kanai, 2004). This is a very small family of type II (with C-ter outside the cells) membrane glycoproteins comprising SLC3A1 and SLC3A2, named rBAT and 4F2hc, respectively. The main function of SLC3 proteins consists in routing the transporters to plasma membrane forming the HAT complex. 4F2hc (heavy chain of the cell surface antigen 4F2) is a multifunctional protein involved, besides transport, in immuno system regulation, cell activation, growth and adhesion (Palacin and Kanai, 2004; Cantor and Ginsberg, 2012; Fotiadis et al., 2013). Furthermore, 4F2hc has been suggested to play a role in integrin signaling and activation linking this protein to human malignancies (Cantor and Ginsberg, 2012).

So far, six human light subunits have been identified as partners of 4F2hc: LAT1, LAT2, y+LAT1, y+LAT2, asc1 and xCT all belonging to the *SLC7* family. This broad spectrum of interaction justifies the ubiquitous expression of 4F2hc and the absence of pathology linked to its complete loss which is, indeed, not compatible with life (Palacin and Kanai, 2004). Interestingly, an heterodimer complex formed with the glucose transporter GLUT1 has been reported, confirming the regulatory role of 4F2hc in general transport mechanism (Ohno et al., 2011). As stated above, the main interaction between LATs and heavy chain occurs via disulphide formation; this, however, is not essential since mutation of Cys does not abolish completely the presence of HATs in membrane (Pfeiffer et al., 1998; Broer et al., 2001; Palacin and Kanai, 2004; Franca et al., 2005; Rosell et al., 2014). The following paragraph deals with the members of SLC7 involved in glutamine disposition.

SLC7A5: LAT1
GENE AND TISSUE LOCALIZATION

The rat and human cDNAs encoding LAT1 protein have been isolated and identified in 1998 and confirmed in 1999

(Kanai et al., 1998; Prasad et al., 1999). The human gene has been annotated in the chromosome 16q24.3 and is characterized by 10 exons. Three transcripts are reported on ensemble database, but only one protein has been characterized. The longest (4537 bp) codes for a protein of 507 amino acid. Other two isoforms of 241 and 334 amino acids are also predicted (**Table 1**), but without evidence of transport activity. LAT1 is expressed in brain, ovary, testis, placenta, spleen, colon, blood-brain barrier, fetal liver, activated lymphocytes, skeletal muscle, heart, lung, thymus, and kidney (Kanai et al., 1998; Yoon et al., 2005). It is mainly localized in the basolateral membranes of polarized epithelia (Verrey et al., 2004; Fotiadis et al., 2013) (**Figure 2**).

FUNCTION

The first report describing hLAT1 activity (Prasad et al., 1999) conducted in cell systems, revealed that LAT1 works as heterodimer. In several studies it has been assessed that LAT1 mediates obligatory antiport of tryptophan, phenylalanine, leucine and histidine with higher affinity (Km in humans ranging from 5 to 50 μM) (Del Amo et al., 2008), glutamine and threonine with lower affinity; alanine, proline and charged amino acids are not recognized as substrates. The non-metabolizable analog BCH is a transportable inhibitor of LAT1(Mastroberardino et al., 1998). Interestingly, LAT1/4F2hc activity has been linked also to transport of L-dopamine across the blood-brain barrier suggesting a potential role of LAT1 in Parkinson's disease (Kageyama et al., 2000). Transport of thyroid hormones by LAT1 was also described being important in neurological development (Kinne et al., 2011). The covalent interaction between LAT1 and 4F2hc requires the conserved C164 of human LAT1(C 165 in rat) and the conserved C109 of human 4F2hc (C103 in rat) (Fort et al., 2007); either the N-ter and the extracellular parts of 4F2hc are essential for non-covalent interaction as demonstrated by experiments with truncated mutants (Broer et al., 2001; Franca et al., 2005). The transport mode has been described as Na$^+$ and pH independent (**Table 1**); this allows the relative concentration of large neutral amino acids acting synergistically with other Na$^+$-coupled amino acid transporters, such as ASCT2 or unidirectional transporters such as system A and N (Verrey, 2003; Del Amo et al., 2008). LAT1 is also involved in regulation of signaling via mTOR, which is responsible of cell growth and survival. This feature shed light on the link between LAT1 and cancer (see Section "Glutamine transporters in human pathology"). In OMIM databank no pathologies are associated to inherited mutations of LAT1 gene. Further information on unknown functional aspects of human isoform of LAT1 will come from the availability of the large scale over-expressed protein (Galluccio et al., 2013). The achievement of such objective represents a great challenge in protein field since amino acid transporters are known to be toxic for bacteria (Miroux and Walker, 1996). hLAT1 represents, in fact, the only case of a mammalian amino acid transporter expressed in bacteria (Galluccio et al., 2013). A strategy previously used for OCTN subfamily was adopted for LAT1and 4F2hc over-expression (Indiveri et al., 2013) (**Figure 3**). After screening of plasmid and cell combinations, the pH6EX3 plasmid revealed suitable for significant over-expression of the protein in *E. coli* Rosetta DE3pLysS (Brizio et al., 2006; Torchetti et al., 2011).

Optimization of some parameters, such as temperature growth and IPTG concentrations (Brizio et al., 2006) was performed and followed by an affinity chromatography procedures (Galluccio et al., 2013). The importance of this study relies on the unique possibility of handling the human protein alone or in combination with the accessory protein 4F2hc (CD98) to characterize function and kinetics.

SLC7A8: LAT2
GENE AND TISSUE LOCALIZATION

LAT2 gene from human and murine source was reported in 1999 by different groups (Pineda et al., 1999; Segawa et al., 1999). The human gene is located in the chromosome 14q11.2 and encodes for a protein of 535 amino acids with 12 putative TMDs. Human LAT2 is mainly expressed in kidney, placenta, brain and at lower extent in spleen, skeletal muscle, small intestine, lung (Pineda et al., 1999; Fraga et al., 2005; Yoon et al., 2005; Fotiadis et al., 2013) but also in prostate, ovaries, testis and fetal liver (Park et al., 2005). As in the case of LAT1, LAT2 is mainly localized at the basolateral membrane of polarized epithelia cells (**Figure 2**). Alignment with hLAT1 shows 50% identity. Alternative transcripts are annotated in GeneBank coding for three additional but uncharacterized proteins (**Table 1**).

FUNCTION

In cell systems, it has been shown that only the co-expression of hLAT2 and 4F2hc mediates uptake of neutral amino acids which is highly trans-stimulated and Na$^+$-independent. LAT2/4F2hc is inhibited by the amino acid analog BCH as LAT1. The main difference with LAT1 is in the substrate choice: while LAT1 prefers large neutral amino acids, LAT2 has a broader specificity mediating transport of small neutral amino acids such as alanine, glycine and serine as well. Glutamine, for which LAT1 has low affinity, is one of the main substrates of LAT2. However, LAT2 shows a general lower affinity for substrates respect to LAT1, with Km ranging in humans from 0.2 to 1 mM (Del Amo et al., 2008). Differently from LAT1, LAT2 is sensitive to more acidic pH (Christensen, 1990). The expression of LAT2 in brain, liver and skeletal muscle suggests a role in mediating release of glutamine. An important advance in LAT2 functional knowledge came from the successful over-expression in *P. pastoris* of the heterodimer 4F2hc/LAT2 (Costa et al., 2013). Soon after the structural bases of the interaction between the two subunits were revealed. This was confirmed by reconstitution in proteoliposomes of the heterodimer. These experiments indicated that 4F2hc is required for LAT2 stabilization in membrane but not for modulating substrate affinity (Rosell et al., 2014).

REGULATORY ASPECTS OF LAT1 AND LAT2 HETERODIMERS

The role played by LAT1 and LAT2 consists in equilibrating glutamine and other amino acids pool into the cells, while their net uptake derives from other transporters (Del Amo et al., 2008). The study on the regulation of LAT1/4F2hc is of great importance due to the key role played by this protein in several human cancers (See Section "Glutamine transporters in human pathology"). One of the first reports concerning this issue has been conducted in lymphocytes, where the basal expression of

LAT1/4F2hc has been found increased upon activation signals. Studying the promoter regions of the two genes by 5′ RACE analysis and Run on assay suggested that control of LAT1 relates to transcriptional activity not to the half-life of mRNA (Nii et al., 2001). More recently, the LAT1 promoter has been analyzed in human pancreatic cancer cells in which LAT1 plays key function in promoting cell growth. The study revealed that LAT1 promoter harbors a canonical binding sequence for the proto-oncogene c-myc (Hayashi et al., 2012). This protein is an important transcription factor whose expression is often altered in cancers since it regulates expression of several genes controlling metabolism and cell cycle (Daye and Wellen, 2012). Silencing c-myc by siRNA leads to down-regulation of LAT1 expression in prostate cancer cells with subsequent decrease of proliferation, due to impairment of neutral amino acids uptake. Mutations of c-myc binding sequence on LAT1 promoter causes loss of responsiveness to c-myc expression. However, the mechanism of control c-myc-mediated may be more complicated since this protein can exert both epigenetic control and can regulate mRNA translation efficiency by microRNA expression (Hayashi et al., 2012). A recent study described that ingestion of essential amino acids resulted in humans in a transient increase of LAT1/4F2hc heterodimer; this event has been linked to activation of mTORC1 signaling and protein synthesis. The augmented expression might be an adaptive response to long term amino acid availability and/or to anabolic stimulus (Drummond et al., 2010). In line with this, a recent finding shows that embryo of null mice for LAT1 gene has a phenotype incompatible with life (Poncet et al., 2014). Glucose deprivation, due to pathological conditions such ischemia, has been also linked to up-regulation of LAT1 via cis-activation of an E-box on its promoter region in retina (Matsuyama et al., 2012). Low insulin concentration increases *SLC7A5*/LAT1 mRNA abundance in an mTORC1-dependent manner in skeletal muscle cells (Walker et al., 2014). Lastly, LAT1 protein expression has been shown to be increased upon chronic treatment with aldosterone (Amaral et al., 2008). Much less is known about regulation of hLAT2 that responds like LAT1 to aldosterone stimulus (Amaral et al., 2008). A regulation of LAT2/4F2hc expression by mTORC1 has been described in glomerular epithelial cells: in glomerulonephritis conditions, mTORC1 integrates signals from inflammatory cytokines stimulating LAT2 translocation to plasma membrane (Kurayama et al., 2011). Increased surface LAT2 expression by DHT, via EGF receptor involving the ERK1/2 cascade, has been described (Hamdi and Mutungi, 2011). The information above reported are summarized in **Table 1**.

STRUCTURAL ASPECTS OF LAT1 AND LAT2

HATs seem to consist of a light chain (*SLC7A5–11*) with 12 putative TMD with the N- and C-terminals localized intracellularly and an heavy chain with the C-terminal localized extracellularly (Verrey et al., 2004; Costa et al., 2013). As stated above, the 12 TMD of the light chains show considerable similarity to the first 12 TMD of the CAT transporters. The heavy chain 4F2hc is a N-glycosylated protein (4 putative N-glycosylation sites: N264,280,323,405) with one transmembrane domain. The structure of the extracellular domain of human 4F2hc has been

solved by X-ray diffraction (Fort et al., 2007). On the contrary, the structure of human LATs is not known, but Cys-scanning mutagenesis showed that the Cys residue between TMD III and IV is conserved being responsible of inter-subunit disulfide bridge with the heavy chain. LATs show significant similarity to the arginine/agmantine antiporter AdiC, to the broad-specificity amino acid transporter ApcT and to the glutamate/GABA antiporter from *E. coli* whose structures have been solved by X-ray crystallography (Shaffer et al., 2009; Ma et al., 2012). A recent study confirmed this hypothesis for LAT1 suggesting a translocation mechanism in which the role of Na^+ is mimicked by a proton with an alternating transport mechanism (Forrest et al., 2011; Geier et al., 2013). The availability of this model allowed screening and molecular docking of potential anti-tumor drugs, revealing new substrates for LAT1 (Geier et al., 2013). Structural studies on LAT2 have been conducted upon over-expression of the heterodimer in *P. pastoris* with subsequent large scale purification (Costa et al., 2013; Rosell et al., 2014). The study shows, by different experimental approaches, that the extracellular domain of 4F2hc covers the surface of human LAT2 modeled on the basis of AdiC transporter. The docking model fitted the steric hindrance derived by cross-linking experiments and highlighted that the Cys residue involved in the formation of disulfide is the C154 between TMD III and IV. Interestingly, the interaction mode described shed new light on the role of 4F2hc on the path cycle of LAT2: the strong interaction of the heavy chain with a specific domain, named "hash domain," of LAT2 makes this part of the heterodimer the more static (close to TMD 11 and 12), while the "boundle domain" including helices 1, 2, 6, 7, and loops 7–8 is the one moving during the transport cycle opening and closing the transporter (Rosell et al., 2014). This mechanism might be shared also by the other HATs.

Further structural features were already described for LAT1 and LAT2 transporters: potential casein kinase II-dependent phosphorylation site(s), PKC phosphorylation motifs, tyrosine kinase dependent phosphorylation sites (Prasad et al., 1999; Segawa et al., 1999). The alignment of the rat and human LAT2 sequences shows high percentage of identity (92%, not shown). This data has been confirmed by the superimposition of the two structural model (RMSD 0.66) obtained by using the structure of ADiC transporter as template (PDB 3OB6; **Figure 6**). Differently from the other glutamine transporters, Scan-Prosite analysis predicts lack of glycosylation sites. For these transporters the targeting to the membrane is guaranteed by their binding partner (Costa et al., 2013; Rosell et al., 2014). This interaction is mediated by disulfide bridge involving C154 for the human isoform and C155 for the rat isoform exposed toward the extracellular side of the proteins (**Figure 6**). Both the isoforms contain 11 Cys residues. No metal binding motif has been found. On the opposite side the human model contains 3 serine residues (S179, S337, S487), putative PKC phosphorylation sites and a threonine residue (T363), putative PKA phosphorylation site. These sites are exposed toward the cytosolic side of the protein. The rat and human proteins show nearly overlapping predicted structures (**Figure 6**).

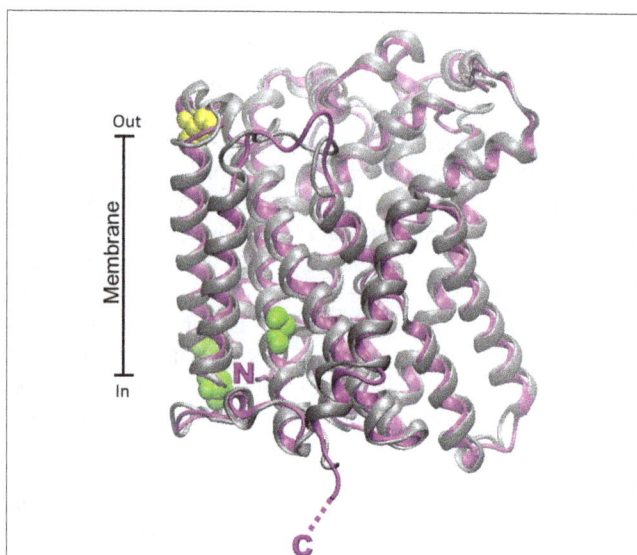

FIGURE 6 | Homology models of LAT2 human and rat transporters. The homology structural models of rat (Chaudhry et al.) and human (purple) LAT2 were constructed as described in **Figure 4** using as template the structure of the arginine/agmantine antiporter AdiC from *E. coli* (PDB 3OB6). Superposition of the rat and human structural models was performed by VMD 1.9.1. C154 of the human protein involved in disulfide bridge with 4F2hc is highlighted in yellow; putative PKC and PKA phosphorylation sites are highlighted in green. N- and C- terminals of rat and human proteins are nearly coincident and highlighted by single N and C.

SLC7A6: Y+LAT2

Together with LAT1 and LAT2, this transporter belongs to the *SLC7* family. The gene encoding the human isoform of y+LAT2 has been isolated in 1999 and annotated on the chromosome 16q22.1 (Pfeiffer et al., 1999). The protein is constituted by 515 amino acids and is mainly expressed in brain, small intestine, testis, parotids, heart, kidney, lung. y+LAT2 works as heterodimer with 4F2hc like the other members of *SLC7* family. It has been identified as arginine/glutamine exchanger: this role would be particularly important in brain where the protein might supply cells with important amino acids. Very interestingly, the catalytic mechanism of this transporter is unique: y+LAT2, indeed, mediates uptake of neutral amino acids in a Na^+-dependent fashion, while cationic amino acid transport is Na^+-independent (**Table 1**). The exchange occurs with a 1:1 stoichiometry. The antiport mechanism is expected to be electroneutral because the positive charge of the cationic amino acid is compensated for the Na^+ ion accompanying the neutral amino acid (Broer et al., 2000).

SLC7A15: ArpAT

The gene encoding this transporter has been annotated by *in silico* approach. BLAST analysis identified on mouse genome a region containing a putative ORF fulfilling the requirements of a transporter belonging to *SLC7* family: it has 64% similarity with $b^{0,+}$AT, 12 TMD domains, 2 putative N-glycosylation sites present also in LAT1 and the conserved Cys residue involved in disulfide with *SLC3* member. This mouse transporter was then

cloned and functionally characterized in HeLa cells; its overexpression in the presence of rBAT or 4F2hc increased neutral amino acid uptake. The transporter mediates uptake of alanine, serine, glutamine and cystine in Na^+-and pH-independent manner. Analogously to LATs, ArpAT showed trans-stimulation. This transporter is expressed in intestine and, at much lower levels, in brain. The main physiological role played by ArpAT may be linked to providing dopamine to the non-neuronal dopaminergic system which regulates intestinal Na^+ absorption (Fernandez et al., 2005). The most intriguing finding concerning this transporter is its evolutionary history: arpAT gene is, in fact, conserved in rat, dog, and chicken while it has been silenced in humans and chimpanzee. This suggests that its role became, along with primates evolution, not essential. The group which identified the transporter speculated that this may find explanation in the different diet of primates compared to other vertebrates (Fernandez et al., 2005).

SLC38: SNATs
GENES AND TISSUE LOCALIZATION

The *SLC38* family belongs to the Amino acid Polyamine-organo Cation family (APC family) and includes 11 membrane transporters most of which are known to mediate the net uptake of glutamine, alanine, asparagine, histidine, arginine, and some other neutral amino acids in those tissues where they are expressed (Hagglund et al., 2011; Schioth et al., 2013). The most acknowledged transporters from this family are 6 proteins, originally classified as systems A and N and subsequently as SNATs or SATs and SNs, respectively. In this more recent classification, the numbers assigned to SNATs (SAT or SN) correspond to the identification numbers of the *SLC38* members. The other five members are still at early stage of characterization. SNAT1, 2, and 4 (*SLC38A1, 2* and *4*) correspond to the old systems A. System A was firstly isolated in 1965 (Christensen et al., 1965) and was defined by the inhibition with the amino acid analog MeAIB. In 2000 the three transporters belonging to such family were identified and the coding genes annotated on chromosome 12. The proteins SNAT1 and SNAT2 are 486 and 505 amino acids in length, respectively. The system N includes SNAT3, 5, and 7 (*SLC38A3, 5* and *7*). SNAT3 gene was identified in 1999 and annotated on the chromosome 3p21.31 (Chaudhry et al., 1999) encoding a protein of 503 amino acids. Then, SNAT5 gene was cloned in 2001 and annotated on the chromosome Xp11.23 (Nakanishi et al., 2001) encoding a protein of 471 amino acids. SNAT7 gene is localized on chromosome 16q21 and encodes a protein of 461 amino acids. According to GeneBank, different transcripts exist for these three genes, reported in **Table 1**. The tissue distribution of SNATs is wide and specific for each member: SNAT1 and SNAT2 are ubiquitous (Albers et al., 2001; Chaudhry et al., 2002); SNAT3 is expressed in liver, skeletal muscle, kidney and pancreas (Chaudhry et al., 1999); SNAT 5 is expressed in stomach, brain, liver, lung, small intestine, spleen, colon and kidney (Nakanishi et al., 2001); SNAT7 is ubiquitous (Hagglund et al., 2011). Regarding subcellular localization, SNATs are mostly localized in the basolateral membrane of absorptive epithelia (Schioth et al., 2013; Broer, 2014) (**Figure 2**). SNAT1 seems to be also localized in intracellular organelles in GABAergic neurons (Solbu et al., 2010). In nervous system, SNAT1 and SNAT2 are expressed in GABAergic and glutamatergic neurons (Solbu et al., 2010) while SNAT3 and SNAT5 are expressed in astrocytes (Chaudhry et al., 2002). In pancreas SNAT2 is localized in the α-cell membranes while SNAT3 in the β-cell ones. In kidney SNAT3 is localized on the basolateral membranes of the S3 segment of proximal tubules. In intestine an apical sub-cellular localization has been proposed for SNAT5. *SLC38* family comprises also other five additional members whose function is still orphan. *SLC38A6* (SNAT6) gene is localized on the chromosome 14q23.1 and is expressed mainly in brain and liver (Nakanishi et al., 2001). *SLC38A8* (SNAT8) gene is localized on the chromosome 16q23.3 and the protein has a distribution similar to that of SNAT7. The transcripts of the other members *SLC38A9, SLC38A10* and *SLC38A11* derive from genes annotated on chromosome 5q11.1, 17q25.3 and 2q24.3 respectively (Sundberg et al., 2008).

FUNCTION

SNAT1 and 2 show a broad substrate specificity with preference for glutamine besides methionine, proline, serine, asparagine, glycine and histidine, while SNAT 4 is not considered as glutamine transporter (Schioth et al., 2013; Broer, 2014) and refs herein). SNAT 7, which was characterized later than SNAT 3 and 5, recognizes also arginine as substrate. SNAT7, considered member of system N, shows also characteristics of system A, such as broad substrate specificity (Hagglund et al., 2011). The known members of the *SLC38* family work as co-transporters for Na^+, in which amino acid uptake is driven by the Na^+ electrochemical gradient, inwardly directed (Mackenzie and Erickson, 2004) (**Table 1**). For system A members (SNAT1, 2) the uptake of glutamine results to be electrogenic, since each transport cycle causes a net movement of a positive charge with a stoichiometry 1:1. On the contrary, the transporters belonging to system N are also able to counter transport Na^+ and H^+ giving rise to an electroneutral transport mechanism. SNAT3 and 5, but not SNAT7, have a unique hallmark to tolerate substitution of Na^+ with Li^+ (Kilberg and Christensen, 1980; Mackenzie et al., 2003; Hagglund et al., 2011). All the known SLC38 transporters are activated by increasing pH from 6 to 8 (Mackenzie and Erickson, 2004). In addition to Na^+ or H^+ coupled transport, SNAT3 mediates uncoupled ion movement across the membrane (Schneider et al., 2007). Thus, protons and other cations can pass through the transporter by substrate dependent mode, while other ion conductances are substrate independent. Whether this uncoupled transport has a physiological function it is still unclear as above described for B^0AT1. Interestingly, a single residue, N76, is critical for coupled and uncoupled ion flows in the glutamine transporter SNAT3 (Broer, 2009). Moreover, it has been reported that a mutation of residue T380 affects the ability of Li^+ to replace Na^+ (Chaudhry et al., 2001).

Under a physiological point of view, SNATs play different role according to their localization (**Figure 2**). In nervous tissue, members of system a take up glutamine released by members of system N giving rise to the glutamine/glutamate cycle in neurons. In astrocytes, glutamine is synthesized via glutamine synthetase, using glutamate released by synapses and taken up via EAAT1 (Glutamate transporter *SLC1A3*) or EAAT2 (Glutamate transporter *SLC1A2*) (Schioth et al., 2013). Another example of concerted action among SNATs occurs in pancreas among alpha and beta-cells. At low plasmatic glutamine concentrations, SNAT3 releases glutamine supplying substrate for SNAT2, stimulating glucagon secretion. In the opposite condition, in the presence of high glutamine concentrations, differently by what above described, SNAT3 will accumulate glutamine which is then converted to glutamate. This reaction provides intermediate for TCA cycle, producing ATP that, similarly to what occurs in the presence of glucose accumulated in the β-cells via GLUT2, will inhibit KC potassium channels causing depolarization. This event facilitates fusion of insulin containing secretory granules with the plasma membrane (Jenstad and Chaudhry, 2013). Thus, SNAT3 which catalyzes both glutamine uptake and efflux can be considered a sensor of the nutritional state of the cell. However, glutamine affinity at the intracellular face is still not known, due to the difficulty in accessing the intracellular space in experiments performed with intact cells. In kidney SNAT3 contributes to glutamine absorption from the circulation; this is involved in regulation of acid-base homeostasis (Busque and Wagner, 2009). In liver SNAT1, SNAT2, SNAT3, and SNAT5 contribute to net import of amino acids, particularly alanine and glutamine regulating gluconeogenesis and giving rise to the glutamine/alanine cycle between liver and muscle (Kondou et al., 2013). In intestine, where glutamine uptake from lumen is mediated by B^0AT1 (**Figure 2**), transporters from system A have been proposed as the main responsible of glutamine transfer to blood (Broer, 2008). Interestingly, in this body district SNAT5 would be involved in glutamine metabolism in crypts cells, even though further investigations are needed (Saha et al., 2012).

The functional and kinetic characterizations described in the cited papers, have been conducted in cell systems; so far, no example of over-expression and reconstitution in proteoliposomes of SNAT members is available. Thus, given their complex interplay, some aspects are still unknown.

REGULATORY ASPECTS

The regulation of SNAT expression is a complex issue, not completely solved; the available information have been included in **Table 1**. SNAT1 expression is stimulated by PKA (Ogura et al., 2007). For SNAT2, a regulation by amino acid deprivation has been described and the mechanism linked to the presence in the first intron of a amino acid response element (AARE) acting as enhancer (Palii et al., 2004). Moreover, in liver SNAT2 is upregulated by glucagon (Ortiz et al., 2011). For SNAT3 a regulation by PKC has been reported and linked to the glutamate/GABA-glutamine of the central nervous system. Derangement of SNAT3 regulation may cause pathological states both in the central nervous system and in peripheral organs (Nissen-Meyer and Chaudhry, 2013). Moreover, SNAT5 expression has been found regulated by c-myc in cancer cells (Deberardinis et al., 2007).

STRUCTURAL ASPECTS

The structure of the *SLC38* members is characterized by a 5+5 inverted repeat fold typical of the bacterial LeuT (Broer, 2014). Recently, by bioinformatics and labeling experiments, the homology model of SNAT4 (built on LeuT and AdiC) has been validated

(Shi et al., 2011). In this model the N- and C-terminals of the protein are extracellular, as well as three loops containing two N-glycosylation sites (N260 and N264). Interestingly, a disulphide bridge linking C249 and C321 is essential for SNAT4 function, which, however, is not a glutamine transporter. (Padmanabhan Iyer et al., 2013). The structural model of SNAT7 has been built as a representative of the *SLC38* glutamine transporters, given that it was the last characterized and, hence, an homology model was not yet built (**Figure 7**). Interestingly, the alignment of SNAT7 with the other *SLC38* members shows the lowest percentage of identity from 19 to 21% (not shown). In addition, SNAT7 has unique properties resembling those of both system A and N. Differently by *SLC1* and *SLC6* transporters, homology models built for human and rat SNAT7 proteins by Modeler do not have the same fidelity of the others due to the low sequence similarity to the available crystallized bacterial homologs (AdiC and ApcT). SNAT2 model was already built on the basis of LeuT crystal structure (Zhang et al., 2009). SNAT7 has the same topology of the homology model previously described for SNAT4 (Shi et al., 2011) with extracellular C- and N- terminals. Differently by SNAT4 no N-glycosylation sites have been predicted by bioinformatics for the human and rat SNAT7 proteins. Concerning the phosphorylation sites, a PKC binding site (T174) and a PKA binding site (T179) have been found in the human isoform. The same phosphorylation sites have been found in rat shifted of one residue.

THE MITOCHONDRIAL GLUTAMINE TRANSPORTER

A mystery in the glutamine transporter ensemble is still represented by the mitochondrial member. This transporter should

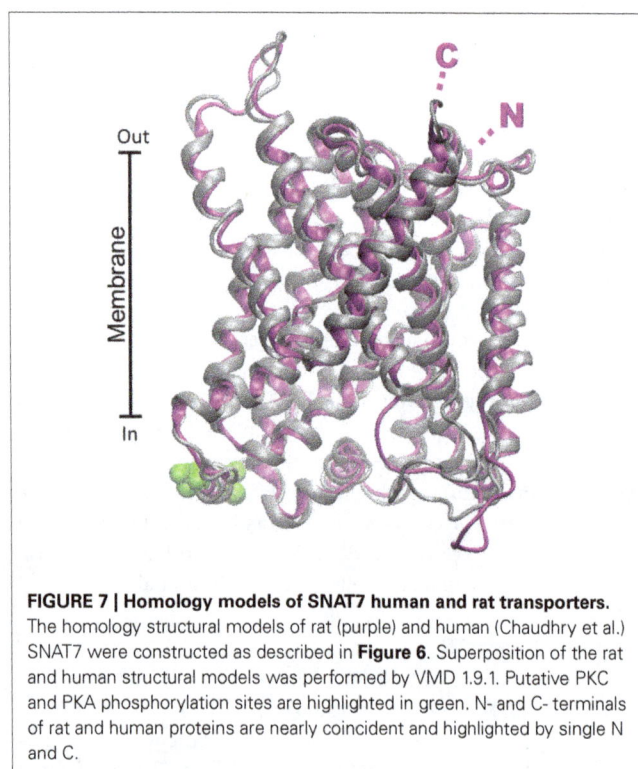

FIGURE 7 | Homology models of SNAT7 human and rat transporters. The homology structural models of rat (purple) and human (Chaudhry et al.) SNAT7 were constructed as described in **Figure 6**. Superposition of the rat and human structural models was performed by VMD 1.9.1. Putative PKC and PKA phosphorylation sites are highlighted in green. N- and C- terminals of rat and human proteins are nearly coincident and highlighted by single N and C.

be essential in metabolism since release of ammonia by glutamine deriving from extra hepatic tissues, should occur inside the mitochondrial matrix, where the ubiquitous form of glutaminase, i.e., derived from GLS1 gene, is localized (**Figure 2**). Glutaminase deriving from GLS1 gene exists in two splicing variants, both localized in mitochondria (Katt and Cerione, 2014). Interestingly and in line with the role of glutamine in cancer cell metabolism (see Section "Glutamine transporters in human pathology"), the GLS1expression is altered in cancer. This has been linked with suppression of miRNA23a and b via both c-myc and NFkB pathways (Gao et al., 2009; Rathore et al., 2012). Indeed, the metabolic flux of ammonia has been well demonstrated to finish in mitochondria also on the basis of a recent work on pancreatic cancer cells, in which glutamine transfer to mitochondria, via transporters, is hypothesized due to the impairment of the cell growth when GLS1 is knocked down (Son et al., 2013). In spite of the great importance of the mitochondrial glutamine transporter, only preliminary evidences on this transport function were provided by old studies in intact mitochondria (Sastrasinh and Sastrasinh, 1989) and then the protein responsible for this function was isolated and purified from mitochondrial extracts (Indiveri et al., 1998). Afterwards no other advances on the identification of the protein and/or the gene encoding the mitochondrial glutamine transporter have been achieved. Interestingly, the mentioned glutamine transporter has a higher apparent molecular mass, 41.5 kDa, than canonical mitochondrial carriers, i.e., 28–34 kDa. Therefore, the mitochondrial glutamine transporter may or not belong to the mitochondrial carrier family. Indeed, a mitochondrial carrier with a longer N-terminal extension has been described (Palmieri, 2008) as well as a mitochondrial transporter which does not belong to the mitochondrial carrier family (Herzig et al., 2012).

GLUTAMINE TRANSPORTERS IN HUMAN PATHOLOGY

The involvement of glutamine transporters in human pathology has been investigated and, with the only exception of the autosomal recessive disorder Hartnup disease, the only pathology linked to alteration of glutamine homeostasis, is cancer. Glutamine plays pivotal role in cell proliferation which requires nutrients, energy and increased biosynthetic activity leading to overall metabolic changes. In this scenario, cells shift to high glycolysis rate, lactate production and increased lipid biosynthesis. These pathways have been firstly described in lymphocytes which represent an useful model for studying cell proliferation (Deberardinis et al., 2008; Vander Heiden, 2011). Interestingly, a similar metabolic shift is a typical feature of cancer cells that need glucose and glutamine to sustain the high rate of proliferation (Fuchs and Bode, 2005; Ganapathy et al., 2009; Chen and Russo, 2012; Daye and Wellen, 2012; Son et al., 2013). Differential expression of specific genes, in particular those coding for plasma membrane transporters, allows increased uptake of the two nutrients. Normal proliferating cells need to integrate signals coming from extracellular growth factors, while cancer cells have increased autonomy. The main metabolic changes occurring in cancer are collectively known as Warburg effect (Ganapathy et al., 2009; Vander Heiden et al., 2009; Dang, 2010; Ohh, 2012). This phenomenon relies, besides increased glucose utilization, also on glutamine, canonically classified as "non-essential" amino acid that gives rise to a truncated form of TCA, terminating with malate and producing ATP at the substrate level (**Figure 8**). Malate is converted to pyruvate and lactate allowing cancer cells to produce ATP and NADPH necessary for anabolic processes (**Figure 8**). Lactate is exported by MCTs which are also over-expressed in tumors (Halestrap, 2013). Moreover, glutamine sustains cytosolic citrate synthesis via IDH1 and lipogenesis (Metallo, 2012) (**Figure 8**).

Therefore, it is not surprising that plasma membrane transporters responsible of glutamine disposition are altered in several human cancers. Among the transporters described in the above Sections, LAT1 (*SLC7A5*) and ASCT2 (*SLC1A5*) are those for which over expression in several tumors has been documented since 2005 (Fuchs and Bode, 2005). Specific examples of human cancers are listed in **Table 2**. In nearly all cases, over-expression of ASCT2 and LAT1 correlates with high grade of malignancies of these tumors. However, the role of ASCT2 in providing glutamine to cells (Nicklin et al., 2009) must be revisited in view of the well assessed obligatory antiport and functional asymmetry (Pingitore et al., 2013). Energy supply from glutamine will not result from the entire glutamine molecule (five carbon atoms) but from the difference between the carbon atoms taken up as glutamine and those exported as

smaller amino acids such as serine and threonine (**Figure 8**). This difference accounts for one or two carbon atoms which are oxidized in the truncated TCA. The membrane transporters SNAT1, SNAT3, and SNAT 5 have been also shown over-expressed in some cancers (**Table 2**), to accomplish glutamine supply; these findings opened perspectives on SNATs as pharmacological targets, even though lack of suitable experimental tool made this issue still difficult to approach. Noteworthy, the low affinity glutamine transporter $ATB^{0,+}$ is also up-regulated in some tumors, even though this is linked with its primary function of providing arginine to cells (Gupta et al., 2005).

Only one inherited pathology, named Hartnup disorder (OMIM 234500), is clearly associated to a glutamine transporter, i.e., $B^0 AT1$. So far, 21 mutations have been described as causative of such pathology onset. This metabolic syndrome is complex and characterized by pellagra-like rash, cerebellar ataxia, emotional instability and strong aminoaciduria. Even though $B^0 AT1$ is responsible also of glutamine absorption and reabsorption, the amino acids for which underlies its role in Hartnup disease are mainly histidine and tryptophan (Broer, 2009). The dramatic loss of such amino acids in urine is responsible of decreased serotonin and melatonin syntheses, explaining the neurological and skin problems, respectively.

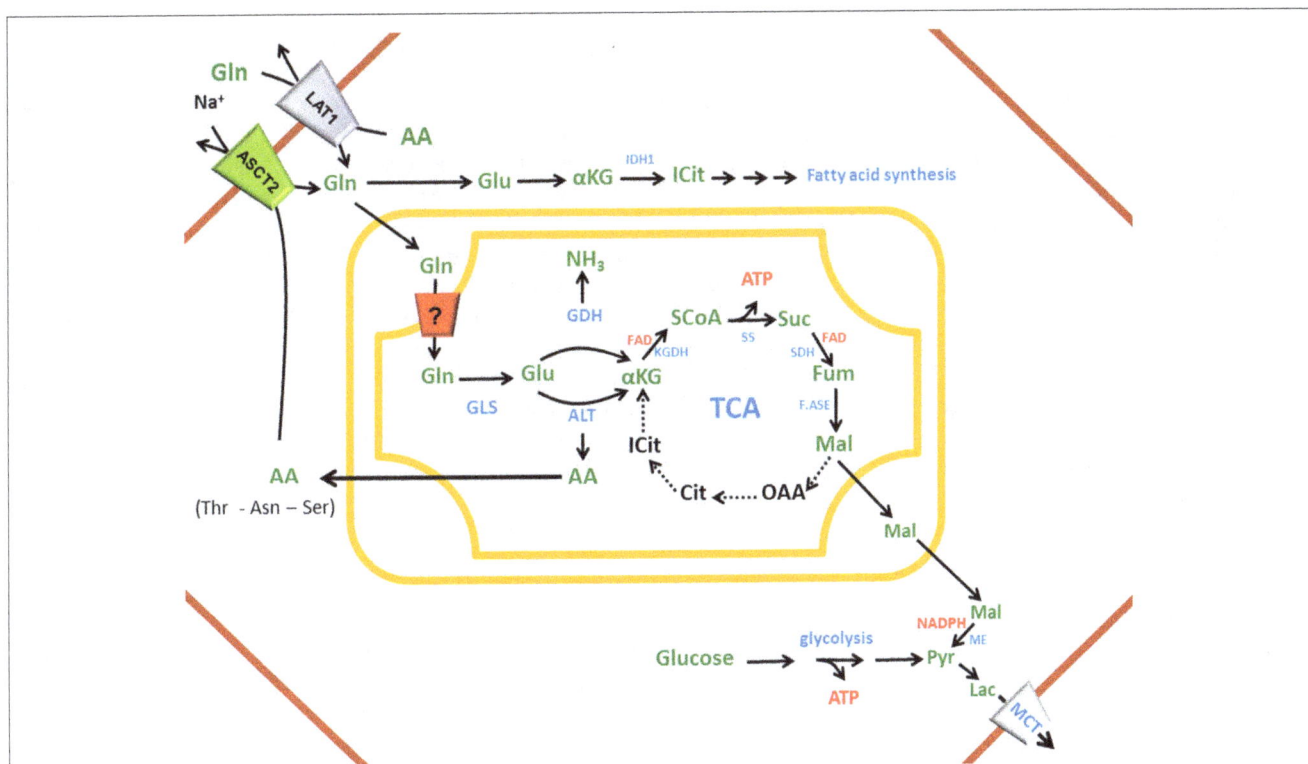

FIGURE 8 | **Network of transporters involved in cancer metabolic switch.** In cell membrane (red), ASCT2 and LAT1: glutamine plasma membrane transporters; MCT: Monocarboxylate Transporter. In cytosol (upper part of the figure), Gln: glutamine, Glu: glutamate, αKG: α Ketoglutarate, ICIT: isocitrate, IDH1: Isocitrate dehydrognase 1 and simplification of reactions to fatty acid synthesis; (lower part of the figure) Mal: Malate, Lac: lactate and simplification of glycolysis with the end product pyruvate (Pyr). In the Inner Mitochondrial Membrane, putative glutamine transporter (?). In mitochondrial matrix, TCA (Tricarboxylic Acid Cycle) with enzymes, GLS: Glutaminase, GDH: Glutamate dehyfrogenase, ALT: Alanine Amino Transferase. AA: Amino Acid. Dotted arrows indicate metabolic pathways depressed in cancers.

Table 2 | SNATs, ASCT2, and LAT1 associated cancers.

Cancer type	SLC38 (SNATs)	SLC1A5 (ASCT2)	SLC7A5 (LAT1)
Prostate cancer	Okudaira et al., 2011	Busque et al., 2009; Li et al., 2013	Patel et al., 2013; Segawa et al., 2013
Colorectal cancer		Witte et al., 2002	Ebara et al., 2010
Hepato Cell Carcinoma	Kondoh et al., 2007	Bode et al., 2002; Fuchs et al., 2004	Li et al., 2013
Lung cancer		Hassanein et al., 2013; Shimizu et al., 2014	Kaira et al., 2010; Imai et al., 2010
Breast cancer	Wang et al., 2013	Kim et al., 2012	Furuya et al., 2012
Neuroblastoma and glioma	Sidoryk et al., 2004	Wasa et al., 2002; Dolinska et al., 2003	
Endometrioid carcinoma			Watanabe et al., 2014
Ovarian Cancer			Kaji et al., 2010
Renal Cell Carcinoma			Betsunoh et al., 2013
Pancreatic andbiliary tract cancer			Yanagisawa et al., 2012; Kaira et al., 2013
Gastric cancer			Shi et al., 2013
Pleural Mesothelioma			Kaira et al., 2011

List of cancer tissues in which SNATs, ASCT2 and/or LAT1 have been found over-expressed with related references.

TOXICOLOGY AND PHARMACOLOGY: ASCT2 AND LAT1 AND DRUG INTERACTIONS

The results above described highlighted the importance of glutamine membrane transporters in metabolic changes for cancer development and/or progression. This shed light, since years, on the pharmacological relevance of ASCT2 and LAT1 which became hot targets of drug discovery. Moreover, the pivotal role played by plasma membrane transporters in interaction with drugs is becoming more and more important and several evidences link these proteins to drug disposition among tissues and among different subcellular compartments (Giacomini et al., 2010; Pochini et al., 2013; Scalise et al., 2013). The challenge of advancing in this research field is represented by the possibility of designing good inhibitors able to specifically target the protein of interest. These studies started since some years and different groups reported attempts of designing and identifying specific inhibitors of ASCT2 (Grewer and Grabsch, 2004; Albers et al., 2012) and LAT1 (Geier et al., 2013). The design of specific and selective molecules requires the knowledge of the tridimensional structures of the target. However, so far, nearly no structures of human secondary membrane transporters have been solved due to the difficulty in handling them and to the lack of methodologies for over-expressing and folding. Only very recently the human isoform of GLUT1 transporter has been solved by X-ray diffraction, even though in an inactive form obtained by site-directed mutagenesis (Deng et al., 2014). The possibility to identify new compounds is therefore mostly based on integration between experimental procedures and bioinformatics, through homology modeling and molecular docking. The homology model structures previously obtained (Oppedisano et al., 2010, 2011; Pingitore et al., 2013; Pochini et al., 2014) and now improved (**Figures 4, 5**) have been used for several toxicological studies (see below). In the meantime, a lot of efforts have been made to obtain human isoforms of ASCT2 and LAT1 by over-expressing them in heterologous systems, following the work flow depicted in **Figure 3**. These procedures are useful to recover high yield of purified protein needed for both functional and structural studies. The two amino acid transporters hASCT2 and hLAT1/CD98 have been recently cloned and over-expressed in the yeast *P. pastoris* and in the bacterium

E. coli, respectively (Galluccio et al., 2013; Pingitore et al., 2013). Even though the two proteins are both glutamine transporters, a common strategy could not be applied. This was not surprising given the experience already acquired with the three members of OCTN family (Indiveri et al., 2013).

The availability of these proteins in purified and in native form opens the possibility of obtaining X-ray structure for better defining or validating the already available homology models depicted in **Figures 4, 6**. Combining the over-expression systems and the procedures of proteoliposomes reconstitution, it will be soon possible to develop assays for high throughput screening of chemical compounds for drug discovery and design before animal experimentation (Pochini et al., 2013; Scalise et al., 2013), as depicted in the work flow of **Figure 3**. These kind of studies can be performed also in cell systems stably or transiently over-expressing the protein of interest; however, given the pleiotropy of glutamine and the broad substrate specificity of amino acid transporters, a strategy allowing the study of isolated proteins in artificial membranes is indispensable as suggested by several authors (McGivan and Bungard, 2007; Scalise et al., 2013; Schioth et al., 2013). By mean of proteoliposome reconstitution, some studies evaluating toxicological aspects of glutamine transporters have been already performed on the rat protein, extracted in native form from kidney, being the human isoforms not available at that time (see above in ASCT2 and B0AT1 Sections). Such studies gave the possibility of confirming also structure/function relationships speculated on the basis of homology models presented in **Figures 4, 5**. The works moved from the observation that some amino acid residues, such as Cys, increased their relative concentrations during protein evolution (Jordan et al., 2005). This makes Cys a potential target of different SH-reactive molecules; noteworthy the glutamine transporters dealt with the present review harbor several Cys residues. ASCT2 and B0AT1, indeed, were shown to be inhibited by several SH-reagents. Under a toxicological point of view, an interesting class of SH-reagents are the mercurial compounds. These molecules are potent environmental contaminants coming from industries with high toxicity for humans and in general for mammals since they enter the food chain as bio-product of bacterial metabolism (Rooney, 2007).

A first work has been performed with ASCT2; this protein showed to be potently inhibited by $HgCl_2$, methyl-mercury and other heavy metals (Oppedisano et al., 2010). The compounds interact very probably with a CXXC metal binding motif located at the end of the large hydrophilic loop of the transporter in a basin surrounded by hydrophobic residues (**Figure 4A**). The IC50 measured for mercury compounds are in the range of toxicity threshold. These results explain at the molecular level the toxic effect of methyl-mercury in brain (Boado et al., 2005). A large scale screening of drugs has been performed on ASCT2 in the same system (Oppedisano et al., 2012). In particular, potential anti-tumor drugs, dithiazoles, where designed specifically for rat ASCT2 and tested to identify the structural features of the most potent ones; the study revealed again that the CXXC motif is involved in covalent interaction with those drugs.

Another study in the proteoliposomes system has been conducted on B0AT1 extracted from rat kidney. Even though the structure of rat B0AT1 is not related with that of rat ASCT2, the transporters share the common feature of carrying CXXC metal binding motifs. In the case of B0AT1, the CXXC motif has been localized far from the putative binding site, correlating with the non-competitive inhibition by mercury compounds (Oppedisano et al., 2011) (**Figure 5**). B0AT1 contains, in addition, a CXXXC motif which has been described previously as target of metals such as Cu^{2+}, Pb^{2+}, Hg^{2+}, Cd^{2+}, and Zn^{2+} (Abajian and Rosenzweig, 2006; Tonazzi and Indiveri, 2011). This seconds second motif might explain the effect exerted by the prototype of mercury agent, mersalyl. In this case, in fact, substrate protection of the inhibition was found (Oppedisano et al., 2011). Interestingly, anti-oxidants are able to reverse the inhibition in both ASCT2 and B0AT1 proteins (Scalise et al., 2013). Very recently, a large scale screening of drugs has been performed on rat B0AT1 describing a specific inhibitory effect exerted by the common anti-inflammatory drug nimesulide (Pochini et al., 2014). Experimental data were supported by molecular docking of nimesulide on the rat B0AT1 showing that the site of drug interaction corresponds to that for antidepressants in SERT and GABA transporters (Krishnamurthy and Gouaux, 2012; Penmatsa et al., 2013). This feature, together with the high degree of identity between rat and human, makes reconstituted B0AT1 an important model for pharmacological and toxicological screening with applicability in humans. On the contrary, among the glutamine transporters described in this review, ASCT2 has the unique feature of being locally different between rat and human orthologs, making the availability of the human over-expressed transporter essential for applications in human health.

CONCLUSION

Glutamine is a "conditionally essential" amino acid which homeostasis is maintained owing to the different mode of transport performed by several membrane transporters. Thus, on the one hand, concentrative transporters allow glutamine absorption from diet in intestine and reabsorption from urine in kidney. On the other hand, both co-transporters and antiporters allow delivery of glutamine in all the tissues and equilibration with other amino acids. The redundancy of transporters with overlapping functional properties is very probably related to the need of avoiding derangements from homeostasis even in case of malfunctioning of one transporter. The lack of known defects of genes encoding these transporters, with very few exception, is in some instances an evidence that derangements of glutamine homeostasis is incompatible with life. Notwithstanding all the transporters dealt with share specificity for glutamine, they display different tridimensional structures. This is in line with their classification in different protein families; however, within the same family, the orthologs are very conserved from rat to human. The only exception is represented by ASCT2 (*SLC1A5*) which may be related with some specific signaling role of the transporter arisen later in evolution. While the function of several transporters has been widely described, no structures are available and the molecular mechanisms of regulation are still not completely solved. Thus, a lot remains to be done for a comprehensive knowledge of glutamine fluxes through tissues. This is very important not only for the biochemical knowledge of the glutamine homeostasis but also, and more importantly, for the involvement of this special amino acid in the hottest topic of human health: cancer biology.

AUTHOR CONTRIBUTIONS

Lorena Pochini, Mariafrancesca Scalise contributed in collecting bibliography, preparing figures and writing; Michele Galluccio contributed in drawing structure figures and writing; Cesare Indiveri contributed in writing and supervision of all the activities.

ACKNOWLEDGMENTS

This work was supported by funds from: Programma Operativo Nazionale [01_00937] - MIUR "Modelli sperimentali biotecnologici integrati per lo sviluppo e la selezione di molecole di interesse per la salute dell'uomo" to Cesare Indiveri.

REFERENCES

Abajian, C., and Rosenzweig, A. C. (2006). Crystal structure of yeast Sco1. *J. Biol. Inorg. Chem.* 11, 459–466. doi: 10.1007/s00775-006-0096-7

Albers, A., Broer, A., Wagner, C. A., Setiawan, I., Lang, P. A., Kranz, E. U., et al. (2001). Na+ transport by the neural glutamine transporter ATA1. *Pflugers Arch.* 443, 92–101. doi: 10.1007/s004240100663

Albers, T., Marsiglia, W., Thomas, T., Gameiro, A., and Grewer, C. (2012). Defining substrate and blocker activity of alanine-serine-cysteine transporter 2 (ASCT2) ligands with novel serine analogs. *Mol. Pharmacol.* 81, 356–365. doi: 10.1124/mol.111.075648

Amaral, J. S., Pinho, M. J., and Soares-Da-Silva, P. (2008). Genomic regulation of intestinal amino acid transporters by aldosterone. *Mol. Cell. Biochem.* 313, 1–10. doi: 10.1007/s11010-008-9735-3

Antony, J. M., Deslauriers, A. M., Bhat, R. K., Ellestad, K. K., and Power, C. (2011). Human endogenous retroviruses and multiple sclerosis: innocent bystanders or disease determinants? *Biochim. Biophys. Acta* 1812, 162–176. doi: 10.1016/j.bbadis.2010.07.016

Avissar, N. E., Sax, H. C., and Toia, L. (2008). In human entrocytes, GLN transport and ASCT2 surface expression induced by short-term EGF are MAPK, PI3K, and Rho-dependent. *Dig. Dis. Sci.* 53, 2113–2125. doi: 10.1007/s10620-007-0120-y

Bak, L. K., Schousboe, A., and Waagepetersen, H. S. (2006). The glutamate/GABA-glutamine cycle: aspects of transport, neurotransmitter homeostasis and ammonia transfer. *J. Neurochem.* 98, 641–653. doi: 10.1111/j.1471-4159.2006.03913.x

Betsunoh, H., Fukuda, T., Anzai, N., Nishihara, D., Mizuno, T., Yuki, H., et al. (2013). Increased expression of system large amino acid transporter (LAT)-1 mRNA is associated with invasive potential and unfavorable prognosis of human clear cell renal cell carcinoma. *BMC Cancer* 13:509. doi: 10.1186/1471-2407-13-509

Bhavsar, S. K., Hosseinzadeh, Z., Merches, K., Gu, S., Broer, S., and Lang, F. (2011). Stimulation of the amino acid transporter SLC6A19 by JAK2. *Biochem. Biophys. Res. Commun.* 414, 456–461. doi: 10.1016/j.bbrc.2011.09.074

Boado, R. J., Li, J. Y., Chu, C., Ogoshi, F., Wise, P., and Pardridge, W. M. (2005). Site-directed mutagenesis of cysteine residues of large neutral amino acid transporter LAT1. *Biochim. Biophys. Acta* 1715, 104–110. doi: 10.1016/j.bbamem.2005.07.007

Bode, B. P. (2001). Recent molecular advances in mammalian glutamine transport. *J. Nutr.* 131, 2475S–2485S. discussion: 2486S–2477S. doi: 10.1146/annurev.nutr. 13.1.137

Bode, B. P., Fuchs, B. C., Hurley, B. P., Conroy, J. L., Suetterlin, J. E., Tanabe, K. K., et al. (2002). Molecular and functional analysis of glutamine uptake in human hepatoma and liver-derived cells. *Am. J. Physiol. Gastrointest. Liver Physiol.* 283, G1062–G1073. doi: 10.1152/ajpgi.00031.2002

Bodoy, S., Fotiadis, D., Stoeger, C., Kanai, Y., and Palacin, M. (2013). The small SLC43 family: facilitator system l amino acid transporters and the orphan EEG1. *Mol. Aspects Med.* 34, 638–645. doi: 10.1016/j.mam.2012.12.006

Bogatikov, E., Munoz, C., Pakladok, T., Alesutan, I., Shojaiefard, M., Seebohm, G., et al. (2012). Up-regulation of amino acid transporter SLC6A19 activity and surface protein abundance by PKB/Akt and PIKfyve. *Cell. Physiol. Biochem.* 30, 1538–1546. doi: 10.1159/000343341

Bohmer, C., Broer, A., Munzinger, M., Kowalczuk, S., Rasko, J. E., Lang, F., et al. (2005). Characterization of mouse amino acid transporter B0AT1 (slc6a19). *Biochem. J.* 389, 745–751. doi: 10.1042/BJ20050083

Bohmer, C., Sopjani, M., Klaus, F., Lindner, R., Laufer, J., Jeyaraj, S., et al. (2010). The serum and glucocorticoid inducible kinases SGK1-3 stimulate the neutral amino acid transporter SLC6A19. *Cell. Physiol. Biochem.* 25, 723–732. doi: 10.1159/000315092

Boudker, O., Ryan, R. M., Yernool, D., Shimamoto, K., and Gouaux, E. (2007). Coupling substrate and ion binding to extracellular gate of a sodium-dependent aspartate transporter. *Nature* 445, 387–393. doi: 10.1038/nature05455

Brizio, C., Galluccio, M., Wait, R., Torchetti, E. M., Bafunno, V., Accardi, R., et al. (2006). Over-expression in *Escherichia coli* and characterization of two recombinant isoforms of human FAD synthetase. *Biochem. Biophys. Res. Commun.* 344, 1008–1016. doi: 10.1016/j.bbrc.2006.04.003

Broer, A., Brookes, N., Ganapathy, V., Dimmer, K. S., Wagner, C. A., Lang, F., et al. (1999). The astroglial ASCT2 amino acid transporter as a mediator of glutamine efflux. *J. Neurochem.* 73, 2184–2194.

Broer, A., Friedrich, B., Wagner, C. A., Fillon, S., Ganapathy, V., Lang, F., et al. (2001). Association of 4F2hc with light chains LAT1, LAT2 or y+LAT2 requires different domains. *Biochem. J.* 355, 725–731.

Broer, A., Juelich, T., Vanslambrouck, J. M., Tietze, N., Solomon, P. S., Holst, J., et al. (2011). Impaired nutrient signaling and body weight control in a Na+ neutral amino acid cotransporter (Slc6a19)-deficient mouse. *J. Biol. Chem.* 286, 26638–26651. doi: 10.1074/jbc.M111.241323

Broer, A., Klingel, K., Kowalczuk, S., Rasko, J. E., Cavanaugh, J., and Broer, S. (2004). Molecular cloning of mouse amino acid transport system B0, a neutral amino acid transporter related to Hartnup disorder. *J. Biol. Chem.* 279, 24467–24476. doi: 10.1074/jbc.M400904200

Broer, A., Wagner, C. A., Lang, F., and Broer, S. (2000). The heterodimeric amino acid transporter 4F2hc/y+LAT2 mediates arginine efflux in exchange with glutamine. *Biochem. J.* 349(Pt 3), 787–795.

Broer, S. (2006). The SLC6 orphans are forming a family of amino acid transporters. *Neurochem. Int.* 48, 559–567. doi: 10.1016/j.neuint.2005.11.021

Broer, S. (2008). Apical transporters for neutral amino acids: physiology and pathophysiology. *Physiology (Bethesda)* 23, 95–103. doi: 10.1152/physiol.00045.2007

Broer, S. (2009). The role of the neutral amino acid transporter B0AT1 (SLC6A19) in Hartnup disorder and protein nutrition. *IUBMB Life* 61, 591–599. doi: 10.1002/iub.210

Broer, S. (2014). The SLC38 family of sodium-amino acid co-transporters. *Pflugers Arch.* 466, 155–172. doi: 10.1007/s00424-013-1393-y

Broer, S., and Brookes, N. (2001). Transfer of glutamine between astrocytes and neurons. *J. Neurochem.* 77, 705–719. doi: 10.1046/j.1471-4159.2001.00322.x

Bungard, C. I., and McGivan, J. D. (2004). Glutamine availability upregulates expression of the amino acid transporter protein ASCT2 in HepG2 cells and stimulates the ASCT2 promoter. *Biochem. J.* 382, 27–32. doi: 10.1042/BJ20040487

Bungard, C. I., and McGivan, J. D. (2005). Identification of the promoter elements involved in the stimulation of ASCT2 expression by glutamine availability in HepG2 cells and the probable involvement of FXR/RXR dimers. *Arch. Biochem. Biophys.* 443, 53–59. doi: 10.1016/j.abb.2005.08.016

Busque, S., Leventhal, J., Brennan, D. C., Steinberg, S., Klintmalm, G., Shah, T., et al. (2009). Calcineurin-inhibitor-free immunosuppression based on the JAK inhibitor CP-690,550: a pilot study *in de novo* kidney allograft recipients. *Am. J. Transplant.* 9, 1936–1945. doi: 10.1111/j.1600-6143.2009.02720.x

Busque, S. M., and Wagner, C. A. (2009). Potassium restriction, high protein intake, and metabolic acidosis increase expression of the glutamine transporter SNAT3 (Slc38a3) in mouse kidney. *Am. J. Physiol. Renal Physiol.* 297, F440–F450. doi: 10.1152/ajprenal.90318.2008

Camargo, S. M., Makrides, V., Virkki, L. V., Forster, I. C., and Verrey, F. (2005). Steady-state kinetic characterization of the mouse B(0)AT1 sodium-dependent neutral amino acid transporter. *Pflugers Arch.* 451, 338–348. doi: 10.1007/s00424-005-1455-x

Cantor, J. M., and Ginsberg, M. H. (2012). CD98 at the crossroads of adaptive immunity and cancer. *J. Cell Sci.* 125, 1373–1382. doi: 10.1242/jcs.096040

Chaudhry, F. A., Krizaj, D., Larsson, P., Reimer, R. J., Wreden, C., Storm-Mathisen, J., et al. (2001). Coupled and uncoupled proton movement by amino acid transport system N. *EMBO J.* 20, 7041–7051. doi: 10.1093/emboj/20. 24.7041

Chaudhry, F. A., Reimer, R. J., Krizaj, D., Barber, D., Storm-Mathisen, J., Copenhagen, D. R., et al. (1999). Molecular analysis of system N suggests novel physiological roles in nitrogen metabolism and synaptic transmission. *Cell* 99, 769–780.

Chaudhry, F. A., Schmitz, D., Reimer, R. J., Larsson, P., Gray, A. T., Nicoll, R., et al. (2002). Glutamine uptake by neurons: interaction of protons with system a transporters. *J. Neurosci.* 22, 62–72.

Chen, J. Q., and Russo, J. (2012). Dysregulation of glucose transport, glycolysis, TCA cycle and glutaminolysis by oncogenes and tumor suppressors in cancer cells. *Biochim. Biophys. Acta* 1826, 370–384. doi: 10.1016/j.bbcan.2012.06.004

Christensen, H. N. (1990). Role of amino acid transport and countertransport in nutrition and metabolism. *Physiol. Rev.* 70, 43–77.

Christensen, H. N., Oxender, D. L., Liang, M., and Vatz, K. A. (1965). The use of N-methylation to direct route of mediated transport of amino acids. *J. Biol. Chem.* 240, 3609–3616.

Conti, F., and Melone, M. (2006). The glutamine commute: lost in the tube? *Neurochem. Int.* 48, 459–464. doi: 10.1016/j.neuint.2005.11.016

Costa, M., Rosell, A., Alvarez-Marimon, E., Zorzano, A., Fotiadis, D., and Palacin, M. (2013). Expression of human heteromeric amino acid transporters in the yeast *Pichia pastoris*. *Protein Expr. Purif.* 87, 35–40. doi: 10.1016/j.pep.2012.10.003

Curi, R., Lagranha, C. J., Doi, S. Q., Sellitti, D. F., Procopio, J., Pithon-Curi, T. C., et al. (2005). Molecular mechanisms of glutamine action. *J. Cell. Physiol.* 204, 392–401. doi: 10.1002/jcp.20339

Cynober, L. A. (2002). Plasma amino acid levels with a note on membrane transport: characteristics, regulation, and metabolic significance. *Nutrition* 18, 761–766. doi: 10.1016/S0899-9007(02)00780-3

Dang, C. V. (2010). p32 (C1QBP) and cancer cell metabolism: is the Warburg effect a lot of hot air? *Mol. Cell. Biol.* 30, 1300–1302. doi: 10.1128/MCB.01661-09

Danilczyk, U., Sarao, R., Remy, C., Benabbas, C., Stange, G., Richter, A., et al. (2006). Essential role for collectrin in renal amino acid transport. *Nature* 444, 1088–1091. doi: 10.1038/nature05475

Daye, D., and Wellen, K. E. (2012). Metabolic reprogramming in cancer: unraveling the role of glutamine in tumorigenesis. *Semin. Cell Dev. Biol.* 23, 362–369. doi: 10.1016/j.semcdb.2012.02.002

Deberardinis, R. J., Lum, J. J., Hatzivassiliou, G., and Thompson, C. B. (2008). The biology of cancer: metabolic reprogramming fuels cell growth and proliferation. *Cell Metab.* 7, 11–20. doi: 10.1016/j.cmet.2007.10.002

Deberardinis, R. J., Mancuso, A., Daikhin, E., Nissim, I., Yudkoff, M., Wehrli, S., et al. (2007). Beyond aerobic glycolysis: transformed cells can engage in glutamine metabolism that exceeds the requirement for protein and nucleotide synthesis. *Proc. Natl. Acad. Sci. U.S.A.* 104, 19345–19350. doi: 10.1073/pnas.0709747104

Deitmer, J. W., Broer, A., and Broer, S. (2003). Glutamine efflux from astrocytes is mediated by multiple pathways. *J. Neurochem.* 87, 127–135. doi: 10.1046/j.1471-4159.2003.01981.x

Del Amo, E. M., Urtti, A., and Yliperttula, M. (2008). Pharmacokinetic role of L-type amino acid transporters LAT1 and LAT2. *Eur. J. Pharm. Sci.* 35, 161–174. doi: 10.1016/j.ejps.2008.06.015

Deng, D., Xu, C., Sun, P., Wu, J., Yan, C., Hu, M., et al. (2014). Crystal structure of the human glucose transporter GLUT1. *Nature* 510, 121–125. doi: 10.1038/nature13306

Dolinska, M., Dybel, A., Zablocka, B., and Albrecht, J. (2003). Glutamine transport in C6 glioma cells shows ASCT2 system characteristics. *Neurochem. Int.* 43, 501–507. doi: 10.1016/S0197-0186(03)00040-8

Drummond, M. J., Glynn, E. L., Fry, C. S., Timmerman, K. L., Volpi, E., and Rasmussen, B. B. (2010). An increase in essential amino acid availability upregulates amino acid transporter expression in human skeletal muscle. *Am. J. Physiol. Endocrinol. Metab.* 298, E1011–E1018. doi: 10.1152/ajpendo.00690.2009

Ducroc, R., Sakar, Y., Fanjul, C., Barber, A., Bado, A., and Lostao, M. P. (2010). Luminal leptin inhibits L-glutamine transport in rat small intestine: involvement of ASCT2 and B0AT1. *Am. J. Physiol. Gastrointest. Liver Physiol.* 299, G179–G185. doi: 10.1152/ajpgi.00048.2010

Ebara, T., Kaira, K., Saito, J., Shioya, M., Asao, T., Takahashi, T., et al. (2010). L-type amino-acid transporter 1 expression predicts the response to preoperative hyperthermo-chemoradiotherapy for advanced rectal cancer. *Anticancer Res.* 30, 4223–4227.

El-Gebali, S., Bentz, S., Hediger, M. A., and Anderle, P. (2013). Solute carriers (SLCs) in cancer. *Mol. Aspects Med.* 34, 719–734. doi: 10.1016/j.mam.2012.12.007

Fairweather, S. J., Broer, A., O'mara, M. L., and Broer, S. (2012). Intestinal peptidases form functional complexes with the neutral amino acid transporter B(0)AT1. *Biochem. J.* 446, 135–148. doi: 10.1042/BJ20120307

Fernandez, E., Torrents, D., Zorzano, A., Palacin, M., and Chillaron, J. (2005). Identification and functional characterization of a novel low affinity aromatic-preferring amino acid transporter (arpAT). One of the few proteins silenced during primate evolution. *J. Biol. Chem.* 280, 19364–19372. doi: 10.1074/jbc.M412516200

Forrest, L. R., Kramer, R., and Ziegler, C. (2011). The structural basis of secondary active transport mechanisms. *Biochim. Biophys. Acta* 1807, 167–188. doi: 10.1016/j.bbabio.2010.10.014

Fort, J., De La Ballina, L. R., Burghardt, H. E., Ferrer-Costa, C., Turnay, J., Ferrer-Orta, C., et al. (2007). The structure of human 4F2hc ectodomain provides a model for homodimerization and electrostatic interaction with plasma membrane. *J. Biol. Chem.* 282, 31444–31452. doi: 10.1074/jbc.M704524200

Fotiadis, D., Kanai, Y., and Palacin, M. (2013). The SLC3 and SLC7 families of amino acid transporters. *Mol. Aspects Med.* 34, 139–158. doi: 10.1016/j.mam.2012.10.007

Fraga, S., Pinho, M. J., and Soares-Da-Silva, P. (2005). Expression of LAT1 and LAT2 amino acid transporters in human and rat intestinal epithelial cells. *Amino Acids* 29, 229–233. doi: 10.1007/s00726-005-0221-x

Franca, R., Veljkovic, E., Walter, S., Wagner, C. A., and Verrey, F. (2005). Heterodimeric amino acid transporter glycoprotein domains determining functional subunit association. *Biochem. J.* 388, 435–443. doi: 10.1042/BJ20050021

Fuchs, B. C., and Bode, B. P. (2005). Amino acid transporters ASCT2 and LAT1 in cancer: partners in crime? *Semin. Cancer Biol.* 15, 254–266. doi: 10.1016/j.semcancer.2005.04.005

Fuchs, B. C., Finger, R. E., Onan, M. C., and Bode, B. P. (2007). ASCT2 silencing regulates mammalian target-of-rapamycin growth and survival signaling in human hepatoma cells. *Am. J. Physiol. Cell Physiol.* 293, C55–C63. doi: 10.1152/ajpcell.00330.2006

Fuchs, B. C., Perez, J. C., Suetterlin, J. E., Chaudhry, S. B., and Bode, B. P. (2004). Inducible antisense RNA targeting amino acid transporter ATB0/ASCT2 elicits apoptosis in human hepatoma cells. *Am. J. Physiol. Gastrointest. Liver Physiol.* 286, G467–G478. doi: 10.1152/ajpgi.00344.2003

Furuya, M., Horiguchi, J., Nakajima, H., Kanai, Y., and Oyama, T. (2012). Correlation of L-type amino acid transporter 1 and CD98 expression with triple negative breast cancer prognosis. *Cancer Sci.* 103, 382–389. doi: 10.1111/j.1349-7006.2011.02151.x

Galluccio, M., Pingitore, P., Scalise, M., and Indiveri, C. (2013). Cloning, large scale over-expression in *E. coli* and purification of the components of the human LAT 1 (SLC7A5) amino acid transporter. *Protein J.* 32, 442–448. doi: 10.1007/s10930-013-9503-4G7A

Ganapathy, M. E., and Ganapathy, V. (2005). Amino Acid Transporter ATB0,+ as a delivery system for drugs and prodrugs. *Curr. Drug Targets Immune Endocr. Metabol. Disord.* 5, 357–364. doi: 10.2174/156800805774912953

Ganapathy, V., Thangaraju, M., and Prasad, P. D. (2009). Nutrient transporters in cancer: relevance to Warburg hypothesis and beyond. *Pharmacol. Ther.* 121, 29–40. doi: 10.1016/j.pharmthera.2008.09.005

Gao, P., Tchernyshyov, I., Chang, T. C., Lee, Y. S., Kita, K., Ochi, T., et al. (2009). c-Myc suppression of miR-23a/b enhances mitochondrial glutaminase expression and glutamine metabolism. *Nature* 458, 762–765. doi: 10.1038/nature07823

Geier, E. G., Schlessinger, A., Fan, H., Gable, J. E., Irwin, J. J., Sali, A., et al. (2013). Structure-based ligand discovery for the large-neutral amino acid transporter 1, LAT-1. *Proc. Natl. Acad. Sci. U.S.A.* 110, 5480–5485. doi: 10.1073/pnas.1218165110

Giacomini, K. M., Huang, S. M., Tweedie, D. J., Benet, L. Z., Brouwer, K. L., Chu, X., et al. (2010). Membrane transporters in drug development. *Nat. Rev. Drug Discov.* 9, 215–236. doi: 10.1038/nrd3028

Gliddon, C. M., Shao, Z., Lemaistre, J. L., and Anderson, C. M. (2009). Cellular distribution of the neutral amino acid transporter subtype ASCT2 in mouse brain. *J. Neurochem.* 108, 372–383. doi: 10.1111/j.1471-4159.2008.05767.x

Grewer, C., and Grabsch, E. (2004). New inhibitors for the neutral amino acid transporter ASCT2 reveal its Na+-dependent anion leak. *J. Physiol.* 557, 747–759. doi: 10.1113/jphysiol.2004.062521

Gupta, N., Miyauchi, S., Martindale, R. G., Herdman, A. V., Podolsky, R., Miyake, K., et al. (2005). Upregulation of the amino acid transporter ATB0,+ (SLC6A14) in colorectal cancer and metastasis in humans. *Biochim. Biophys. Acta* 1741, 215–223. doi: 10.1016/j.bbadis.2005.04.002

Gupta, N., Prasad, P. D., Ghamande, S., Moore-Martin, P., Herdman, A. V., Martindale, R. G., et al. (2006). Up-regulation of the amino acid transporter ATB(0,+) (SLC6A14) in carcinoma of the cervix. *Gynecol. Oncol.* 100, 8–13. doi: 10.1016/j.ygyno.2005.08.016

Hagglund, M. G., Sreedharan, S., Nilsson, V. C., Shaik, J. H., Almkvist, I. M., Backlin, S., et al. (2011). Identification of SLC38A7 (SNAT7) protein as a glutamine transporter expressed in neurons. *J. Biol. Chem.* 286, 20500–20511. doi: 10.1074/jbc.M110.162404

Halestrap, A. P. (2013). The SLC16 gene family—structure, role and regulation in health and disease. *Mol. Aspects Med.* 34, 337–349. doi: 10.1016/j.mam.2012.05.003

Hamdi, M. M., and Mutungi, G. (2011). Dihydrotestosterone stimulates amino acid uptake and the expression of LAT2 in mouse skeletal muscle fibres through an ERK1/2-dependent mechanism. *J. Physiol.* 589, 3623–3640. doi: 10.1113/jphysiol.2011.207175

Hansen, I. A., Boudko, D. Y., Shiao, S. H., Voronov, D. A., Meleshkevitch, E. A., Drake, L. L., et al. (2011). AaCAT1 of the yellow fever mosquito, *Aedes aegypti*: a novel histidine-specific amino acid transporter from the SLC7 family. *J. Biol. Chem.* 286, 10803–10813. doi: 10.1074/jbc.M110.179739

Hassanein, M., Hoeksema, M. D., Shiota, M., Qian, J., Harris, B. K., Chen, H., et al. (2013). SLC1A5 mediates glutamine transport required for lung cancer cell growth and survival. *Clin. Cancer Res.* 19, 560–570. doi: 10.1158/1078-0432.CCR-12-2334

Hayashi, K., Jutabha, P., Endou, H., and Anzai, N. (2012). c-Myc is crucial for the expression of LAT1 in MIA Paca-2 human pancreatic cancer cells. *Oncol. Rep.* 28, 862–866. doi: 10.3892/or.2012.1878

Herzig, S., Raemy, E., Montessuit, S., Veuthey, J. L., Zamboni, N., Westermann, B., et al. (2012). Identification and functional expression of the mitochondrial pyruvate carrier. *Science* 337, 93–96. doi: 10.1126/science.1218530

Holecek, M. (2013). Side effects of long-term glutamine supplementation. *JPEN J. Parenter. Enteral Nutr.* 37, 607–616. doi: 10.1177/0148607112460682

Imai, H., Kaira, K., Oriuchi, N., Shimizu, K., Tominaga, H., Yanagitani, N., et al. (2010). Inhibition of L-type amino acid transporter 1 has antitumor activity in non-small cell lung cancer. *Anticancer Res.* 30, 4819–4828.

Indiveri, C., Abruzzo, G., Stipani, I., and Palmieri, F. (1998). Identification and purification of the reconstitutively active glutamine carrier from rat kidney mitochondria. *Biochem. J.* 333(Pt 2), 285–290.

Indiveri, C., Galluccio, M., Scalise, M., and Pochini, L. (2013). Strategies of bacterial over expression of membrane transporters relevant in human health: the successful case of the three members of OCTN subfamily. *Mol. Biotechnol.* 54, 724–736. doi: 10.1007/s12033-012-9586-8

Indiveri, C., Palmieri, L., and Palmieri, F. (1994). Kinetic characterization of the reconstituted ornithine carrier from rat liver mitochondria. *Biochim. Biophys. Acta* 1188, 293–301. doi: 10.1016/0005-2728(94)90048-5

Indiveri, C., Pochini, L., Galluccio, M., and Scalise, M. (2014). SLC1A5 (solute carrier family 1 (neutral amino acid transporter), member 5). *Atlas Genet. Cytogenet. Oncol. Haematol.* Available online at: http://AtlasGeneticsOncology.org/Genes/SLC1A5ID42313ch19q13.html

Jenstad, M., and Chaudhry, F. A. (2013). The amino acid transporters of the glutamate/GABA-glutamine cycle and their impact on insulin and glucagon secretion. *Front. Endocrinol. (Lausanne)* 4:199. doi: 10.3389/fendo.2013.00199

Jordan, I. K., Kondrashov, F. A., Adzhubei, I. A., Wolf, Y. I., Koonin, E. V., Kondrashov, A. S., et al. (2005). A universal trend of amino acid gain and loss in protein evolution. *Nature* 433, 633–638. doi: 10.1038/nature03306

Kageyama, T., Nakamura, M., Matsuo, A., Yamasaki, Y., Takakura, Y., Hashida, M., et al. (2000). The 4F2hc/LAT1 complex transports L-DOPA across the blood-brain barrier. *Brain Res.* 879, 115–121. doi: 10.1016/S0006-8993(00)02758-X

Kaira, K., Oriuchi, N., Imai, H., Shimizu, K., Yanagitani, N., Sunaga, N., et al. (2010). Prognostic significance of L-type amino acid transporter 1 (LAT1) and 4F2 heavy chain (CD98) expression in surgically resectable stage III non-small cell lung cancer. *Exp. Ther. Med.* 1, 799–808. doi: 10.3892/etm.2010.117

Kaira, K., Oriuchi, N., Takahashi, T., Nakagawa, K., Ohde, Y., Okumura, T., et al. (2011). L-type amino acid transporter 1 (LAT1) expression in malignant pleural mesothelioma. *Anticancer Res.* 31, 4075–4082.

Kaira, K., Sunose, Y., Ohshima, Y., Ishioka, N. S., Arakawa, K., Ogawa, T., et al. (2013). Clinical significance of L-type amino acid transporter 1 expression as a prognostic marker and potential of new targeting therapy in biliary tract cancer. *BMC Cancer* 13:482. doi: 10.1186/1471-2407-13-482

Kaji, M., Kabir-Salmani, M., Anzai, N., Jin, C. J., Akimoto, Y., Horita, A., et al. (2010). Properties of L-type amino acid transporter 1 in epidermal ovarian cancer. *Int. J. Gynecol. Cancer* 20, 329–336. doi: 10.1111/IGC.0b013e3181d28e13

Kanai, Y., Clemencon, B., Simonin, A., Leuenberger, M., Lochner, M., Weisstanner, M., et al. (2013). The SLC1 high-affinity glutamate and neutral amino acid transporter family. *Mol. Aspects Med.* 34, 108–120. doi: 10.1016/j.mam.2013.01.001

Kanai, Y., Segawa, H., Miyamoto, K., Uchino, H., Takeda, E., and Endou, H. (1998). Expression cloning and characterization of a transporter for large neutral amino acids activated by the heavy chain of 4F2 antigen (CD98). *J. Biol. Chem.* 273, 23629–23632. doi: 10.1074/jbc.273.37.23629

Karunakaran, S., Ramachandran, S., Coothankandaswamy, V., Elangovan, S., Babu, E., Periyasamy-Thandavan, S., et al. (2011). SLC6A14 (ATB0,+) protein, a highly concentrative and broad specific amino acid transporter, is a novel and effective drug target for treatment of estrogen receptor-positive breast cancer. *J. Biol. Chem.* 286, 31830–31838. doi: 10.1074/jbc.M111.229518

Katt, W. P., and Cerione, R. A. (2014). Glutaminase regulation in cancer cells: a druggable chain of events. *Drug Discov. Today* 19, 450–457. doi: 10.1016/j.drudis.2013.10.008

Kekuda, R., Prasad, P. D., Fei, Y. J., Torres-Zamorano, V., Sinha, S., Yang-Feng, T. L., et al. (1996). Cloning of the sodium-dependent, broad-scope, neutral amino acid transporter Bo from a human placental choriocarcinoma cell line. *J. Biol. Chem.* 271, 18657–18661. doi: 10.1074/jbc.271.31.18657

Kekuda, R., Torres-Zamorano, V., Fei, Y. J., Prasad, P. D., Li, H. W., Mader, L. D., et al. (1997). Molecular and functional characterization of intestinal Na(+)-dependent neutral amino acid transporter B0. *Am. J. Physiol.* 272, G1463–G1472.

Kilberg, M. S., and Christensen, H. N. (1980). The relation between membrane potential and the transport activity of system A and L in plasma membrane vesicles of the Ehrlich cell. *Membr. Biochem.* 3, 155–168. doi: 10.3109/09687688009063883

Kim, M. S., Park, Y. S., Kim, S. H., Kim, S. Y., Lee, M. H., Kim, Y. H., et al. (2012). Quantification of nimesulide in human plasma by high-performance liquid chromatography with ultraviolet detector (HPLC-UV): application to pharmacokinetic studies in 28 healthy Korean subjects. *J. Chromatogr. Sci.* 50, 396–400. doi: 10.1093/chromsci/bms014

Kinne, A., Schulein, R., and Krause, G. (2011). Primary and secondary thyroid hormone transporters. *Thyroid Res.* 4(Suppl. 1), S7. doi: 10.1186/1756-6614-4-S1-S7

Kleta, R., Romeo, E., Ristic, Z., Ohura, T., Stuart, C., Arcos-Burgos, M., et al. (2004). Mutations in SLC6A19, encoding B0AT1, cause Hartnup disorder. *Nat. Genet.* 36, 999–1002. doi: 10.1038/ng1405

Kondoh, N., Imazeki, N., Arai, M., Hada, A., Hatsuse, K., Matsuo, H., et al. (2007). Activation of a system A amino acid transporter, ATA1/SLC38A1, in human hepatocellular carcinoma and preneoplastic liver tissues. *Int. J. Oncol.* 31, 81–87. doi: 10.3892/ijo.31.1.81

Kondou, H., Kawai, M., Tachikawa, K., Kimoto, A., Yamagata, M., Koinuma, T., et al. (2013). Sodium-coupled neutral amino acid transporter 4 functions as a regulator of protein synthesis during liver development. *Hepatol. Res.* 43, 1211–1223. doi: 10.1111/hepr.12069

Krishnamurthy, H., and Gouaux, E. (2012). X-ray structures of LeuT in substrate-free outward-open and apo inward-open states. *Nature* 481, 469–474. doi: 10.1038/nature10737

Kurayama, R., Ito, N., Nishibori, Y., Fukuhara, D., Akimoto, Y., Higashihara, E., et al. (2011). Role of amino acid transporter LAT2 in the activation of mTORC1 pathway and the pathogenesis of crescentic glomerulonephritis. *Lab. Invest.* 91, 992–1006. doi: 10.1038/labinvest.2011.43

Li, J., Qiang, J., Chen, S. F., Wang, X., Fu, J., and Chen, Y. (2013). The impact of L-type amino acid transporter 1 (LAT1) in human hepatocellular carcinoma. *Tumour Biol.* 34, 2977–2981. doi: 10.1007/s13277-013-0861-5

Ma, D., Lu, P., Yan, C., Fan, C., Yin, P., Wang, J., et al. (2012). Structure and mechanism of a glutamate-GABA antiporter. *Nature* 483, 632–636. doi: 10.1038/nature10917

Mackenzie, B., and Erickson, J. D. (2004). Sodium-coupled neutral amino acid (System N/A) transporters of the SLC38 gene family. *Pflugers Arch.* 447, 784–795. doi: 10.1007/s00424-003-1117-9

Mackenzie, B., Schafer, M. K., Erickson, J. D., Hediger, M. A., Weihe, E., and Varoqui, H. (2003). Functional properties and cellular distribution of the system a glutamine transporter SNAT1 support specialized roles in central neurons. *J. Biol. Chem.* 278, 23720–23730. doi: 10.1074/jbc.M212718200

Marin, M., Lavillette, D., Kelly, S. M., and Kabat, D. (2003). N-linked glycosylation and sequence changes in a critical negative control region of the ASCT1 and ASCT2 neutral amino acid transporters determine their retroviral receptor functions. *J. Virol.* 77, 2936–2945. doi: 10.1128/JVI.77.5.2936-2945.2003

Mastroberardino, L., Spindler, B., Pfeiffer, R., Skelly, P. J., Loffing, J., Shoemaker, C. B., et al. (1998). Amino-acid transport by heterodimers of 4F2hc/CD98 and members of a permease family. *Nature* 395, 288–291. doi: 10.1038/26246

Matsuyama, R., Tomi, M., Akanuma, S., Tabuchi, A., Kubo, Y., Tachikawa, M., et al. (2012). Up-regulation of L-type amino acid transporter 1 (LAT1) in cultured rat retinal capillary endothelial cells in response to glucose deprivation. *Drug Metab. Pharmacokinet.* 27, 317–324. doi: 10.2133/dmpk.DMPK-11-RG-122

McGivan, J. D., and Bungard, C. I. (2007). The transport of glutamine into mammalian cells. *Front. Biosci.* 12:874–882. doi: 10.2741/2109

Metallo, C. M. (2012). Expanding the reach of cancer metabolomics. *Cancer Prev. Res. (Phila.)* 5, 1337–1340. doi: 10.1158/1940-6207.CAPR-12-0433

Miroux, B., and Walker, J. E. (1996). Over-production of proteins in *Escherichia coli*: mutant hosts that allow synthesis of some membrane proteins and globular proteins at high levels. *J. Mol. Biol.* 260, 289–298. doi: 10.1006/jmbi.1996.0399

Nakanishi, T., Hatanaka, T., Huang, W., Prasad, P. D., Leibach, F. H., Ganapathy, M. E., et al. (2001). Na+- and Cl−-coupled active transport of carnitine by the amino acid transporter ATB(0,+) from mouse colon expressed in HRPE cells and Xenopus oocytes. *J. Physiol.* 532, 297–304. doi: 10.1111/j.1469-7793.2001.0297f.x

Newsholme, P., Procopio, J., Lima, M. M., Pithon-Curi, T. C., and Curi, R. (2003). Glutamine and glutamate–their central role in cell metabolism and function. *Cell Biochem. Funct.* 21, 1–9. doi: 10.1002/cbf.1003

Nicklin, P., Bergman, P., Zhang, B., Triantafellow, E., Wang, H., Nyfeler, B., et al. (2009). Bidirectional transport of amino acids regulates mTOR and autophagy. *Cell* 136, 521–534. doi: 10.1016/j.cell.2008.11.044

Nii, T., Segawa, H., Taketani, Y., Tani, Y., Ohkido, M., Kishida, S., et al. (2001). Molecular events involved in up-regulating human Na+-independent neutral amino acid transporter LAT1 during T-cell activation. *Biochem. J.* 358, 693–704. doi: 10.1042/0264-6021:3580693

Nissen-Meyer, L. S., and Chaudhry, F. A. (2013). Protein kinase C phosphorylates the system N glutamine transporter SN1 (Slc38a3) and regulates its membrane trafficking and degradation. *Front. Endocrinol. (Lausanne)* 4:138. doi: 10.3389/fendo.2013.00138

Ogura, M., Taniura, H., Nakamichi, N., and Yoneda, Y. (2007). Upregulation of the glutamine transporter through transactivation mediated by cAMP/protein kinase a signals toward exacerbation of vulnerability to oxidative stress in rat neocortical astrocytes. *J. Cell. Physiol.* 212, 375–385. doi: 10.1002/jcp.21031

Ohh, M. (2012). Tumor strengths and frailties: cancer SUMmOns Otto's metabolism. *Nat. Med.* 18, 30–31. doi: 10.1038/nm.2631

Ohno, H., Nakatsu, Y., Sakoda, H., Kushiyama, A., Ono, H., Fujishiro, M., et al. (2011). 4F2hc stabilizes GLUT1 protein and increases glucose transport activity. *Am. J. Physiol. Cell Physiol.* 300, C1047–C1054. doi: 10.1152/ajpcell.004 16.2010

Okudaira, H., Shikano, N., Nishii, R., Miyagi, T., Yoshimoto, M., Kobayashi, M., et al. (2011). Putative transport mechanism and intracellular fate of trans-1-amino-3-18F-fluorocyclobutanecarboxylic acid in human prostate cancer. *J. Nucl. Med.* 52, 822–829. doi: 10.2967/jnumed.110.086074

O'mara, M., Oakley, A., and Broer, S. (2006). Mechanism and putative structure of B(0)-like neutral amino acid transporters. *J. Membr. Biol.* 213, 111–118. doi: 10.1007/s00232-006-0879-3

Oppedisano, F., Catto, M., Koutentis, P. A., Nicolotti, O., Pochini, L., Koyioni, M., et al. (2012). Inactivation of the glutamine/amino acid transporter ASCT2 by 1,2,3-dithiazoles: proteoliposomes as a tool to gain insights in the molecular mechanism of action and of antitumor activity. *Toxicol. Appl. Pharmacol.* 265, 93–102. doi: 10.1016/j.taap.2012.09.011

Oppedisano, F., Galluccio, M., and Indiveri, C. (2010). Inactivation by Hg2+ and methylmercury of the glutamine/amino acid transporter (ASCT2) reconstituted in liposomes: Prediction of the involvement of a CXXC motif by homology modelling. *Biochem. Pharmacol.* 80, 1266–1273. doi: 10.1016/j.bcp.2010.06.032

Oppedisano, F., and Indiveri, C. (2008). Reconstitution into liposomes of the B degrees -like glutamine-neutral amino acid transporter from renal cell plasma membrane. *Biochim. Biophys. Acta* 1778, 2258–2265. doi: 10.1016/j.bbamem.2008.05.011

Oppedisano, F., Pochini, L., Broer, S., and Indiveri, C. (2011). The B degrees AT1 amino acid transporter from rat kidney reconstituted in liposomes: kinetics and inactivation by methylmercury. *Biochim. Biophys. Acta* 1808, 2551–2558. doi: 10.1016/j.bbamem.2011.05.011

Oppedisano, F., Pochini, L., Galluccio, M., Cavarelli, M., and Indiveri, C. (2004). Reconstitution into liposomes of the glutamine/amino acid transporter from renal cell plasma membrane: functional characterization, kinetics and activation by nucleotides. *Biochim. Biophys. Acta* 1667, 122–131. doi: 10.1016/j.bbamem.2004.09.007

Oppedisano, F., Pochini, L., Galluccio, M., and Indiveri, C. (2007). The glutamine/amino acid transporter (ASCT2) reconstituted in liposomes: transport mechanism, regulation by ATP and characterization of the glutamine/glutamate antiport. *Biochim. Biophys. Acta* 1768, 291–298. doi: 10.1016/j.bbamem.2006.09.002

Ortiz, V., Aleman, G., Escamilla-Del-Arenal, M., Recillas-Targa, F., Torres, N., and Tovar, A. R. (2011). Promoter characterization and role of CRE in the basal transcription of the rat SNAT2 gene. *Am. J. Physiol. Endocrinol. Metab.* 300, E1092–E1102. doi: 10.1152/ajpendo.00459.2010

Oxender, D. L., and Christensen, H. N. (1963). Evidence for two types of mediation of neutral and amino-acid transport in Ehrlich cells. *Nature* 197, 765–767. doi: 10.1038/197765a0

Padmanabhan Iyer, R., Gu, S., Nicholson, B. J., and Jiang, J. X. (2013). Identification of a disulfide bridge important for transport function of SNAT4 neutral amino acid transporter. *PLoS ONE* 8:e56792. doi: 10.1371/journal.pone.0056792

Palacin, M., and Kanai, Y. (2004). The ancillary proteins of HATs: SLC3 family of amino acid transporters. *Pflugers Arch.* 447, 490–494. doi: 10.1007/s00424-003-1062-7

Palii, S. S., Chen, H., and Kilberg, M. S. (2004). Transcriptional control of the human sodium-coupled neutral amino acid transporter system a gene by amino acid availability is mediated by an intronic element. *J. Biol. Chem.* 279, 3463–3471. doi: 10.1074/jbc.M310483200

Palmada, M., Speil, A., Jeyaraj, S., Bohmer, C., and Lang, F. (2005). The serine/threonine kinases SGK1, 3 and PKB stimulate the amino acid transporter ASCT2. *Biochem. Biophys. Res. Commun.* 331, 272–277. doi: 10.1016/j.bbrc.2005.03.159

Palmieri, F. (2008). Diseases caused by defects of mitochondrial carriers: a review. *Biochim. Biophys. Acta* 1777, 564–578. doi: 10.1016/j.bbabio.2008.03.008

Park, S. Y., Kim, J. K., Kim, I. J., Choi, B. K., Jung, K. Y., Lee, S., et al. (2005). Reabsorption of neutral amino acids mediated by amino acid transporter LAT2 and TAT1 in the basolateral membrane of proximal tubule. *Arch. Pharm. Res.* 28, 421–432. doi: 10.1007/BF02977671

Patel, M., Dalvi, P., Gokulgandhi, M., Kesh, S., Kohli, T., Pal, D., et al. (2013). Functional characterization and molecular expression of large neutral amino acid transporter (LAT1) in human prostate cancer cells. *Int. J. Pharm.* 443, 245–253. doi: 10.1016/j.ijpharm.2012.12.029

Penmatsa, A., Wang, K. H., and Gouaux, E. (2013). X-ray structure of dopamine transporter elucidates antidepressant mechanism. *Nature* 503, 85–90. doi: 10.1038/nature12533

Pfeiffer, R., Rossier, G., Spindler, B., Meier, C., Kuhn, L., and Verrey, F. (1999). Amino acid transport of y+L-type by heterodimers of 4F2hc/CD98 and members of the glycoprotein-associated amino acid transporter family. *EMBO J.* 18, 49–57. doi: 10.1093/emboj/18.1.49

Pfeiffer, R., Spindler, B., Loffing, J., Skelly, P. J., Shoemaker, C. B., and Verrey, F. (1998). Functional heterodimeric amino acid transporters lacking cysteine residues involved in disulfide bond. *FEBS Lett.* 439, 157–162. doi: 10.1016/S0014-5793(98)01359-3

Pineda, M., Fernandez, E., Torrents, D., Estevez, R., Lopez, C., Camps, M., et al. (1999). Identification of a membrane protein, LAT-2, that Co-expresses with 4F2 heavy chain, an L-type amino acid transport activity with broad specificity for small and large zwitterionic amino acids. *J. Biol. Chem.* 274, 19738–19744. doi: 10.1074/jbc.274.28.19738

Pingitore, P., Pochini, L., Scalise, M., Galluccio, M., Hedfalk, K., and Indiveri, C. (2013). Large scale production of the active human ASCT2 (SLC1A5) transporter in *Pichia pastoris*–functional and kinetic asymmetry revealed in proteoliposomes. *Biochim. Biophys. Acta* 1828, 2238–2246. doi: 10.1016/j.bbamem.2013.05.034

Pochini, L., Scalise, M., Galluccio, M., and Indiveri, C. (2013). OCTN cation transporters in health and disease: role as drug targets and assay development. *J. Biomol. Screen.* 18, 851–867. doi: 10.1177/1087057113493006

Pochini, L., Seidita, A., Sensi, C., Scalise, M., Eberini, I., and Indiveri, C. (2014). Nimesulide binding site in the B0AT1 (SLC6A19) amino acid transporter. mechanism of inhibition revealed by proteoliposome transport assay and molecular modelling. *Biochem. Pharmacol.* 89, 422–430. doi: 10.1016/j.bcp.2014.03.014

Poncet, N., Mitchell, F. E., Ibrahim, A. F., McGuire, V. A., English, G., Arthur, J. S., et al. (2014). The catalytic subunit of the system L1 amino acid transporter (slc7a5) facilitates nutrient signalling in mouse skeletal muscle. *PLoS ONE* 9:e89547. doi: 10.1371/journal.pone.0089547

Pramod, A. B., Foster, J., Carvelli, L., and Henry, L. K. (2013). SLC6 transporters: structure, function, regulation, disease association and therapeutics. *Mol. Aspects Med.* 34, 197–219. doi: 10.1016/j.mam.2012.07.002

Prasad, P. D., Wang, H., Huang, W., Kekuda, R., Rajan, D. P., Leibach, F. H., et al. (1999). Human LAT1, a subunit of system L amino acid transporter: molecular cloning and transport function. *Biochem. Biophys. Res. Commun.* 255, 283–288. doi: 10.1006/bbrc.1999.0206

Rask-Andersen, M., Masuram, S., Fredriksson, R., and Schioth, H. B. (2013). Solute carriers as drug targets: current use, clinical trials and prospective. *Mol. Aspects Med.* 34, 702–710. doi: 10.1016/j.mam.2012.07.015

Rathore, M. G., Saumet, A., Rossi, J. F., De Bettignies, C., Tempe, D., Lecellier, C. H., et al. (2012). The NF-kappaB member p65 controls glutamine metabolism through miR-23a. *Int. J. Biochem. Cell Biol.* 44, 1448–1456. doi: 10.1016/j.biocel.2012.05.011

Reyes, N., Ginter, C., and Boudker, O. (2009). Transport mechanism of a bacterial homologue of glutamate transporters. *Nature* 462, 880–885. doi: 10.1038/nature08616

Reynolds, M. R., Lane, A. N., Robertson, B., Kemp, S., Liu, Y., Hill, B. G., et al. (2014). Control of glutamine metabolism by the tumor suppressor Rb. *Oncogene* 33, 556–566. doi: 10.1038/onc.2012.635

Rooney, J. P. (2007). The role of thiols, dithiols, nutritional factors and interacting ligands in the toxicology of mercury. *Toxicology* 234, 145–156. doi: 10.1016/j.tox.2007.02.016

Rosell, A., Meury, M., Alvarez-Marimon, E., Costa, M., Perez-Cano, L., Zorzano, A., et al. (2014). Structural bases for the interaction and stabilization of the human amino acid transporter LAT2 with its ancillary protein 4F2hc. *Proc. Natl. Acad. Sci. U.S.A.* 111, 2966–2971. doi: 10.1073/pnas.1323779111

Saha, P., Arthur, S., Kekuda, R., and Sundaram, U. (2012). Na-glutamine cotransporters B(0)AT1 in villus and SN2 in crypts are differentially altered in chronically inflamed rabbit intestine. *Biochim. Biophys. Acta* 1818, 434–442. doi: 10.1016/j.bbamem.2011.11.005

Sali, A., and Blundell, T. L. (1993). Comparative protein modelling by satisfaction of spatial restraints. *J. Mol. Biol.* 234, 779–815. doi: 10.1006/jmbi.1993.1626

Sastrasinh, S., and Sastrasinh, M. (1989). Glutamine transport in submitochondrial particles. *Am. J. Physiol.* 257, F1050–F1058.

Savas, S., Schmidt, S., Jarjanazi, H., and Ozcelik, H. (2006). Functional nsSNPs from carcinogenesis-related genes expressed in breast tissue: potential breast cancer risk alleles and their distribution across human populations. *Hum. Genomics* 2, 287–296. doi: 10.1186/1479-7364-2-5-287

Scalise, M., Pochini, L., Giangregorio, N., Tonazzi, A., and Indiveri, C. (2013). Proteoliposomes as tool for assaying membrane transporter functions and interactions with xenobiotics. *Pharmaceutics* 5, 472–497. doi: 10.3390/pharmaceutics5030472

Schioth, H. B., Roshanbin, S., Hagglund, M. G., and Fredriksson, R. (2013). Evolutionary origin of amino acid transporter families SLC32, SLC36 and SLC38 and physiological, pathological and therapeutic aspects. *Mol. Aspects Med.* 34, 571–585. doi: 10.1016/j.mam.2012.07.012

Schneider, H. P., Broer, S., Broer, A., and Deitmer, J. W. (2007). Heterologous expression of the glutamine transporter SNAT3 in Xenopus oocytes is associated with four modes of uncoupled transport. *J. Biol. Chem.* 282, 3788–3798. doi: 10.1074/jbc.M609452200

Segawa, A., Nagamori, S., Kanai, Y., Masawa, N., and Oyama, T. (2013). L-type amino acid transporter 1 expression is highly correlated with Gleason score in prostate cancer. *Mol. Clin. Oncol.* 1, 274–280. doi: 10.3892/mco.2012.54

Segawa, H., Fukasawa, Y., Miyamoto, K., Takeda, E., Endou, H., and Kanai, Y. (1999). Identification and functional characterization of a Na+-independent neutral amino acid transporter with broad substrate selectivity. *J. Biol. Chem.* 274, 19745–19751. doi: 10.1074/jbc.274.28.19745

Seow, H. F., Broer, S., Broer, A., Bailey, C. G., Potter, S. J., Cavanaugh, J. A., et al. (2004). Hartnup disorder is caused by mutations in the gene encoding the neutral amino acid transporter SLC6A19. *Nat. Genet.* 36, 1003–1007. doi: 10.1038/ng1406

Shaffer, P. L., Goehring, A., Shankaranarayanan, A., and Gouaux, E. (2009). Structure and mechanism of a Na+-independent amino acid transporter. *Science* 325, 1010–1014. doi: 10.1126/science.1176088

Shanware, N. P., Mullen, A. R., Deberardinis, R. J., and Abraham, R. T. (2011). Glutamine: pleiotropic roles in tumor growth and stress resistance. *J. Mol. Med. (Berl)* 89, 229–236. doi: 10.1007/s00109-011-0731-9

Shi, L., Luo, W., Huang, W., Huang, S., and Huang, G. (2013). Downregulation of L-type amino acid transporter 1 expression inhibits the growth, migration and invasion of gastric cancer cells. *Oncol. Lett.* 6, 106–112. doi: 10.3892/ol.2013.1342

Shi, Q., Padmanabhan, R., Villegas, C. J., Gu, S., and Jiang, J. X. (2011). Membrane topological structure of neutral system N/A amino acid transporter 4 (SNAT4) protein. *J. Biol. Chem.* 286, 38086–38094. doi: 10.1074/jbc.M111.220277

Shimizu, K., Kaira, K., Tomizawa, Y., Sunaga, N., Kawashima, O., Oriuchi, N., et al. (2014). ASC amino-acid transporter 2 (ASCT2) as a novel prognostic marker in non-small cell lung cancer. *Br. J. Cancer* 110, 2030–2039. doi: 10.1038/bjc.2014.88

Sidoryk, M., Matyja, E., Dybel, A., Zielinska, M., Bogucki, J., Jaskolski, D. J., et al. (2004). Increased expression of a glutamine transporter SNAT3 is a marker of malignant gliomas. *Neuroreport* 15, 575–578. doi: 10.1097/00001756-200403220-00001

Sloan, J. L., Grubb, B. R., and Mager, S. (2003). Expression of the amino acid transporter ATB 0+ in lung: possible role in luminal protein removal. *Am. J. Physiol. Lung Cell. Mol. Physiol.* 284, L39–L49. doi: 10.1152/ajplung.00164.2002

Sloan, J. L., and Mager, S. (1999). Cloning and functional expression of a human Na(+) and Cl(-)-dependent neutral and cationic amino acid transporter B(0+). *J. Biol. Chem.* 274, 23740–23745. doi: 10.1074/jbc.274.34.23740

Solbu, T. T., Bjorkmo, M., Berghuis, P., Harkany, T., and Chaudhry, F. A. (2010). SAT1, A Glutamine transporter, is preferentially expressed in GABAergic neurons. *Front. Neuroanat.* 4:1. doi: 10.3389/neuro.05.001.2010

Son, J., Lyssiotis, C. A., Ying, H., Wang, X., Hua, S., Ligorio, M., et al. (2013). Glutamine supports pancreatic cancer growth through a KRAS-regulated metabolic pathway. *Nature* 496, 101–105. doi: 10.1038/nature12040

Souba, W. W., Pan, M., and Stevens, B. R. (1992). Kinetics of the sodium-dependent glutamine transporter in human intestinal cell confluent monolayers. *Biochem. Biophys. Res. Commun.* 188, 746–753. doi: 10.1016/0006-291X(92)91119-B

Stumvoll, M., Perriello, G., Meyer, C., and Gerich, J. (1999). Role of glutamine in human carbohydrate metabolism in kidney and other tissues. *Kidney Int.* 55, 778–792. doi: 10.1046/j.1523-1755.1999.055003778.x

Sundberg, B. E., Waag, E., Jacobsson, J. A., Stephansson, O., Rumaks, J., Svirskis, S., et al. (2008). The evolutionary history and tissue mapping of amino acid transporters belonging to solute carrier families SLC32, SLC36, and SLC38. *J. Mol. Neurosci.* 35, 179–193. doi: 10.1007/s12031-008-9046-x

Tonazzi, A., and Indiveri, C. (2003). Chemical modification of the mitochondrial ornithine/citrulline carrier by SH reagents: effects on the transport activity and transition from carrier to pore-like function. *Biochim. Biophys. Acta* 1611, 123–130. doi: 10.1016/S0005-2736(03)00033-6

Tonazzi, A., and Indiveri, C. (2011). Effects of heavy metal cations on the mitochondrial ornithine/citrulline transporter reconstituted in liposomes. *Biometals* 24, 1205–1215. doi: 10.1007/s10534-011-9479-5

Torchetti, E. M., Bonomi, F., Galluccio, M., Gianazza, E., Giancaspero, T. A., Iametti, S., et al. (2011). Human FAD synthase (isoform 2): a component of the machinery that delivers FAD to apo-flavoproteins. *FEBS J.* 278, 4434–4449. doi: 10.1111/j.1742-4658.2011.08368.x

Torres-Zamorano, V., Leibach, F. H., and Ganapathy, V. (1998). Sodium-dependent homo- and hetero-exchange of neutral amino acids mediated by the amino acid transporter ATB degree. *Biochem. Biophys. Res. Commun.* 245, 824–829. doi: 10.1006/bbrc.1998.8434

Utsunomiya-Tate, N., Endou, H., and Kanai, Y. (1996). Cloning and functional characterization of a system ASC-like Na+-dependent neutral amino acid transporter. *J. Biol. Chem.* 271, 14883–14890. doi: 10.1074/jbc.271.25.14883

Vander Heiden, M. G. (2011). Targeting cancer metabolism: a therapeutic window opens. *Nat. Rev. Drug Discov.* 10, 671–684. doi: 10.1038/nrd3504

Vander Heiden, M. G., Cantley, L. C., and Thompson, C. B. (2009). Understanding the Warburg effect: the metabolic requirements of cell proliferation. *Science* 324, 1029–1033. doi: 10.1126/science.1160809

Verrey, F. (2003). System L: heteromeric exchangers of large, neutral amino acids involved in directional transport. *Pflugers Arch.* 445, 529–533. doi: 10.1007/s00424-002-0973-z

Verrey, F., Closs, E. I., Wagner, C. A., Palacin, M., Endou, H., and Kanai, Y. (2004). CATs and HATs: the SLC7 family of amino acid transporters. *Pflugers Arch.* 447, 532–542. doi: 10.1007/s00424-003-1086-z

Verrey, F., Ristic, Z., Romeo, E., Ramadan, T., Makrides, V., Dave, M. H., et al. (2005). Novel renal amino acid transporters. *Annu. Rev. Physiol.* 67, 557–572. doi: 10.1146/annurev.physiol.67.031103.153949

Verrey, F., Singer, D., Ramadan, T., Vuille-Dit-Bille, R. N., Mariotta, L., and Camargo, S. M. (2009). Kidney amino acid transport. *Pflugers Arch.* 458, 53–60. doi: 10.1007/s00424-009-0638-2

Wagner, C. A., Lang, F., and Broer, S. (2001). Function and structure of heterodimeric amino acid transporters. *Am. J. Physiol. Cell Physiol.* 281, C1077–C1093.

Walker, D. K., Drummond, M. J., Dickinson, J. M., Borack, M. S., Jennings, K., Volpi, E., et al. (2014). Insulin increases mRNA abundance of the amino acid transporter SLC7A5/LAT1 via an mTORC1-dependent mechanism in skeletal muscle cells. *Physiol. Rep.* 2, e00238. doi: 10.1002/phy2.238

Wang, K., Cao, F., Fang, W., Hu, Y., Chen, Y., Ding, H., et al. (2013). Activation of SNAT1/SLC38A1 in human breast cancer: correlation with p-Akt overexpression. *BMC Cancer* 13:343. doi: 10.1186/1471-2407-13-343

Wasa, M., Wang, H. S., and Okada, A. (2002). Characterization of L-glutamine transport by a human neuroblastoma cell line. *Am. J. Physiol. Cell Physiol.* 282, C1246–C1253. doi: 10.1152/ajpcell.00324.2001

Watanabe, J., Yokoyama, Y., Futagami, M., Mizunuma, H., Yoshioka, H., Washiya, K., et al. (2014). L-type amino acid transporter 1 expression increases in well-differentiated but decreases in poorly differentiated endometrial endometrioid adenocarcinoma and shows an inverse correlation with p53 expression. *Int. J. Gynecol. Cancer* 24, 659–663. doi: 10.1097/IGC.0000000000000123

Witte, D., Ali, N., Carlson, N., and Younes, M. (2002). Overexpression of the neutral amino acid transporter ASCT2 in human colorectal adenocarcinoma. *Anticancer Res.* 22, 2555–2557.

Xu, D., and Hemler, M. E. (2005). Metabolic activation-related CD147-CD98 complex. *Mol. Cell. Proteomics* 4, 1061–1071. doi: 10.1074/mcp.M400207-MCP200

Yamashita, A., Singh, S. K., Kawate, T., Jin, Y., and Gouaux, E. (2005). Crystal structure of a bacterial homologue of Na+/Cl–dependent neurotransmitter transporters. *Nature* 437, 215–223. doi: 10.1038/nature03978

Yanagisawa, N., Ichinoe, M., Mikami, T., Nakada, N., Hana, K., Koizumi, W., et al. (2012). High expression of L-type amino acid transporter 1 (LAT1) predicts poor prognosis in pancreatic ductal adenocarcinomas. *J. Clin. Pathol.* 65, 1019–1023. doi: 10.1136/jclinpath-2012-200826

Yernool, D., Boudker, O., Jin, Y., and Gouaux, E. (2004). Structure of a glutamate transporter homologue from Pyrococcus horikoshii. *Nature* 431, 811–818. doi: 10.1038/nature03018

Yoon, J. H., Kim, I. J., Kim, H., Kim, H. J., Jeong, M. J., Ahn, S. G., et al. (2005). Amino acid transport system L is differently expressed in human normal oral keratinocytes and human oral cancer cells. *Cancer Lett.* 222, 237–245. doi: 10.1016/j.canlet.2004.09.040

Zander, C. B., Albers, T., and Grewer, C. (2013). Voltage-dependent processes in the electroneutral amino acid exchanger ASCT2. *J. Gen. Physiol.* 141, 659–672. doi: 10.1085/jgp.201210948

Zhang, Z., Albers, T., Fiumera, H. L., Gameiro, A., and Grewer, C. (2009). A conserved Na(+) binding site of the sodium-coupled neutral amino acid transporter 2 (SNAT2). *J. Biol. Chem.* 284, 25314–25323. doi: 10.1074/jbc.M109. 038422

Ziegler, T. R., Bazargan, N., Leader, L. M., and Martindale, R. G. (2000). Glutamine and the gastrointestinal tract. *Curr. Opin. Clin. Nutr. Metab. Care* 3, 355–362. doi: 10.1097/00075197-200009000-00005

Conflict of Interest Statement: The authors declare that the research was conducted in the absence of any commercial or financial relationships that could be construed as a potential conflict of interest.

7

Microwave induced synthesis of graft copolymer of binary vinyl monomer mixtures onto delignified *Grewia optiva* fiber: application in dye removal

Vinod Kumar Gupta[1]*, Deepak Pathania[2]*, Bhanu Priya[2], Amar Singh Singha[3] and Gaurav Sharma[2]

[1] *Department of Chemistry, Indian Institute of Technology Roorkee, Roorkee, India*
[2] *Department of Chemistry, Shoolini University of Biotechnology and Management Sciences, Solan, India*
[3] *Department of Applied Chemistry, National Institute of Technology Hamirpur, Hamirpur, India*

Edited by:
Shusheng Zhang, Qingdao University of Science and Technology, Linyi University, China

Reviewed by:
Raquel Santos Mauler, Federal University of Rio Grande do Sul, Brazil
Daisuke Takeuchi, Tokyo Institute of Technology, Japan

*Correspondence:
Vinod Kumar Gupta, Department of Chemistry, Indian Institute of Technology Roorkee, Roorkee 247667, India
e-mail: vinodfcy@gmail.com;
Deepak Pathania, Department of Chemistry, Shoolini University of Biotechnology and Management Sciences, Solan, India
e-mail: dpathania74@gmail.com

Grafting method, through microwave radiation technique is very effective in terms of time consumption, cost effectiveness and environmental friendliness. Via this method, delignified *Grewia optiva* identified as a waste biomass, was graft copolymerized with methylmethacrylate (MMA) as an principal monomer in a binary mixture of ethyl methacrylate (EMA) and ethyl acrylate (EA) under microwave irradiation (MWR) using ascorbic acid/H_2O_2 as an initiator system. The concentration of the comonomer was optimized to maximize the graft yield with respect to the primary monomer. Maximum graft yield (86.32%) was found for dGo-poly(MMA-co-EA) binary mixture as compared to other synthesized copolymer. The experimental results inferred that the optimal concentrations for the comonomers to the optimized primary monomer was observed to be 3.19 mol/L $\times$ 10^{-1} for EMA and 2.76 mol/L $\times$ 10^{-1} for EA. Delignified and graft copolymerized fiber were subjected to evaluation of physicochemical properties such as swelling behavior and chemical resistance. The synthesized graft copolymers were characterized with Fourier transform infrared spectroscopy (FTIR), scanning electron microscopy (SEM), thermogravimetric analysis (TGA) and X-ray diffraction techniques. Thermal stability of dGo-poly(MMA-co-EA) was found to be more as compared to the delignified *Grewia optiva* fiber and other graft copolymers. Although the grafting technique was found to decrease percentage crystallinity and crystallinity index among the graft copolymers but there was significant increase in their acid/base and thermal resistance properties. The grafted samples have been explored for the adsorption of hazardous methylene dye from aqueous system.

Keywords: comonomer, microwave, delignified *Grewia optiva* fiber, physicochemical properties, methylene blue

INTRODUCTION

Natural biomasses such as natural fibers are utilized by humans for household or other conventional applications (Necula et al., 2010; Ramanaiah et al., 2011a; Sharma et al., 2013). However, during the past few decades, natural polymers have found various applications in different fields such as building materials, sports equipment, automobiles, electrolytes, energy storage, aerospace, and as adsorbent for toxic metal ion from different resources (Kiani et al., 2011; Ramanaiah et al., 2011b; Sis et al., 2013).

The wider applicability of natural fibers has been due to the exhibition of diverse properties like low density, low health hazards, biodegradability, better wear resistance, and a high degree of flexibility, low cost, renewability, and high specific strength. These fibers have been found to be sensitive to moisture, chemicals, water, and their properties are consequently degraded when they come in contact with harsh environmental conditions. A variety of chemical treatments and modifications have been employed onto natural fibers to enhance their application in different areas including composite materials; thereby necessitating

the improvement of their existing properties (Singha and Rana, 2010a).

Graft copolymerization of vinyl monomers onto natural and synthetic polymers has the advantage of incorporation of additional properties of the monomer. A considerable number of studies on graft copolymerization of single monomers onto cellulose using different methods of initiation have been reported. But graft copolymerization of binary mixtures of vinyl monomers has special importance in comparison to simple grafting of individual monomers. This technique of grafting of monomer mixtures has the advantage of creating grafted chains with tailor made properties for specific applications. The synergistic effect of the comonomers in grafting mixtures plays an important role in controlling the composition and graft yield onto cellulose (Singha et al., 2013; Thakur et al., 2014).

Graft copolymerization of methyl methacrylate (MMA) onto cellulose by chemical and radiation methods is well investigated (Singha et al., 2014). Microwave irradiation (MWR) gives rapid energy transfer and high-energy efficiency. Microwave radiation

(MWR) assists in direct heating of solvents and reactants. Owing to this interesting heating mechanism, which is clearly different from other conventional heating, selective heating can be accomplished and many reactions can be accelerated (Mansour and Nagaty, 1975). MWR technique reduces the extent of physicochemical stresses to which the fibers are exposed during the conventional techniques. Microwave technology uses electromagnetic waves, which passes through material and causes its molecules to oscillate. Microwave energy is not observed by non-polar materials to any degree while polar water molecules held within a polymer matrix do absorb energy very proficiently, thus becoming heated (Kaith and Kalia, 2008a). A few workers have studied the grafting of vinyl monomers onto various natural polymers under MWRs inorder to improve the properties of the backbone polymer (Kaith and Kalia, 2007, 2008b).

The grafting with binary mixture of monomers provides an opportunity to prepare the material for specific applications (Singha and Rana, 2010b, 2012a). However, in these investigations no systematic analysis of grafting parameters has been reported, as has been carried out using binary mixture of monomers. In these investigations, the concentration dependant monomer-monomer interactions in the reaction mixture have been found responsible for controlling the graft yield and the composition of the grafted chains. Because of synergistic effect of the added comonomer, the graft yield and other grafting parameters have shown improvement, in comparison with graft copolymerization carried out with individual monomers. Investigations have revealed that grafted chains of desired properties can be obtained by using suitable combination of monomers and their compositions in the feed (Kitagawa and Tokiwa, 2006; Singha and Rana, 2012b,c).

Lignocellulosic fibers have drawn considerable attention for environment protection as they are abundant in nature, inexhaustible, inexpensive, renewable, stable, hydrophilic, biodegradable and modifiable biopolymers. The fibers are generally composed of cellulose embedded in a matrix of other structural biopolymers like hemicelluloses, lignin, pectin, waxy substances, nitrogen-containing substances, minerals, organic acids etc. Cellulose is the highly functionalized, linear stiff chain homopolymer. It is known to be hydophillic, biodegradable and has broad chemical modifying capacity. These materials have been used in many industrial applications and as an adsorbent due to their heterogeneous nature (Jonathan and Chen, 2010; Kalia and Averous, 2011). Therefore, efforts have been made to convert this biomass into inexpensive and effective adsorbent.

Methylene blue is known to cause problems in humans ranging from nausea, vomiting, profuse sweating, mental disorders, hemoglobinemia or blue baby syndrome etc. (Cowling, 1975; Lynd et al., 1999). Hence, wastewater containing dyes need at source remediation before being discharged into natural water bodies. Many physiochemical methods like coagulation, flocculation, ion exchange, membrane separation, photodegradation, electrochemical oxidation etc. have been used for the treatment of contaminated water (Bhattacharyya and Sharma, 2005; Wang and Zhu, 2007; Gupta et al., 2013b,c, 2014; Pathania and Rathore, 2014). Among these methods, adsorption

is known to be efficient, and is widely used in wastewater treatment because of its simplicity, economic viability, technical feasibility, and social acceptability. This has led to the development of cheaper, effective, easily available and biodegradable materials for the adsorption of pollutants from water system. Several researches have made significant contributions in this area, utilizing a number of natural materials like chitosan, pectin, rice-husk, mango seed kernel powder, peanut hull, cross-linked chitosan beads, natural fibers etc. for the removal of the dyes from water (Pathania and Sharma, 2012; Gupta et al., 2013a).

In the present study, efforts have been made to modify the surface of deilgnified *Grewia optiva* fiber through graft copolymerization method of binary monomer mixture of MMA-co-EMA and MMA-co-EA using ascorbic acid (ASC)/H_2O_2 redox initiator system with MMA as the principal monomer. Fiber was extracted from the *Grewia optiva* tree and delignified by chemical method. *Grewia optiva* fiber had a composition of approximately 58–62% cellulose, 22–24% hemicelluloses, and 14–16% lignin. Its composition showed a dependence on the source, age and separating techniques. The extracted fibers due to their excellent mechanical properties are utilized by villagers for domestic purposes in the making of ropes, bags, mats etc. These fibers have excellent potential as textile fiber and are a promising candidate as fiber reinforcement for polymer matrix based composites. The potential of the grafted fiber for the removal of methylene blue dye from water system was also explored.

EXPERIMENTAL
MATERIALS AND METHODS

The monomers, MMA, ethyl acrylate (EA) and ethyl methacrylate (EMA) were received from S.D. Fine, India and used as received. Acetone of 99% purity supplied by Rankem, India was used for removal of homopolymer. ASC and hydrogen peroxide (H_2O_2) were supplied by E. Merck Pvt. Ltd., India, used as initiator, Sodium chlorite ($NaClO_2$) (Himedia Pvt. Ltd., India) was used as received. Weighing of the samples was done on Libror AEG-220 (Shimadzu, Japan) electronic balance. Microwave equipment (Grill Microwave Oven 20PGI) was used for graft copolymerization.

Delignification of Grewia optiva fiber

Grewia optiva tree branches were collected from Shivalik region of Himachal Pradesh, India. After collection, these branches were immersed in continuously flowing fresh water for 30 days at temperature between 25 and 30°C. The branches were then taken out of water and fibers were gently separated from sticks by dissolving the cementing and gummy material through beating. The obtained fibers were washed with detergent to remove the impurities. The fibers were then dried in a hot air oven maintained at 80°C for 12 h. Fibers were treated with 0.7% $NaClO_2$ at pH 4 and maintained to liquor ratio of 1:50. This mixture was boiled for 2 h with continuous stirring. After treatment, the fibers were washed and dried at 80°C for 72 h. The treated fibers were designated as delignified *Grewia optiva* fiber. The delignified *Grewia optiva* fibers were then cut into pieces of length 85–100 mm and 500 mg of the fibers was used for grafting reaction.

Microwave radiations induced grafting onto delignified Grewia optiva fiber

Initially optimization of different reaction parameters such as reaction time, microwave power, initiator concentration and monomer concentration was carried out for graft copolymerization of the principal monomer MMA onto delignified *Grewia optiva* fiber backbone prior to carry out grafting with a binary mixture under MWR radiation. Delignified *Grewia optiva* fiber (500 mg) was immersed in 100 ml of distilled water for 24 h prior to its grafting under the influence of MWRs. A definite amount of ASC ($3.74\,mol/L \times 10^{-2}$), H_2O_2 ($0.97\,mol/L \times 10^{-1}$), and monomer ($1.87\,mol/L \times 10^{-1}$) was added to the reaction flask. The co-monomers were then added to the reaction mixture with concentration ranging from 2.39 to $3.99 \times 10^{-1}\,mol/L$ for EMA and 1.84 to $3.68 \times 10^{-1}\,mol/L$ for EA under continuous stirring. The mixture was then subjected to MWR for 10 min at MWR power of 110 W to get maximum percent graft yield. The reaction was stopped after 10 min and the graft copolymer obtained was taken out and subjected to the removal of homopolymer formed during the grafting reaction.

Percent graft yield (% P_g), percent graft efficiency (% P_e), percent monomer conversion (%C), and percent homocopolymers (% H_{cp}) were calculated by following methods (Kaith and Kalia, 2007).

$$\% \ Grafting \ (p_g) = \frac{W_g - W}{W} \times 100 \qquad (1)$$

$$\% \ Efficiency \ (p_e) = \frac{W_g - W}{W_m} \times 100 \qquad (2)$$

$$\% \ Homocopolymer \ (\% \ H_{cp}) = \frac{W_{Hcp}}{W_m} \times 100 \qquad (3)$$

where W, W_g, W_m, and W_{Hcp} are the weights of delignified fiber, grafted fiber, monomer and homocopolymers.

Swelling behavior

The swelling behavior of the grafted and ungrafted fibers was studied in polar and non-polar solvents such as water, n-butanol, DMSO and carbon tetrachloride. Dry samples of grafted and ungrafted fibers were subjected to the evaluation of swelling behavior by immersing the known weights of the fibers in certain amounts of different solvents for 24 h. The samples were then taken out and excess solvent was removed by pressing between the folds of the filter paper. The samples were weighed again to obtain the final weight. The degree of swelling was calculated by using the following formula (Singha and Rana, 2012c):

$$\% \ Swelling = \frac{W_f - W_i}{W_i} \times 100 \qquad (4)$$

where, W_i is the initial weight of the dried fiber and W_f is the final weight after the swelling.

Acid and base resistance

The acid and base resistance of the fibers was studied as a function of percentage weight loss when treated with different chemicals. A known amount (W_i) of the ungrafted and grafted fibers was separately treated with a fixed volume of hydrochloric acid and sodium hydroxide of different strengths for 24 h. The fibers were then washed 2–3 times with distilled water and finally dried in an oven at 70°C to a constant weight. The samples were weighed again to obtain the final weight (W_f). The percentage of weight loss was determined by using the following formula (Kitagawa and Tokiwa, 2006).

$$\% \ Weight \ loss = \frac{W_i - W_f}{W_i} \times 100. \qquad (5)$$

CHARACTERIZATION

Scanning electron microscopy (SEM)

Scanning electron microscopic studies of raw, delignified and grafted *Grewia optiva* fibers were carried out on Quanta FEG 450 electron microscope. All the samples were gold plated to make them conducting. The images were taken at a resolution of 1000×.

X-ray diffraction studies (XRD)

X-ray diffraction studies were performed on a Philip 1710 X-ray diffractometer using CuKα (1.5418 A) radiation, a Ni-Filter and a scintillation counter as a detector at 40 kV and 40 mA on rotation from 2 to 60° at 2θ scale. The counter reading of peak intensity close to 22 and 15° represents the crystalline and amorphous materials in cellulose, respectively. The percentage crystallinity (%Cr) and crystallinity index (C.I.) were calculated according to Equations (6, 7) as follow (Singha and Rana, 2012c):

$$\% \ Cr = \frac{I_{22}}{I_{22} + I_{15}} \times 100 \qquad (6)$$

$$C.I. = \frac{I_{22} - I_{15}}{I_{22}} \qquad (7)$$

Fourier transforms infrared studies (FTIR)

Fourier transform infrared spectra of raw *Grewia optiva* fiber, delignified *Grewia optiva* fiber, dGo-poly(MMA), dGo-poly(MMA-co-EA), and dGo-poly(MMA-co-EMA) were recorded on Perkin Elmer FT-IR spectrophotometer using KBr pellets. The samples were exposed to IR radiations in the range of 4000 to 400 cm^{-1} with a resolution of 2 cm^{-1}.

Thermogravimetric analysis

Thermogravimetric analysis of the raw *Grewia optiva* fiber, delignified *Grewia optiva* fiber, dGo-poly(MMA), dGo-poly(MMA-co-EA), and dGo-poly(MMA-co-EMA) were performed using EXSTAR TG/DTA 6300 at a heating rate of 10°C/min under nitrogen atmosphere. The temperature ranged from 25 to 800°C.

DYE ADSORPTION STUDY

The adsorption experiments were carried out using simple batch process. In these experiments, 0.1 g of raw, dGo-poly(MMA), dGo-poly(MMA-co-EA), and dGo-poly(MMA-co-EMA) fibers were used as adsorbent and kept in 50 mL of methylene dye solution for adsorption at 35°C. The initial methylene blue dye concentrations were varied between 5 and 30 mgL^{-1}. The resultant reaction mixture was stirred for 60 min and amount of

Generation of free radical

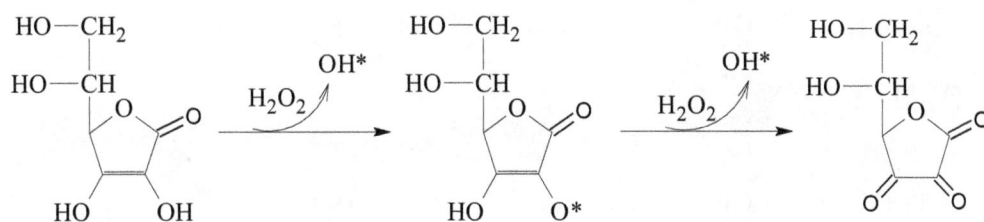

$Cell + OH^*$ $\longrightarrow$ $Cell^* + H_2O$

$M_1 + OH^*$ $\longrightarrow$ $^*M_1 - OH$

$M_2 + OH^*$ $\longrightarrow$ $^*M_2 - OH$

Chain Initiation

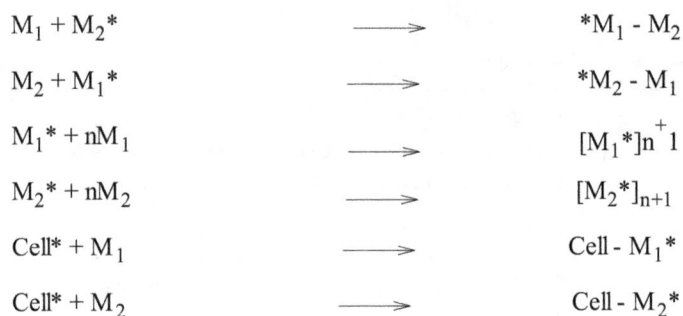

$M_1 + M_2^*$ $\longrightarrow$ $^*M_1 - M_2$

$M_2 + M_1^*$ $\longrightarrow$ $^*M_2 - M_1$

$M_1^* + nM_1$ $\longrightarrow$ $[M_1^*]_{n+1}$

$M_2^* + nM_2$ $\longrightarrow$ $[M_2^*]_{n+1}$

$Cell^* + M_1$ $\longrightarrow$ $Cell - M_1^*$

$Cell^* + M_2$ $\longrightarrow$ $Cell - M_2^*$

Chain Propagation

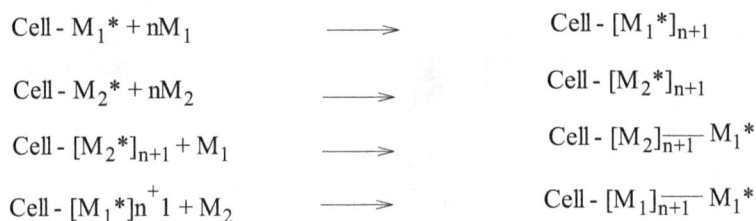

$Cell - M_1^* + nM_1$ $\longrightarrow$ $Cell - [M_1^*]_{n+1}$

$Cell - M_2^* + nM_2$ $\longrightarrow$ $Cell - [M_2^*]_{n+1}$

$Cell - [M_2^*]_{n+1} + M_1$ $\longrightarrow$ $Cell - [M_2]_{\overline{n+1}} M_1^*$

$Cell - [M_1^*]_{n+1} + M_2$ $\longrightarrow$ $Cell - [M_1]_{\overline{n+1}} M_1^*$

Termination Reaction

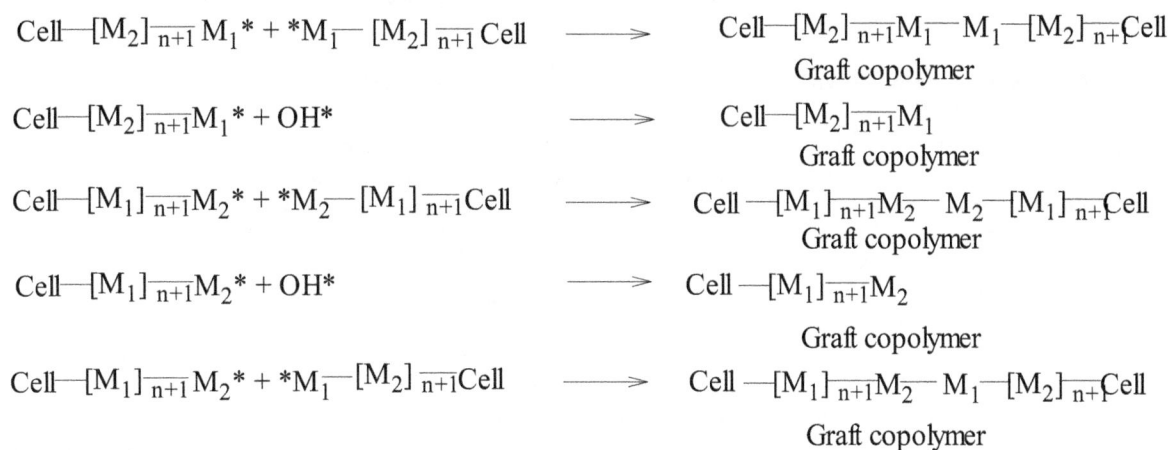

$Cell-[M_2]_{\overline{n+1}} M_1^* + {}^*M_1 - [M_2]_{\overline{n+1}} Cell$ $\longrightarrow$ $Cell-[M_2]_{\overline{n+1}}M_1-M_1-[M_2]_{\overline{n+1}}Cell$
Graft copolymer

$Cell-[M_2]_{\overline{n+1}}M_1^* + OH^*$ $\longrightarrow$ $Cell-[M_2]_{\overline{n+1}}M_1$
Graft copolymer

$Cell-[M_1]_{\overline{n+1}}M_2^* + {}^*M_2 - [M_1]_{\overline{n+1}}Cell$ $\longrightarrow$ $Cell -[M_1]_{\overline{n+1}}M_2-M_2-[M_1]_{\overline{n+1}}Cell$
Graft copolymer

$Cell-[M_1]_{\overline{n+1}}M_2^* + OH^*$ $\longrightarrow$ $Cell -[M_1]_{\overline{n+1}}M_2$

Graft copolymer

$Cell-[M_1]_{\overline{n+1}} M_2^* + {}^*M_1 - [M_2]_{\overline{n+1}}Cell$ $\longrightarrow$ $Cell -[M_1]_{\overline{n+1}}M_2-M_1-[M_2]_{\overline{n+1}}Cell$

Graft copolymer

Scheme 1 | Continued

Homopolymer formation

$$\text{Cell—}[M_1{}^*]_{n+1} + OH^* \longrightarrow \text{Cell—}[M_1]_{n+1}$$

$$[M_1{}^*]_{n+1} + OH^* \longrightarrow [M_1]_{n+1}$$
Homopolymer

$$\text{Cell—}[M_2{}^*]_{n+1} + OH^* \longrightarrow \text{Cell—}[M_2]_{n+1}$$

$$[M_2{}^*]_{n+1} + OH^* \longrightarrow [M_2]_{n+1}$$
Homopolymer

Cell*- Delignified Grewia optiva fibre free radical,
$M_1{}^*$ - Monomer (EMA) free radical,
$M_2{}^*$ - Monomer (EA) free radical

Scheme 1 | Probable mechanism of graft copolymerization of binary monomer mixture.

dye adsorbed was determined. The amount of dye adsorbed q_e (mg g^{-1}) was calculated by using following formula:

$$q_e = \frac{(C_0 - C_e)\, V}{M}$$

where C_0 is the initial concentration of dye, C_e is the concentration time at equilibrium in solution, V is the volume and M is the adsorbed mass.

RESULTS AND DISCUSSIONS
MECHANISM

The interaction of ASC with H_2O_2 generates OH* (**Scheme 1**) (Kitagawa and Tokiwa, 2006), and these free radicals are known to be responsible for free radical generation on polymer backbone and monomer as well as for further chain propagation, thereby resulting in the formation of a graft copolymer along with a homopolymer. On the other hand, MWR also produces free radicals on polymeric backbone and monomer, which can be explained through the following mechanism:

EFFECT OF THE CONCENTRATION OF BINARY VINYL MONOMER MIXTURES ON PERCENTAGE GRAFTING

Initially optimizations of different reaction parameters were carried out under MWR irradiation for graft copolymerization of MMA onto backbone. The experimental results showed that the optimal conditions for grafting were: exposure time—10 min, microwave power—110 W, ASC conc.—3.74 mol/L × 10^{-2}, H_2O_2 conc.—0.97 mol/L × 10^{-1}, monomer conc.—1.87 mol/L × 10^{-1}. The maximum P_g of dGo-poly(MMA) was found to be (26.54%). Graft copolymerization of MMA-co-EMA (ethyl methacrylate) and MMA-co-EA (ethyl acrylate) binary mixtures onto delignified Grewia optiva fiber under optimized reaction conditions using MMA—1.87 mol/L × 10^{-1} the principal monomer and EMA—3.19 mol/L × 10^{-1} showed 51.56%; and MMA—3.19 mol/L × 10^{-1} and EA—2.76 mol/L × 10^{-1}

Table 1 | Evaluation of optimum reaction parameter for grafting of binary vinyl monomer mixture onto delignified *Grewia optiva* fiber.

Sr. No	× 10^{-1} mol/L	P_g(%)	P_e(%)	%H_{cp}
MMA + EMA				
1	1.87 + 2.39	33.62	6.14	19.63
2	1.87 + 2.79	44.68	6.99	20.43
3	**1.87 + 3.19**	**51.56**	**7.06**	**22.84**
4	1.87 + 3.59	44.16	5.38	20.13
5	1.87 + 3.99	29.74	3.26	17.38
MMA + EA				
6	1.87 + 1.84	40.38	10.93	17.26
7	1.87 + 2.31	55.28	11.98	18.34
8	**1.87 + 2.76**	**86.32**	**15.58**	**20.73**
9	1.87 + 3.22	64.18	9.93	19.35
10	1.87 + 3.68	36.72	4.97	16.67

Optimal conditions for grafting: exposure time = 10 min, microwave power = 110 W, ASC conc. = 3.74 mol/L × 10^{-2}, H_2O_2 conc. = 0.97 mol/L × 10^{-1} (Singha et al., 2014).

showed 86.32% (**Table 1**). Higher percentage graft yield observed in the case of binary vinyl mixtures could be explained on the basis of the monomer reactivity ratio. The reactivity ratios in the case of different binary vinyl monomer mixtures have been found to be MMA-co-EMA: $r_1 = 2.15$, $r_2 = 0.42$; MMA-co-EA: $r_1 = 1.322$, $r_2 = 0.138$ (Brandrup, 1975). The formation of copolymer between the two different monomeric moieties takes place, which suppresses the comonomer homopolymerization due to small r_2 values in all the binary mixtures. On the other hand r_1 values clearly indicate the formation of a copolymer between the different monomers in the binary mixtures, which suppresses the formation of principal monomer homopolymerization, which resulting in higher graft yields. In the case of principal monomer, lower graft yield could be due to the higher reactivity ratio of MMA with MMA which result in more homopolymerization. Higher percentage graft yield obtained in case of MMA-co-EA (86.32%) than in the case of MMA-co-EMA (51.56%) binary mixtures which is higher than grafting of MMA (26.54%). It could be due to the more reactivity of EA that MMA and EA radical is less hysterical hindered. Then it can form more graft sites. The low graft yield with MMA-co-EMA binary mixture was due to the less reactivity of EMA then EA.

CHEMICAL RESISTANCE

Chemical resistance of delignified, MMA grafted fibers, dGo-g-poly(MMA-co-EMA) and dGo-g-poly(MMA-co-EA) samples was studied in HCl and NaOH of different normalities at 0.5 and 1.0 N for 24 h as shown in **Figures 1A,B**. It has been evident that grafted fibers were more resistant to the attack of acids/base and hence lesser weight loss occurred as compared to ungrafted fibers.

SWELLING STUDIES

Figure 2 Shows the swelling behavior of delignified *Grewia optiva* fiber, dGo-poly(MMA), dGo-poly(MMA-co-EA), and dGo-poly(MMA-co-EMA) at P_g 0, 26.54, 86.32, 51.56%, respectively were performed in water, n-butanol, DMSO and CCl_4. The swelling behavior of ungrafted fiber in different solvents followed the trend as H_2O > DMSO > n-butanol > CCl_4. Where as in grafted fiber it varied with P_g and followed the trend as CCl_4 > n-butanol > DMSO > H_2O. Delignified *Grewia optiva* fiber possesses hydrophilic -OH groups at C2, C3, and C6 of the glucose unit, and hence has strong affinity with water as compared to other solvents. In case of grafted fibers containing poly(MMA), poly(MMA-co-EMA), and poly(MMA-co-EA) monomer chains, the extent of interaction with water and alcohols is different as compared with ungrafted fibers. This may be due to the blockage of active sites on the main polymeric backbone

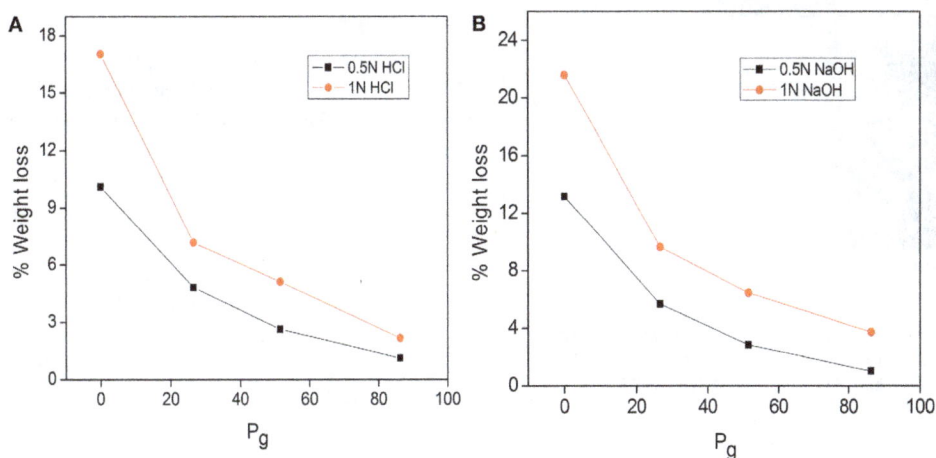

Figure 1 | (A) Acid resistance (B) base resistance of delignified *Grewia optiva* fibre, dGo-poly(MMA), dGo-poly(MMA-co-EA), and dGo-poly(MMA-co-EMA) at maximum P_g.

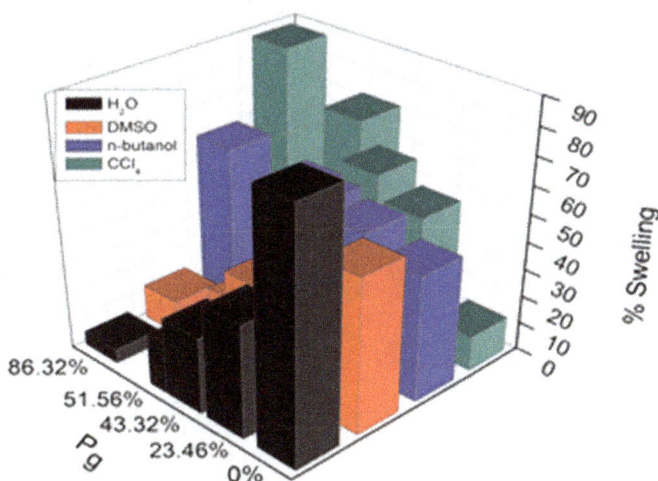

Figure 2 | Swelling behavior of delignified *Grewia optiva* fiber, dGo-poly(MMA), dGo-poly(MMA-co-EA), and dGo-poly(MMA-co-EMA) at maximum P_g in different solvents.

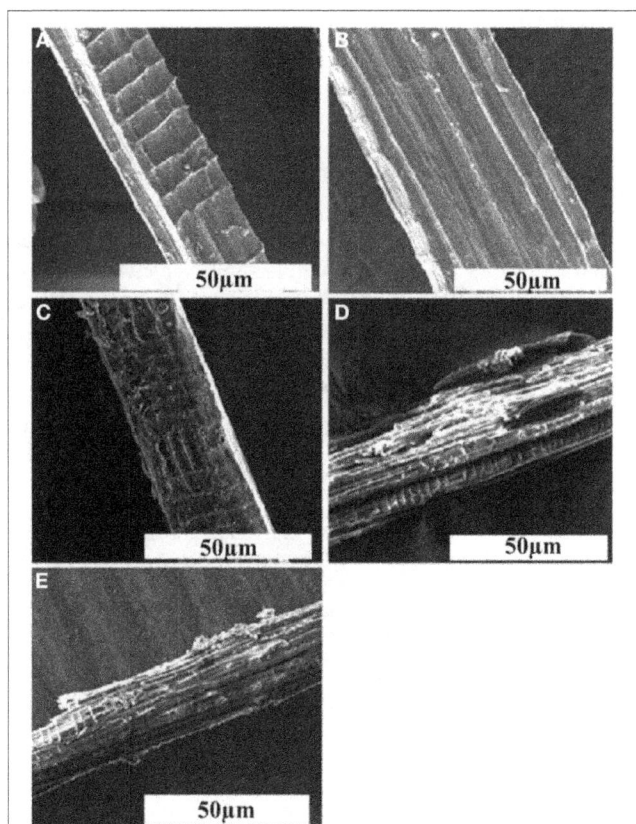

Figure 3 | Scanning electron micrographs of (A) raw *Grewia optiva* fibre, (B) delignified *Grewia optiva* fibre, (C) dGo-g-poly(MMA), (D) dGo-g-poly(MMA-co-EMA), and (E) dGo-g-poly(MMA-co-EA).

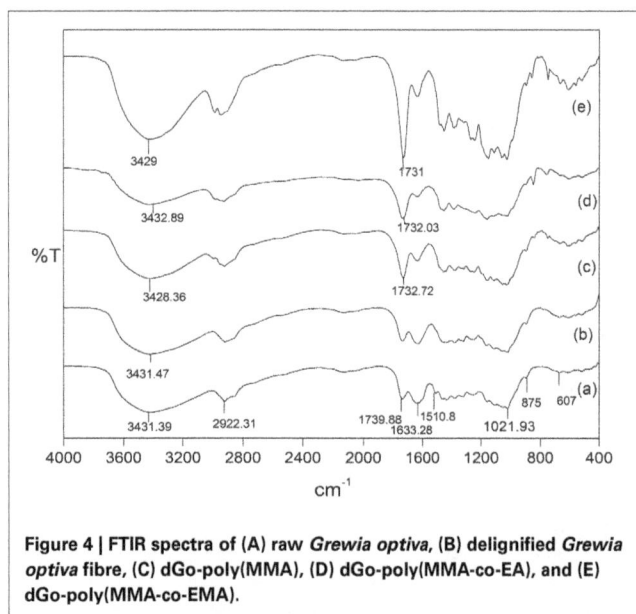

Figure 4 | FTIR spectra of (A) raw *Grewia optiva*, (B) delignified *Grewia optiva* fibre, (C) dGo-poly(MMA), (D) dGo-poly(MMA-co-EA), and (E) dGo-poly(MMA-co-EMA).

Table 2 | Percentage crystallinity (%Cr) and crystalline index (C.I.) of raw, delignified and graft copolymers prepared under the influence of MWR.

Sample name	P_g	At 2θ scale		% Cr	C.I.
		I_{22}	I_{15}		
Raw *Grewia optiva* fiber	–	903	432	67.64	0.52
Delignified *Grewia optiva* fiber	–	396	158	71.48	0.60
dGo-g-poly(MMA)	26.54	467	216	68.37	0.53
dGo-g-poly(MMA-co-EMA)	51.56	511	428	54.41	0.48
dGo-g-poly(MMA-co-EA)	86.32	508	472	51.31	0.43

non-polar solvent (CCl_4) than by polar aprotic solvent (DMSO) as compared to that with water or alcohol. Hence more swelling is observed in CCl_4 as compared to that with other solvents.

SEM ANALYSIS

The scanning electron micrographs of raw, delignified and poly (MMA) grafted delignified *Grewia optiva* fibers show a clear cut distinction in, **Figures 3A–E**. This provides a strong evidence for the change in the surface morphology of the fibers as a result of delignification and grafting of the monomer onto the cellulosic backbone. It is quite evident from the micrographs that delignified fibers (**Figure 3B**) are smoother than raw fiber (**Figure 3A**) and surface of graft copolymerized fibers (**Figures 3C–E**) is visibly rough due to the incorporation of polymeric chains onto the fiber backbone (Singha et al., 2014). This change in morphology ultimately causes changes in the properties of the raw *Grewia optiva* fibers onto graft copolymerization.

FTIR ANALYSIS

The comparison of the FTIR spectra of raw, delignified, and MMA grafted *Grewia optiva* fibers has been reported earlier (Singha et al., 2014). **Figures 4A–E** shows the FT-IR spectrum of raw, delignified, MMA grafted fibers, dGo-g-poly(MMA-co-EMA), and dGo-g-poly(MMA-co-EA). FT-IR spectrum of raw *Grewia optiva* fiber (**Figure 4A**) shows a broad peak at 607.25 cm^{-1} (due to out of plane –OH bending), 875 cm^{-1} (due to β-glycosidic linkage), 1021.93 cm^{-1} (due to stretching of C-O and -OH), peaks at $1431.01–1451.99 \text{ cm}^{-1}$ (due to -CH, -CH$_2$, and –CH$_3$ bending), 1510.8 cm^{-1} (lignin aromatic ring vibration and stretching). The peak due to lignin at 1510.8 cm^{-1} is significantly disappeared in delignified *Grewia optiva* fiber (**Figure 4B**). The peaks are observed at 1633.27 cm^{-1} (due to H-O-H bending of absorbed water and for lignin C-H deformation), 2922.31 cm^{-1} (for C-H stretching vibration of aliphatic methylene groups), and 3431.39 cm^{-1} (due to bonded -OH group) in all spectra. An additional peak at 1734.94 cm^{-1} (**Figure 4C**) is observed due to the carbonyl group ($>C=O$) of MMA. This confirms the grafting of MMA onto delignified *Grewia optiva* fiber. poly(MMA-co-EMA) and poly(MMA-co-EA) copolymers grafted onto fiber has shown absorption bands at 1732.03 cm^{-1} and 1731 cm^{-1} which correspond to ester carbonyl group of EMA and EA (**Figures 4D,E**) (Gupta et al., 2002; Gupta and Khandekar, 2006).

by poly(MMA), poly(MMA-co-EMA), and poly(MMA-co-EA) chains which resulted in the sorption behavior with different solvents. Further poly(MMA), poly(MMA-co-EMA), and poly(MMA-co-EA) chains on grafted fiber is more solvolysed by

Table 3 | Thermogravimetric analysis of grafted and ungrafted samples.

Sample name	P_g	IDT (°C)	FDT (°C)	DT (°C) at every 20% weight loss			Residual left (%)
				20	40	60	
Raw fiber	–	241.18	356.38	287.74	326.35	348.77	18.93
Delignified fiber	–	250.13	375.39	302.45	348.31	368.13	18.03
dGo-g-poly(MMA)	26.54	261.13	432.18	294.11	333.62	401.11	13.84
dGo-g-poly(MMA-co-EMA)	51.56	270.23	490.24	304.45	345.65	413.43	7.63
dGo-g-poly(MMA-co-EA)	86.32	273.63	523.52	325.54	357.84	434.84	4.57

Table 4 | q_e values for dye adsorption of raw, MMA grafted, MMA-co-EMA, and MMA-co-EA grafted *Grewia optiva* fiber.

Dye concentration (mg/L)	q_e (mg/g)			
	Raw fiber	dGo-g-poly (MMA)	dGo-g-poly (MMA-co-EMA)	dGo-g-poly (MMA-co-EA)
5	2.92	3.07	3.22	3.40
10	3.45	3.78	3.92	4.16
15	4.77	4.82	5.25	5.48
20	7.05	7.35	7.50	7.75
25	7.12	7.46	7.82	7.98
30	7.18	7.64	7.92	8.20

CRYSTALLINITY OF COPOLYMERS

It is evident from **Table 2** that raw *Grewia optiva* fiber and delignified *Grewia optiva* fiber showed 66.15 and 71.48% crystallinity. In case of delignified *Grewia optiva*, the incorporation of monomers chains to fiber backbone had impaired the crystallinity of fiber. %Cr decreased rapidly with reduction in its stiffness and hardness (Singha and Rana, 2012c). Whereas, graft copolymers prepared under the influence of MWRs showed fewer disturbances in the crystalline lattice. This is due to the reason that optimum reaction time for grafting under MWR was very low and thus the fiber underwent a fewer disturbances in its crystalline structure. Moreover, fiber faces less surface deformations during grafting process under the influence of MWRs, thereby retaining better crystalline structure. dGo-g-poly(MMA-co-EMA) and dGo-g-poly(MMA-co-EA) showed the %Cr values 54.41 and 51.38, respectively. The grafted fibers show lower percent crystallinity (%Cr) as well as C.I. Lower crystallinity index of the grafted fiber indicates that there may be the disorientation of the cellulose crystals when poly(MMA-co-EA) and poly(MMA-co-EMA) chains are incorporated in the fiber.

THERMAL ANALYSIS

The TGA of raw, delignified, MMA grafted fibers has been reported earlier (Singha et al., 2014). TGA of raw, delignified, MMA grafted fibers, dGo-g-poly(MMA-co-EMA), and dGo-g-poly(MMA-co-EA) were studied as a function of percentage weight loss vs. temperature and results are shown in **Table 3**. The initial decomposition temperature (IDT) of the delignified *Grewia optiva* fiber has been found to be 250.13°C, while the final decomposition temperature (FDT) was recorded at 375.39°C. The IDT and FDT for delignified *Grewia optiva* fiber were higher than that of the raw fiber. The grafted sample exhibits two-stage decomposition behavior, IDT and FDT were found to be 261.13 and 432.18°C, respectively, which were much higher than that of the raw and delignified *Grewia optiva* fiber. It may be due to the decomposition of the cellulosic material in the first stage and decomposition of poly MMA in second stage. From the table it is also evident that there has been an increase in the IDT of MMA grafted fiber upon graft copolymerization with different binary monomer mixture. The increase in the IDT of grafted fibers could be attributed to the incorporation of the polymer chains of comonomers on the surface of fiber. Further IDT of fibers with binary mixture of MMA-co-EA was higher than that of other binary monomer mixture which probably may be due to more thermal stability of EA. FDT values for dGo-g-poly(MMA-co-EMA) and dGo-g-poly(MMA-co-EA) were found to be 490.24 and 523.52°C, respectively.

DYE ADSORPTION STUDY

The grafted and ungrafted fiber has been used as adsorbent for removal of methylene blue from water system. At a fixed adsorbent dosage of fibers, the amount of dye adsorbed increased with the concentration of dye. It may be due to the increase in the driving force of the concentration gradient at higher initial dye concentration. It has been revealed that the percentage of adsorption increased initially and then became almost constant. Thus, dye concentration is one of the most important factors controlling the adsorption of dye onto adsorbent. Adsorption is a mass transfer process that can generally be defined as the accumulation of material at the interface between two phases (**Table 4**). The q_e value for grafted fiber was much higher than that of ungrafted fiber.

CONCLUSION

MWR induced grafting is an effective method for modifying the properties of natural fibers in terms of graft yield, time consumption, and cost effectiveness. The grafting of MMA-co-EMA, MMA-co-EA binary mixtures onto delignified *Grewia optiva* fiber in the presence of ASC/H_2O_2 as a redox initiator has been found to have physicochemical, thermal, as well as morphological impact. Although with an increase in grafting, percentage crystallinity and CI decreased, but incorporation of Poly(MMA-co-EMA), and Poly(MMA-co-EA) chains on the backbone polymer resulted in higher acid, base, and thermal resistance properties as compared to those of delignified sample. Moreover, on grafting morphological changes with respect to surface topography have taken place and graft copolymers have been found to exhibit different physical and chemical properties. Therefore, the cellulosic fibers graft copolymerized with vinyl monomers from their binary mixtures have improved properties which can ensure the utilization of these fibers in various industrial applications. Moreover, the removal of methylene dye was found more for grafted samples as compared to that with ungrafted samples.

ACKNOWLEDGMENTS

The authors wish to thank vice chancellor Shoolini University Solan (H.P.) and Director NIT Hamirpur (H.P.) India for providing laboratory facility to complete this work.

REFERENCES

Bhattacharyya, K. G., and Sharma, A. (2005). Kinetics and thermodynamics of Methylene Blue adsorption on Neem (*Azadirachta indica*) leaf powder. *Dyes Pigments* 65, 51–59. doi: 10.1016/j.dyepig.2004.06.016

Brandrup, J. E. H. (eds.). (1975). *Polymer Handbook Second Edition*, New York, NY: Wiley.

Cowling, E. B. (1975). Physical and chemical constraints in the hydrolysis of cellulose and lignocellulosic materials. *J. Biotech. Bioeng.* 5, 163–181.

Gupta, K. C., and Khandekar, K. J. (2006). Graft copolymerization of acrylamide onto cellulose in presence of comonomer using ceric ammonium nitrate as initiator. *Appl. Polym. Sci.* 101, 2546–2558. doi: 10.1002/app.23919

Gupta, K. C., Sahoo, S., and Khandekar, K. (2002). Graft copolymerization of ethyl acrylate onto cellulose using ceric ammonium nitrate as initiator in aqueous medium. *Biomacromolecules* 3, 1087–1094. doi: 10.1021/bm020060s

Gupta, V. K., Pathania, D., Agarwal, S., and Sharma, S. (2013b). Removal of Cr (VI) onto Ficus carica biosorbent from water. *Environ. Sci. Pollut. Res. Int.* 20, 2632–2644. doi: 10.1007/s11356-012-1176-6

Gupta, V. K., Pathania, D., Agarwal, S., and Sharma, S. (2014). Amputation of congo red dye from waste water using microwave induced grafted Luffa cylindrica cellulosic fiber. *Carbohydr. Polym.* 111, 556–566. doi: 10.1016/j.carbpol.2014.04.032

Gupta, V. K., Pathania, D., Sharma, S., Agarwal, S., and Singh, P. (2013a). Remediation of noxious chromium (VI) utilizing acrylic acid grafted lignocellulosic adsorbent. *J. Mol. Liq.* 177, 343–352. doi: 10.1016/j.molliq.2012.10.017

Gupta, V. K., Pathania, D., Sharma, S., Agarwal, S., and Singh, P. (2013c). Remediation and recovery of methyl orange from aqueous solution onto acrylic acid grafted *Ficus carica* fiber: isotherms, kinetics and thermodynamics. *J. Mol. Liq.* 177, 325–334. doi: 10.1016/j.molliq.2012.10.007

Jonathan, Y., and Chen, L. F. (2010). "Bast fibres: from plants to products," in *Industrial Crops and Uses, 2nd Edn.*, ed B. P. Singh (Oxon, UK: CAB International), 308–325.

Kaith, B. S., and Kalia, S. (2007). Grafting of flax fiber (*Linum usitatissimum*) with vinyl monomer for enhancement of properties of flax-phenolic composires. *Polym. J.* 39, 1319–1327. doi: 10.1295/polymj.PJ2007073

Kaith, B. S., and Kalia, S. (2008a). Graft copolymerization of MMA onto flax under different reaction conditions: a comparative study. *Express Polym. Lett.* 2, 93–100. doi: 10.3144/expresspolymlett.2008.13

Kaith, B. S., and Kalia, S. (2008b). Preparation of microwave radiation induced graft copolymers and their applications as reinforcing material in phenolic composites. *Polym. Compos.* 29, 791–797. doi: 10.1002/pc.20445

Kalia, S., and Averous, L. (2011). *Biopolymers: Biomedical and Environmental Applications, 1st Edn.* New Jersey, NJ: Wiley.

Kiani, G. R., Sheikhloie, H., and Arsalani, N. (2011). Heavy metal ion removal from aqueous solutions by functionalized polyacrylonitrile. *Deselination* 269, 266–270. doi: 10.1016/j.desal.2010.11.012

Kitagawa, M., and Tokiwa, Y. (2006). Polymerization of vinyl sugar ester using ascorbic acid and hydrogen peroxide as a redox reagent. *Carbohydr. Polym.* 64, 218–223. doi: 10.1016/j.carbpol.2005.11.029

Lynd, L. R., Wyman, C. E., and Gerngross, T. U. J. (1999). Biocommodity engineering. *Biotech. Prog.* 15, 777–793. doi: 10.1021/bp990109e

Mansour, O. Y., and Nagaty, A. J. (1975). Graft polymerization of vinyl monomers onto cellulose in presence of soda lime glass. *J. Polym. Sci. Pol. Chem.* 13, 2785–2793. doi: 10.1002/pol.1975.170131213

Necula, A. M., Olaru, N., Olaru, L., Homocianu, M., and Ioan, S. (2010). Influence of the substitution degrees on the optical properties of cellulose acetates. *J. Appl. Polym. Sci.* 115, 1751–1757. doi: 10.1002/app.31276

Pathania, D., and Rathore, B. S. (2014). Styrene-tin(IV) phosphate nanocomposite for pfotocatalytic degradation of organic dye in presence of visible light. *J. Alloy. Comp.* 606, 105–111. doi: 10.1016/j.jallcom.2014.03.160

Pathania, D., and Sharma, S. (2012). Effect of surfactants and electrolyte on removal and recovery of basic dye by using *Ficus Carica* cellulosic fibers. *Tenside Surfact. Det.* 49, 306–314. doi: 10.3139/113.110196

Ramanaiah, K., Ratnaprasad, A. V., Hema, K., and Reddy, C. (2011a). Thermal and mechanical properties of sansevieria green fiber reinforcement. *Int. J. Polym. Anal. Char.* 16, 602–608. doi: 10.1080/1023666X.2011.622358

Ramanaiah, K. A. V., Prasad, R., Hema, K., and Reddy, C. (2011b). Mechanical properties and thermal conductivity of typha angustifolia natural fiber-reinforced polyester composites. *Int. J. Polym. Anal. Char.* 16, 496–503. doi: 10.1080/1023666X.2011.598528

Sharma, S., Pathania, D., and Singh, P. (2013). Preparation, characterization and Cr(VI) adsorption behavior study of poly(acrylic acid) grafted *Ficus carica* bast fiber. *Adv. Mat. Lett.* 4, 271–276. doi: 10.5185/amlett.2012.8409

Singha, A. S., Guleria, A., and Rana, R. K. (2013). Ascorbic acid/H_2O_2-initiated graft copolymerization of Methyl Methacrylate onto abelmoschus esculentus fiber: a kinetic approach. *Int. J. Polym. Anal. Char.* 18, 1–8. doi: 10.1080/1023666X.2012.723852

Singha, A. S., and Rana, A. K. (2012b). Ce(IV) ion initiated and microwave radiation induced graft copolymerization of acrylic acid onto lignocellulosic fibers. *Int. J. Polym. Anal. Char.* 17, 72–84. doi: 10.1080/1023666X.2012.638753

Singha, A. S., and Rana, A. K. (2012c). Functionalisation of cellulosic fibers by graft copolymerisation of acrylonitrile and ethyl acrylate from their binary mixtures. *J. Carbohydr. Polym.* 87, 500–511. doi: 10.1016/j.carbpol.2011.08.011

Singha, A. S., and Rana, R. K. (2010a). Enhancement of hydrophobic character of lignocellulosic fibers through graft-copolymerization. *Adv. Mat. Lett.* 1, 156–163. doi: 10.5185/amlett.2010.6134

Singha, A. S., and Rana, R. K. (2010b). Microwave induced graft copolymerization of methyl methacrylate onto lignocellulosic fibers. *Int. J. Polym. Anal. Char.* 5, 370–386. doi: 10.1080/1023666X.2010.500539

Singha, A. S., and Rana, R. K. (2012a). Chemically induced graft copolymerization of acrylonitrile onto lignocellulosic fibers. *J Appl. Polym. Sci.* 124, 1891–1898. doi: 10.1002/app.35221

Singha, A. S., Thakur, B. P., and Pathania, D. (2014). Analysis and characterization of microwave irradiation–induced graft copolymerization of methyl methacrylate onto delignified *Grewia optiva* fiber. *Int. J. Polym. Anal. Char.* 19, 115–119. doi: 10.1080/1023666X.2014.872817

Sis, A. L. M., Ibrahim, N. A., and Yunus, W. M. Z. W. (2013). Effect of (3-aminopropyl) trimethoxysilane on mechanical properties of PLA/PBAT blend reinforced kenaf fiber. *Iran. Polym. J.* 22, 101–108. doi: 10.1007/s13726-012-0108-0

Thakur, V. K., Singha, A. S., and Thakur, M. K. (2014). Pressure induced synthesis of EA grafted *Saccaharum cilliare* fibers. *Int. J. Polym. Mater. Polym. Biomater.* 63, 17–22. doi: 10.1080/00914037.2013.769243

Wang, S., and Zhu, Z. (2007). Effect of acidic treatment of activated carbon on dye absorption. *Dyes Pigments* 75, 306–314. doi: 10.1016/j.dyepig.2006.06.005

Conflict of Interest Statement: The authors declare that the research was conducted in the absence of any commercial or financial relationships that could be construed as a potential conflict of interest.

8

Nanoparticles from renewable polymers

Frederik R. Wurm[1]* and Clemens K. Weiss[2]*

[1] *Physical Chemistry of Polymers, Max Planck Institute for Polymer Research, Mainz, Germany*
[2] *Life Sciences and Engineering, University of Applied Sciences Bingen, Bingen, Germany*

Edited by:
Frederic Jacquemin, Université de Nantes, France

Reviewed by:
Domenico Larobina, National Research Council of Italy, Italy
Vincent LEGRAND, Institut de Recherche en Génie Civil et Mécanique (UMR CNRS 6183), France

*Correspondence:
Frederik R. Wurm, Physical Chemistry of Polymers, Max Planck Institute for Polymer Research, Ackermannweg 10, 55128 Mainz, Germany
e-mail: wurm@mpip-mainz.mpg.de;
Clemens K. Weiss, Life Sciences and Engineering, University of Applied Sciences Bingen, Berlinstrasse 109, 55411 Bingen, Germany
e-mail: c.weiss@fh-bingen.de

The use of polymers from natural resources can bring many benefits for novel polymeric nanoparticle systems. Such polymers have a variety of beneficial properties such as biodegradability and biocompatibility, they are readily available on large scale and at low cost. As the amount of fossil fuels decrease, their application becomes more interesting even if characterization is in many cases more challenging due to structural complexity, either by broad distribution of their molecular weights (polysaccharides, polyesters, lignin) or by complex structure (proteins, lignin). This review summarizes different sources and methods for the preparation of biopolymer-based nanoparticle systems for various applications.

Keywords: nanoparticles, polymers, biodegradation, formulation of nanoparticles

INTRODUCTION

Nanoparticles have become omnipresent in science but also have found their ways in a plethora of everyday consumer goods. Synthetic inorganic nanoparticles find application from catalysts to sunscreens and are made from a wide variety of materials, ranging from metals or alloys, to semiconductors, oxides or other ceramic materials (Daniel and Astruc, 2004; Astruc et al., 2005; Kamat, 2007; Vallet-Regi et al., 2007; Na et al., 2009; Becker et al., 2010; Muñoz-Espí et al., 2012). In contrast to these materials, polymeric nanoparticles are designed synthetically for manifold applications. Typically, common vinyl monomer-based polymers, such as polystyrene, polyalkyl(meth)acrylates, few polyesters and polyurethanes are used. Most of them are of synthetic origin. Functionality is added to the particles by chemical modification of the surface. In addition to the general discussion about sustainability (Musto, 2013), and the drive to use "green" processes or raw materials, the naturally occurring polymers or monomers-derived from natural resources- offer great potential for broadening the scope of materials for the preparation of polymeric nanoparticles. In this review, we will briefly introduce polymers and monomers, which are of natural origin or derived from renewable sources, and present some of their unique properties. Subsequently, suitable formulation techniques are discussed and we conclude with recent examples and an outlook to potential future applications.

We will only discuss and present *nano*particle systems and, although there are several examples, not refer to *micro*particles, including microgels.

POLYMERS AND MONOMERS FROM NATURAL RESOURCES

Polymers from natural resources accompanied mankind throughout history. There is archeological evidence for the use of flax fibers, consisting of cellulose, other polysaccharides, and lignin, already 30,000 years ago (Kvavadze et al., 2009). Although animal derived materials had found their use before, evidence for the use of wool, fur, silk or leather, which are animal derived, protein (keratin, collagen, silk protein) materials, is dating back almost 7000 years (Good et al., 2009). Ancient Mesoamerican people used natural rubber (polyisoprene) in its liquid or colloidal form as medicines, and in its solid form for creating decorative items, coatings, and solid balls to be used in ritual ball games (Hosler, 1999). Widespread application for rubber, however, was enabled by the invention of vulcanization of natural rubber by Goodyear in the nineteenth century. Although polymers from renewable resources have been used for millennia, the recent years witnessed enormous efforts to use plants, animals or microbes as sources for polymers or monomers as alternatives to fossil oil-derived systems, especially in pursuit to match physical, chemical and mechanical properties of synthetic polymers. Extensive literature with excellent reviews presenting details on the different materials have been published recently (Gandini and Belgacem, 1997; Lora and Glasser, 2002; Luengo et al., 2003; Yu et al., 2006; Gandini, 2008, 2011; Kim et al., 2008; Sharma and Kundu, 2008; Vemula and John, 2008; Calvo-Flores and Dobado, 2010; Madhavan Nampoothiri et al., 2010; Mutlu and Meier, 2010; Raquez et al., 2010, 2013; Biermann et al., 2011; Montero de Espinosa and Meier, 2011; Fernandes et al., 2013; Lligadas et al., 2013; Mosiewicki and Aranguren, 2013;

Auvergne et al., 2014; Miao et al., 2014). Here, we present a brief overview of some prominent types of polymers and monomers, highlighting their properties and their potential benefits for potential applications of nanoparticulate materials.

POLYSACCHARIDES

Natural polysaccharides are ideal materials for the preparation of nanoparticles for drug or protein delivery due to many advantages over synthetic polymers such as natural abundance, generally low cost, ease of manipulation, facile derivatization due to the presence of several nucleophilic groups and, in most cases, biocompatibility. Polysaccharides have a high number of functional groups, a wide range of molecular weights with varying chemical and structural compositions, which contribute to their diversity in structure and in physical property. Due to the presence of various derivable groups on the molecular chains, polysaccharides can be easily modified chemically and biochemically, resulting in many kinds of polysaccharide derivatives. As these biomaterials occur naturally, they are typically very stable, hydrophilic, non-toxic and biodegradable. Particularly, most of natural polysaccharides have hydrophilic groups such as hydroxyl (**Figure 1**), carboxyl and amino groups, which could form non-covalent bonds with biological tissues (mainly epithelia and mucous membranes).

POLYESTERS

Polyesters such as polylactide (PLA) (**Figure 2**) or copolymers of poly(lactide-*co*-glycolide) (**Figure 3**) are biocompatible and biodegradable materials that are already used as commodity plastics (Gupta and Kumar, 2007; Madhavan Nampoothiri et al., 2010; Kfoury et al., 2013; Raquez et al., 2013) and also widely used in biomedical applications such as drug delivery or in tissue engineering.

This is mainly due to their safe degradation profile and their commercial availability. They have been approved by the United States Food and Drug Administration (FDA) and European Medicine Agency (EMA) for diverse biomedical applications (drug delivery devices, sutures, and implants). As polylactide is a readily available thermoplastic, biodegradable polyester based on renewable resources, it is a promising material for present and future applications, which substitute fossil materials. The monomer lactic acid (or lactide, the cyclic dilactone) is chiral, having two possible enantiomers (D and L), which can be polymerized enantiomerically pure or in their racemic form: poly(L-lactide) (PLLA) or poly(D-lactide) (PDLA), and the racemic mixture, i.e., poly(D,L-lactide) (PDLLA). Polylactides with an L-content above 90% are crystalline materials, while PDLLA is an amorphous polymer due to the random orientation of the pendant methyl side chain. For applications, semicrystalline PLA is often preferred over the amorphous materials especially when higher mechanical stability is necessary (tensile strength ca. 50–70 MPa). PLA has a glass transition temperature (T_g) between 50 and 65°C and a melting temperature (T_m) of ca. 175–180°C depending on the ratio of D:L in the monomer mixture. Copolymers with glycolide are also widely applied due to the varied mechanical and degradation profile. Besides the broad application several factors may limit their use, such as high crystallinity resulting in (too) slow release, or acidic degradation products, i.e., carboxylic acids, which hamper cell growth in some cases.

Other polyesters, such as poly(hydroxy butyrate), which is produced by microbes, have so far not found many applications in nanotechnology.

POLYPEPTIDES/PROTEINS

Proteins/peptides as building blocks for nanoparticles are obvious due to their degradability, usually non-toxicity and easy

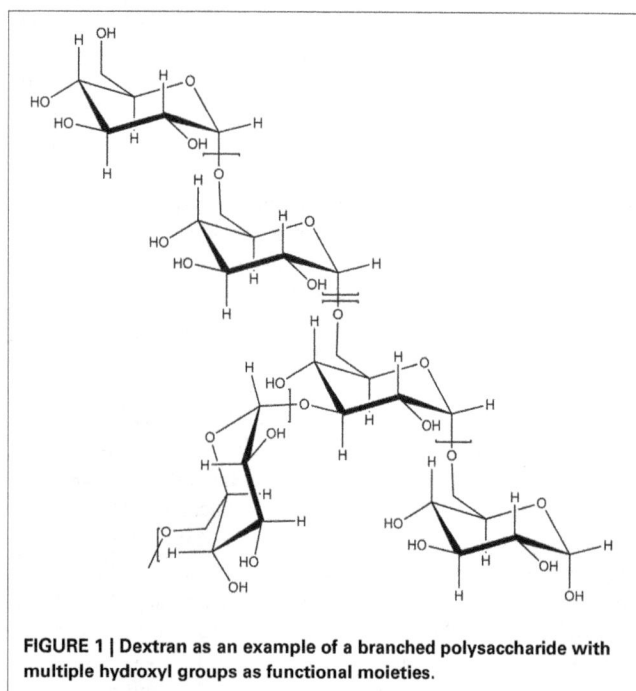

FIGURE 1 | Dextran as an example of a branched polysaccharide with multiple hydroxyl groups as functional moieties.

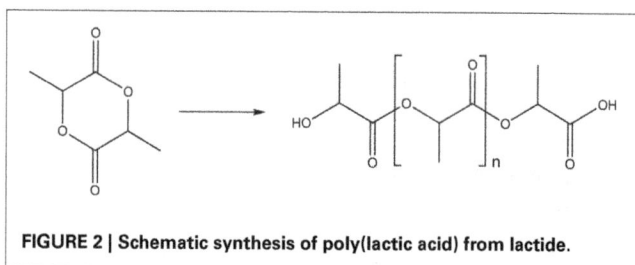

FIGURE 2 | Schematic synthesis of poly(lactic acid) from lactide.

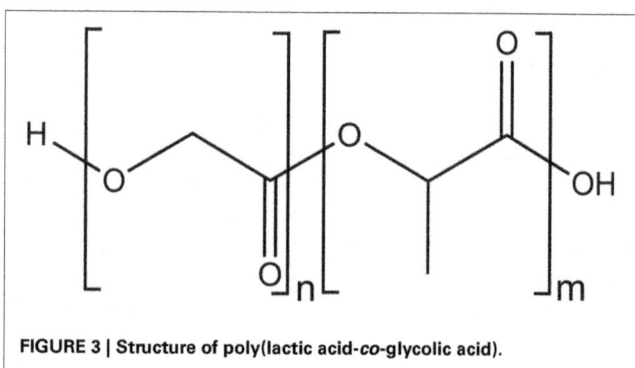

FIGURE 3 | Structure of poly(lactic acid-*co*-glycolic acid).

availability. Proteins are polymers from amino acids, produced by living organisms. They constitute tissues, have regulatory and transport properties, and catalyze metabolic reactions. So far, gelatin (a degradation product of collagen) and albumin have been used for the preparation of nanoparticles. The first commercialized protein nanoparticle-based product was albumin-bound paclitaxel (nab™-paclitaxel; Abraxane®) with a mean diameter of ca. 130 nm. It was approved by the FDA in 2005 for the treatment of breast cancer in patients who fail combination chemotherapy for metastases, or relapse within 6 months of adjuvant chemotherapy (Hawkins et al., 2008). Gelatin nanoparticles were proposed for drug delivery and, in combination with mineralized calcium phosphate, as bone substitute or bone-regrowth substrate (Balthasar et al., 2005; Kommareddy and Amiji, 2007; Ethirajan et al., 2008a,b; Gan et al., 2012; Khan and Schneider, 2013; Xu et al., 2014).

LIGNIN

Lignin is one of the most abundant renewable biopolymers. It is extracted from plants and represents approximately 15–30% of their total biomass besides cellulose as the major component. In spite of the huge amounts of lignin, which are readily available, lignin is normally considered as a non-preferred byproduct from the paper industry for example with over 30 mio t/a (Hatakeyama and Hatakeyama, 2010). A potential reason for considering lignin as a waste product is probably the very limited solubility of the native material and the complexity of the lignin structure with very broad molecular weight distributions and a random microstructure. Thus only 2% of lignin (mainly lignosulfonates) are applied in agricultural uses or other industries as binders, e.g., for animal feed pellets, bricks, ceramics, or road dust, dispersants for oil well drilling products, etc. (Lora and Glasser, 2002; Calvo-Flores and Dobado, 2010). Lignin is a highly branched polyphenolic polyether, with the main structural elements being three "monomeric units," i.e., 4-hydroxyphenyl, guaiacyl, and syringyl derivatives, which are connected via aromatic and aliphatic ether bonds that build up a hyperbranched, i.e., irregularly branched polymer (Calvo-Flores and Dobado, 2010). Lignin is highly functional and carries both phenolic and aliphatic hydroxyl groups that can be used for further modification or polymerization. Several publications deal with the use of lignin in materials science, however only very few example are present that use lignin in nanoparticle formulations. Recently, lignin was doped with multi-walled carbon nanotubes and used as a macromonomer in a step-growth polymerization with toluene-2,4-diisocyanate (TDI) terminated poly(propylene glycol) for chemical sensing applications (Faria et al., 2012). In addition, the hydroxyl groups of lignin films were modified with PNIPAM through ATRP under aqueous conditions to prepare ion-responsive nanofibers (Gao et al., 2012). We recently developed a straightforward strategy to generate hollow nanocapsules by interfacial polyaddition of lignin with TDI in an inverse miniemulsion. Lignin derivatives were dissolved in water and dispersed in organic solvent. Upon addition of TDI a polyaddition reaction with lignins' hydroxyl groups occurred at the droplet surface (**Figure 4**). By this approach hydrophilic substances can be incorporated into biodegradable lignin nanocontainers (Yiamsawas et al., 2014).

BIOGENIC MONOMERS
Amino acids

Amino acids are ubiquitous in nature as they are the monomeric units of proteins. Although recent studies describe the presence and importance of α-D-amino acids in proteins, the predominant natural form is the L-enantiomer. In addition to the proteinogenic amino acids, chemical functionality can easily be synthetically introduced into amino acids offering an extremely wide variety of possible applications. Amino acids are typically commercially available with fluorenylmethoxycarbonyl (Fmoc) or t-butoxycarbonyl (Boc) protected amino groups. In these forms they are suitable for controlled solid phase peptide synthesis. Peptides are sequentially constructed on a resin, offering structural and chemical control over the generated peptides. The obtained peptides can be used for pharmaceutical purposes, but also as structural motif in peptide polymer conjugates, as intermolecular and intramolecular interactions can be tuned by the peptide sequence. This determines the peptide structure and can eventually control the self-assembly behavior of peptide-polymer conjugates (Gauthier and Klok, 2008; Lutz and Börner, 2008; Krishna and Kiick, 2010). Introduced in nanoparticulate systems, they were used as cleavage site for release of encapsulated materials, but also for the recognition and optical detection of enzymes (Maier et al., 2011; Andrieu et al., 2012; Fuchs et al., 2013).

From polysaccharides

Polysaccharides can also serve as raw materials for the generation of fine chemicals, such as monomers for polymerization. Examples are bio-ethanol, which can be converted to ethylene glycol for polyaddition (PET) (Jang et al., 2012), or furan-based systems (Gandini and Belgacem, 1997; Gandini, 2008, 2011; Pranger and Tannenbaum, 2008; Gandini et al., 2009).

From triglycerides

Phospholipid vesicles, i.e., liposomes, have found increasing attention as nano-drug carriers since several decades. Additionally, oils and fatty acids derived from plants are currently discussed as versatile sources for the preparation of different products, including polymers and polymeric nanoparticles (Biermann et al., 2011). Since the beginning of the 1990s solid lipid nanoparticles (SLN) have been discussed as alternatives to synthetic polymer particles, due to their good biocompatibility, large-scale production and cost-effective formulation. The use of lipid pellets was already known for oral drug delivery, e.g., Mucosolvan(R) retard capsules, which was followed by lipid microparticles made by spray congealing (Eldem et al., 1991) and nanoparticles from microemulsion (Schwarz et al., 1994). A typical argument as an advantage of SLNs over polymeric nanoparticles is that they can be produced by high pressure homogenization identical to conventional oil in water emulsions (Müller et al., 2000), a process, which can be easily conducted industrially on large scale. Loading capacities are reported for several drugs and vary strongly with drug administered from a few percent for retinol (Westesen et al., 1997) for example up to 50% for ubidecarenone (i.e., Coenzym Q10) which is mainly related to the solubility of the drug in the lipid (Jenning et al., 2000). This loading capacity is also reflected in the drug release,

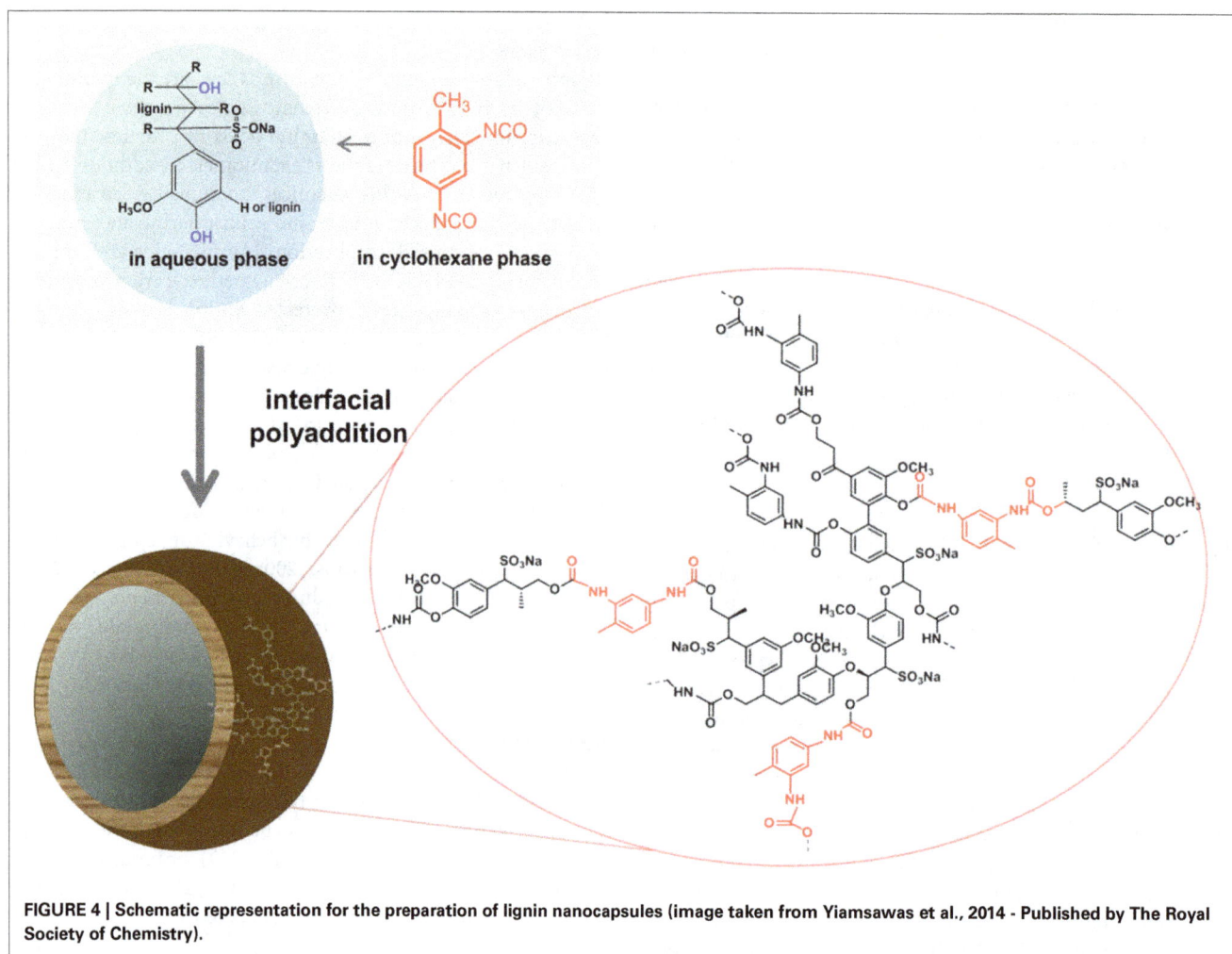

FIGURE 4 | Schematic representation for the preparation of lignin nanocapsules (image taken from Yiamsawas et al., 2014 - Published by The Royal Society of Chemistry).

which is often found as a burst release. Some examples present a release up to several weeks (5–7 weeks) by variation of the lipid matrix, surfactant concentration and production parameters (such as temperature) (Zur Mühlen and Mehnert, 1998).

Fatty acids can be derived from vegetable oils by hydrolysis (Mutlu and Meier, 2010; De Lima et al., 2013). They have been intensively studied for the use as building blocks in polymer science and derivatized to a variety of functional monomers. Recently, metathesis polymerization and used to prepare nanoparticles by miniemulsion. A strong dependence on the nature of the catalyst and surfactant was found; however the removal of Ru-based catalysts may be challenging in such systems (Cardoso et al., 2014). Also the encapsulation of several vegetable oils by emulsion techniques has been investigated (Cardoso et al., 2013). Jojoba and andiroba oil were encapsulated into polymeric nanostructures by miniemulsion polymerization. The effects of the addition of different hydrophilic monomers such as acrylic or methacrylic acid with the oils being co-stabilizers were evaluated. The formation of hollow (oil-filled) nanocapsules was investigated. Fatty acids were also used to prepare nanoparticles from protein-fatty acid ionic, i.e., salt, complexes (Yoo et al., 2001). Lysozyme was modified with hydrophobic fatty acid salt, i.e., sodium oleate, via

ionic binding between the anionic carboxylate and the basic amino groups in lysozyme. The lysozyme-oleate complex dissolved in an organic solvent exhibited much higher conformational stability at elevated temperature than free lysozyme in the same solvent. The complex was formulated into biodegradable nanoparticles by a spontaneous emulsion and solvent diffusion method. The resultant formulation showed near 100% encapsulation efficiency of lysozyme within nanoparticles with <100 nm in diameter with a narrow size distribution. These biodegradable nanoparticles showed efficient encapsulation of proteins which are potentially useful for oral protein delivery including mucosal vaccination.

OTHERS

Glycolipids are complex materials that can be produced by fermentation by yeast (such as *Candida bombicola*) from a mixture of carbohydrates and lipids. One class, the so called sophorolipids can be produced by fermentation of *C. bombicola* on glucose/oleic acid mixtures resulting in a mixture of several components including lactonic and acidic forms. Natural sophorolipids and selected derivatives are currently discussed for surfactants, emulsifiers, and therapeutic agents (antibacterial, antifungal, septic shock, anticancer, etc.) (Guilmanov et al.,

2002; Zini et al., 2008). They have been polymerized via various polymerization techniques such as ring-opening metathesis polymerization and represents a promising class of materials for future development of drug-delivery vehicles for example.

Other plant derived chemicals as, e.g., terpenes have also been proposed for application in polymer science (Wilbon et al., 2013).

PREPARATION TECHNIQUES

Nanoparticles can be formulated by basically three ways: (1) the monomer is polymerized during the preparation process to eventually form a nanoparticulate system, (2) an insoluble polymer is subjected to a physical process resulting in nanoparticles, (3) a soluble polymer is cross-linked in a suitable way. (1) Comprises classic or emulsifier free emulsion polymerization, miniemulsion polymerization and microemulsion polymerization, dispersion polymerization usually leads to microspheres. (2) Comprises (mini)emulsion solvent evaporation, emulsion solvent shifting or nanoprecipitation (Ouzo-effect). (3) Describes the formation of nanogels or crosslinked polymeric particles.

NANOPRECIPITATION/OUZO-EFFECT

Several terms can be found in the literature for the process, where a solution of an organic compound in a water miscible organic solvent is mixed with an excess of water and nanoparticles are generated by a phase separation process (**Figure 5**). Most widely used are the terms: nanoprecipitation, solvent shifting, or "Ouzo-effect" (Thioune et al., 1997; Vitale and Katz, 2003; Ganachaud and Katz, 2005; Hornig et al., 2009; Schubert et al., 2011; Aschenbrenner et al., 2013).

This phenomenon is well known from alcoholic beverages containing the oil of the anise seed, such as Ouzo or Pastis. These drinks are, chemically spoken, a solution of anethol (and some other components) in water/ethanol. When additional water is added to this solution, the solubility limit of anethol is passed and phase separation occurs. Interestingly, no macroscopic phases are initially generated. In contrast, submicron droplets with high stability are generated, resulting in the cloudy appearance of the drink after water addition. In other words, an aqueous dispersion of nanoparticles/droplets is generated without the necessity of additional energy input, which makes the process mild and cheap.

This process also finds application in the formulation of pigments (Horn and Rieger, 2001; Van Keuren et al., 2003, 2008; Chen et al., 2009), and has massively been used for the preparation of polymeric nanoparticles, especially polyester based, for

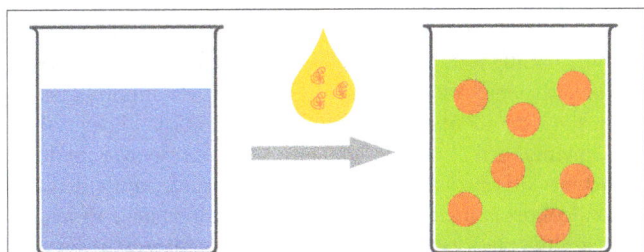

FIGURE 5 | Schematic representation of the polymeric Ouzo-effect. A polymeric solution is added to water. Nanophaseseparation leads to polymeric nanoparticles.

pharmaceutical applications (Brick et al., 2003; Liebert et al., 2005; Miller, 2006; Hornig et al., 2009; Beck-Broichsitter et al., 2010; Mora-Huertas et al., 2010, 2011). Especially for the encapsulation of sensitive drugs, the absence of energy input is a great benefit.

Although a variety of solvent/polymer combinations have been evaluated, the mechanism of formation and stability is not completely elucidated. Despite the wide applicability and the ease of the system, the major drawback may be the low (around 1 wt%) solids contents which can be obtained (Brick et al., 2003; Vitale and Katz, 2003; Ganachaud and Katz, 2005; Aubry et al., 2009).

EMULSION BASED PROCESSES

Classic emulsion polymerization finds major application in the industrial production of (meth)acrylic and vinyl polymers via radical polymerization (Harkins, 1947). As the monomer has to diffuse through an aqueous phase to reach the place of polymerization, very hydrophobic monomers are difficult to process. Additionally, other polyreactions besides the free radical polymerization are difficult to handle. Preformed polymers cannot be formulated to nanoparticles using the emulsion polymerization technique. In contrast to the above mentioned phase separation stable preformed emulsions can be used as templates or enclosed reaction vessels of the dimensions of only few microns or several hundreds of nanometers for polyreactions or phase separation processes.

The stability of droplets in emulsions is governed by two processes: First, droplets fuse by collision and coalescence, and second, droplets degrade by diffusion (Ostwald Ripening). Coalescence can be controlled by the addition of surfactants to the emulsions. These create an electrostatic or steric barrier around the droplets, creating a repulsive force between the droplets. Ostwald Ripening is caused by the imbalance of Laplace pressures between droplets of different sizes. Higher curvature leads to higher Laplace pressure, forcing the contents of smaller droplets to diffuse into larger droplets. This can be counteracted by an as-narrow-as-possible droplet size distribution, and by the addition of a so-called costabilizer to the droplets. The presence of this costabilizer, which is a component with a very low solubitiy in the continuous phase, creates an osmotic pressure, opposite to the Laplace pressure. Consequently, diffusion is limited. Such systems are typically referred to as miniemulsions. The droplets stabilized in this manner can be used for various polyreactions and physical processes for nanoparticle formulation (**Figure 6**).

The stability of the droplets in miniemulsions can be used in several ways for the formation of nanoparticluate systems. Depending on the formulation technique, nanoparticles, nanocapsules, or nanogels can be created. Detailed reviews about the mechanism and the preparative possibilities have recently been published (Landfester et al., 1999a,b; Landfester, 2000, 2009; Asua, 2002, 2004; Schork et al., 2005; Crespy and Landfester, 2010; Landfester and Weiss, 2010; Weiss and Landfester, 2010).

EMULSION/SOLVENT EVAPORATION

In contrast to the above mentioned technique of nanoprecipitation, where phase separation for the formation of nanoparticles is induced in the bulk of the solution, phase separation can

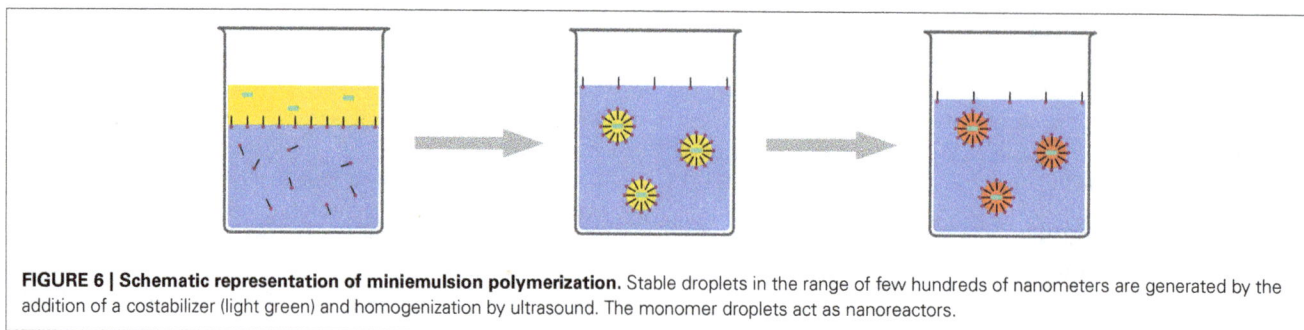

FIGURE 6 | Schematic representation of miniemulsion polymerization. Stable droplets in the range of few hundreds of nanometers are generated by the addition of a costabilizer (light green) and homogenization by ultrasound. The monomer droplets act as nanoreactors.

also be induced in droplets (**Figure 7**). Such is typically achieved by dispersing a polymer solution in a surfactant solution (typically aqueous) to form stable droplets. Either the polymer is precipitated by a water soluble organic solvent, diffusing from the droplet to the aqueous phase or by evaporating the solvent from the droplets (Mora-Huertas et al., 2011; Staff et al., 2013). Both lead to nucleation of the solid polymer, which starts at the droplet/solvent interface (Anton et al., 2008). As the solvent from the droplets is removed through the continuous phase, the miscibility of the solvents is crucial for evaporation and consequently for the (nano)particle formation. Solvent, temperature, pressure, and stirring speed are parameters which need to be controlled. Eventually, when the entire solvent has been removed, solid polymeric particles-stabilized by the surfactant which was initially added for stabilizing the droplets-form a stable dispersion.

The process is conceptually easy and is principally suitable for all polymers, for which a good solvent can be found. Furthermore, the process is quite fast (few hours) and, in contrast to polymerization techniques, no residual monomer or initiator can be found in the dispersion. Drawbacks are mainly the broad particle size distribution, the low solids contents (few %) and residual surfactants. Concentrating and cleaning may overcome the latter both issues; they are however, time consuming discontinuous steps, which may prevent large scale application.

The use of not only one dissolved component in the dispersed phase reveals another exciting topic in this area. Depending of the wetting properties (interfacial energies) (Torza and Mason, 1970) of the dissolved components, the phase separation can be directed in such a way, that complex particle morphologies can be generated. Given enough time for complete phase separation, the morphology adopted presents a minimum of the Gibbs Free Enthalpy. These ideal equilibrium morphologies are core-shell, partially engulfed (snowman-like), and separated. In real systems, several factors such as the viscosity of the solution, the temperature, and the amount of surfactant, which influences interfacial energies, determine the final morphology of the particle. This can be one of the ideal morphologies, capsules if one component is liquid, but also multi-compartment structures and other kinetically controlled structures. A review about the features, the mechanism and the scope of the emulsion solvent evaporation process was recently published (Staff et al., 2013).

EMULSION/POLYREACTIONS

In the classic emulsion polymerization process, acrylic or vinylic monomers are dispersed in an aqueous surfactant solution by

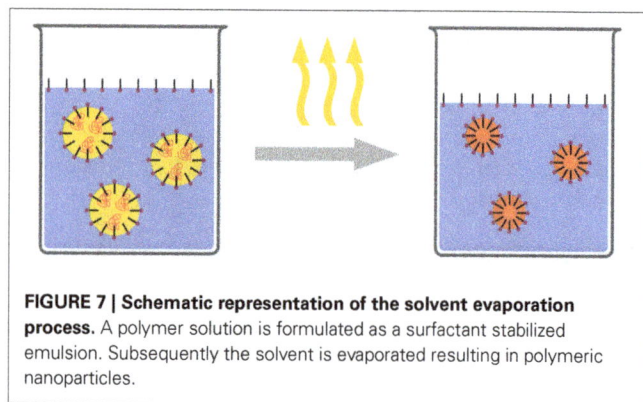

FIGURE 7 | Schematic representation of the solvent evaporation process. A polymer solution is formulated as a surfactant stabilized emulsion. Subsequently the solvent is evaporated resulting in polymeric nanoparticles.

mechanical stirring. A water-soluble initiator for free radical polymerization is also added to the system. The surfactant concentration is higher than the critical micelle concentration (cmc). Polymerization is initiated in the aqueous phase, leading to water-insoluble oligomers, entering micelles containing the monomer. Here, polymerization is maintained, as further monomer is diffusing through the aqueous phase to the locus of polymerization (**Figure 8**). A modification is the so called soap-free emulsion polymerization (Goodwin et al., 1974). Here, the monomer is dispersed in an aqueous solution of initiator, which is typically a peroxodisulfate. Thus, polymerization is also initiated in the aqueous phase. The sulfate terminated oligomers act as surfactants, generating micelles, which then act as locus for polymerization. Eventually, polymeric particles of few hundreds of nanometers or smaller are generated. To initiate polymerization and sustain diffusion in a sufficient way, a certain water solubility of the monomers is required for both techniques. Although there is a report about polycondenstation in emulsion (Jönsson et al., 2013), leading to polyester particles, classic emulsion polymerization was mainly found to be suitable for free radical polymerization of sufficiently water-soluble acrylic or vinylic monomers.

In addition to microemulsion polymerization (Pavel, 2004; McClements, 2012) where aqueous solutions with very high surfactant concentration and additional co-surfactant (medium chain alcohol) is used (**Figure 9**), miniemulsions have shown to be a versatile platform for a variety of polyreactions (Crespy and Landfester, 2010, 2011).

The above described features of miniemulsions allow among others conducting free radical polymerization of "normal" monomers and very hydrophobic monomers, even fluorinated

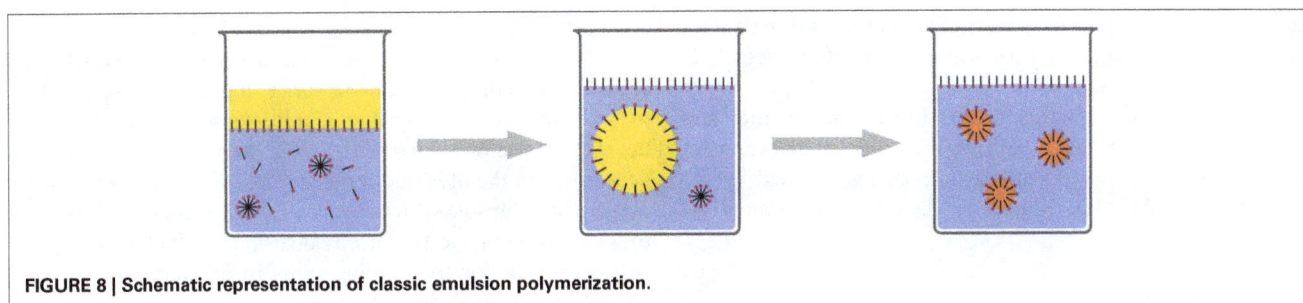

FIGURE 8 | Schematic representation of classic emulsion polymerization.

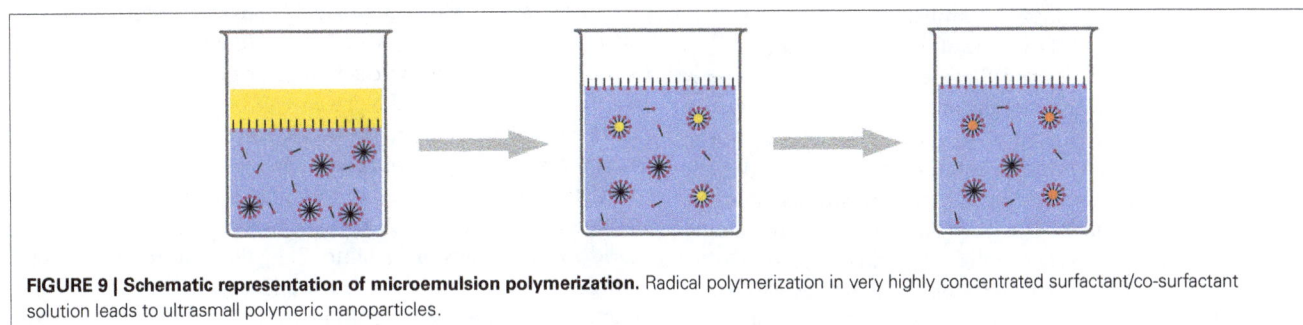

FIGURE 9 | Schematic representation of microemulsion polymerization. Radical polymerization in very highly concentrated surfactant/co-surfactant solution leads to ultrasmall polymeric nanoparticles.

monomers (Landfester et al., 2002), controlled radical polymerization (Bombalski et al., 2007; Esteves et al., 2007; Oh et al., 2007; Siegwart et al., 2009), anionic (Limouzin et al., 2003; Weiss et al., 2007a,b) and cationic polymerization (Cauvin and Ganachaud, 2004), catalytic polymerization (Cardoso et al., 2014; Malzahn et al., 2014b), as well as polyaddition and polycondesation reactions (Landfester et al., 2000; Tiarks et al., 2001; Barrère and Landfester, 2003). Each of these reactions can be used for the preparation of aqueous dispersions of polymeric nanoparticles. The major benefit of emulsion polymerization techniques is clearly the high achievable solid content, and the fact that this technique is already established in industry.

EMULSION/INTERFACIAL REACTIONS/CROSSLINKING

Polymerization can be conducted at interfaces-a very famous example is the "nylon rope trick." Two immiscible solutions are overlaid, one containing a diamine the other one a diacid chloride. The product of the polycondensation, which can only proceed at the interface, i.e., a Schotten–Baumann reaction, can be drawn out of the biphasic system to produce a nylon thread. Translating such a system to the nanoscale can be used for creating a polymeric shell or membrane at the interface of a droplet in an emulsion. Aiming at nanocapsules with aqueous core, this classic reaction is not suitable, as the diacid chloride reacts to the unreactive diacid, upon contact with water which is a "dead end" and cannot participate in polymerization. Besides a very special monomer pair (hydrophilic vinyl ether and hydrophobic maleates) (Scott et al., 2005; Wu et al., 2006) for strictly alternating radical polymerization and the metathesis polymerization of acrylated dextran and unsaturated organophosphates (Malzahn et al., 2014b), mainly polyaddition reactions of a polyol or a polyamine with diisocyanates to

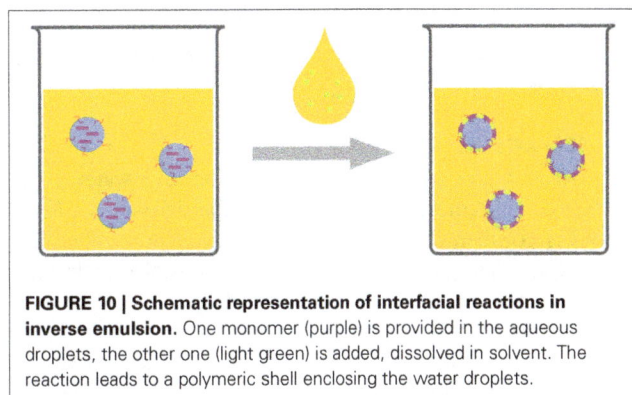

FIGURE 10 | Schematic representation of interfacial reactions in inverse emulsion. One monomer (purple) is provided in the aqueous droplets, the other one (light green) is added, dissolved in solvent. The reaction leads to a polymeric shell enclosing the water droplets.

generate polyurethane or polyurea are reported (Crespy et al., 2006). An aqueous solution of the water soluble component, such as low molecular compounds (Paiphansiri et al., 2009), polysaccharides (Crespy et al., 2006; Baier et al., 2012, 2013), or peptides (Andrieu et al., 2012), is dispersed in a solution of a surfactant in an inert solvent, typically a hydrocarbon, such as cyclohexane or isooctane. The other reactive component, e.g., toluene diisocyanate is added to the mixture, diffuses to the droplets and reacts at the interface with the alcohol or amine comonomer, or with water (**Figure 10**). The reaction with water generates a diamine, which can also act as comonomer and, in contrast to the diacid chloride, is not lost for polymer generation.

If each monomer is insoluble in the other solvent, the reaction proceeds only at the interface. If the organo-soluble monomer shows certain solubility in the dispersed phase, polymerization can also proceed in the droplet. Once the molecular weight is high enough the polymer can become insoluble and precipitate at the interface. Thus, changing the quality of the solvent can be used for tuning the morphology from particle to capsule (Andrieu et al., 2012).

Creating droplets of a preformed functional polymer and adding a crosslinker can be used for the preparation of highly crosslinked organic nanoparticles or nano-scaled hydrogels (Maier et al., 2011; Klinger et al., 2012; Malzahn et al., 2014a).

APPLICATIONS

The benefits of polymers from renewable resources are at the moment probably most prominent in applications in biomedicine. As outlined above, one of the major benefits of such systems is biocompatibility and the specific biochemical (enzymatic) or chemical (hydrolytic) degradation, which are of great importance, when such a system is proposed for application in living organisms. In addition to the use of proteins, especially gelatin, for the formulation of nanoparticles, the most of the relevant systems found in literature are based on polysaccharides and polyesters, especially poly(lactic acid). Indeed, besides commercially available polystyrene beads, nanoparticles made from poly(lactic acid) or derivatives are the most used polymeric nanoparticulate systems in biomedicine. Only part of the huge number of publications can be accounted for in this section. We hope to give an idea about the possibilities and refer the reader to comprehensive reviews about nanoparticles for biomedical applications.

POLYSACCHARIDES

As polysaccharides are typically hydrophilic and water soluble, formulating these polymers as nanoparticles in aqueous dispersion requires either hydrophobization prior to formulation, or crosslinking the polymer to form nanoscaled hydrogels. When the crosslinking reaction is conducted at the interface of a droplet, nanocapsules with liquid core can be generated (see above). Typically, the crosslinking with hydrophobic agents leads to impermeable shells. If the crosslinking reaction is conducted in a droplet containing the polysaccharide, nanogels are created. Rather recent reviews by Janes et al. (2001), Prabaharan et al. (Prabaharan and Mano, 2005), and Liu et al. (2008) describe many polysaccharide nanoparticle carriers in detail with respect to synthesis and applications. Herein, the fundamental and the most recent reports on nanoparticle preparation based on polysaccharides are addressed. Hydroxy ethyl starch (HES) based on natural starch, which is modified with ethylene glycol units, has proven to exhibit a stealth effect similar to PEG (see below, polyesters). The reduced protein adsorption of HES was already proven by several pharmaceuticals that are modified with it (Noga et al., 2012; Liebner et al., 2014). As a typical cross-linker for polysaccharides in early reports glutaraldehyde was frequently applied, e.g., for chitosan-based nanoparticles (Zhi et al., 2005), however the toxicity of glutaraldehyde limits its application in drug delivery as residues are difficult to remove from the crosslinked particles but may leak out during application. A more promising approach was established by the use of water-soluble carbodiimides, condensing the polysaccharide with natural di- or tricarboxylic acids, such as succinic acid, malic

acid, tartaric acid, and citric acid. This condensation reaction was very efficient for chitosan particles as the reaction product is an amide. Such nanoparticles were stable in aqueous media at low pH, neutral, and mild alkaline conditions. In the swollen state, the average size of the particles was in the range of 270–370 nm depending on the pH (Bodnar et al., 2005). A convenient way to prepare nanocapsule structures, i.e., an aqueous liquid core, surrounded by a polysaccharide based shell, uses the OH-groups of polysaccharides for crosslinking with di-isocyanates at the interface of an aqueous droplet in an inverse emulsion. With the procedure described above, stable or degradable capsules were created. Dextran, hydroxyethyl starch, or hyaluronic acid was used and typically crosslinked with toluenediisocyanate. Baier et al. recently reported hyaluronic acid based nanocapsules containing an antibacerial agent. In the presence of bacteria (*Staphylococcus aureus* and *Escherichia coli*), which secrete hyaluronidase, an enzyme degrading hyaluronic acid, the capsules were opened and the bactericidal agent released, resulting in an inhibition in bacterial growth (Baier et al., 2013). The hydroxyl groups of polysaccharides can be transformed into many different reactive groups in order to use them for a subsequent polymerization/crosslinking step in a dispersed medium. Acrylation or Methacrylation are typical ways for the introduction of polymerizable groups, which additionally alter the solubility profile of the polysaccharide and allow their dispersion with an organic solvent in water or -if only a few crosslinking units are attached- in an inverse system, i.e., water in oil. We recently prepared hollow nanocapsules based on acrylated dextran which was crosslinked via olefin cross metathesis at the interface of water droplets in an organic continuous phase containing a diolefin phosphate as the comonomer. The olefin-cross metathesis was tailored to the interface of a miniemulsion for the design of novel fully degradable polysaccharide-*co*-polyphosphate nanoparticles. Dextran was modified with acrylate side groups which reacted at the water-oil-interface with hydrophobic diolefin phosphates to generate hollow nanocapsules. The Ru-catalyst remained in the organic continuous phase making a purification to obtain metal-free nanocapsules feasible (Malzahn et al., 2014b). Hydrophobically modified polysaccharides, e.g., by esterification, have been used for the preparation of water dispersible nanoparticles, typically by nanoprecipitation processes (Liebert et al., 2005; Hornig and Heinze, 2007, 2008; Hornig et al., 2008). The possibility for encapsulating small molecules was shown and confirmed by fluorescence correlation spectroscopy (Aschenbrenner et al., 2013).

POLYESTERS

As mentioned above polyesters are typically hydrophobic polymers which are degraded hydrolytically and are well tolerated by biological systems. Most of the work found in literature uses poly(lactic acid) "polylactide," and copolymers with glycolic acid, PEG, or polysaccharides. Nanoparticles from poly(ε-caprolactone) are also reported. Typically, the material is used in its polymeric state. Thus, phase separation techniques for the generation of nanoparticles or nanopcapsules are mostly used. Hydrophobic and hydrophilic fluorophores (Bourges, 2003; Musyanovych et al., 2008), or drugs were encapsulated in pure PLA or in more complex copolymers (Niwa et al., 1993;

Jeong et al., 2006a,b; Beck-Broichsitter et al., 2010). Capsules with diclofenac dissolved in an oil core were reported by Guterres et al. (1995) prepared with the nanoprecipitation technique. The preparation techniques also allow the protection of large biomolecules, as plasmid DNA (Perez et al., 2001) and proteins (Pillay et al., 2013; Zaric et al., 2013) as well as the encapsulation of hydrophobized magnetic iron oxides in PLA based nanoparticles. Using the emulsion/solvent evaporation technique, Urban et al. (2009) prepared poly-L-lactic acid (PLLA) nanoparticles of around 100 nm with almost 40 wt% of iron oxide. Encapsulation did not alter the magnetic properties of the iron oxide. Larger particles of 300 nm to 3 μm containing iron oxide were used by Xu et al. (2012) for tracking mesenchymal stem cells with MRI for almost 1 month. Cell functions and properties were not disturbed by the presence of the particles, allowing long term tracking, e.g., of cell migration. Nanoparticles prepared from hydrophobic polyesters are usually readily uptaken by cells via endocytosis (Mailander and Landfester, 2009). If the material is released from the endosome or partly degraded in it, an efficient drug delivery is observed. A typical drawback for all nanoparticles is the adsorption of plasma proteins after the injection in a living organism, resulting in an altered cell uptake or faster degradation of the nanoparticles before reaching the target site (Torchilin and Trubetskoy, 1995; Sahil et al., 1997; Owens and Peppas, 2006). To overcome this problem many particles are modified with poly(ethylene glycol) (PEG), which reduces the amount of adsorbed protein due to the so called "stealth effect," which increases blood half life time by several orders of magnitude. The so called "PEGylation" is well-known for nanoparticles or molecularly dissolved drugs as the highly hydrated polyether reduces protein adsorption dramatically, (Bazile et al., 1995; Torchilin and Trubetskoy, 1995; Peracchia et al., 1999a,b; Andrieux and Couvreur, 2009) however, PEG is not degraded in vivo making a search for degradable alternatives in combination with degradable drug carriers necessary (Liebner et al., 2014) (see polysaccharides). Encapsulated material can either be released by diffusion from the polymeric matrix of the nanoparticles, and seems to depend on several factors as the preparation conditions, and the solubility of the drug in the dispersion medium. The reports range from few minutes to several days (Bourges, 2003; Pillay et al., 2013). The degradation of the polymeric matrix seems to play a minor role, only, as it is quite slow at physiological pH. In fact, diclofenac could be almost completely retained in PDLA for 8 months (Guterres et al., 1995). Control of the degradation properties was achieved by using graft copolymers, consisting of a polysaccharide backbone and grafted PLA (Nouvel et al., 2009) or PLGA chains. Polysaccharide specific enzymes degraded the backbone leading to decomposition of the whole nanostructure, leading to faster release than in the absence of the enzyme (Jeong et al., 2006a,b). Zaric et al. report a delivery system based on a combination of microneedles and antigen loaded PLGA nanoparticles. Microneedles from quickly degrading poly(methylvinylether-co-maleic anhydride) are used to deliver PLGA nanoparticles loaded with ovalbumin as model antigen. Encapsulation increases antigen stability in the microneedles and facilitates the targeting and activation of dendritic cells in the skin, opening up a promising route for

vaccination (Zaric et al., 2013). Another, so far neglected, class of polyesters found recently more attention, namely polyesters based on phosphoric acid. Similar to DNA or RNA, these materials are connected by main-chain phosphates that are typically derived from phosphorus oxychloride but can also be made from natural occurring sodium phosphate by enzymatic esterification. Several synthesis strategies have been reported ranging from classical polycondensation, via anionic ring-opening polymerization to metathesis polymerization. Moreover, the materials have been investigated in various emulsion and micellar approaches, also in combination with polyesters such as poly lactide or DNA in cation/anion complexes for gene delivery (Mao and Leong, 2005; Zhang et al., 2012; Shen et al., 2013; Steinbach et al., 2013; Alexandrino et al., 2014; Baier et al., 2014).

CONCLUSION AND OUTLOOK

The properties of nanoparticles from renewable resources are currently mainly exploited in biomedical sciences. Specific or unspecific degradation of the polymeric matrix is used for sustained release of encapsulated drugs or as probe for specific environments, e.g., the presence of enzymes or pH regimes. Despite all benefits and promising results, most of these applications, however, are far from being suitable for the industrial scale or a market-ready product. Environmental benignity does not play a major role here, as the prospective quantities are relatively low. This is probably due to the still low costs and established procedures for common vinyl monomers, etc., which are applied in industry today. Furthermore, the structural variety of many natural polymers makes their selective application challenging, broad molecular weight distributions, many different electrophilic and nucleophilic groups-sometimes in the same material- make their chemical modification and further postmodification more difficult compared to fully synthetic polymers. Additionally, the solubility profiles are usually different compared to industrial products [they dissolve in water (proteins, saccharides) or are hardly soluble at all (lignin)] making selective chemistry challenging as all other components must tolerate the aqueous conditions for example.

A prospective large scale application for polymeric nanoparticles from renewable resources may be found as major constituent of paints. For this application, the performance of the polymers has to match these of synthetic polymers and a suitable formulation technique has to be found. For these reasons, monomers derived from plant oils and formulated in conventional, emulsion polymerization based processes are probably most promising, as raw materials are readily available and polymerization techniques are established and allow the generation of dispersions with high solids contents.

REFERENCES
Alexandrino, E. M., Ritz, S., Marsico, F., Baier, G., Mailander, V., Landfester, K., et al. (2014). Paclitaxel-loaded polyphosphate nanoparticles: a potential strategy for bone cancer treatment. J. Mater. Chem. B 2, 1298–1306. doi: 10.1039/c3tb21295e

Andrieu, J., Kotman, N., Maier, M., Mailänder, V., Strauss, W. S. L., Weiss, C. K., et al. (2012). Live monitoring of cargo release from peptide-based hybrid nanocapsules induced by enzyme cleavage. Macromol. Rapid Commun. 33, 248–253. doi: 10.1002/marc.201100729

Andrieux, K., and Couvreur, P. (2009). Polyalkylcyanoacrylate nanoparticles for delivery of drugs across the blood-brain barrier. *Wiley Interdiscip. Rev. Nanomed. Nanobiotechnol.* 1, 463–474. doi: 10.1002/wnan.5

Anton, N., Benoit, J.-P., and Saulnier, P. (2008). Design and production of nanoparticles formulated from nano-emulsion templates—a review. *J. Control. Release* 128, 185–199. doi: 10.1016/j.jconrel.2008.02.007

Aschenbrenner, E., Bley, K., Koynov, K., Makowski, M., Kappl, M., Landfester, K., et al. (2013). Using the polymeric ouzo effect for the preparation of polysaccharide-based nanoparticles. *Langmuir* 29, 8845–8855. doi: 10.1021/la4017867

Astruc, D., Lu, F., and Aranzaes, J. R. (2005). Nanoparticles as recyclable catalysts: the frontier between homogeneous and heterogeneous catalysis. *Angew. Chem. Int. Ed. Engl.* 44, 7852–7872. doi: 10.1002/anie.200500766

Asua, J. M. (2002). Miniemulsion polymerization. *Prog. Polym. Sci.* 27, 1283–1346. doi: 10.1016/S0079-6700(02)00010-2

Asua, J. M. (2004). Emulsion polymerization: from fundamental mechanisms to process developments. *J. Polym. Sci. A Polym. Chem.* 42, 1025–1041. doi: 10.1002/pola.11096

Aubry, J., Ganachaud, F., Addad, J. P. C., and Cabane, B. (2009). Nanoprecipitation of polymethylmethacrylate by solvent shifting: 1. Boundaries. *Langmuir* 25, 1970–1979. doi: 10.1021/la803000e

Auvergne, R., Caillol, S., David, G., Boutevin, B., and Pascault, J.-P. (2014). Biobased thermosetting epoxy: present and future. *Chem. Rev.* 114, 1082–1115. doi: 10.1021/cr3001274

Baier, G., Baki, A., Tomcin, S., Mailänder, V., Alexandrino, E., Wurm, F., et al. (2014). Stabilization of nanoparticles synthesized by miniemulsion polymerization using "green" amino-acid based surfactants. *Macromol. Symp.* 337, 9–17. doi: 10.1002/masy.201450301

Baier, G., Baumann, D., Siebert, J. M., Musyanovych, A., Mailänder, V., and Landfester, K. (2012). Suppressing unspecific cell uptake for targeted delivery using hydroxyethyl starch nanocapsules. *Biomacromolecules* 13, 2704–2715. doi: 10.1021/bm300653v

Baier, G., Cavallaro, A., Vasilev, K., Mailänder, V., Musyanovych, A., and Landfester, K. (2013). Enzyme responsive hyaluronic acid nanocapsules containing polyhexanide and their exposure to bacteria to prevent infection. *Biomacromolecules* 14, 1103–1112. doi: 10.1021/bm302003m

Balthasar, S., Michaelis, K., Dinauer, N., Von Briesen, H., Kreuter, J., and Langer, K. (2005). Preparation and characterization of antibody modified gelatin nanoparticles as drug carrier system for uptake in lymphocytes. *Biomater* 26, 2723–2732. doi: 10.1016/j.biomaterials.2004.07.047

Barrère, M., and Landfester, K. (2003). High molecular weight polyurethane and polymer hybrid particles in aqueous miniemulsion. *Macromolecules* 36, 5119–5125. doi: 10.1021/ma025981+

Bazile, D., Prud'homme, C., Bassoullet, M.-T., Marlard, M., Spenlehauer, G., and Veillard, M. (1995). Stealth me.PEG-PLA nanoparticles avoid uptake by the mononuclear phagocytes system. *J. Pharm. Sci.* 84, 493–498. doi: 10.1002/jps.2600840420

Beck-Broichsitter, M., Rytting, E., Lebhardt, T., Wang, X. Y., and Kissel, T. (2010). Preparation of nanoparticles by solvent displacement for drug delivery: a shift in the "ouzo region" upon drug loading. *Eur. J. Pharm. Sci.* 41, 244–253. doi: 10.1016/j.ejps.2010.06.007

Becker, J., Trugler, A., Jakab, A., Hohenester, U., and Sonnichsen, C. (2010). The optimal aspect ratio of gold nanorods for plasmonic bio-sensing. *Plasmonics* 5, 161–167. doi: 10.1007/s11468-010-9130-2

Biermann, U., Bornscheuer, U., Meier, M. A., Metzger, J. O., and Schäfer, H. J. (2011). Oils and fats as renewable raw materials in chemistry. *Angew. Chem. Int. Ed. Engl.* 50, 3854–3871. doi: 10.1002/anie.201002767

Bodnar, M., Hartmann, J. F., and Borbely, J. (2005). Preparation and characterization of chitosan-based nanoparticles. *Biomacromolecules* 6, 2521–2527. doi: 10.1021/bm0502258

Bombalski, L., Min, K., Dong, H., Tang, C., and Matyjaszewski, K. (2007). Preparation of well-defined hybrid materials by atrp in miniemulsion. *Macromolecules* 40, 7429–7432. doi: 10.1021/ma071408k

Bourges, J. L. (2003). Ocular drug delivery targeting the retina and retinal pigment epithelium using polylactide nanoparticles. *Invest. Ophthalmol. Vis. Sci.* 44, 3562–3569. doi: 10.1167/iovs.02-1068

Brick, M. C., Palmer, H. J., and Whitesides, T. H. (2003). Formation of colloidal dispersions of organic materials in aqueous media by solvent shifting. *Langmuir* 19, 6367–6380. doi: 10.1021/la034173o

Calvo-Flores, F. G., and Dobado, J. A. (2010). Lignin as renewable raw material. *ChemSusChem* 3, 1227–1235. doi: 10.1002/cssc.201000157

Cardoso, P. B., Araújo, P. H. H., and Sayer, C. (2013). Encapsulation of jojoba and andiroba oils by miniemulsion polymerization. effect on molar mass distribution. *Macromol. Symp.* 324, 114–123. doi: 10.1002/masy.201200075

Cardoso, P. B., Musyanovych, A., Landfester, K., Sayer, C., De Araújo, P. H. H., and Meier, M. A. R. (2014). ADMET reactions in miniemulsion. *J. Polym. Sci. A Polym. Chem.* 52, 1300–1305. doi: 10.1002/pola.27118

Cauvin, S., and Ganachaud, F. (2004). On the preparation and polymerization of p-methoxystyrene miniemulsions in the presence of excess ytterbium triflate. *Macromol. Symp.* 215, 179–190. doi: 10.1002/masy.200451115

Chen, Q., Schönherr, H., and Vancso, G. J. (2009). Mechanical properties of block copolymer vesicle membranes by atomic force microscopy. *Soft Matter* 5, 4944. doi: 10.1039/b903110c

Crespy, D., and Landfester, K. (2010). Miniemulsion polymerization as a versatile tool for the synthesis of functionalized polymers. *Beilstein J. Org. Chem.* 6, 1132–1148. doi: 10.3762/bjoc.6.130

Crespy, D., and Landfester, K. (2011). Making dry fertile: a practical tour of non-aqueous emulsions and miniemulsions, their preparation and some applications. *Soft Matter* 7, 11054–11064. doi: 10.1039/c1sm06156a

Crespy, D., Musyanovych, A., and Landfester, K. (2006). Synthesis of polymer particles and nanocapsules stabilized with PEO/PPO containing polymerizable surfactants in miniemulsion. *Colloid Polym. Sci.* 284, 780–787. doi: 10.1007/s00396-005-1446-7

Daniel, M. C., and Astruc, D. (2004). Gold nanoparticles: assembly, supramolecular chemistry, quantum-size-related properties, and applications toward biology, catalysis, and nanotechnology. *Chem. Rev.* 104, 293–346. doi: 10.1021/cr030698+

De Lima, A. P. D., Aschenbrenner, E. M., Oliveira, S. D. N., Doucet, J.-B., Weiss, C. K., Ziener, U., et al. (2013). Towards regioselective enzymatic hydrolysis and glycerolysis of tricaprylin in miniemulsion and the direct preparation of polyurethane from the hydrolysis products. *J. Mol. Catal. B Enzym.* 98, 127–137. doi: 10.1016/j.molcatb.2013.10.013

Eldem, T., Speiser, P., and Hincal, A. (1991). Optimization of spray-dried and -congealed lipid micropellets and characterization of their surface morphology by scanning electron microscopy. *Pharm. Res.* 8, 47–54. doi: 10.1023/A:1015874121860

Esteves, A. C. C., Bombalski, L., Trindade, T., Matyjaszewski, K., and Barros-Timmons, A. (2007). Polymer grafting from Cds quantum dots via AGET ATRP in miniemulsion. *Small* 3, 1230–1236. doi: 10.1002/smll.200600510

Ethirajan, A., Schoeller, K., Musyanovych, A., Ziener, U., and Landfester, K. (2008a). Synthesis and optimization of gelatin nanoparticles using the miniemulsion process. *Biomacromolecules* 9, 2383–2389. doi: 10.1021/bm800377w

Ethirajan, A., Ziener, U., Chuvilin, A., Kaiser, U., Colfen, H., and Landfester, K. (2008b). Biomimetic hydroxyapatite crystallization in gelatin nanoparticles synthesized using a miniemulsion process. *Adv. Funct. Mater.* 18, 2221–2227. doi: 10.1002/adfm.200800048

Faria, F. A. C., Evtuguin, D. V., Rudnitskaya, A., Gomes, M. T. S. R., Oliveira, J. A. B. P., and Costa, L. C. (2012). Lignin-based polyurethane doped with carbon nanotubes for sensor applications. *Polym. Int.* 61, 788–794. doi: 10.1002/pi.4140

Fernandes, E. M., Pires, R. A., Mano, J. F., and Reis, R. L. (2013). Bionanocomposites from lignocellulosic resources: properties, applications and future trends for their use in the biomedical field. *Prog. Polym. Sci.* 38, 1415–1441. doi: 10.1016/j.progpolymsci.2013.05.013

Fuchs, A. V., Kotman, N., Andrieu, J., Mailänder, V., Weiss, C. K., and Landfester, K. (2013). Enzyme cleavable nanoparticles from peptide based triblock copolymers. *Nanoscale* 5, 4829–4839. doi: 10.1039/c3nr00706e

Gan, Z., Zhang, T., Liu, Y., and Wu, D. (2012). Temperature-triggered enzyme immobilization and release based on cross-linked gelatin nanoparticles. *PLoS ONE* 7:e47154. doi: 10.1371/journal.pone.0047154

Ganachaud, F., and Katz, J. L. (2005). Nanoparticles and nanocapsules created using the ouzo effect: spontaneous emulsification as an alternative to ultrasonic and high-shear devices. *Chemphyschem* 6, 209–216. doi: 10.1002/cphc.200400527

Gandini, A. (2008). Polymers from renewable resources: a challenge for the future of macromolecular materials. *Macromolecules* 41, 9491–9504. doi: 10.1021/ma801735u

Gandini, A. (2011). The irruption of polymers from renewable resources on the scene of macromolecular science and technology. *Green Chem.* 13, 1061–1083. doi: 10.1039/c0gc00789g

Gandini, A., and Belgacem, M. N. (1997). Furans in polymer chemistry. *Prog. Polym. Sci.* 22, 1203–1379. doi: 10.1016/S0079-6700(97)00004-X

Gandini, A., Silvestre, A. J. D., Neto, C. P., Sousa, A. F., and Gomes, M. (2009). The furan counterpart of poly(ethylene terephthalate): an alternative material based on renewable resources. *J. Polym. Sci. A Polym. Chem.* 47, 295–298. doi: 10.1002/pola.23130

Gao, G., Dallmeyer, J. I., and Kadla, J. F. (2012). Synthesis of lignin nanofibers with ionic-responsive shells: water-expandable lignin-based nanofibrous mats. *Biomacromolecules* 13, 3602–3610. doi: 10.1021/bm301039f

Gauthier, M. A., and Klok, H.-A. (2008). Peptide/protein-polymer conjugates: synthetic strategies and design concepts. *Chem. Commun.* 2591–2611. doi: 10.1039/b719689j

Good, I. L., Kenoyer, J. M., and Meadow, R. H. (2009). New evidence for early silk in the indus civilization. *Archaeometry* 51, 457–466. doi: 10.1111/j.1475-4754.2008.00454.x

Goodwin, J. W., Hearn, J., Ho, C. C., and Ottewill, R. H. (1974). Studies on the preparation and characterisation of monodisperse polystyrene laticee. *Colloid Polym. Sci.* 252, 464–471. doi: 10.1007/BF01554752

Guilmanov, V., Ballistreri, A., Impallomeni, G., and Gross, R. A. (2002). Oxygen transfer rate and sophorose lipid production by Candida bombicola. *Biotechnol. Bioeng.* 77, 489–494. doi: 10.1002/bit.10177

Gupta, A. P., and Kumar, V. (2007). New emerging trends in synthetic biodegradable polymers—polylactide: a critique. *Eur. Polym. J.* 43, 4053–4074. doi: 10.1016/j.eurpolymj.2007.06.045

Guterres, S. S., Fessi, H., Barratt, G., Devissaguet, J. P., and Puisieux, F. (1995). Poly(DL-lactide) nanocapsules containing diclofenac: I. Formulation and stability study. *Int. J. Pharm.* 113, 57–63. doi: 10.1016/0378-5173(94)00177-7

Harkins, W. D. (1947). A general theory of the mechanism of emulsion polymerization1. *J. Am. Chem. Soc.* 69, 1428–1444. doi: 10.1021/ja01198a053

Hatakeyama, H., and Hatakeyama, T. (2010). "Lignin, proteins, bioactive nanocomposites," in *Biopolymers*, eds. A. Abe, K. Dusek, and S. Kobayashi. (Berlin: Springer Verlag), 1–63.

Hawkins, M. J., Soon-Shiong, P., and Desai, N. (2008). Protein nanoparticles as drug carriers in clinical medicine. Adv. Drug Deliv. Rev. 60, 876–885. doi: 10.1016/j.addr.2007.08.044

Horn, D., and Rieger, J. (2001). Organic nanoparticles in the aqueous phase—theory, experiment, and use. *Angew. Chem.Int. Ed.Engl.* 40, 4331–4361. doi: 10.1002/1521-3773(20011203)40:23<4330::AID-ANIE4330>3.0.CO;2-W

Hornig, S., Biskup, C., Gräfe, A., Wotschadlo, J., Liebert, T., Mohr, G. J., et al. (2008). Biocompatible fluorescent nanoparticles for pH-sensing. *Soft Matter* 4, 1169–1172. doi: 10.1039/b800276b

Hornig, S., and Heinze, T. (2007). Nanoscale structures of dextran esters. *Carbohydr. Polym.* 68, 280–286. doi: 10.1016/j.carbpol.2006.12.007

Hornig, S., and Heinze, T. (2008). Efficient approach to design stable water-dispersible nanoparticles of hydrophobic cellulose esters. *Biomacromolecules* 9, 1487–1492. doi: 10.1021/bm8000155

Hornig, S., Heinze, T., Becer, C. R., and Schubert, U. S. (2009). Synthetic polymeric nanoparticles by nanoprecipitation. *J. Mater. Chem.* 19, 3838–3840. doi: 10.1039/b906556n

Hosler, D. (1999). Prehistoric polymers: rubber processing in ancient mesoamerica. *Science* 284, 1988–1991. doi: 10.1126/science.284.5422.1988

Janes, K. A., Calvo, P., and Alonso, M. J. (2001). Polysaccharide colloidal particles as delivery systems for macromolecules. *Adv. Drug Deliv. Rev.* 47, 83–97. doi: 10.1016/S0169-409X(00)00123-X

Jang, Y.-S., Kim, B., Shin, J. H., Choi, Y. J., Choi, S., Song, C. W., et al. (2012). Bio-based production of C2–C6 platform chemicals. *Biotechnol. Bioeng.* 109, 2437–2459. doi: 10.1002/bit.24599

Jenning, V., Gysler, A., Schäfer-Korting, M., and Gohla, S. H. (2000). Vitamin A loaded solid lipid nanoparticles for topical use: occlusive properties and drug targeting to the upper skin. *Eur. J. Pharm. Biopharm.* 49, 211–218. doi: 10.1016/S0939-6411(99)00075-2

Jeong, Y.-I., Choi, K.-C., and Song, C.-E. (2006a). Doxorubicin release from coreshell type nanoparticles of poly(DL-lactide-co-glycolide)-grafted dextran. *Arch. Pharm. Res.* 29, 712–719. doi: 10.1007/BF02968257

Jeong, Y.-I., Na, H.-S., Oh, J.-S., Choi, K.-C., Song, C.-E., and Lee, H.-C. (2006b). Adriamycin release from self-assembling nanospheres of poly(dl-lactide-co-glycolide)-grafted pullulan. *Int. J. Pharm.* 322, 154–160. doi: 10.1016/j.ijpharm.2006.05.020

Jönsson, J. B., Müllner, M., Piculell, L., and Karlsson, O. J. (2013). Emulsion condensation polymerization in dispersed aqueous media. interfacial reactions and nanoparticle formation. *Macromolecules* 46, 9104–9113. doi: 10.1021/ma401799g

Kamat, P. V. (2007). Meeting the clean energy demand: nanostructure architectures for solar energy conversion. *J. Phys. Chem. C* 111, 2834–2860. doi: 10.1021/jp066952u

Kfoury, G., Raquez, J.-M., Hassouna, F., Odent, J., Toniazzo, V., Ruch, D., et al. (2013). Recent advances in high performance poly(lactide): from "green" plasticization to super-tough materials via (reactive) compounding. *Front. Chem.* 1:32 doi: 10.3389/fchem.2013.00032

Khan, S. A., and Schneider, M. (2013). Improvement of nanoprecipitation technique for preparation of gelatin nanoparticles and potential macromolecular drug loading. *Macromol. Biosci.* 13, 455–463. doi: 10.1002/mabi.201200382

Kim, I.-Y., Seo, S.-J., Moon, H.-S., Yoo, M.-K., Park, I.-Y., Kim, B.-C., et al. (2008). Chitosan and its derivatives for tissue engineering applications. *Biotechnol. Adv.* 26, 1–21. doi: 10.1016/j.biotechadv.2007.07.009

Klinger, D., Aschenbrenner, E. M., Weiss, C. K., and Landfester, K. (2012). Enzymatically degradable nanogels by inverse miniemulsion copolymerization of acrylamide with dextran methacrylates as crosslinkers. *Polym. Chem.* 3, 204–216. doi: 10.1039/c1py00415h

Kommareddy, S., and Amiji, M. (2007). Poly(Ethylene Glycol)-Modified thiolated gelatin nanoparticles for glutathione-responsive intracellular DNA delivery. *Nanomedicine* 3, 32–42. doi: 10.1016/j.nano.2006.11.005

Krishna, O. D., and Kiick, K. L. (2010). Protein- and peptide-modified synthetic polymeric biomaterials. *Biopolymers* 94, 32–48. doi: 10.1002/bip.21333

Kvavadze, E., Bar-Yosef, O., Belfer-Cohen, A., Boaretto, E., Jakeli, N., Matskevich, Z., et al. (2009). 30,000-year-old wild flax fibers. *Science* 325, 1359–1359. doi: 10.1126/science.1175404

Landfester, K. (2000). Recent developments in miniemlsions—formation and stability mechanisms. *Macromol. Symp.* 150, 171–178. doi: 10.1002/1521-3900(200002)150:1<171::AID-MASY171>3.0.CO;2-D

Landfester, K. (2009). Miniemulsion polymerization and the structure of polymer and hybrid nanoparticles13. *Angew. Chem. Int. Ed. Engl.* 48, 4488–4507. doi: 10.1002/anie.200900723

Landfester, K., Bechthold, N., Tiarks, F., and Antonietti, M. (1999a). Formulation and stability mechanisms of polymerizable miniemulsions. *Macromolecules* 32, 5222–5228. doi: 10.1021/ma990299+

Landfester, K., Bechthold, N., Tiarks, F., and Antonietti, M. (1999b). Miniemulsion polymerization with cationic and nonionic surfactants: a very efficient use of surfactants for heterophase polymerization. *Macromolecules* 32, 2679–2683. doi: 10.1021/ma9819438

Landfester, K., Rothe, R., and Antonietti, M. (2002). Convenient synthesis of fluorinated latexes and core-shell structures by miniemulsion polymerization. *Macromolecules* 35, 1658–1662. doi: 10.1021/ma011608a

Landfester, K., Tiarks, F., Hentze, H.-P., and Antonietti, M. (2000). Polyaddition in miniemulsions: a new route to polymer dispersions. *Macromol. Chem. Phys.* 201, 1–5. doi: 10.1002/(SICI)1521-3935(20000101)201:1<1::AID-MACP1>3.0.CO;2-N

Landfester, K., and Weiss, C. K. (2010). Encapsulation by miniemulsion polymerization. *Adv. Polym. Sci.* 229, 1–49. doi: 10.1007/12_2009_43

Liebert, T., Hornig, S., Hesse, S., and Heinze, T. (2005). Nanoparticles on the basis of highly functionalized dextrans. *J. Am. Chem. Soc.* 127, 10484–10485. doi: 10.1021/ja052594h

Liebner, F., Mathaes, R., Meyer, M., Hey, T., Winter, G., and Besheer, A. (2014). Protein HESylation for half-life extension: synthesis, characterization and pharmacokinetics of HESylated anakinra. *Eur. J. Pharm. Biopharm.* 87, 378–385. doi: 10.1016/j.ejpb.2014.03.010

Limouzin, C., Caviggia, A., Ganachaud, F., and Hémery, P. (2003). Anionic polymerization of n-butyl cyanoacrylate in emulsion and miniemulsion. *Macromolecules* 36, 667–674. doi: 10.1021/ma0257402

Liu, Z., Jiao, Y., Wang, Y., Zhou, C., and Zhang, Z. (2008). Polysaccharides-based nanoparticles as drug delivery systems. *Adv. Drug Deliv. Rev.* 60, 1650–1662. doi: 10.1016/j.addr.2008.09.001

Lligadas, G., Ronda, J. C., Galià, M., and Cádiz, V. (2013). Renewable polymeric materials from vegetable oils: a perspective. *Mater. Today* 16, 337–343. doi: 10.1016/j.mattod.2013.08.016

Lora, J. H., and Glasser, W. G. (2002). Recent industrial applications of lignin: a sustainable alternative to nonrenewable materials. *J. Polym. Environ.*10, 39–48. doi: 10.1023/A:1021070006895

Luengo, J. M., GarcíA. B., Sandoval, A., Naharro, G., and Olivera, E. A. R. (2003). Bioplastics from microorganisms. *Curr. Opin. Microbiol.* 6, 251–260. doi: 10.1016/S1369-5274(03)00040-7

Lutz, J.-F., and Börner, H. G. (2008). Modern trends in polymer bioconjugates design. *Prog. Polym. Sci.* 33, 1–39. doi: 10.1016/j.progpolymsci.2007.07.005

Madhavan Nampoothiri, K., Nair, N. R., and John, R. P. (2010). An overview of the recent developments in polylactide (PLA) research. *Bioresour. Technol.* 101, 8493–8501. doi: 10.1016/j.biortech.2010.05.092

Maier, M., Kotman, N., Friedrichs, C., Andrieu, J., Wagner, M., Graf, R., et al. (2011). Highly site specific, protease cleavable, hydrophobic peptide–polymer nanoparticles. *Macromolecules* 44, 6258–6267. doi: 10.1021/ma 201149b

Mailander, V., and Landfester, K. (2009). Interaction of nanoparticles with cells. *Biomacromolecules* 10, 2379–2400. doi: 10.1021/bm900266r

Malzahn, K., Jamieson, W. D., Dröge, M., Mailänder, V., Jenkins, A. T. A., Weiss, C. K., et al. (2014a). Advanced dextran based nanogels for fighting *Staphylococcus aureus* infections by sustained zinc release. *J. Mater. Chem. B* 2, 2175–2183. doi: 10.1039/c3tb21335h

Malzahn, K., Marsico, F., Koynov, K., Landfester, K., Weiss, C. K., and Wurm, F. R. (2014b). Selective interfacial olefin cross metathesis for the preparation of hollow nanocapsules. *ACS Macro Lett.* 3, 40–43. doi: 10.1021/mz400578e

Mao, H. Q., and Leong, K. W. (2005). "Design of polyphosphoester-DNA nanoparticles for non-viral gene delivery," in *Advances in Genetics*, eds M.-C. H. Leaf Huang and W. Ernst (San Diego, CA; London: Academic Press), 275–306.

McClements, D. J. (2012). Nanoemulsions versus microemulsions: terminology, differences, and similarities. *Soft Matter* 8, 1719. doi: 10.1039/c2sm06903b

Miao, S., Wang, P., Su, Z., and Zhang, S. (2014). Vegetable-oil-based polymers as future polymeric biomaterials. *Acta Biomater.* 10, 1692–1704. doi: 10.1016/j.actbio.2013.08.040

Miller, C. A. (2006). "Spontaneous emulsification: recent developments with emphasis on self-emulsification," in *Emulsion and Emulsion Stability, Surfactant Science Series*, Vol. 61, ed J. Sjöblom (Boca Raton, FL: CRC Press; Taylor & Francis).

Montero de Espinosa, L., and Meier, M. A. R. (2011). Plant oils: the perfect renewable resource for polymer science?! *Eur. Polym. J.* 47, 837–852. doi: 10.1016/j.eurpolymj.2010.11.020

Mora-Huertas, C. E., Fessi, H., and Elaissari, A. (2010). Polymer-based nanocapsules for drug delivery. *Int. J. Pharm.* 385, 113–142. doi: 10.1016/j.ijpharm.2009.10.018

Mora-Huertas, C. E., Fessi, H., and Elaissari, A. (2011). Influence of process and formulation parameters on the formation of submicron particles by solvent displacement and emulsification-diffusion methods Critical comparison. *Adv. Colloid Interface Sci.* 163, 90–122. doi: 10.1016/j.cis.2011.02.005

Mosiewicki, M. A., and Aranguren, M. I. (2013). A short review on novel biocomposites based on plant oil precursors. *Eur. Polym. J.* 49, 1243–1256. doi: 10.1016/j.eurpolymj.2013.02.034

Müller, R. H., Mäder, K., and Gohla, S. (2000). Solid lipid nanoparticles (SLN) for controlled drug delivery—a review of the state of the art. *Eur. J. Pharm. Biopharm.* 50, 161–177. doi: 10.1016/S0939-6411(00)00087-4

Muñoz-Espí, R., Weiss, C. K., and Landfester, K. (2012). Inorganic nanoparticles prepared in miniemulsion. *Curr. Opin. Colloid Interface Sci.* 17, 212–224. doi: 10.1016/j.cocis.2012.04.002

Musto, P. (2013). Grand challenges in polymer chemistry: energy, environment, health. *Front. Chem.* 1:31. doi: 10.3389/fchem.2013.00031

Musyanovych, A., Schmitz-Wienke, J., Mailänder, V., Walther, P., and Landfester, K. (2008). Preparation of biodegradable polymer nanoparticles by miniemulsion technique and their cell interactions. *Macromol. Biosci.* 8, 127–139. doi: 10.1002/mabi.200700241

Mutlu, H., and Meier, M. A. R. (2010). Castor oil as a renewable resource for the chemical industry. *Eur. J. Lipid Sci. Technol.* 112, 10–30. doi: 10.1002/ejlt.200900138

Na, H. B., Song, I. C., and Hyeon, T. (2009). Inorganic nanoparticles for MRI contrast agents. *Adv. Mater.* 21, 2133–2148. doi: 10.1002/adma.200802366

Niwa, T., Takeuchi, H., Hino, T., Kunou, N., and Kawashima, Y. (1993). Preparations of biodegradable nanospheres of water-soluble and insoluble drugs with D, L-lactide/glycolide copolymer by a novel spontaneous emulsification solvent diffusion method, and the drug release behavior. *J. Control. Release* 25, 89–98. doi: 10.1016/0168-3659(93)90097-O

Noga, M., Edinger, D., Rödl, W., Wagner, E., Winter, G., and Besheer, A. (2012). Controlled shielding and deshielding of gene delivery polyplexes using hydroxyethyl starch (HES) and alpha-amylase. *J. Control. Release* 159, 92–103. doi: 10.1016/j.jconrel.2012.01.006

Nouvel, C., Raynaud, J., Marie, E., Dellacherie, E., Six, J. L., and Durand, A. (2009). Biodegradable nanoparticles made from polylactide-grafted dextran copolymers. *J. Colloid Interface Sci.* 330, 337–343. doi: 10.1016/j.jcis.2008.10.069

Oh, J. K., Siegwart, D. J., Lee, H. I., Sherwood, G., Peteanu, L., Hollinger, J. O., et al. (2007). Biodegradable nanogels prepared by atom transfer radical polymerization as potential drug delivery carriers: synthesis, biodegradation, *in vitro* release, and bioconjugation. *J. Am. Chem. Soc.* 129, 5939–5945. doi: 10.1021/ja069150l

Owens, D. E. III., and Peppas, N. A. (2006). Opsonization, biodistribution, and pharmacokinetics of polymeric nanoparticles. *Int. J. Pharm.* 307, 93–102. doi: 10.1016/j.ijpharm.2005.10.010

Paiphansiri, U., Dausend, J., Musyanovych, A., Mailänder, V., and Landfester, K. (2009). Fluorescent polyurethane nanocapsules prepared via inverse miniemulsion: surface functionalization for use as biocarriers. *Macromol. Biosci.* 9, 575–584. doi: 10.1002/mabi.200800293

Pavel, F. M. (2004). Microemulsion polymerization. *J. Disper. Sci. Technol.* 25, 1–16. doi: 10.1081/DIS-120027662

Peracchia, M. T., Fattal, E., Desmaële, D., Besnard, M., Noel, J.-P., Gomis, J. M., et al. (1999a). Stealth PEGylated polycyanoacrylate nanoparticles for intravenous administratio and splenic targeting. *J. Control. Release* 60, 121–128. doi: 10.1016/S0168-3659(99)00063-2

Peracchia, M. T., Harnisch, S., Pinto-Alphandary, H., Gulik, A., Dedieu, J.-C., Desmaële, D., et al. (1999b). Visualization of *in vitro* protein-rejecting properties of PEGylated stealth(R) polycyanoacrylate nanoparticles. *Biomaterials* 20, 1269–1275. doi: 10.1016/S0142-9612(99)00021-6

Perez, C., Sanchez, A., Putnam, D., Ting, D., Langer, R., and Alonso, M. J. (2001). Poly (lactic acid)-poly (ethylene glycol) nanoparticles as new carriers for the delivery of plasmid DNA. *J. Control. Release* 75, 211–224. doi: 10.1016/S0168-3659(01)00397-2

Pillay, V., Tomar, T., Kumar, M., Kumar, P., Singh, H., and Choonara, Y. E. (2013). *In vivo* evaluation of a conjugated poly(lactide-ethylene glycol) nanoparticle depot formulation for prolonged insulin delivery in the diabetic rabbit model. *Int. J. Nanomedicine.* 8, 505–520. doi: 10.2147/IJN.S38011

Prabaharan, M., and Mano, J. F. (2005). Chitosan-based particles as controlled drug delivery systems. *Drug Deliv.* 12, 41–57. doi: 10.1080/10717540590889781

Pranger, L., and Tannenbaum, R. (2008). Biobased nanocomposites prepared by *in situ* polymerization of furfuryl alcohol with cellulose whiskers or montmorillonite clay. *Macromolecules* 41, 8682–8687. doi: 10.1021/ma8020213

Raquez, J. M., Deléglise, M., Lacrampe, M. F., and Krawczak, P. (2010). Thermosetting (bio)materials derived from renewable resources: a critical review. *Prog. Polym. Sci.* 35, 487–509. doi: 10.1016/j.progpolymsci.2010.01.001

Raquez, J.-M., Habibi, Y., Murariu, M., and Dubois, P. (2013). Polylactide (PLA)-based nanocomposites. *Prog. Polym. Sci.* 38, 1504–1542. doi: 10.1016/j.progpolymsci.2013.05.014

Sahil, H., Tapon-Bretaudière, J., Fischer, A.-M., Sternberg, C., Spenlehauer, G., Verrecchia, T., et al. (1997). Interactions of poly(lactic acid) and poly(lactic acid-co-ethylene oxide) nanoparticles with the plasma factors of the coagulation system. *Biomaterials* 18, 281–288. doi: 10.1016/S0142-9612(96)00146-9

Schork, F. J., Luo, Y., Smulders, W., Russum, J. P., Butté, A., and Fontenot, K. (2005). Miniemulsion polymerization. *Adv. Polym. Sci.* 175, 129–255. doi: 10.1007/b100115

Schubert, S., Delaney, J. J. T., and Schubert, U. S. (2011). Nanoprecipitation and nanoformulation of polymers: from history to powerful possibilities beyond poly(lactic acid). *Soft Matter* 7, 1581–1588. doi: 10.1039/c0sm00862a

Schwarz, C., Mehnert, W., Lucks, J. S., and Müller, R. H. (1994). Solid lipid nanoparticles (SLN) for controlled drug delivery. I. Production, characterization and sterilization. *J. Control. Release* 30, 83–96. doi: 10.1016/0168-3659(94)90047-7

Scott, C., Wu, D., Ho, C. C., and Co, C. C. (2005). Liquid-core capsules via interfacial polymerization: a free-radical analogy of the nylon rope trick. *J. Am. Chem. Soc.* 127, 4160–4161. doi: 10.1021/ja044532h

Sharma, V., and Kundu, P. P. (2008). Condensation polymers from natural oils. *Prog. Polym. Sci.* 33, 1199–1215. doi: 10.1016/j.progpolymsci.2008.07.004

Shen, Y., Zhang, S., Zhang, F., Loftis, A., Pavía-Sanders, A., Zou, J., et al. (2013). Polyphosphoester-based cationic nanoparticles serendipitously release integral biologically-active components to serve as novel degradable inducible nitric oxide synthase inhibitors. *Adv. Mater.* 25, 5609–5614. doi: 10.1002/adma.201302842

Siegwart, D. J., Srinivasan, A., Bencherif, S. A., Karunanidhi, A., Oh, J. K., Vaidya, S., et al. (2009). Cellular uptake of functional nanogels prepared by inverse miniemulsion atrp with encapsulated proteins, carbohydrates, and gold nanoparticles. *Biomacromolecules* 10, 2300–2309. doi: 10.1021/bm9004904

Staff, R. H., Landfester, K., and Crespy, D. (2013). Recent advances in the emulsion solvent evaporation technique for the preparation of nanoparticles and nanocapsules. *Adv. Polym. Sci.* 262, 329–344. doi: 10.1007/12_2013_233

Steinbach, T., Alexandrino, E. M., and Wurm, F. R. (2013). Unsaturated poly(phosphoester)s via ring-opening metathesis polymerization. *Polym. Chem.* 4, 3800–3806. doi: 10.1039/c3py00437f

Thioune, O., Fessi, H., Devissaguet, J. P., and Puisieux, F. (1997). Preparation of pseudolatex by nanoprecipitation: influence of the solvent nature on intrinsic viscosity and interaction constant. *Int. J. Pharm.* 146, 233–238. doi: 10.1016/S0378-5173(96)04830-2

Tiarks, F., Landfester, K., and Antonietti, M. (2001). One-step preparation of polyurethane dispersions by miniemulsion polyaddition. *J. Polym. Sci. A: Polym. Chem.* 39, 2520–2524. doi: 10.1002/pola.1228

Torchilin, V. P., and Trubetskoy, V. S. (1995). Which polymers can make nanoparticulate drug carriers long-circulating? *Adv. Drug Deliv. Rev.* 16, 141–155. doi: 10.1016/0169-409X(95)00022-Y

Torza, S., and Mason, S. G. (1970). Three-phase interactions in shear and electrical fields. *J. Colloid Interface Sci.* 33, 67–83. doi: 10.1016/0021-9797(70)90073-1

Urban, M., Musyanovych, A., and Landfester, K. (2009). Fluorescent superparamagnetic polylactide nanoparticles by combination of miniemulsion and emulsion/solvent evaporation techniques. *Macromol. Chem. Phys.* 210, 961–970. doi: 10.1002/macp.200900071

Vallet-Regi, M., Balas, F., and Arcos, D. (2007). Mesoporous materials for drug delivery. *Angew. Chem. Int. Ed. Engl.* 46, 7548–7558. doi: 10.1002/anie.200604488

Van Keuren, E., Bone, A., and Ma, C. B. (2008). Phthalocyanine nanoparticle formation in supersaturated solutions. *Langmuir* 24, 6079–6084. doi: 10.1021/la800290s

Van Keuren, E., Georgieva, E., and Durst, M. (2003). Kinetics of the growth of anthracene nanoparticles. *J. Disper. Sci. Technol.* 24, 721–729. doi: 10.1081/DIS-120023819

Vemula, P. K., and John, G. (2008). Crops: a green approach toward self-assembled soft materials. *Acc. Chem. Res.* 41, 769–782. doi: 10.1021/ar7002682

Vitale, S. A., and Katz, J. L. (2003). Liquid droplet dispersions formed by homogeneous liquid-liquid nucleation: "The Ouzo Effect." *Langmuir* 19, 4105–4110. doi: 10.1021/la026842o

Weiss, C. K., and Landfester, K. (2010). Miniemulsion polymerization as a means to encapsulate organic and inorganic materials. *Adv. Polym. Sci.* 233, 185–236. doi: 10.1007/12_2010_61

Weiss, C. K., Lorenz, M. R., Mailänder, V., and Landfester, K. (2007a). Cellular uptake behavior of unfunctionalized and functionalized pbca particles prepared in a miniemulsion. *Macromol. Biosci.* 7, 883–896. doi: 10.1002/mabi.200700046

Weiss, C. K., Ziener, U., and Landfester, K. (2007b). A route to nonfunctionalized and functionalized poly (n-butylcyanoacrylate) nanoparticles: preparation in miniemulsion. *Macromolecules* 40, 928–938. doi: 10.1021/ma061865l

Westesen, K., Bunjes, H., and Koch, M. H. J. (1997). Physicochemical characterization of lipid nanoparticles and evaluation of their drug loading capacity and sustained release potential. *J. Control. Release* 48, 223–236. doi: 10.1016/S0168-3659(97)00046-1

Wilbon, P. A., Chu, F., and Tang, C. (2013). Progress in renewable polymers from natural terpenes, terpenoids, and rosin. *Macromol. Rapid Commun.* 34, 8–37. doi: 10.1002/marc.201200513

Wu, D., Scott, C., Ho, C. C., and Co, C. C. (2006). Aqueous-core capsules via interfacial free radical alternating copolymerization. *Macromolecules* 39, 5848–5853. doi: 10.1021/ma060951i

Xu, C., Miranda-Nieves, D., Ankrum, J. A., Matthiesen, M. E., Phillips, J. A., Roes, I., et al. (2012). Tracking mesenchymal stem cells with iron oxide nanoparticle loaded poly(lactide-co-glycolide) microparticles. *Nano Lett.* 12, 4131–4139. doi: 10.1021/nl301658q

Xu, J., Singh, A., and Amiji, M. M. (2014). Redox-responsive targeted gelatin nanoparticles for delivery of combination wt-p53 expressing plasmid DNA and gemcitabine in the treatment of pancreatic cancer. *BMC Cancer* 14:75. doi: 10.1186/1471-2407-14-75

Yiamsawas, D., Baier, G., Thines, E., Landfester, K., and Wurm, F. R. (2014). Biodegradable lignin nanocontainers. *RSC Adv.* 4, 11661–11663. doi: 10.1039/c3ra47971d

Yoo, H. S., Choi, H.-K., and Park, T. G. (2001). Protein–fatty acid complex for enhanced loading and stability within biodegradable nanoparticles. *J. Pharm. Sci.* 90, 194–201. doi: 10.1002/1520-6017(200102)90:2<194::AID-JPS10>3.0.CO;2-Q

Yu, L., Dean, K., and Li, L. (2006). Polymer blends and composites from renewable resources. *Prog. Polym. Sci.* 31, 576–602. doi: 10.1016/j.progpolymsci.2006.03.002

Zaric, M., Lyubomska, O., Touzelet, O., Poux, C., Al-Zahrani, S., Fay, F., et al. (2013). Skin dendritic cell targeting via microneedle arrays laden with antigen-encapsulated polyL-lactide-*co*-glycolide nanoparticles induces efficient antitumor and antiviral immune responses. *ACS Nano* 7, 2042–2055. doi: 10.1021/nn304235j

Zhang, S., Zou, J., Zhang, F., Elsabahy, M., Felder, S. E., Zhu, J., et al. (2012). Rapid and versatile construction of diverse and functional nanostructures derived from a polyphosphoester-based biomimetic block copolymer system. *J. Am. Chem. Soc.* 134, 18467–18474. doi: 10.1021/ja309037m

Zhi, J., Wang, Y., and Luo, G. (2005). Adsorption of diuretic furosemide onto chitosan nanoparticles prepared with a water-in-oil nanoemulsion system. *React. Funct. Polym.* 65, 249–257. doi: 10.1016/j.reactfunctpolym.2005.06.009

Zini, E., Gazzano, M., Scandola, M., Wallner, S. R., and Gross, R. A. (2008). Glycolipid biomaterials: solid-state properties of a poly(sophorolipid). *Macromolecules* 41, 7463–7468. doi: 10.1021/ma800496f

Zur Mühlen, A., and Mehnert, W. (1998). Drug release and release mechanisms of prednisolone loaded solid lipid nanoparticles. *Pharmazie* 2, 552–555.

Conflict of Interest Statement: The authors declare that the research was conducted in the absence of any commercial or financial relationships that could be construed as a potential conflict of interest.

9

Protein disulfide isomerase a multifunctional protein with multiple physiological roles

Hyder Ali Khan and Bulent Mutus *

Chemistry and Biochemistry Department, University of Windsor, Windsor, ON, Canada

Edited by:
Fernando Antunes, Universidade de Lisboa, Portugal

Reviewed by:
Joao Batista Teixeira Rocha, Universidade Federal de Santa Maria, Brazil
Laura De Gara, Università Campus Bio-Medico di Roma, Italy

*Correspondence:
Bulent Mutus, Chemistry and Biochemistry Department, University of Windsor, 401 Sunset Avenue, Windsor, ON N9B 3P4, Canada
e-mail: mutusb@uwindsor.ca

Protein disulfide isomerase (PDI), is a member of the thioredoxin superfamily of redox proteins. PDI has three catalytic activities including, thiol-disulfide oxireductase, disulfide isomerase and redox-dependent chaperone. Originally, PDI was identified in the lumen of the endoplasmic reticulum and subsequently detected at additional locations, such as cell surfaces and the cytosol. This review will provide an overview of the recent advances in relating the structural features of PDI to its multiple catalytic roles as well as its physiological and pathophysiological functions related to redox regulation and protein folding.

Keywords: PDI, protein disulfide isomerase, chaperone, oxidoreductase, disulfides, endoplasmic reticulum, cell surface

INTRODUCTION

In 1963, microsomal preparations of rat liver were shown to reactivate reduced ribonuclease A (Goldberger et al., 1963). The enzyme that catalyzed this reaction was eventually identified as protein disulfide isomerase (PDI; EC 5.3.4.1). PDI is one of 20 proteins belonging to the PDI family (Kozlov et al., 2010). The proteins in this family all contain at least one domain with a thioredoxin-like fold($\beta\alpha\beta\alpha\beta\beta\alpha$), but may vary in length and domain arrangement (Kozlov et al., 2010).

Edman and coworkers were able to determine the active site sequence of rat PDI as WCGHCK through sequencing of cDNA (Edman et al., 1985). This sequence was homologous to the active site of thioredoxin and was found to catalyze the reduction and isomerization of disulfide bonds and the oxidation of thiols (**Figure 1**) (Holmgren, 1968). In 1994 Cai and coworkers discovered PDI also had chaperone activity in addition to its redox activity (Cai et al., 1994). In 1998, McLaughlin and Bulleid showed that the chaperone activity was independent of the redox status of active site thiols (McLaughlin and Bulleid, 1998). The three catalytic activities of PDI, thiol redox, disulfide exchange, and chaperone are central to endoplasmic reticulum (ER) function (Maattanen et al., 2010).

PDI is expressed in almost all mammalian tissues (Marcus et al., 1996; Noiva, 1999). Although PDI has a C-terminal KDEL ER retention sequence, significant amounts of this protein were shown to escape the ER and were detected in the nucleus, cytosol, cell surface, and extracellularly (Edman et al., 1985; Koch, 1987; Yoshimori et al., 1990).

This review will focus on recent discoveries on PDI structure-function and physiological and pathophysiological roles arising from its localization in tissues including hemostasis, facilitation of pathogen entry, and reactive nitrogen and oxygen signaling. For more information on the role PDI in cancer, lipid homeostasis,

infertility etc., the readers are directed to the reviews by Bulleid and Ellgaard (2011), Andreu et al. (2012), Laurindo et al. (2012), and Benham (2012).

PDI STRUCTURE

All members of the PDI family share the thioredoxin-like domain structure characterized by the $\beta\alpha\beta\alpha\beta\beta\alpha$ fold (Kemmink et al., 1997). PDI contains four thioredoxin-like domains **abb'a'**. The two redox active sites containing the CXXC motif are found in the **a** and **a'** domains (**Figure 2**). The active site domains are linked by the **b** and **b'** domains. There is also a small interdomain region known as the **x**-linker located between **b'** and **a'** (Freedman et al., 1998; Alanen et al., 2003). For an in depth review on the structure of PDI the reader is directed to a recent review by Kozlov et al. (2010).

Recent advances in NMR and x-ray crystallography have given further insight into PDI structure and function, by identifying the **b'** as the chaperone domain. Through NMR, the structure and amino acid residues of the **b'** domain were observed to interact with unfolded RNase A, an oft used enzyme to assay the chaperone activity of PDI. The **b'** domain contains a large multivalent hydrophobic surface allowing for a structurally promiscuous binding site (Denisov et al., 2009). In addition, computational analysis indicates that the **bb'** domains contain 4 cavities allowing for the possible binding of a variety of ligands (see Section PDI Chaperone Activity). Recently human PDI is observed to dimerize *in vivo* through the binding of **bb'** (Bastos-Aristizabal et al., 2014).

PDI CHAPERONE ACTIVITY

The chaperone activity of PDI is an essential area of study to further understand the protein folding related roles of PDI in the ER as well as neuronal tissues (see Section PDI and Coagulation).

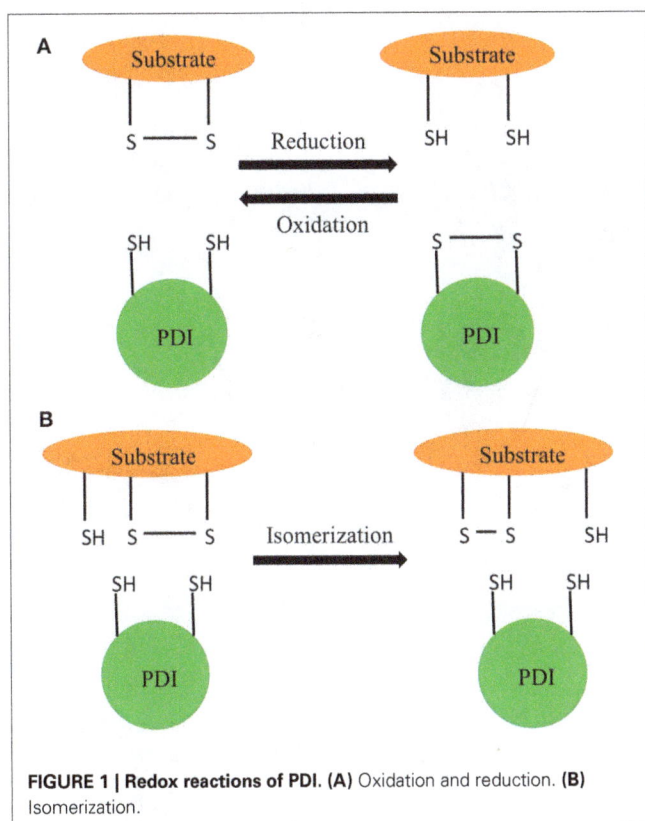

FIGURE 1 | Redox reactions of PDI. (A) Oxidation and reduction. **(B)** Isomerization.

Historically the chaperone activity of PDI was assessed by a variety of methods depending on renaturation of denatured proteins monitored by activity-gain or loss-of-aggregates (Shao et al., 2000; Ben Khalaf et al., 2012; Wang et al., 2012; Hashimoto and Imaoka, 2013). A recent addition to this list utilizes acid-denatured green fluorescent protein (GFP), which interacts with a chaperone protein like PDI, and refold to yield the proper configuration and fluorescent properties (Mares et al., 2011). This technology allows for a high-throughput assays for chaperone activity and its inhibitors (Mares et al., 2011).

Recent work using NMR, indicated that PDI is able to distinguish between unfolded, partly folded, and fully folded proteins. In these studies, it was observed that the dissociation constant (K_D) for fully unfolded basic pancreatic trypsin inhibitor was $\sim$1.5 μM (Irvine et al., 2014). On the other hand the fully folded protein had a K_D that was $\sim$10 fold higher. Partially unfolded protein had a K_D that was $\sim$3 fold higher. These data lead the authors to conclude that PDI can distinguish between unfolded, partially unfolded, and folded proteins (Irvine et al., 2014).

A major recent discovery shows that the chaperone activity of PDI is regulated by its redox status. Wang and coworkers were the first to obtain the crystal structure of human PDI in both the reduced and oxidized forms (Wang et al., 2012). In oxidized form of PDI the active site of *a* and *a'* are 40.3 Å apart and the thioredoxin domains *abb'a'* were all in the same plane (Wang et al., 2013). In the reduced state of PDI the active sites are 27.6 Å, however only *abb'* are in the same plane where *a'* is twisted 45° (**Figure 2**) (Wang et al., 2013), illustrating that the oxidized state has a more open conformation allowing for the entry of

chaperone substrates (i.e., unfolded peptides) and the reduced state has a closed conformation inhibiting their entry. This further illustrates long range conformational changes induced by redox status of the active sites and further suggests redox regulation of chaperone activity.

Another interesting observation came from the work of Fu et al., where the hormone 17β-estradiol was able to bind to the hydrophobic cavity formed between the *bb'* domains (Fu et al., 2011). It should be noted that this site is different from the putative chaperone binding site in the *b'* domain. The key amino acid residue in the interaction of PDI with 17β-estradiol was shown to H256 where it is believed that nitrogen of the histidine forms a hydrogen bond with the 3-hydroxyl group (Fu et al., 2011). The authors, based on this data, postulate yet another physiological role for PDI: a reservoir for hormones (Fu et al., 2011).

PDI REDOX ACTIVITY AND ENDOPLASMIC RETICULUM OXIDOREDUCTIN-1

When PDI catalyzes the oxidation of thiols and the reduction and isomerization of disulfides, the catalytic vicinal active site thiols (-CXXC-) undergo sequential oxidation and reduction reactions. It is suggested that an oxidized *a*-domain catalyzes the oxidation of the reduced substrates and in the process becomes reduced. The N-terminus cysteine (C53) on the catalytic motif reacts with the substrate to form the mixed heterodimer while the C-terminus cysteine (C56) releases the substrate (Walker and Gilbert, 1997). The *a*-domain is then subsequently oxidized by the *a'*-domain back to a disulfide through intramolecular reactions (Araki et al., 2013). The re-oxidation of the *a'*-domain is catalyzed by the protein endoplasmic reticulum oxidoreductin-1 (Ero1) in the process reducing O_2 to produce H_2O_2 (Araki and Nagata, 2011). It should be noted that this enzyme has two isoforms α and β (Wang et al., 2011). The binding affinity and catalytic activity of Ero1α with PDI was observed to be better at pH 7, and less so at 7.5 and 8 (Araki and Nagata, 2011). The interaction of Ero1α with PDI occurs through the interaction of the βhairpin in Ero1α with the *b'*-domain of reduced PDI only (Masui et al., 2011). The non-covalent interaction was found to be between the aromatic residues, the W272 of Ero1α and F240, F249, and F304 of PDI (Masui et al., 2011). This interaction allows for Ero1α to be positioned in a manner that allows the oxidation of the *a'*-domain of PDI but not the *a*-domain (Araki and Nagata, 2011; Masui et al., 2011). NMR studies on Ero1α indicate that it is inactivated when disulfides are formed between C94–C104 and C99–C131 (Araki and Nagata, 2011). This conclusion was reached since C99 is the residue responsible for oxidizing the *a'*-domain of PDI. Therefore, it is thought the formation of disulfide between C94 and C104 results in the C99 close C131. The subsequent formation of the disulfide C99–C131 thus inactivates Ero1α (Araki et al., 2013). The other isoform, Ero1β was observed to also preferentially oxidize the *a'* domain of PDI (Nguyen et al., 2011; Wang et al., 2011).

A further consequence of Ero1 activity is the oxidation of the *a*-domain of PDI. This is modulated through three other ER resident proteins: peroxiredoxin 4 (Prx IV), glutathione peroxidase 7 (GPx7), and glutathione peroxidase 8 (GPx8) (Kakihana et al., 2013; Wang et al., 2014). GPx7 and GPx8 catalyzes the reduction

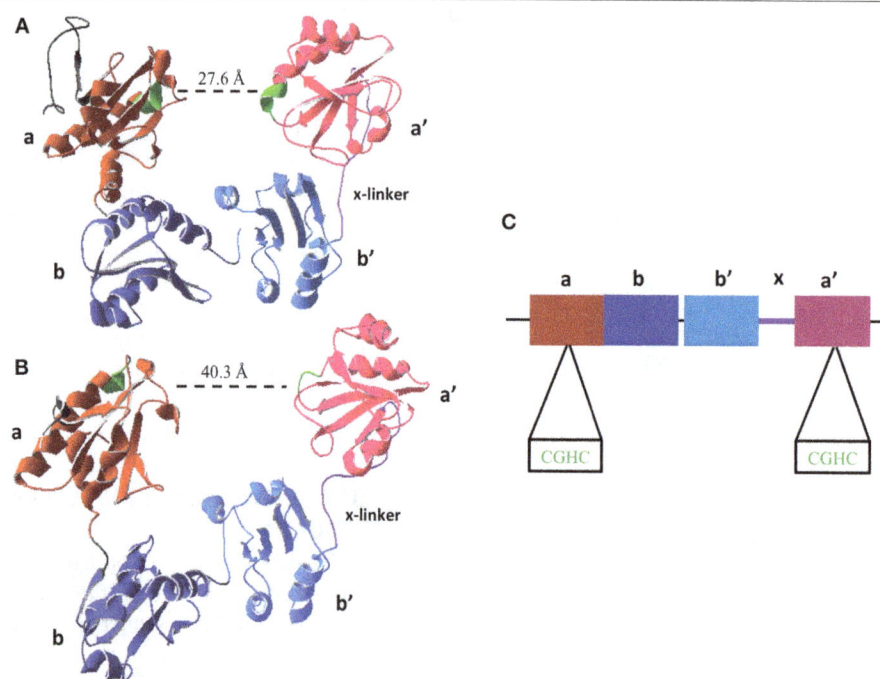

FIGURE 2 | Crystal Structure of Reduced and Oxidized PDI. (A) Reduced crystal structure of human PDI (PDB ID: 4EKZ). **(B)** Oxidized crystal structure of human PDI (PDB ID: 4EL1). **(C)** Schematic of PDI domains present in the crystal structure, as well as the two active site CXXC motifs (Crystal structures created using Swiss PDBViewer) (Wang et al., 2013).

of Ero1-produced H_2O_2 to water. In the process, the C57 and C86 of GPx7 are converted to sulfenic acid and disulfide respectively. The PDI *a*-domain then reduces these thiols, becoming oxidized in the process (Wang et al., 2014). When GPx7 or GPx8 was replaced with Prx IV this caused a slower oxidative folding of protein in comparison with either GPx7 or GPx8 (Nguyen et al., 2011). It is still controversial whether proteins such as, Ero1α, are the primary cause of the oxidation of PDI, since oxidized glutathione (GSSG) can also oxidize PDI active sites (Lappi and Ruddock, 2011). The second-order rate constant for the formation of mixed disulfide from reduced PDI and GSSG was observed to be $191 \pm 3\ M^{-1}s^{-1}$ (Lappi and Ruddock, 2011). *In vivo* studies of the rate of oxidation of the *a* domain of PDI by GSSG were found to be ~0.3 s^{-1}, indicating that this rate is too fast to be overlooked (Lappi and Ruddock, 2011). The oxidation of PDI by excess H_2O_2 is also observed and has a pseudo-first-order reaction with a rate constant of $9.2\ M^{-1}s^{-1}$ (Karala et al., 2009). However this study was conducted solely on the *a* domain on PDI, further studies of glutathione and glutathione disulfide on the oxidation of PDI are required to determine how much affect these molecules have on PDI. It should also be noted that more research must be conducted on how much affect PDI has on controlling the equilibrium of glutathione and glutathione disulfide.

PDI INHIBITORS

With the importance of PDI in many different cellular functions, there has been an increased interest in the modulation of its activity with small molecule inhibitors. A commonly used inhibitor of PDI is bacitracin, which is a mixture of cyclic polypeptides.

Recently, the analogs of bacitracin were individually tested to observe their inhibitory effect on the reductive activity of PDI. The analogs H and F were 25-fold more effective than the A and B analogs (Dickerhof et al., 2011). These analogs bind to the **b'** domain and interestingly do not seem to inhibit the oxidative and isomerase activity of PDI (Roth, 1981; Dickerhof et al., 2011).

Another inhibitor of PDI reductase activity is quercetin-3-rutinoside. This compound inhibits PDI mediated platelet aggregation as a result has been suggested as antithrombic agent (Jasuja et al., 2012). Phenyl vinyl sulfonate containing molecules were also showed to inhibit PDI (Ge et al., 2013). The inhibitor designated P1 was observed to inhibit PDI in mammalian cells with an IC_{50} of $1.7 \pm 0.4\ \mu m$ (Ge et al., 2013).

Propynoic acid carbamoyl methyl amides (PACMA) also inhibit PDI (Xu et al., 2012). Of these, the one designated PACMA31 was observed to be an irreversible inhibitor, by forming a covalent (carbon-sulfur) bond to the active site cysteine (Xu et al., 2012). Intriguingly, this inhibitor was observed to target ovarian tumors, inhibiting their growth but not causing harm to normal cells (Xu et al., 2012).

PDI AND NADPH OXIDASE

NADPH oxidase complex (Nox) is the major contributor of reactive oxygen species (ROS) in cells which can act as molecular signal and when produced in excess can cause oxidative stress (Janiszewski et al., 2005; Clempus and Griendling, 2006; Bedard and Krause, 2007; Marchetti et al., 2007). The NADPH oxidase complex is comprised of many subunits, that are expressed differently depending on the cell type (Janiszewski et al., 2005; Clempus and Griendling, 2006; Bedard and Krause, 2007; Marchetti et al., 2007). Laurindo's group were the first to report a role for PDI in the regulation of Nox. PDI is shown to interact with Nox within the ER as well as outside the ER in the cytosol (Laurindo et al., 2008, 2014; Santos et al., 2009b).

More recently, Fernandes et al. showed that in vascular smooth muscle cells, the overexpression of PDI increased Nox1 mRNA which encodes for NOX1 in NADPH oxidase complex (Fernandes et al., 2009). The PDI overexpression also produced an increase in the activity of NADPH oxidase, leading to an increase in the steady-state levels of ROS (Fernandes et al., 2009). Silencing of PDI triggered a decrease in mRNA and protein of Nox1 (Fernandes et al., 2009). Overexpression of a mutant PDI where the catalytic cysteines were mutated resulted in the increase in Nox1 mRNA and protein indicating that PDI redox activity is not required and it is most likely the chaperone activity of PDI that is modulating this effect (Fernandes et al., 2009).

In other studies, PDI silencing has shown to greatly decrease platelet derived growth factor (PDGF)-induced VSMC migration (Pescatore et al., 2012). In the same study, it was observed that PDI co-immunoprecipitated with Rac1, RhoA, and RhoGDI (Pescatore et al., 2012). In macrophages, PDI plays a similar role where interaction with NADPH oxidase causes an increase in ROS (Santos et al., 2009a). Laurindo's group has shown that PDI co-immuneprecipitated and co-localized with p22phox subunit of NADPH oxidase in J774 cells (Santos et al., 2009a). They also used a leukocyte cell free system to observe the interaction of PDI with NADPH oxidase. They found that oxidized PDI stimulates ROS production however reduced PDI inhibited the production of ROS (De et al., 2011). This group further suggests that PDI possibly interacts with p47phox subunit of NADPH oxidase through hydrophobic effects and not through cysteines (Santos et al., 2009a). From all this data it can be suggested that PDI plays an important role in the regulation of ROS.

PDI AND INFECTION

The internalization of some pathogens has been shown to be modulated by PDI. Ryser and coworkers were the first to test the potential role of PDI in the internalization of pathogens (Ryser et al., 1994). In mouse macrophage J774 cells, Laurindo's group observed that increased levels of PDI would increase the phagocytosis of *L. chagasi* promastigote but not the amastigote (Santos et al., 2009a). Phagocytosis was decreased by inhibition of expression levels of PDI as well as through the addition of thiol inhibitors such as *p*-phenylarsine oxide (Santos et al., 2009a). Santos and coworkers hypothesized that PDI plays a role in the reduction of the disulfide bonds present on the parasite, which may help with internalization of the parasite (Santos et al., 2009a). They also suggest that PDI's role in the increase of ROS through the NADPH oxidase interaction resulting in an "intraphagosomal-oxidizing milieu" and thus favoring infection of promastigotes (Santos et al., 2009a).

Reiser and coworkers have examined the role of PDI in HIV infections. Both PDI and thioredoxin 1 (Trx1) have been shown to reduce disulfides on the viral glycoprotein gp120 causing the internalization of HIV-1 (Reiser et al., 2012). Through semi-quantitative ELISA analysis it was observed that Trx1 was more efficient at reducing disulfides on gp120, however the authors hypothesize that the two proteins may be acting at different stages of internalization (Reiser et al., 2012). Interestingly, galectin-9 was observed to bind to PDI on Th2 cells this helped with the retention of PDI to the cell surface as well as increased the entry of HIV into these cells (Bi et al., 2011). It is suggested that inhibitors of gp120 that prevent the reduction of the disulfides maybe a better drug target than reductases since reductases are widely used in cells (Reiser et al., 2012).

In endothelial cells surface PDI possibly reduces β1 and β3 integrins allowing for the entry of dengue virus (Wan et al., 2012). Also in MA104 cells thiol blockers and PDI inhibitors decreased the entry of rotavirus, indicating the involvement of thiols for infectivity (Calderon et al., 2012).

Recently, PDI was observed to regulate the cytoskeleton reorganization by its interaction with β-actin (Sobierajska et al., 2014). PDI was observed to bind to Cys374 of β-actin through a disulfide bond (Sobierajska et al., 2014). In MEG-01 cells the down regulation of PDI resulted in the prevention of adhesion to fibronectin (Sobierajska et al., 2014).

During the bacterial infection of the cholera toxin it is observed that PDI does not unfold the CT protein into its subunit CTA1 for further infection (Taylor et al., 2011). Nevertheless, PDI does disassemble the holotoxin allowing for the dissociation of CTA1 from CTA2 and CTB_5 and its spontaneous unfolding, allowing CTA1 to exit the ER (Taylor et al., 2011). Interestingly, Taylor and coworkers recently discovered that PDI when interacting with the holotoxin, will itself unfold, causing a wedge to form between the holotoxin and causing the dissociation of CTA1 (Taylor et al., 2014). PDI locked in the folded state would not dissociate the holotoxin (Taylor et al., 2014). This appears not to required the oxireductase activity of PDI since redox inhibitors had no effect on toxin activation. On the other hand, chaperone inhibitors prevent the release of CTA1 thus indicating that PDI unfolding in the presence of holotoxin requires PDI chaperone activity (Taylor et al., 2014).

PDI AND COAGULATION

The activation of platelets is a complex series of proteolytic, protein-protein, and protein-ligand interactions that is not yet fully understood. A quick summary of this process is, that damaged endothelial cells will cause the exposure of collagen allowing for the glycoprotein, von Willebrand factor to attract platelets (Lopez et al., 1992; Modderman et al., 1992; Ruggeri, 2007; Herr and Farndale, 2009). The platelets form a monolayer, which in turns produces thrombin for additional platelet aggregation and causes a series of cascades (Monroe et al., 2002; Clemetson, 2012). The platelets also start to secrete other activators these will bind to receptor on the platelets and increase the intracellular calcium levels (Purvis et al., 2008). This causes receptors on platelets to bind fibrinogen resulting in the increase adhesion of the platelets and the formation of thrombi(Ma et al., 2007; Furie and Furie, 2008). Many of the integrin receptors associated with platelet activation contain a cysteine rich subunit, possibly allowing for PDI to regulate the redox state of the thiols.

Platelets deficient of PDI display irregular and improper aggregation, when wild-type PDI is added the aggregation of platelets returns to normal (Kim et al., 2013). PDI on the surface of platelets plays a crucial role in the formation of thrombus in collagen-coated platelets (Kim et al., 2013). In monocytic cells, cell surface PDI is required for antithymocyte globulin decryption of tissue factor (Langer et al., 2013).

Recent studies have shown that free thiols and isomerization are associated with the process of coagulation (Jurk et al., 2011). *In vitro* studies by Jurk and coworkers have shown that PDI plays a role in the feedback activation of thrombin in platelets that have already been stimulated by thrombin (Jurk et al., 2011). In the same paper, they showed that PDI may play a role in the interaction of coagulation factors with platelets (Jurk et al., 2011). With the use of CHO cells expressing different integrin factors, $\alpha_V\beta_3$ and $\alpha_{IIb}\beta_3$, PDI was observed to bind to these integrin molecules (Cho et al., 2012). This binding of PDI with the integrin was determined to be through the β_3 subunit and not the α subunit (Cho et al., 2012). The binding of PDI to integrins was not inhibited by the presence of PDI redox inhibitors, the authors however do not hypothesize how this interaction is occurring (Cho et al., 2012). In a mice model, where there is a deficiency in β_3 integrins, it is observed that fibrin generation was diminished from the lack of accumulation of PDI at the platelet surface (Cho et al., 2012). In neutrophil cells, PDI may mediate the adhesive activity during vasculature inflammation through its interaction with $\alpha_M\beta_2$ integrin (Hahm et al., 2013). With the use of flow cytometry, it was observed that exogenous PDI interacted with neutrophil cell surface by electrostatic interactions (Hahm et al., 2013). The authors propose that PDI catalyzes thiol exchange through electrostatic interactions of $\alpha_M\beta_2$ integrin thus regulating the clustering of the integrin within lipid rafts (Hahm et al., 2013). It should also be noted that vitamin K epoxide reductase, an important enzyme in blood coagulation, requires the help of a thioredoxin-like protein to regenerate vitamin K hydroquinone from the epoxide form. It is speculated that the PDI is the thioredoxin-like protein that provides the reductive equivalent to perform the regeneration (Schulman et al., 2010).

In endothelial cells, when PDI was silenced or inhibited on the cell surface, the coagulation was increased (Popescu et al., 2010). In agreement with this finding, it was observed that addition of exogenous PDI resulted in the decrease in coagulation activity (Popescu et al., 2010). The authors suggest that coagulation of endothelial cells is negatively regulated by PDI, specifically its oxidoreductase activity (Popescu et al., 2010). In the same study, it was also observed that PDI may play a role in the regulation of phosphatidylserine exposure through the activity of flippase and floppase (Popescu et al., 2010).

For more information of previous knowledge on the effects of redox on coagulation can be found in reviews by Cho (2013), Flaumenhaft (2013), Langer and Ruf (2014), and Murphy et al. (2014).

PDI AND SUPEROXIDE DISMUTASE 1

Protein aggregation has been shown to occur in a variety of neurodegenerative diseases and disorders, such as cerebral ischemia and amyotrophic lateral sclerosis. Cai and coworkers were one of the first to show that PDI alleviates the aggregation of proteins (Cai et al., 1994). Cerebral ischemia is caused by decrease

blood flow to the brain resulting in the loss of oxygen to neurons and astrocytes, which has shown to cause an increase in nitric oxide (NO) production specifically by inducible nitric oxide synthase (iNOS) (Cherian et al., 2000; Greve and Zink, 2009). Elevated NO production can lead to the inactivation of proteins by *S*-nitrosylation. In astrocytes, recent studies by Chen have shown the presence of S-nitrosylated PDI (SNO-PDI), which has previously been shown to inhibit the activity of PDI (Ramachandran et al., 2001; Chen et al., 2012). When the cells were oxygen deprived, this caused the formation of ubiquitinated protein aggregates, of proteins like superoxide dismutase 1 (SOD1), which is crucial to cell survival (Chen et al., 2012).

When NO production in these cells was attenuated, the formation of both SNO-PDI and protein aggregates decreased (Chen et al., 2012). Interestingly, in SH-SY5Y cells the overexpression of mutant SOD1 caused an increase expression of iNOS, where wild type SOD1 did not cause any increase (Chen et al., 2013). SOD1 is a protein that alleviates oxidative stress by converting free radicals to H_2O_2 (Chen et al., 2012). The mutant SOD1 expressing cells were observed to contain SNO-PDI and protein aggregation while the wild type SOD1 did not show any increase in either, this was also observed *in vivo*, in transgenic mice (Chen et al., 2013). Furthermore, the expression of PDI increased when protein aggregates accumulated in cells, and the cells became apoptotic (Hoffstrom et al., 2010; Chen et al., 2013).

The aggregation of mutant SOD1 has shown to cause an increase in neuronal cell death in amyotrophic lateral sclerosis, it has been hypothesized that denitrosylation of PDI can be considered as a therapeutic intervention for this disease (Jeon et al., 2014). In both HEK293A cells and NSC-34 motor neuron cells, the overexpression of PDI lead to a decrease in the aggregation of mutant SOD1 (Jeon et al., 2014). The interaction of PDI with SOD1 appears to be through the chaperone activity of PDI rather than the isomerase activity since aggregation was also reduced with mutant PDI lacking active site thiols (Jeon et al., 2014). Intriguingly, the overexpression of PDI caused by mutant SOD1 has shown to cause the activation of NADPH oxidase causing an increase in ROS (Jaronen et al., 2013). In murine microglial BV-2 cells Nox activation was PDI-dependent suggesting that PDI inhibitors may help reduce oxidative stress in cells (Jaronen et al., 2013). However, further studies must be conducted to determine PDI's role in the interplay of protein aggregation and the production of ROS.

ROLE OF PDI IN NITRIC OXIDE SIGNALING

Our group was the first to show that cell surface PDI mediated the entry of *S*-nitrosothiols (SNO) into cells (Ramachandran et al., 2001). Cell surface PDI was observed to release NO from *S*-nitrosylated glutathione (GSNO), one of the main nitric oxide carriers in tissues (Root et al., 2004). PDI would denitrosylate GSNO when NO levels were low, however during high levels of NO, PDI would act as a carrier either through the formation of SNO-PDI (Sliskovic et al., 2005).

Recently our group, showed that PDI plays a role in the transport of nitric oxide from red blood cells to endothelium cells (Kallakunta et al., 2013). This process was observed to be oxygen-dependent where nitric oxide bound hemoglobin

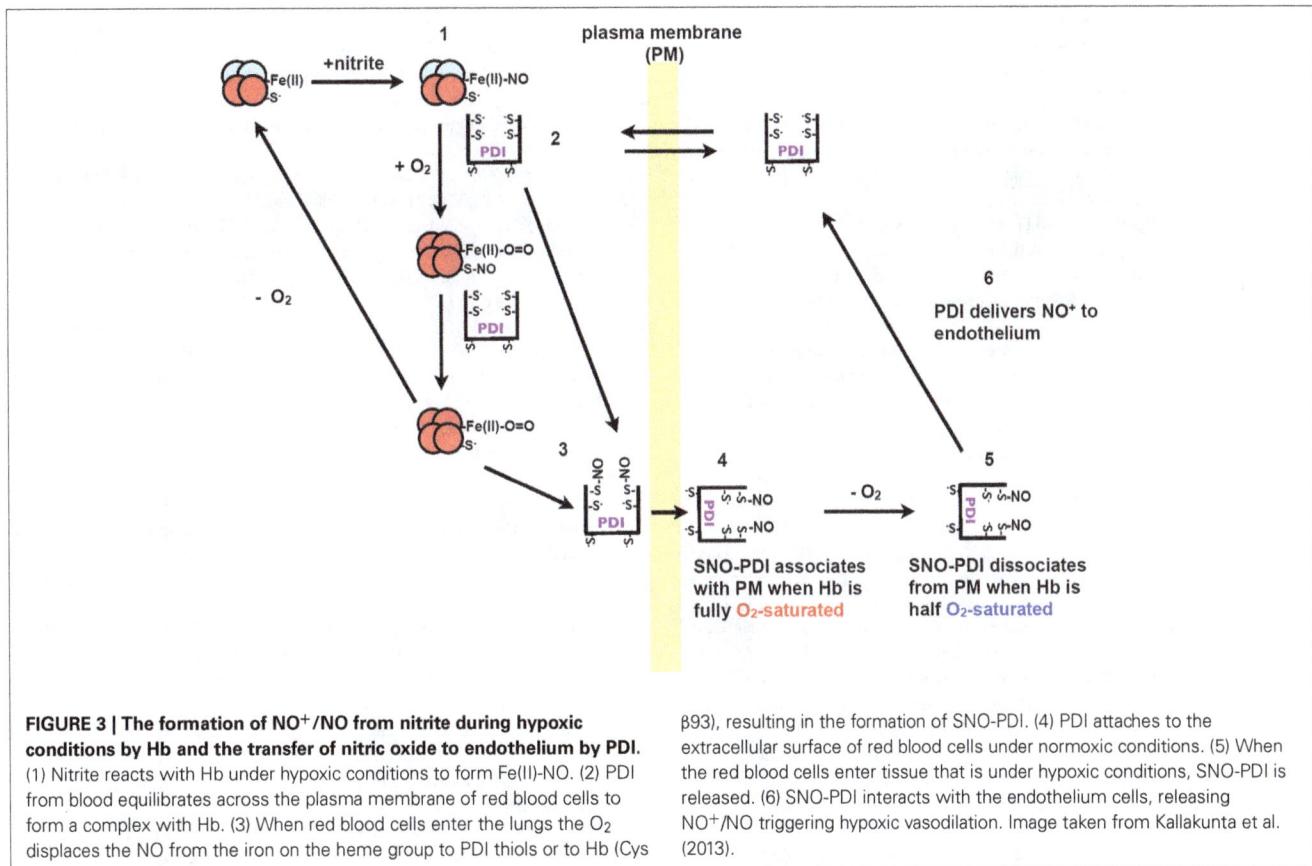

FIGURE 3 | The formation of NO⁺/NO from nitrite during hypoxic conditions by Hb and the transfer of nitric oxide to endothelium by PDI. (1) Nitrite reacts with Hb under hypoxic conditions to form Fe(II)-NO. (2) PDI from blood equilibrates across the plasma membrane of red blood cells to form a complex with Hb. (3) When red blood cells enter the lungs the O_2 displaces the NO from the iron on the heme group to PDI thiols or to Hb (Cys β93), resulting in the formation of SNO-PDI. (4) PDI attaches to the extracellular surface of red blood cells under normoxic conditions. (5) When the red blood cells enter tissue that is under hypoxic conditions, SNO-PDI is released. (6) SNO-PDI interacts with the endothelium cells, releasing NO⁺/NO triggering hypoxic vasodilation. Image taken from Kallakunta et al. (2013).

(NO-Hb) required hypoxic conditions. Oxygen is then required for the formation of SNO-Hb, which then transfers the nitric oxide to PDI to form SNO-PDI (Kallakunta et al., 2013). When hemoglobin is half oxygen saturated it was observed that nitric oxide would dissociate from extracellular PDI and be transferred to endothelium cells (**Figure 3**) (Kallakunta et al., 2013).

CONCLUSION

During the early years of research on PDI it was well thought of that PDI was required for the maintenance of healthy cells and tissues, in view of the catalytic roles of PDI in reducing and isomerizing disulfides into native confirmations of proteins and to alleviate protein aggregates through its chaperone activity. However recent studies indicate that this may not be the case. PDI through its thiol redox activity, is observed to play a role for the internalization of some pathogens, as well as it is shown to promote increased ROS in cells. Furthermore, in neuronal cells post transnationally modified SNO-PDI etc. has been shown to promote protein aggregation commonly associated with the pathophysiology of neurodegenerative diseases. These observations show that PDI is a complex protein that can play a role in both physiology and pathophysiology. Therefore, further studies are warranted in order to solve the complex relationships of PDI in health and disease.

ACKNOWLEDGMENT

This work was supported by a NSERC Discovery Grant to Bulent Mutus and was supported by the University of Windsor tuition scholarship.

REFERENCES

Alanen, H. I., Salo, K. E., Pekkala, M., Siekkinen, H. M., Pirneskoski, A., and Ruddock, L. W. (2003). Defining the domain boundaries of the human protein disulfide isomerases. *Antioxid. Redox Signal.* 5, 367–374. doi: 10.1089/152308603768295096

Andreu, C. I., Woehlbier, U., Torres, M., and Hetz, C. (2012). Protein disulfide isomerases in neurodegeneration: from disease mechanisms to biomedical applications. *FEBS Lett.* 586, 2826–2834. doi: 10.1016/j.febslet.2012.07.023

Araki, K., Iemura, S., Kamiya, Y., Ron, D., Kato, K., Natsume, T., et al. (2013). Ero1-alpha and PDIs constitute a hierarchical electron transfer network of endoplasmic reticulum oxidoreductases. *J. Cell Biol.* 202, 861–874. doi: 10.1083/jcb.201303027

Araki, K., and Nagata, K. (2011). Functional *in vitro* analysis of the ERO1 protein and protein-disulfide isomerase pathway. *J. Biol. Chem.* 286, 32705–32712. doi: 10.1074/jbc.M111.227181

Bastos-Aristizabal, S., Kozlov, G., and Gehring, K. (2014). Structural insight into the dimerization of human protein disulfide isomerase. *Protein Sci.* 23, 618–626. doi: 10.1002/pro.2444

Bedard, K., and Krause, K.-H. (2007). The NOX family of ROS-generating NADPH oxidases: physiology and pathophysiology. *Physiol. Rev.* 87, 245–313. doi: 10.1152/physrev.00044.2005

Benham, A. M. (2012). The protein disulfide isomerase family: key players in health and disease. *Antioxid. Redox Signal.* 16, 781–789. doi: 10.1089/ars.2011.4439

Ben Khalaf, N., De Muylder, G., Louzir, H., Mckerrow, J., and Chenik, M. (2012). Leishmania major protein disulfide isomerase as a drug target: enzymatic and functional characterization. *Parasitol. Res.* 110, 1911–1917. doi: 10.1007/s00436-011-2717-5

Bi, S., Hong, P. W., Lee, B., and Baum, L. G. (2011). Galectin-9 binding to cell surface protein disulfide isomerase regulates the redox environment to enhance T-cell migration and HIV entry. *Proc. Natl. Acad. Sci. U.S.A.* 108, 10650–10655. doi: 10.1073/pnas.1017954108

Bulleid, N. J., and Ellgaard, L. (2011). Multiple ways to make disulfides. *Trends Biochem. Sci.* 36, 485–492. doi: 10.1016/j.tibs.2011.05.004

Cai, H., Wang, C. C., and Tsou, C. L. (1994). Chaperone-like activity of protein disulfide isomerase in the refolding of a protein with no disulfide bonds. *J. Biol. Chem.* 269, 24550–24552.

Calderon, M. N., Guerrero, C. A., Acosta, O., Lopez, S., and Arias, C. F. (2012). Inhibiting rotavirus infection by membrane-impermeant thiol/disulfide exchange blockers and antibodies against protein disulfide isomerase. *Intervirology* 55, 451–464. doi: 10.1159/000335262

Chen, X., Guan, T., Li, C., Shang, H., Cui, L., Li, X. M., et al. (2012). SOD1 aggregation in astrocytes following ischemia/reperfusion injury: a role of NO-mediated S-nitrosylation of protein disulfide isomerase (PDI). *J. Neuroinflammation* 9, 237. doi: 10.1186/1742-2094-9-237

Chen, X., Zhang, X., Li, C., Guan, T., Shang, H., Cui, L., et al. (2013). S-nitrosylated protein disulfide isomerase contributes to mutant SOD1 aggregates in amyotrophic lateral sclerosis. *J. Neurochem.* 124, 45–58. doi: 10.1111/jnc.12046

Cherian, L., Goodman, J. C., and Robertson, C. S. (2000). Brain nitric oxide changes after controlled cortical impact injury in rats. *J. Neurophysiol.* 83, 2171–2178. doi: 10.1097/00003246-199901001-00091

Cho, J. (2013). Protein disulfide isomerase in thrombosis and vascular inflammation. *J. Thromb. Haemost.* 11, 2084–2091. doi: 10.1111/jth.12413

Cho, J., Kennedy, D. R., Lin, L., Huang, M., Merrill-Skoloff, G., Furie, B. C., et al. (2012). Protein disulfide isomerase capture during thrombus formation *in vivo* depends on the presence of beta3 integrins. *Blood* 120, 647–655. doi: 10.1182/blood-2011-08-372532

Clemetson, K. J. (2012). Platelets and primary haemostasis. *Thromb. Res.* 129, 220–224. doi: 10.1016/j.thromres.2011.11.036

Clempus, R. E., and Griendling, K. K. (2006). Reactive oxygen species signaling in vascular smooth muscle cells. *Cardiovasc. Res.* 71, 216–225. doi: 10.1016/j.cardiores.2006.02.033

De, A. P. A. M., Verissimo-Filho, S., Guimaraes, L. L., Silva, A. C., Takiuti, J. T., and Lopes, L. R. (2011). Protein disulfide isomerase redox-dependent association with p47(phox): evidence for an organizer role in leukocyte NADPH oxidase activation. *J. Leukoc. Biol.* 90, 799–810. doi: 10.1189/jlb.0610324

Denisov, A. Y., Maattanen, P., Dabrowski, C., Kozlov, G., Thomas, D. Y., and Gehring, K. (2009). Solution structure of the bb' domains of human protein disulfide isomerase. *FEBS J.* 276, 1440–1449. doi: 10.1111/j.1742-4658.2009.06884.x

Dickerhof, N., Kleffmann, T., Jack, R., and Mccormick, S. (2011). Bacitracin inhibits the reductive activity of protein disulfide isomerase by disulfide bond formation with free cysteines in the substrate-binding domain. *FEBS J.* 278, 2034–2043. doi: 10.1111/j.1742-4658.2011.08119.x

Edman, J. C., Ellis, L., Blacher, R. W., Roth, R. A., and Rutter, W. J. (1985). Sequence of protein disulphide isomerase and implications of its relationship to thioredoxin. *Nature* 317, 267–270.

Fernandes, D. C., Manoel, A. H., Wosniak, J. Jr., and Laurindo, F. R. (2009). Protein disulfide isomerase overexpression in vascular smooth muscle cells induces spontaneous preemptive NADPH oxidase activation and Nox1 mRNA expression: effects of nitrosothiol exposure. *Arch. Biochem. Biophys.* 484, 197–204. doi: 10.1016/j.abb.2009.01.022

Flaumenhaft, R. (2013). Protein disulfide isomerase as an antithrombotic target. *Trends Cardiovasc. Med.* 23, 264–268. doi: 10.1016/j.tcm.2013.03.001

Freedman, R. B., Gane, P. J., Hawkins, H. C., Hlodan, R., Mclaughlin, S. H., and Parry, J. W. (1998). Experimental and theoretical analyses of the domain architecture of mammalian protein disulphide-isomerase. *Biol. Chem.* 379, 321–328.

Fu, X. M., Wang, P., and Zhu, B. T. (2011). Characterization of the estradiol-binding site structure of human protein disulfide isomerase (PDI). *PLoS ONE* 6:e27185. doi: 10.1371/journal.pone.0027185

Furie, B., and Furie, B. C. (2008). Mechanisms of thrombus formation. *N. Engl. J. Med.* 359, 938–949. doi: 10.1056/NEJMra0801082

Ge, J., Zhang, C. J., Li, L., Chong, L. M., Wu, X., Hao, P., et al. (2013). Small molecule probe suitable for in situ profiling and inhibition of protein disulfide isomerase. *ACS Chem. Biol.* 8, 2577–2585. doi: 10.1021/cb4002602

Goldberger, R. F., Epstein, C. J., and Anfinsen, C. B. (1963). Acceleration of reactivation of reduced bovine pancreatic ribonuclease by a microsomal system from rat liver. *J. Biol. Chem.* 238, 628–635.

Greve, M. W., and Zink, B. J. (2009). Pathophysiology of traumatic brain injury. *Mt. Sinai J. Med.* 76, 97–104. doi: 10.1002/msj.20104

Hahm, E., Li, J., Kim, K., Huh, S., Rogelj, S., and Cho, J. (2013). Extracellular protein disulfide isomerase regulates ligand-binding activity of alphaMbeta2 integrin and neutrophil recruitment during vascular inflammation. *Blood* 121, 3789–3800, S3781–S3715. doi: 10.1182/blood-2012-11-467985

Hashimoto, S., and Imaoka, S. (2013). Protein-disulfide isomerase regulates the thyroid hormone receptor-mediated gene expression via redox factor-1 through thiol reduction-oxidation. *J. Biol. Chem.* 288, 1706–1716. doi: 10.1074/jbc.M112.365239

Herr, A. B., and Farndale, R. W. (2009). Structural insights into the interactions between platelet receptors and fibrillar collagen. *J. Biol. Chem.* 284, 19781–19785. doi: 10.1074/jbc.R109.013219

Hoffstrom, B. G., Kaplan, A., Letso, R., Schmid, R. S., Turmel, G. J., Lo, D. C., et al. (2010). Inhibitors of protein disulfide isomerase suppress apoptosis induced by misfolded proteins. *Nat. Chem. Biol.* 6, 900–906. doi: 10.1038/nchembio.467

Holmgren, A. (1968). Thioredoxin. 6. The amino acid sequence of the protein from *Escherichia coli* B. *Eur. J. Biochem.* 6, 475–484.

Irvine, A. G., Wallis, A. K., Sanghera, N., Rowe, M. L., Ruddock, L. W., Howard, M. J., et al. (2014). Protein disulfide-isomerase interacts with a substrate protein at all stages along its folding pathway. *PLoS ONE* 9:e82511. doi: 10.1371/journal.pone.0082511

Janiszewski, M., Lopes, L. R., Carmo, A. O., Pedro, M. A., Brandes, R. P., Santos, C. X., et al. (2005). Regulation of NAD(P)H oxidase by associated protein disulfide isomerase in vascular smooth muscle cells. *J. Biol. Chem.* 280, 40813–40819. doi: 10.1074/jbc.M509255200

Jaronen, M., Vehvilainen, P., Malm, T., Keksa-Goldsteine, V., Pollari, E., Valonen, P., et al. (2013). Protein disulfide isomerase in ALS mouse glia links protein misfolding with NADPH oxidase-catalyzed superoxide production. *Hum. Mol. Genet.* 22, 646–655. doi: 10.1093/hmg/dds472

Jasuja, R., Passam, F. H., Kennedy, D. R., Kim, S. H., Van Hessem, L., Lin, L., et al. (2012). Protein disulfide isomerase inhibitors constitute a new class of antithrombotic agents. *J. Clin. Invest.* 122, 2104–2113. doi: 10.1172/JCI61228

Jeon, G. S., Nakamura, T., Lee, J.-S., Choi, W.-J., Ahn, S.-W., Lee, K.-W., et al. (2014). Potential effect of *S*-nitrosylated protein disulfide isomerase on mutant SOD1 aggregation and neuronal cell death in amyotrophic lateral sclerosis. *Mol. Neurobiol.* 49, 796–807. doi: 10.1007/s12035-013-8562-z

Jurk, K., Lahav, J., VAN Aken, H., Brodde, M. F., Nofer, J. R., and Kehrel, B. E. (2011). Extracellular protein disulfide isomerase regulates feedback activation of platelet thrombin generation via modulation of coagulation factor binding. *J. Thromb. Haemost.* 9, 2278–2290. doi: 10.1111/j.1538-7836.2011.04509.x

Kakihana, T., Araki, K., Vavassori, S., Iemura, S., Cortini, M., Fagioli, C., et al. (2013). Dynamic regulation of Ero1alpha and peroxiredoxin 4 localization in the secretory pathway. *J. Biol. Chem.* 288, 29586–29594. doi: 10.1074/jbc.M113.467845

Kallakunta, V. M., Slama-Schwok, A., and Mutus, B. (2013). Protein disulfide isomerase may facilitate the efflux of nitrite derived S-nitrosothiols from red blood cells. *Redox Biol.* 1, 373–380. doi: 10.1016/j.redox.2013.07.002

Karala, A. R., Lappi, A. K., Saaranen, M. J., and Ruddock, L. W. (2009). Efficient peroxide-mediated oxidative refolding of a protein at physiological pH and implications for oxidative folding in the endoplasmic reticulum. *Antioxid. Redox Signal.* 11, 963–970. doi: 10.1089/ARS.2008.2326

Kemmink, J., Darby, N. J., Dijkstra, K., Nilges, M., and Creighton, T. E. (1997). The folding catalyst protein disulfide isomerase is constructed of active and inactive thioredoxin modules. *Curr. Biol.* 7, 239–245.

Kim, K., Hahm, E., Li, J., Holbrook, L. M., Sasikumar, P., Stanley, R. G., et al. (2013). Platelet protein disulfide isomerase is required for thrombus formation but not for hemostasis in mice. *Blood* 122, 1052–1061. doi: 10.1182/blood-2013-03-492504

Koch, G. L. (1987). Reticuloplasmins: a novel group of proteins in the endoplasmic reticulum. *J. Cell Sci.* 87(Pt 4), 491–492.

Kozlov, G., Maattanen, P., Thomas, D. Y., and Gehring, K. (2010). A structural overview of the PDI family of proteins. *FEBS J.* 277, 3924–3936. doi: 10.1111/j.1742-4658.2010.07793.x

Langer, F., and Ruf, W. (2014). Synergies of phosphatidylserine and protein disulfide isomerase in tissue factor activation. *Thromb. Haemost.* 111, 590–597. doi: 10.1160/TH13-09-0802

Langer, F., Spath, B., Fischer, C., Stolz, M., Ayuk, F. A., Kroger, N., et al. (2013). Rapid activation of monocyte tissue factor by antithymocyte globulin is dependent on complement and protein disulfide isomerase. *Blood* 121, 2324–2335. doi: 10.1182/blood-2012-10-460493

Lappi, A. K., and Ruddock, L. W. (2011). Reexamination of the role of interplay between glutathione and protein disulfide isomerase. *J. Mol. Biol.* 409, 238–249. doi: 10.1016/j.jmb.2011.03.024

Laurindo, F. R., Araujo, T. L., and Abrahao, T. B. (2014). Nox NADPH oxidases and the endoplasmic reticulum. *Antioxid. Redox Signal.* 20, 2755–2775. doi: 10.1089/ars.2013.5605

Laurindo, F. R., Fernandes, D. C., Amanso, A. M., Lopes, L. R., and Santos, C. X. (2008). Novel role of protein disulfide isomerase in the regulation of NADPH oxidase activity: pathophysiological implications in vascular diseases. *Antioxid. Redox Signal.* 10, 1101–1113. doi: 10.1089/ars.2007.2011

Laurindo, F. R., Pescatore, L. A., and Fernandes Dde, C. (2012). Protein disulfide isomerase in redox cell signaling and homeostasis. *Free Radic. Biol. Med.* 52, 1954–1969. doi: 10.1016/j.freeradbiomed.2012.02.037

Lopez, J. A., Leung, B., Reynolds, C. C., Li, C. Q., and Fox, J. E. (1992). Efficient plasma membrane expression of a functional platelet glycoprotein Ib-IX complex requires the presence of its three subunits. *J. Biol. Chem.* 267, 12851–12859.

Ma, Y. Q., Qin, J., and Plow, E. F. (2007). Platelet integrin alpha(IIb)beta(3): activation mechanisms. *J. Thromb. Haemost.* 5, 1345–1352. doi: 10.1111/j.1538-7836.2007.02537.x

Maattanen, P., Gehring, K., Bergeron, J. J., and Thomas, D. Y. (2010). Protein quality control in the ER: the recognition of misfolded proteins. *Semin. Cell Dev. Biol.* 21, 500–511. doi: 10.1016/j.semcdb.2010.03.006

Marchetti, P., Bugliani, M., Lupi, R., Marselli, L., Masini, M., Boggi, U., et al. (2007). The endoplasmic reticulum in pancreatic beta cells of type 2 diabetes patients. *Diabetologia* 50, 2486–2494. doi: 10.1007/s00125-007-0816-8

Marcus, N., Shaffer, D., Farrar, P., and Green, M. (1996). Tissue distribution of three members of the murine protein disulfide isomerase (PDI) family. *Biochim. Biophys. Acta* 1309, 253–260.

Mares, R. E., Melendez-Lopez, S. G., and Ramos, M. A. (2011). Acid-denatured Green Fluorescent Protein (GFP) as model substrate to study the chaperone activity of protein disulfide isomerase. *Int. J. Mol. Sci.* 12, 4625–4636. doi: 10.3390/ijms12074625

Masui, S., Vavassori, S., Fagioli, C., Sitia, R., and Inaba, K. (2011). Molecular bases of cyclic and specific disulfide interchange between human ERO1alpha protein and protein-disulfide isomerase (PDI). *J. Biol. Chem.* 286, 16261–16271. doi: 10.1074/jbc.M111.231357

McLaughlin, S. H., and Bulleid, N. J. (1998). Thiol-independent interaction of protein disulphide isomerase with type X collagen during intra-cellular folding and assembly. *Biochem. J.* 331(Pt 3), 793–800.

Modderman, P. W., Admiraal, L. G., Sonnenberg, A., and Von Dem Borne, A. E. (1992). Glycoproteins V and Ib-IX form a noncovalent complex in the platelet membrane. *J. Biol. Chem.* 267, 364–369.

Monroe, D. M., Hoffman, M., and Roberts, H. R. (2002). Platelets and thrombin generation. *Arterioscler. Thromb. Vasc. Biol.* 22, 1381–1389. doi: 10.1161/01.ATV.0000031340.68494.34

Murphy, D. D., Reddy, E. C., Moran, N., and O'neill, S. (2014). Regulation of platelet activity in a changing redox environment. *Antioxid. Redox Signal.* 20, 2074–2089. doi: 10.1089/ars.2013.5698

Nguyen, V. D., Saaranen, M. J., Karala, A. R., Lappi, A. K., Wang, L., Raykhel, I. B., et al. (2011). Two endoplasmic reticulum PDI peroxidases increase the efficiency of the use of peroxide during disulfide bond formation. *J. Mol. Biol.* 406, 503–515. doi: 10.1016/j.jmb.2010.12.039

Noiva, R. (1999). Protein disulfide isomerase: the multifunctional redox chaperone of the endoplasmic reticulum. *Semin. Cell Dev. Biol.* 10, 481–493. doi: 10.1006/scdb.1999.0319

Pescatore, L. A., Bonatto, D., Forti, F. L., Sadok, A., Kovacic, H., and Laurindo, F. R. (2012). Protein disulfide isomerase is required for platelet-derived growth factor-induced vascular smooth muscle cell migration, Nox1 NADPH oxidase

expression, and RhoGTPase activation. *J. Biol. Chem.* 287, 29290–29300. doi: 10.1074/jbc.M112.394551

Popescu, N. I., Lupu, C., and Lupu, F. (2010). Extracellular protein disulfide isomerase regulates coagulation on endothelial cells through modulation of phosphatidylserine exposure. *Blood* 116, 993–1001. doi: 10.1182/blood-2009-10-249607

Purvis, J. E., Chatterjee, M. S., Brass, L. F., and Diamond, S. L. (2008). A molecular signaling model of platelet phosphoinositide and calcium regulation during homeostasis and P2Y1 activation. *Blood* 112, 4069–4079. doi: 10.1182/blood-2008-05-157883

Ramachandran, N., Root, P., Jiang, X. M., Hogg, P. J., and Mutus, B. (2001). Mechanism of transfer of NO from extracellular S-nitrosothiols into the cytosol by cell-surface protein disulfide isomerase. *Proc. Natl. Acad. Sci. U.S.A.* 98, 9539–9544. doi: 10.1073/pnas.171180998

Reiser, K., Francois, K. O., Schols, D., Bergman, T., Jornvall, H., Balzarini, J., et al. (2012). Thioredoxin-1 and protein disulfide isomerase catalyze the reduction of similar disulfides in HIV gp120. *Int. J. Biochem. Cell Biol.* 44, 556–562. doi: 10.1016/j.biocel.2011.12.015

Root, P., Sliskovic, I., and Mutus, B. (2004). Platelet cell-surface protein disulphide-isomerase mediated S-nitrosoglutathione consumption. *Biochem. J.* 382, 575–580. doi: 10.1042/BJ20040759

Roth, R. A. (1981). Bacitracin: an inhibitor of the insulin degrading activity of glutathione-insulin transhydrogenase. *Biochem. Biophys. Res. Commun.* 98, 431–438.

Ruggeri, Z. M. (2007). The role of von Willebrand factor in thrombus formation. *Thromb. Res.* 120(Suppl. 1), S5–S9. doi: 10.1016/j.thromres.2007.03.011

Ryser, H. J., Levy, E. M., Mandel, R., and Disciullo, G. J. (1994). Inhibition of human immunodeficiency virus infection by agents that interfere with thiol-disulfide interchange upon virus-receptor interaction. *Proc. Natl. Acad. Sci. U.S.A.* 91, 4559–4563.

Santos, C. X., Stolf, B. S., Takemoto, P. V., Amanso, A. M., Lopes, L. R., Souza, E. B., et al. (2009a). Protein disulfide isomerase (PDI) associates with NADPH oxidase and is required for phagocytosis of *Leishmania chagasi* promastigotes by macrophages. *J. Leukoc. Biol.* 86, 989–998. doi: 10.1189/jlb.0608354

Santos, C. X., Tanaka, L. Y., Wosniak, J., and Laurindo, F. R. (2009b). Mechanisms and implications of reactive oxygen species generation during the unfolded protein response: roles of endoplasmic reticulum oxidoreductases, mitochondrial electron transport, and NADPH oxidase. *Antioxid. Redox Signal.* 11, 2409–2427. doi: 10.1089/ARS.2009.2625

Schulman, S., Wang, B., Li, W., and Rapoport, T. A. (2010). Vitamin K epoxide reductase prefers ER membrane-anchored thioredoxin-like redox partners. *Proc. Natl. Acad. Sci. U.S.A.* 107, 15027–15032. doi: 10.1073/pnas.1009972107

Shao, F., Bader, M. W., Jakob, U., and Bardwell, J. C. (2000). DsbG, a protein disulfide isomerase with chaperone activity. *J. Biol. Chem.* 275, 13349–13352. doi: 10.1074/jbc.275.18.13349

Sliskovic, I., Raturi, A., and Mutus, B. (2005). Characterization of the S-denitrosation activity of protein disulfide isomerase. *J. Biol. Chem.* 280, 8733–8741. doi: 10.1074/jbc.M408080200

Sobierajska, K., Skurzynski, S., Stasiak, M., Kryczka, J., Cierniewski, C. S., and Swiatkowska, M. (2014). Protein disulfide isomerase directly interacts with beta-actin Cys374 and regulates cytoskeleton reorganization. *J. Biol. Chem.* 289, 5758–5773. doi: 10.1074/jbc.M113.479477

Taylor, M., Banerjee, T., Ray, S., Tatulian, S. A., and Teter, K. (2011). Protein-disulfide isomerase displaces the cholera toxin A1 subunit from the holotoxin without unfolding the A1 subunit. *J. Biol. Chem.* 286, 22090–22100. doi: 10.1074/jbc.M111.237966

Taylor, M., Burress, H., Banerjee, T., Ray, S., Curtis, D., Tatulian, S. A., et al. (2014). Substrate-induced unfolding of protein disulfide isomerase displaces the cholera toxin A1 subunit from its holotoxin. *PLoS Pathog.* 10:e1003925. doi: 10.1371/journal.ppat.1003925

Walker, K. W., and Gilbert, H. F. (1997). Scanning and escape during protein-disulfide isomerase-assisted protein folding. *J. Biol. Chem.* 272, 8845–8848.

Wan, S. W., Lin, C. F., Lu, Y. T., Lei, H. Y., Anderson, R., and Lin, Y. S. (2012). Endothelial cell surface expression of protein disulfide isomerase activates beta1 and beta3 integrins and facilitates dengue virus infection. *J. Cell. Biochem.* 113, 1681–1691. doi: 10.1002/jcb.24037

Wang, C., Li, W., Ren, J., Fang, J., Ke, H., Gong, W., et al. (2013). Structural insights into the redox-regulated dynamic conformations of human protein disulfide isomerase. *Antioxid. Redox Signal.* 19, 36–45. doi: 10.1089/ars.2012.4630

Wang, C., Yu, J., Huo, L., Wang, L., Feng, W., and Wang, C. C. (2012). Human protein-disulfide isomerase is a redox-regulated chaperone activated by oxidation of domain a'. *J. Biol. Chem.* 287, 1139–1149. doi: 10.1074/jbc.M111.303149

Wang, L., Zhang, L., Niu, Y., Sitia, R., and Wang, C. C. (2014). Glutathione peroxidase 7 utilizes hydrogen peroxide generated by Ero1alpha to promote oxidative protein folding. *Antioxid. Redox Signal.* 20, 545–556. doi: 10.1089/ars.2013.5236

Wang, L., Zhu, L., and Wang, C. C. (2011). The endoplasmic reticulum sulfhydryl oxidase Ero1beta drives efficient oxidative protein folding with loose regulation. *Biochem. J.* 434, 113–121. doi: 10.1042/BJ20101357

Xu, S., Butkevich, A. N., Yamada, R., Zhou, Y., Debnath, B., Duncan, R., et al. (2012). Discovery of an orally active small-molecule irreversible inhibitor of protein disulfide isomerase for ovarian cancer treatment. *Proc. Natl. Acad. Sci. U.S.A.* 109, 16348–16353. doi: 10.1073/pnas.1205226109

Yoshimori, T., Semba, T., Takemoto, H., Akagi, S., Yamamoto, A., and Tashiro, Y. (1990). Protein disulfide-isomerase in rat exocrine pancreatic cells is exported from the endoplasmic reticulum despite possessing the retention signal. *J. Biol. Chem.* 265, 15984–15990.

Conflict of Interest Statement: The authors declare that the research was conducted in the absence of any commercial or financial relationships that could be construed as a potential conflict of interest.

10

Nanomaterials for biosensing applications: a review

Michael Holzinger*, Alan Le Goff and Serge Cosnier

Département de Chimie Moléculaire UMR 5250, Biosystèmes Electrochimique and Analytiques, CNRS, University of Grenoble Alpes, Grenoble, France

Edited by:
Margarita Stilianova Stoytcheva,
Universidad Autonoma de Baja
California, Mexico

Reviewed by:
Xiaoyi Xu, Stanford University, USA
Alberto Bianco, Centre National de
la Recherche Scientifique, France

*Correspondence:
Michael Holzinger, Département de
Chimie Moléculaire UMR 5250,
Biosystèmes Electrochimique and
Analytiques, CNRS, University of
Grenoble Alpes, 570 rue de la
Chimie, BP 53F-38041 Grenoble,
France
e-mail: michael.holzinger@
ujf-grenoble.fr

A biosensor device is defined by its biological, or bioinspired receptor unit with unique specificities toward corresponding analytes. These analytes are often of biological origin like DNAs of bacteria or viruses, or proteins which are generated from the immune system (antibodies, antigens) of infected or contaminated living organisms. Such analytes can also be simple molecules like glucose or pollutants when a biological receptor unit with particular specificity is available. One of many other challenges in biosensor development is the efficient signal capture of the biological recognition event (transduction). Such transducers translate the interaction of the analyte with the biological element into electrochemical, electrochemiluminescent, magnetic, gravimetric, or optical signals. In order to increase sensitivities and to lower detection limits down to even individual molecules, nanomaterials are promising candidates due to the possibility to immobilize an enhanced quantity of bioreceptor units at reduced volumes and even to act itself as transduction element. Among such nanomaterials, gold nanoparticles, semi-conductor quantum dots, polymer nanoparticles, carbon nanotubes, nanodiamonds, and graphene are intensively studied. Due to the vast evolution of this research field, this review summarizes in a non-exhaustive way the advantages of nanomaterials by focusing on nano-objects which provide further beneficial properties than "just" an enhanced surface area.

Keywords: biosensors, gold nanoparticles, quantum dots, magnetic nanoparticles, nanostructured carbon

INTRODUCTION

As in many different technological sections, nanomaterials have demonstrated their appropriateness for biosensing applications. The intelligent use of such nano-objects led to clearly enhanced performances with increased sensitivities and lowered detection limits of several orders of magnitudes. One general advantage of all nanomaterials is the high specific surface thus already enabling the immobilization of an enhanced amount of bioreceptor units. However, one of the constant challenges is the immobilization strategy used to conjugate intimately the bio-specific entity onto such nanomaterials. Therefore, the technique used to immobilize the enzyme is one of the key factors in developing a reliable biosensor.

Efficient methods for the biofunctionalization of nanomaterials are summarized in reference (Putzbach and Ronkainen, 2013). Briefly, non-covalent approaches representing electrostatic interaction, π–π stacking, entrapment in polymers, or van der Waals forces between the nanomaterial and the biological entity. These principles preserve all specific properties of both, nanomaterial and biomolecule.

Covalent binding: the strategy to attached covalently biomolecules to nanomaterials has an advantage in terms of stability and reproducibility of the surface functionalization and lowers unspecific physisorption. Covalent links can be formed, e.g., by classic amide coupling reactions, cross-linking, or click chemistry. One drawback is the uncontrolled anchoring of the biomolecule which can affect the domain which is responsible for the recognition event.

The immobilization of biomolecules via supramolecular or coordinative interactions: this technique has achieved wide acceptance in recent years in binding biological species to surfaces. The most famous example used in the field of biosensor engineering is the biotin/avidin (or streptavidin) system (Wilchek and Bayer, 1988). Biotinylated biomolecules can be attached to biotinylated substrates via avidin (or streptavidin) bridges. Other affinity systems have been reported like the nitrilotriacetic acid (NTA)/Cu^{2+}/histidine complex (Haddour et al., 2005) or the host-guest system adamantane/β-cyclodextrin (Holzinger et al., 2009). The advantage of such systems, compared to the other immobilization methods, is the reversibility, enabling the possibility to regenerate the transducer element. Furthermore, all components like the functionalized transducer surface and the modified bio-receptor can be characterized individually assuring the reproducibility of the constructed biosensor.

According to the chemical composition, almost all nanomaterials can be equipped with appropriate functions via direct functionalization (in some cases already during synthesis), or via coating with functional polymers without affecting their specific properties (Biju, 2014). Such functionalization not only allows the reproducible immobilization of bioreceptor units but can also increase the biocompatibility of these materials.

One particular issue in biosensing devices is that the recognition event is not directly detectable by the used transduction technique. This is the case for affinity biosensors like the immunoreaction between an antigen and its antibody or the hybridization of corresponding DNA strands. Here, further biospecific components (secondary antibodies or DNA strands) modified with labels for optical or electrochemical transduction have to be used. The specific properties of some nanomaterials clearly contributed to the development of "label free" transduction techniques or contribute to clear signal amplifications when used as labels.

GOLD NANOPARTICLES

Within the group of noble metal nanoparticles, gold nanoparticles are mostly used for biosensor application (Li et al., 2010a) due to their biocompatibility, their optical and electronic properties, and their relatively simple production and modification (Biju, 2014).

Particular interesting is the optical behavior of gold surfaces where irradiation with light of one specific wavelength causes an oscillation of the electrons in the conduction band, called resonant surface plasmons. When the particle size is much smaller than the incident wavelength, the oscillating electrons cannot propagate along the surface as it is for classic surface plasmon resonance (SPR) setups. The electron density is then polarized on one side of the particle where the plasmons oscillate in resonance with the light frequency (**Figure 1**). This phenomenon was described applying the Mie theory (Mulvaney, 1996; Hao et al., 2004) and is strongly dependent on the size, shape of the nanoparticle and the dielectric constant of its environment (Kelly et al., 2002). This environmental dependency represents a great advantage for (bio)-analytics since the recognition event can result in a change of the oscillation frequency and therefore to a color change of the gold nanoparticles observable with bare eye. In this context, a wide series of efficient colorimetric biosensors were developed for DNA or oligonucleotide detection, or immunosensors (Reynolds et al., 2000; Oldenburg et al., 2002; Liu and Lu, 2004; Xu et al., 2009).

Gold nanoparticles have also demonstrated their advantages in bioanalysis using SPR transduction. This method is usually based on the change of the dielectric constant of propagating surface plasmons' environment of gold films where the detection of the analyte can be recorded in different ways like the changes of the angle, intensity, or phase of the reflected light (Wijaya et al., 2011; Guo, 2012). Beside the use of a pure gold nanoparticle based

SPR transduction replacing the gold film (Pedersen and Duncan, 2005), a clear SPR signal enhancement can be obtained when gold films and gold nanoparticles are used in a sandwich configuration. In fact, the surface plasmons on gold nanoparticles provoke a perturbation of the evanescent field of the gold film in addition to the immobilized bioreceptor unit and the recognized analyte. The optimal configuration of this approach was determined for gold nanoparticles smaller than 40 nm at a distance to the gold film surface of 5 nm (Zeng et al., 2013) as illustrated in **Figure 2**. In this case, gold nanoparticles serve as labels when attached to secondary antibodies or DNA strands. Even when such labeling needs further preparative steps than label less detection, the signal enhancement is of several orders of magnitudes which is a convincing argument for such an approach.

Gold nanoparticles have also shown their ability to form a powerful transduction platform for single molecule detection. By refractive index sensing of localized surface plasmon resonance (LSPR) coupled with enzyme linked immunosorbent assay (ELISA) using isolated gold nanoparticles of 60 nm sizes. The model enzyme horseradish peroxidase (HRP) was immobilized on these gold nanoparticles via biotin streptavidin linkage. HRP is widely used as label in biosensing applications since it can form colored, fluorescent, or redox active molecules while reducing hydrogen peroxide to water (Veitch, 2004). In the work of Chen et al., HRP oxidized the soluble monomer 3, 3′-Diaminobenzidine (DAB) to insoluble colored polybenzimidazole which aggregates around the enzyme. This aggregation led to an additional shift of the LSPR scattering wavelength enabling the detection of few or even single HRP proteins attached to the gold particles (Chen et al., 2011). The beneficial effects of gold nanoparticles could be validated for high sensitive biosensing using surface enhanced Raman spectroscopy (SERS). Based on surface plasmon assisted signal amplification (Moskovits, 1978) of the vibrational spectrum of adsorbed or immobilized compounds, detection limits also down to the single molecule level could be reached (Nie and Emory, 1997; Hossain et al., 2009; Lim et al., 2011; Saha et al., 2012).

Besides the outstanding optical properties, gold nanoparticles also have the ability to transfer electrons between a wide range of electroactive biological species and the electrode. This principle

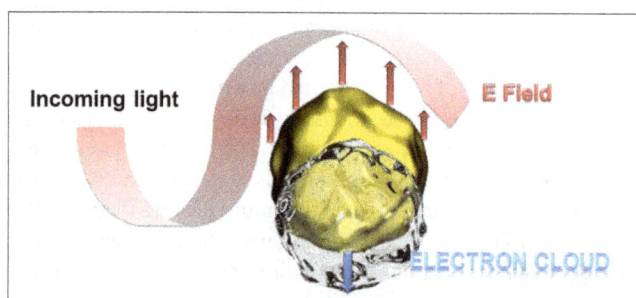

FIGURE 1 | Schematic presentation of the polarization of the electron density at resonant excitation wavelength. At particle sizes smaller than the excitation wavelength, the oscillating electrons (surface plasmons) cannot propagate along the gold surface leading to a polarization of the electron cloud at one side of the particle.

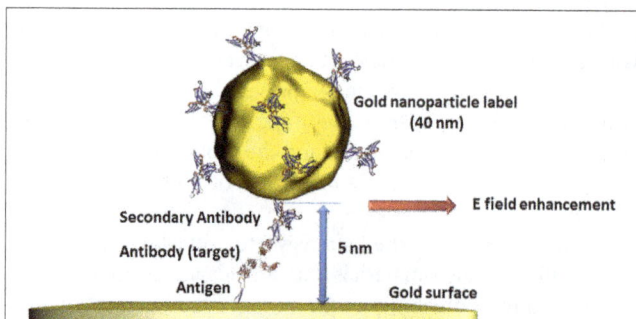

FIGURE 2 | Illustration of the perturbation of propagating surface plasmons on gold surfaces provoked by gold nanoparticles of defined size and distance leading to an additional change of the evanescent field and therefore to an enhanced signal.

is principally used for redox enzyme biosensing where the bioreceptor unit catalyzes the oxidation or reduction of the analyte. In classical electrochemical enzyme biosensors setups, the formed species are oxidized or reduced by the electrode, giving the electrochemical signal. The disadvantage of this approach is that the detectable molecules have to diffuse to the electrode where a non-negligible amount is lost in solution. Gold nanoparticles can act as electron shuttles, i.e., the gold nanoparticles can approach to the redox center of the enzyme regenerating this biocatalyst by transferring the electrons involved in the redox reaction to the electrode. Since such wiring of enzymes prevents the need of enzymatically formed redox species which have to reach the electrode surface, a clear increase of the electrochemical signal is to expect. Particular efficient wiring could be obtained using enzymes where metal ions are involved in the catalytic redox process as it is for HRP (Xu et al., 2006). Nonetheless, the holy grail of enzymes to wire is glucose oxidase (GOx) since its active center is deeply embedded inside the protein structure and contains no metal ion which would facilitate electron transfer (Mena et al., 2005; Willner et al., 2006; Pingarrón et al, 2008). Wiring of GOx is not only a great challenge for glucose sensing applications but also for the design of bioanodes for glucose fuel cells. This is a steady expanding research field since such glucose biofuel cells could generate power to supply implantable medical devices out of body liquids (Cosnier et al., 2014). GOx catalyzes the oxidation of glucose to gluconolactone by releasing 2 electrons. These electrons are generally used to reduce oxygen to hydrogen peroxide. The involved natural electron shuttle is the co-factor flavin adenine dinucleotide (FAD) which is after glucose oxidation in reduced form ($FADH_2$). The goal is therefore to provide an electron carrier which regenerates FAD with favored kinetics to avoid the oxygen reduction process. Xiao et al. proposed an original approach by the reconstitution of apo-GOx (GOx without the cofactor FAD) with a FAD modified gold nanoparticle (**Figure 3**).

This work even demonstrated that the as reactivated enzyme showed a 7-fold improved electron transfer turnover rate compared to the natural electron transfer rate to oxygen (Xiao et al., 2003).

These outstanding properties of gold nanoparticles made them promising candidates not only for bioanalytics but also for many other research fields. The particular properties of such gold nanoparticles can be tuned and adjusted. Whatever the desired application, almost any desired shape or size can be obtained using the appropriate synthesis technique. These different morphologies result in different optical, catalytic, and electronic behavior of these gold nanoparticles (Eustis and El-Sayed, 2006).

QUANTUM DOTS

Another prominent example of nanomaterials used for bioanalytics are luminescent semiconducting nanocrystals called quantum dots (QDs). The most studied colloidal QDs are based on cadmium chalcogenides (S, Se, Te) (Murray et al., 1993; Park et al., 2007; Reiss et al., 2009) which provide a very large absorption spectrum with a size-dependent narrow emission spectrum. This phenomenon is due to the varying band gaps of the semiconductor material for different nanocrystal sizes (the bigger the particle the lower the band gap) which leads to distinct emission wavelengths from the recombination of the electron-hole exiton (Weller, 1993). The availability of a large range of emission wavelengths of QDs with different sizes enables efficient multiplexed analysis using classic optical transduction (Geißler et al., 2010; Petryayeva and Algar, 2014). However, structural defects in the crystal lattice can trap the exited electrons or holes leading to non-radiative relaxation (Murphy, 2002). To overcome this issue, core/shell composited with another semiconductor material with a wider band gap range (generally ZnS) were realized to passivate these surface defects and to enhance quantum yields and photo-stability (Dabbousi et al., 1997; Jaiswal et al., 2003).

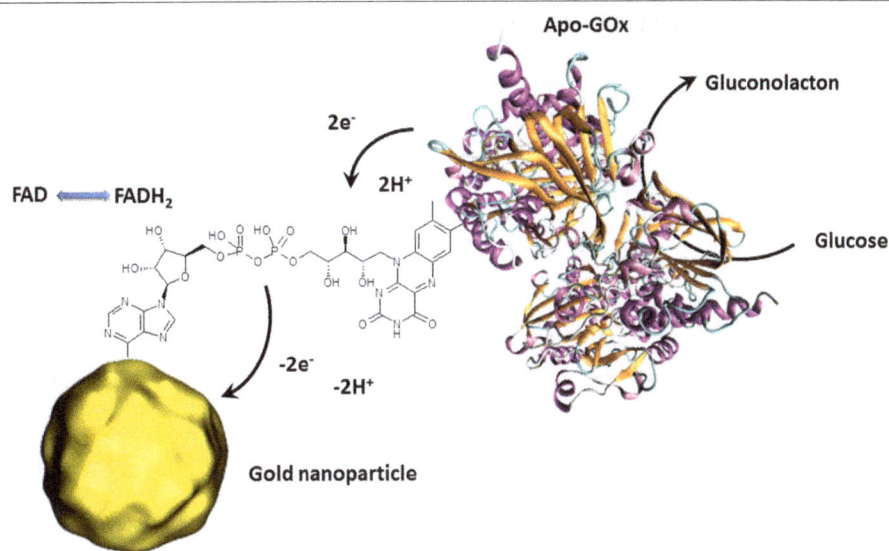

FIGURE 3 | Schematic presentation of the reconstitution of apo-GOx with FAD modified gold nanoparticles enabling the transfer of electron released by the electrocatalytic oxidation of glucose. The collected electrons are then directly transferred to the external circuit giving an electrochemical signal.

Due to this high photochemical stability of core/shell QDs, this material became a promising alternative to organic fluorophores (Resch-Genger et al., 2008).

In order to provide functional groups for bioreceptor immobilization and also to anticipate possible toxicity issues, QDs are nowadays available with inert or biocompatible coatings (Biju et al., 2010). Thus, almost any kind of biomolecule can be attached to these nanocrystals as long as the photophysical recombination event is not affected.

Such so-called fluorescence quenching of QDs, described by the Förster (Fluorescence) Resonance Energy Transfer (FRET) (Clapp et al., 2006), is based on a non-radiative energy transfer between the exited QD (donor) and a quencher (acceptor) such as organic fluorophores. Such FRET quenching revealed original approaches used for optical transduction since QD fluorescence reappears when the quencher is removed. This strategy is particularly used for optical DNA and oligonucleotide sensors (Zhang et al., 2005; Freeman et al., 2013). The challenge is to bring the quenching dye in close contact to the QD and to release it after the recognition event. One example deals with the hybridization of a QD tagged receptor DNA and a short corresponding DNA sequence equipped with a gold nanoparticle. As already described, gold nanoparticles are excellent acceptors and are highly efficient QD quenchers. Due to the more favorable hybridization kinetics of the analyte DNA, the short sequence with the gold nanoparticle is released and the QDs' fluorescence reappears where its intensity is correlated to the analyte concentration (Dyadyusha et al., 2005; Dai et al., 2007) as shown in **Figure 4**.

Another use of non-radiative energy transfer provoking QD fluorescence is called Bioluminescence Resonance Energy Transfer (BRET) (So et al., 2006). Here, a light-emitting protein label transfers the energy to QDs and eliminates the necessity of an external excitation light source. FRET and BRET, together with charge transfer quenching and chemiluminescence resonance energy transfer (CRET) (Huang et al., 2006) are the most common strategies in biosensing applications using QDs as optical

transducers (Frasco and Chaniotakis, 2009; Algar et al., 2010; Petryayeva et al., 2013).

For all cases, the distance between the quencher and QD is one of the key factors in the sensitivity of such detection principles. For instance, an enhancement of QD fluorescence was observed when gold nanoparticles are localized at around 30 nm (**Figure 5**). In this case, the gold nanoparticle acts as antenna (and not as quencher) increasing the excitation rates of the QDs and therefore the fluorescence intensity (Maye et al., 2010). Another effect leading to further example for optical signal enhancement is that QDs can interact with propagating surface plasmons on gold surfaces. These surface plasmons can excite QDs leading to light emission and vice versa - the light induced excited state of QDs are converted to propagating surface plasmons (Wei et al., 2009). Malic et al. demonstrated the beneficial combination of near infrared QDs and gold surfaces for drastic signal enhancement in a SPR imaging biosensor setup (Malic et al., 2011).

QDs have demonstrated their appropriateness in bioanalytics either as transducer unit or as optical labels. A tremendous evolution of QD based biosensors is therefore predictable.

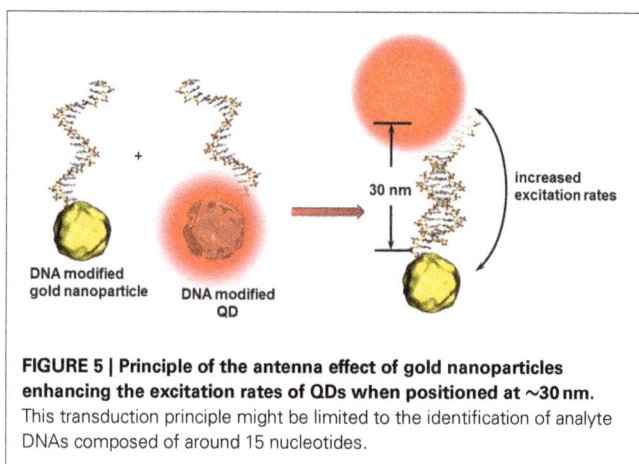

FIGURE 5 | Principle of the antenna effect of gold nanoparticles enhancing the excitation rates of QDs when positioned at ~30 nm. This transduction principle might be limited to the identification of analyte DNAs composed of around 15 nucleotides.

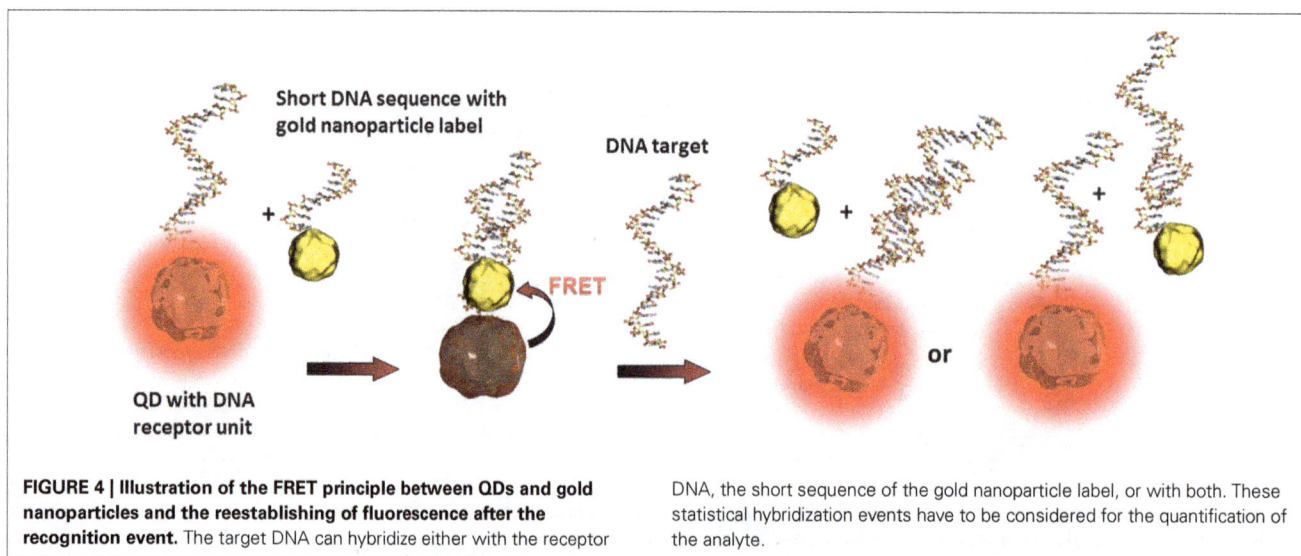

FIGURE 4 | Illustration of the FRET principle between QDs and gold nanoparticles and the reestablishing of fluorescence after the recognition event. The target DNA can hybridize either with the receptor DNA, the short sequence of the gold nanoparticle label, or with both. These statistical hybridization events have to be considered for the quantification of the analyte.

Furthermore, the described principles and examples about the use of QDs in bioanalysis let suggest that the combination of different nano-objects and their specific particularity is a promising way for future biosensors with new and original highly efficient transduction techniques.

MAGNETIC NANOPARTICLES

Magnetic nanoparticles are promising alternatives to fluorescent labels in biosensor devices. Nanosized magnetic nanoparticles show different magnetic behaviors compared to its bulk material due to the reduced number of magnetic domains (regions of parallel oriented magnetic moments caused by interacting unpaired electrons of an atom) leading to so called superparamagnetic behavior. This means that magnetization can flip the direction randomly within very short time (Neel's relaxation time) and the magnetization appears in average zero in absence of an external magnetic field. This temperature dependent phenomenon disappears by applying an external magnetic field aligning the magnetic moments. Even when this effect seems to be similar to this of classic paramagnetic materials, the magnetic susceptibility of superparamagnets are much higher (Bishop et al., 2009). Such superparamagnetic behavior prevents therefore from attractive or repulsive forces between the magnetic nanoparticles as long as no external magnetic field is applied.

Beside a wide range of ferromagnetic materials, iron oxide is mostly used for bioanalytical applications (Haun et al., 2010). The outstanding advantage using magnetic nanoparticles is the possibility to concentrate the analyte before the detection event. Receptor unit modified magnetic nanoparticles can simply be mixed with the analyte solution and interacts specifically with the specific target. After applying an external magnetic field, the nanoparticles agglomerate and can be separated from the solution. Efficient isolation of DNA strands in complex media was achieved in a fast and efficient manner using silica or gold coated core/shell nanoparticles (He et al., 2007; Li et al., 2011; Min et al., 2014).

Besides optical (Bi et al., 2009) or electrochemical (Mejri et al., 2011) detection techniques in combination with other often nanosized labels, magnetic nanoparticles offer a further high sensitive transduction technique in the domain of diagnostic magnetic resonance (**Figure 6**). In this context, high performant giant magnetoresistance, spin valve, or magnetic tunnel junction biosensors could be developed (Wang and Li, 2008; Konry et al., 2012). Magnetic labels are particular interesting for biosensing applications since biological entities do not show any magnetic behavior or susceptibility and therefore, no interferences or noise is to expect during signal capturing (Tamanaha et al., 2008). For instance, an ultra-high sensitive magnetoresistant biosensor was developed for *Escherichia coli* (Mujika et al., 2009) detection or *Salmonella* were identified in skimmed-milk samples with a limit of detection (LOD) of 1 colony forming-unit (cfu)/mL using a magneto-genosensing setup (Liébana et al., 2009).

A further advantage of magnetic nanoparticles is the possibility to carry the analytes to the transduction platform in

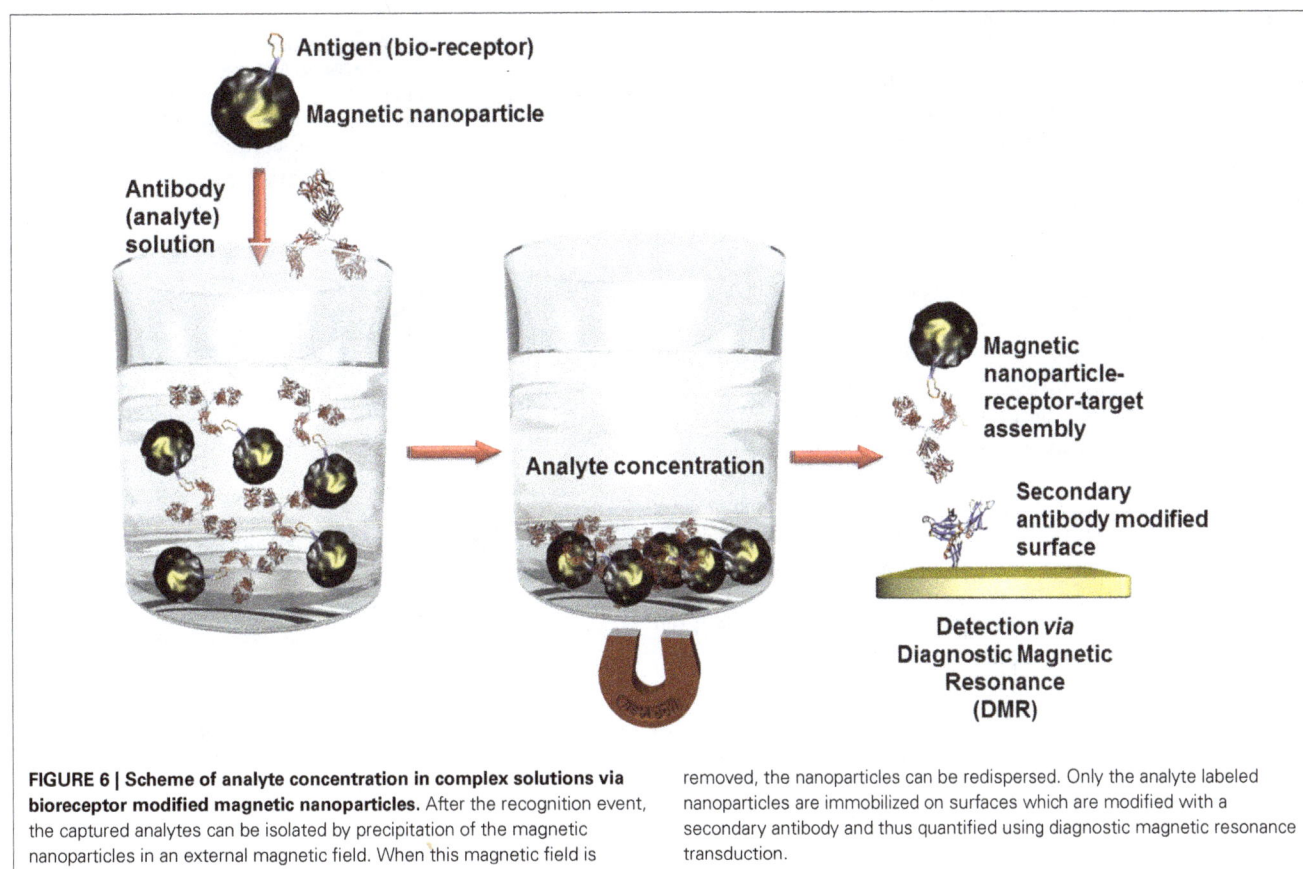

FIGURE 6 | Scheme of analyte concentration in complex solutions via bioreceptor modified magnetic nanoparticles. After the recognition event, the captured analytes can be isolated by precipitation of the magnetic nanoparticles in an external magnetic field. When this magnetic field is removed, the nanoparticles can be redispersed. Only the analyte labeled nanoparticles are immobilized on surfaces which are modified with a secondary antibody and thus quantified using diagnostic magnetic resonance transduction.

microfluidic systems (Konry et al., 2012) enabling even complex multiplexed analyses for the simultaneous detection of different analytes. Nonetheless, these carrier properties are mainly applied for the transport of drugs and genes, or for magnetic resonance imaging (Sun et al., 2008).

CARBON NANOSTRUCTURES

The beneficial properties of nanostructured carbons such as carbon nanotubes or graphene made them a widely used material as electronic or electrochemical transducer in biosensor devices (Vamvakaki and Chaniotakis, 2007; Valentini et al., 2013). In particular, carbon nanotubes possess the outstanding combination of nanowire morphology, biocompatibility and electronic properties (Battigelli et al., 2013). Therefore, carbon nanotube interfaces present clearly enhanced capacities, e.g., to approach the active sites of a redox enzyme and to wire it to the bulk electrode. Furthermore, their ease and well-documented organic functionalization (Ménard-Moyon et al., 2010) brings new properties to nanostructured electrodes such as specific docking sites for biomolecules or redox mediation of bioelectrochemical reactions. Moreover, CNT films exhibit a high electroactive surface areas due to the natural formation of highly porous three-dimensional networks, suitable for the anchoring of a high amount of bioreceptor units, leading consequently to high sensitivities (Wang, 2005; Le Goff et al., 2011).

Nonetheless, since purely the increase of the specific surface is not discussed here, the presented examples of carbon nanomaterials focus the use of their electronic or electrochemical properties for biosensing applications.

As already mentioned, the nanowire morphology of CNTs enables the approach to the active centers of redox enzyme leading to fast and efficient electron transfers. In few cases, this enzyme wiring spontaneously occurs simply by adsorption of the enzyme on the nanocarbon. The driving force is mostly based on hydrophobic interactions. An important condition is that the hydrophobic domain of the enzyme has to be situated close to its active center for efficient electron transfer after immobilization and orientation.

The nitrite reductase *from Desulfovibrio desulfuricans* is a famous example of such an enzyme having the hydrophobic and catalytic domains at favorable localizations. The multihemic catalytic site for the reduction of nitrite ions is closely placed to a hydrophobic domain of the protein structure as illustrated in **Figure 7**. After catalytic reduction of nitrite to ammonia, the enzyme, in oxidized state, is directly regenerated by the carbon electrode. This concept was validated for pyrolitic graphite (Silveira et al., 2010a), and CNTs (Silveira et al., 2010b).

Since nature does not always provide such excellent conditions for electron transfers, appropriate functionalization of the carbon material with anchor molecules and/or with redox active species with corresponding redox potentials is necessary to establish so-called mediated electron transfer (via electron shuttle). One particular case of mediated electron transfer is the use of the enzyme specific co-factor. As already demonstrated by Xiao et al. (2003) for glucose sensors using gold nanoparticles as electron carriers, FAD was also attached to carbon nanotubes enabling the

wiring glucose oxidase that led to a performant glucose biosensor (Patolsky et al., 2004).

However, the main purpose for efficient enzyme wiring on nanostructured carbon clearly targets bioenergy conversion (Holzinger et al., 2012).

The particular electric properties of CNTs were used in field effect transistor (FET) biosensor setups where changes of the conductivity of the CNT channel or the modulation of the Schottky barrier after the bio-recognition event (Gruner, 2006) led to high sensitivities and low detection limits down to single molecules (Besteman et al., 2003).

An original approach for CNT-FET based DNA sensor was proposed by taking advantage of the high affinity of CNTs and DNA strands. The nucleic bases of a ssDNA attach to the CNTs via π-π stacking leading to a wrapping of the DNA around the CNT

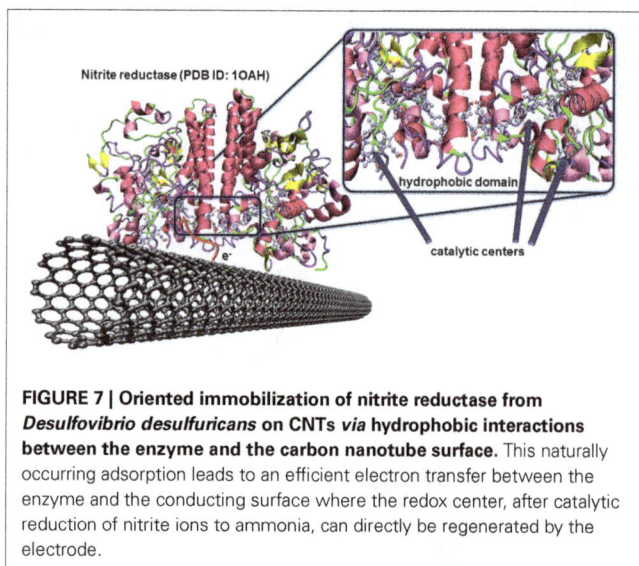

FIGURE 7 | Oriented immobilization of nitrite reductase from *Desulfovibrio desulfuricans* **on CNTs** *via* **hydrophobic interactions between the enzyme and the carbon nanotube surface.** This naturally occurring adsorption leads to an efficient electron transfer between the enzyme and the conducting surface where the redox center, after catalytic reduction of nitrite ions to ammonia, can directly be regenerated by the electrode.

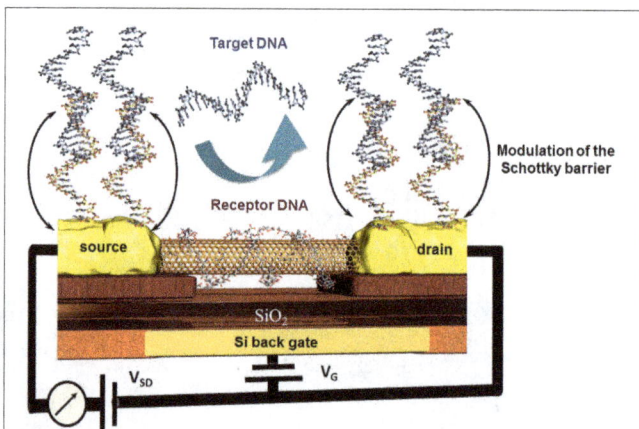

FIGURE 8 | Principle of the modulation of the Schottky barrier at source and drain electrodes of CNT field effect transistor devices after hybridization of the receptor DNA with the target DNA. Here, the wrapping properties of DNA around nanotubes inhibit the capability of this receptor DNA to hybridize with its counterpart and thus deactivate indirectly the CNT channel which remains unaffected toward the recognition event.

(Gigliotti et al., 2006) thus eliminating the recognition capacity of this ssDNA toward its counterpart. Tang et al. reported a clear electrical conductance change due to the modulation of energy level alignment between CNT and gold contacts of the CNT-FET device. The receptor unit was attached to the gold electrodes via thioethers where the wrapping of the ssDNA around the conductive CNT channel was accepted. The recognition event was therefore localized only on source and drain, modulating the Schottky barrier and finally the conduction of the CNT channel as sketched in **Figure 8** (Tang et al., 2006). This is one of very little examples of CNT based DNA-sensors where the CNTs were not blended in composites to avoid this wrapping effect (Zhou et al., 2009). Else CNT-FET biosensors were mainly developed for enzyme and immunosensing applications (Chao et al., 2005; Gruner, 2006). Some other spectroscopic characteristics of CNTs like NIR photoluminescence and Raman scattering features used for biosensing applications are barely described which is most likely due to the low photoluminescence quantum efficiency (Biju, 2014).

CNT-based biosensors are clearly more developed than graphene-based biosensors (Yang et al., 2010) but there is a steady increasing interest in this two-dimensional material for bioanalytical applications. Besides the fabrication of isolation of real monolayer graphene sheets via mechanical cleavage using a scotch tape to exfoliate one sheet from highly oriented pyrolytic graphite (HOPG), by CVD on metal foils, or using epitaxy techniques where the graphene layer is formed out of silicon carbide, a wide series of graphite based bulk materials are also called graphene (Bonaccorso et al., 2012). These materials are mostly obtained after mechanic exfoliation (Coleman, 2012; Paton et al.,

2014) or chemical oxidation of graphite based on the Hummers and Offeman method (Hummers and Offeman, 1958). This initially called graphitic oxide is now generally known as graphene oxide and allows obtaining soluble carbon oxide sheets of undefined layer composition and sizes. The electric conductivity of this isolating material can be reestablished by chemical, thermal, or electrochemical reduction (Kuila et al., 2013). Even when the exceptional conductivity of real monolayer graphene (Novoselov et al., 2004) cannot be obtained, reduced graphene oxide has other beneficial properties in high performance biosensor devices. As for CNTs, graphene-based materials are mostly used in electrochemical biosensors (Shao et al., 2010; Kuila et al., 2011; Ratinac et al., 2011) or field effect transistor setups (Liu and Guo, 2012) where graphene is principally used as electrode material with enhanced specific surface. Graphene materials can also act as transduction element itself in optical or colorimetric biosensor setups. High efficient graphene-based FRET biosensors were developed in combination with organic dyes (Li et al., 2010b) or quantum dots (Dong et al., 2010) modified receptor units for DNA, aptamer, immuno, or protein sensors (Ma et al., 2012). Contrary to CNTs, ssDNA or oligonucleotide receptors adsorb reversibly to graphene oxide and are released after the recognition event. This allows recovering the fluorescence of the dyes as shown in **Figure 9**. An interesting example was the adsorption of several ssDNA receptors, each modified with differently colored dyes, leading to a fast multiplex colorimetric DNA sensor (He et al., 2010).

Graphene and CNTs are the most promising nanostructured carbon materials for biosensing applications where each allotrope has its particular advantage as transducer element.

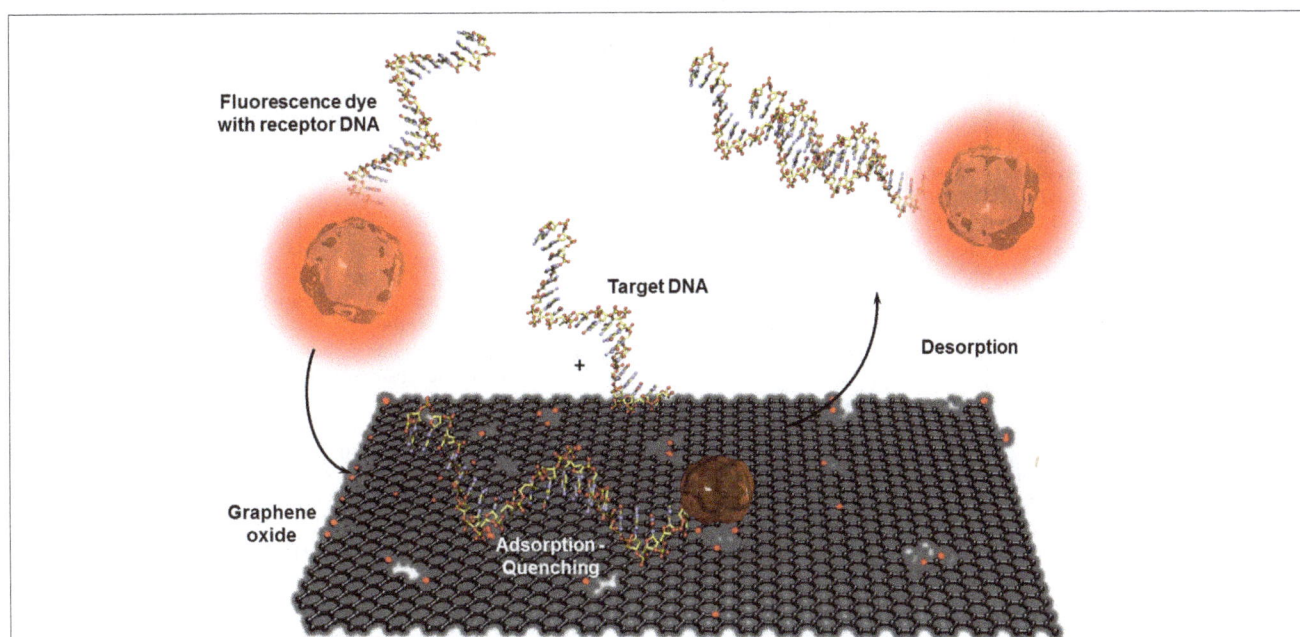

FIGURE 9 | Adsorption and desorption of a fluorescence dye modified receptor DNA on graphene oxide before and after hybridization with the corresponding target DNA. Graphene oxide quenches the fluorescence of the dye when the receptor unit is adsorbed. By hybridization with the analyte DNA, the receptor DNA and therefor the dye is released which leads to the reestablishment of the fluorescence.

CONCLUSION

Nanomaterials became important components in bioanalytical devices since they clearly enhance the performances in terms of sensitivity and detection limits down to single molecules detection. The specific properties of such nano objects also offer alternatives to classic transduction methods. Furthermore, the combination of different nanomaterials, each with its characteristics, to increase even more the performances of biosensors is a well-accepted strategy. Due to the vast number of different nanomaterials all with its own specific properties, only few examples could be mentioned here by emphasizing the principal advantages of such materials. More detailed examples of the use of nanomaterials in biosensor applications are described in the references (Ju et al., 2011; Lei and Ju, 2012), and in the cited review articles.

REFERENCES

Algar, W. R., Tavares, A. J., and Krull, U. J (2010). Beyond labels: a review of the application of quantum dots as integrated components of assays, bioprobes, and biosensors utilizing optical transduction. *Anal. Chim. Acta* 673, 1–25. doi: 10.1016/j.aca.2010.05.026

Battigelli, A., Ménard-Moyon, C., Da Ros, T., Prato, M., and Bianco, A. (2013). Endowing carbon nanotubes with biological and biomedical properties by chemical modifications. *Adv. Drug Deliv. Rev.* 65, 1899–1920. doi: 10.1016/j.addr.2013.07.006

Besteman, K., Lee, J. O., Wiertz, F. G. M., Heering, H. A., and Dekker, C. (2003). Enzyme-coated carbon nanotubes as single-molecule biosensors. *Nano Lett.* 3, 727–730. doi: 10.1021/nl034139u

Bi, S., Yan, Y., Yang, X., and Zhang, S. (2009). Gold nanolabels for new enhanced chemiluminescence immunoassay of alpha-fetoprotein based on magnetic beads. *Chemistry* 15, 4704–4709. doi: 10.1002/chem.200801722

Biju, V. (2014). Chemical modifications and bioconjugate reactions of nanomaterials for sensing, imaging, drug delivery and therapy. *Chem. Soc. Rev.* 43, 744–764. doi: 10.1039/C3CS60273G

Biju, V., Itoh, T., and Ishikawa, M. (2010). Delivering quantum dots to cells: bioconjugated quantum dots for targeted and nonspecific extracellular and intracellular imaging. *Chem. Soc. Rev.* 39, 3031–3056. doi: 10.1039/B926512K

Bishop, K. J. M., Wilmer, C. E., Soh, S., and Grzybowski, B. A. (2009). Nanoscale forces and their uses in self-assembly. *Small* 5, 1600–1630. doi: 10.1002/smll.200900358

Bonaccorso, F., Lombardo, A., Hasan, T., Sun, Z., Colombo, L., and Ferrari, A. C (2012). Production and processing of graphene and 2d crystals. *Mater. Today* 15, 564–589. doi: 10.1016/S1369-7021(13)70014-2

Chao, L. I., Curreli, M., Lin, H., Lei, B., Ishikawa, F. N., Datar, R., et al. (2005). Complementary detection of prostate-specific antigen using In2O3 nanowires and carbon nanotubes. *J. Am. Chem. Soc.* 127, 12484–12485. doi: 10.1021/ja053761g

Chen, S., Svedendahl, M., Duyne, R. P. V., and Kaĺll, M. (2011). Plasmon-enhanced colorimetric ELISA with single molecule sensitivity. *Nano Lett.* 11, 1826–1830. doi: 10.1021/nl2006092

Clapp, A. R., Medintz, I. L., and Mattoussi, H. (2006). Förster resonance energy transfer investigations using quantum-dot fluorophores. *Chemphyschem* 7, 47–57. doi: 10.1002/cphc.200500217

Coleman, J. N. (2012). Liquid exfoliation of defect-free graphene. *Acc. Chem. Res.* 46, 14–22. doi: 10.1021/ar300009f

Cosnier, S., Le Goff, A., and Holzinger, M. (2014). "Enzymatic fuel cells: from design to implantation in mammals," in *Implantable Bioelectronics*, ed E. Katz (Weinheim: Wiley-VCH Verlag GmbH and Co. KGaA), 347–362.

Dabbousi, B. O., Rodriguez-Viejo, J., Mikulec, F. V., Heine, J. R., Mattoussi, H., Ober, R., et al. (1997). (CdSe)ZnS core-shell quantum dots: synthesis and characterization of a size series of highly luminescent nanocrystallites. *J. Phys. Chem. B* 101, 9463–9475. doi: 10.1021/jp971091y

Dai, Z., Zhang, J., Dong, Q., Guo, N., Xu, S., Sun, B., et al. (2007). Adaption of Au nanoparticles and CdTe quantum dots in DNA detection. *Chin. J. Chem. Eng.* 15, 791–794. doi: 10.1016/S1004-9541(08)60004-X

Dong, H., Gao, W., Yan, F., Ji, H., and Ju, H. (2010). Fluorescence resonance energy transfer between quantum dots and graphene oxide for sensing biomolecules. *Anal. Chem.* 82, 5511–5517. doi: 10.1021/ac100852z

Dyadyusha, L., Yin, H., Jaiswal, S., Brown, T., Baumberg, J. J., Booy, F. P., et al. (2005). Quenching of CdSe quantum dot emission, a new approach for biosensing. *Chem. Commun.* 3201–3203. doi: 10.1039/B500664C

Eustis, S., and El-Sayed, M. A. (2006). Why gold nanoparticles are more precious than pretty gold: noble metal surface plasmon resonance and its enhancement of the radiative and nonradiative properties of nanocrystals of different shapes. *Chem. Soc. Rev.* 35, 209–217. doi: 10.1039/B514191E

Frasco, M., and Chaniotakis, N. (2009). Semiconductor quantum dots in chemical sensors and biosensors. *Sensors* 9, 7266–7286. doi: 10.3390/s90907266

Freeman, R., Girsh, J., and Willner, I. (2013). Nucleic acid/Quantum dots (QDs) hybrid systems for optical and photoelectrochemical sensing. *ACS Appl. Mater. Interfaces* 5, 2815–2834. doi: 10.1021/am303189h

Geißler, D., Charbonnière, L. J., Ziessel, R. F., Butlin, N. G., Löhmannsröben, H.-G., and Hildebrandt, N. (2010). Quantum dot biosensors for ultrasensitive multiplexed diagnostics. *Angew. Chem. Int. Ed.* 49, 1396–1401. doi: 10.1002/anie.200906399

Gigliotti, B., Sakizzie, B., Bethune, D. S., Shelby, R. M., and Cha, J. N (2006). Sequence-independent helical wrapping of single-walled carbon nanotubes by long genomic DNA. *Nano Lett.* 6, 159–164. doi: 10.1021/nl0518775

Gruner, G. (2006). Carbon nanotube transistors for biosensing applications. *Anal. Bioanal. Chem.* 384, 322–335. doi: 10.1007/s00216-005-3400-4

Guo, X. (2012). Surface plasmon resonance based biosensor technique: a review. *J. Biophotonics* 5, 483–501. doi: 10.1002/jbio.201200015

Haddour, N., Cosnier, S., and Gondran, C. (2005). Electrogeneration of a poly(pyrrole)-NTA chelator film for a reversible oriented immobilization of histidine-tagged proteins. *J. Am. Chem. Soc.* 127, 5752–5753. doi: 10.1021/ja050390v

Hao, E., Schatz, G. C., and Hupp, J. T. (2004). Synthesis and optical properties of anisotropic metal nanoparticles. *J. Fluoresc.* 14, 331–341. doi: 10.1023/B:JOFL.0000031815.71450.74

Haun, J. B., Yoon, T.-J., Lee, H., and Weissleder, R. (2010). Magnetic nanoparticle biosensors. *Wiley Interdiscip. Rev.* 2, 291–304. doi: 10.1002/wnan.84

He, S., Song, B., Li, D., Zhu, C., Qi, W., Wen, Y., et al. (2010). A graphene nanoprobe for rapid, sensitive, and multicolor fluorescent DNA analysis. *Adv. Funct. Mater.* 20, 453–459. doi: 10.1002/adfm.200901639

He, X., Huo, H., Wang, K., Tan, W., Gong, P., and Ge, J. (2007). Plasmid DNA isolation using amino-silica coated magnetic nanoparticles (ASMNPs). *Talanta* 73, 764–769. doi: 10.1016/j.talanta.2007.04.056

Holzinger, M., Bouffier, L., Villalonga, R., and Cosnier, S. (2009). Adamantane/β-cyclodextrin affinity biosensors based on single-walled carbon nanotubes. *Biosens. Bioelectron.* 24, 1128–1134. doi: 10.1016/j.bios.2008.06.029

Holzinger, M., Le Goff, A., and Cosnier, S. (2012). Carbon nanotube/enzyme biofuel cells. *Electrochim. Acta* 82, 179–190. doi: 10.1016/j.electacta.2011.12.135

Hossain, M. K., Huang, G. G., Kaneko, T., and Ozaki, Y (2009). Characteristics of surface-enhanced Raman scattering and surface-enhanced fluorescence using a single and a double layer gold nanostructure. *Phys. Chem. Chem. Phys.* 11, 7484–7490. doi: 10.1039/B903819C

Huang, X., Li, L., Qian, H., Dong, C., and Ren, J. (2006). A resonance energy transfer between chemiluminescent donors and luminescent quantum-dots as acceptors (CRET). *Angew. Chem.* 118, 5264–5267. doi: 10.1002/ange.200601196

Hummers, W. S, and Offeman, R. E (1958). Preparation of graphitic oxide. *J. Am. Chem. Soc.* 80, 1339–1339. doi: 10.1021/ja01539a017

Jaiswal, J. K., Mattoussi, H., Mauro, J. M., and Simon, S. M. (2003). Long-term multiple color imaging of live cells using quantum dot bioconjugates. *Nat. Biotechnol.* 21, 47–51. doi: 10.1038/nbt767

Ju, H., Zhang, X., and Wang, J. (eds.). (2011). "Nanomaterials for immunosensors and immunoassays," in *NanoBiosensing, Biological and Medical Physics, Biomedical Engineering* (New York, NY: Springer), 425–452.

Kelly, K. L., Coronado, E., Zhao, L. L., and Schatz, G. C (2002). The optical properties of metal nanoparticles:? the influence of size, shape, and dielectric environment. *J. Phys. Chem. B* 107, 668–677. doi: 10.1021/jp026731y

Konry, T., Bale, S., Bhushan, A., Shen, K., Seker, E., Polyak, B., et al. (2012). Particles and microfluidics merged: perspectives of highly sensitive diagnostic detection. *Microchim. Acta* 176, 251–269. doi: 10.1007/s00604-011-0705-1

Kuila, T., Bose, S., Khanra, P., Mishra, A. K., Kim, N. H., and Lee, J. H (2011). Recent advances in graphene-based biosensors. *Biosens. Bioelectron.* 26, 4637–4648. doi: 10.1016/j.bios.2011.05.039

Kuila, T., Mishra, A. K., Khanra, P., Kim, N. H., and Lee, J. H (2013). Recent advances in the efficient reduction of graphene oxide and its application as energy storage electrode materials. *Nanoscale* 5, 52–71. doi: 10.1039/C2NR32703A

Le Goff, A., Holzinger, M., and Cosnier, S. (2011). Enzymatic biosensors based on SWCNT-conducting polymer electrodes. *Analyst* 136, 1279–1287. doi: 10.1039/C0AN00904K

Lei, J., and Ju, H. (2012). Signal amplification using functional nanomaterials for biosensing. *Chem. Soc. Rev.* 41, 2122–2134. doi: 10.1039/C1CS15274B

Li, F., Huang, Y., Yang, Q., Zhong, Z., Li, D., Wang, L., et al. (2010b). A graphene-enhanced molecular beacon for homogeneous DNA detection. *Nanoscale* 2, 1021–1026. doi: 10.1039/B9NR00401G

Li, K., Lai, Y., Zhang, W., and Jin, L. (2011). Fe$_2$O$_3$@Au core/shell nanoparticle-based electrochemical DNA biosensor for Escherichia coli detection. *Talanta* 84, 607–613. doi: 10.1016/j.talanta.2010.12.042

Li, Y., Schluesener, H., and Xu, S. (2010a). Gold nanoparticle-based biosensors. *Gold Bull.* 43, 29–41. doi: 10.1007/BF03214964

Liébana, S., Lermo, A., Campoy, S., Barbé J, Alegret, S., and Pividori, M. I (2009). Magneto immunoseparation of pathogenic bacteria and electrochemical magneto genosensing of the double-tagged amplicon. *Anal. Chem.* 81, 5812–5820. doi: 10.1021/ac9007539

Lim, D.-K., Jeon, K.-S., Hwang, J.-H., Kim, H., Kwon, S., Suh, Y. D., et al. (2011). Highly uniform and reproducible surface-enhanced Raman scattering from DNA-tailorable nanoparticles with 1-nm interior gap. *Nat. Nano* 6, 452–460. doi: 10.1038/nnano.2011.79

Liu, J., and Lu, Y. (2004). Colorimetric biosensors based on DNAzyme-assembled gold nanoparticles. *J. Fluoresc.* 14, 343–354. doi: 10.1023/B:JOFL.0000031816.06134.d3

Liu, S., and Guo, X. (2012). Carbon nanomaterials field-effect-transistor-based biosensors. *NPG Asia Mater.* 4, e23. doi: 10.1038/am.2012.42

Ma, H., Wu, D., Cui, Z., Li, Y., Zhang, Y., Du, B., et al. (2012). Graphene-based optical and electrochemical biosensors: a review. *Anal. Lett.* 46, 1–17. doi: 10.1080/00032719.2012.706850

Malic, L., Sandros, M. G., and Tabrizian, M. (2011). Designed biointerface using near-infrared quantum dots for ultrasensitive surface plasmon resonance imaging biosensors. *Anal. Chem.* 83, 5222–5229. doi: 10.1021/ac200465m

Maye, M. M., Gang, O., and Cotlet, M. (2010). Photoluminescence enhancement in CdSe/ZnS-DNA linked-Au nanoparticle heterodimers probed by single molecule spectroscopy. *Chem. Commun.* 46, 6111–6113. doi: 10.1039/C0CC00660B

Mejri, M. B., Tlili, A., and Abdelghani, A. (2011). Magnetic nanoparticles immobilization and functionalization for biosensor applications. *Int. J. Electrochem.* 2011:421387. doi: 10.4061/2011/421387

Mena, M. L., Yáñez-Sedeño, P., and Pingarrón, J. M. (2005). A comparison of different strategies for the construction of amperometric enzyme biosensors using gold nanoparticle-modified electrodes. *Anal. Biochem.* 336, 20–27. doi: 10.1016/j.ab.2004.07.038

Ménard-Moyon, C., Kostarelos, K., Prato, M., and Bianco, A. (2010). Functionalized carbon nanotubes for probing and modulating molecular functions. *Chem. Biol.* 17, 107–115. doi: 10.1016/j.chembiol.2010.01.009

Min, J. H., Woo, M-K, Yoon, H. Y., Jang, J. W., Wu, J. H., Lim, C.-S., et al. (2014). Isolation of DNA using magnetic nanoparticles coated with dimercaptosuccinic acid. *Anal. Biochem.* 447, 114–118. doi: 10.1016/j.ab.2013.11.018

Moskovits, M. (1978). Surface roughness and the enhanced intensity of Raman scattering by molecules adsorbed on metals. *J. Chem. Phys.* 69, 4159–4161. doi: 10.1063/1.437095

Mujika, M., Arana, S., Castaño, E., Tijero, M., Vilares, R., Ruano-López, J. M., et al. (2009). Magnetoresistive immunosensor for the detection of Escherichia coli O157:H7 including a microfluidic network. *Biosens. Bioelectron.* 24, 1253–1258. doi: 10.1016/j.bios.2008.07.024

Mulvaney, P. (1996). Surface plasmon spectroscopy of nanosized metal particles. *Langmuir* 12, 788–800. doi: 10.1021/la9502711

Murphy, C. J. (2002). Peer reviewed: optical sensing with quantum dots. *Anal. Chem.* 74, 520A–526A. doi: 10.1021/ac022124v

Murray, C. B., Norris, D. J., and Bawendi, M. G (1993). Synthesis and characterization of nearly monodisperse CdE (E = sulfur, selenium, tellurium)

semiconductor nanocrystallites. *J. Am. Chem. Soc.* 115, 8706–8715. doi: 10.1021/ja00072a025

Nie, S., and Emory, S. R (1997). Probing single molecules and single nanoparticles by surface-enhanced raman scattering. *Science* 275, 1102–1106. doi: 10.1126/science.275.5303.1102

Novoselov, K. S., Geim, A. K., Morozov, S. V., Jiang, D., Zhang, Y., Dubonos, S. V., et al. (2004). Electric field effect in atomically thin carbon films. *Science* 306, 666–669. doi: 10.1126/science.1102896

Oldenburg, S. J., Genick, C. C., Clark, K. A., and Schultz, D. A (2002). Base pair mismatch recognition using plasmon resonant particle labels. *Anal. Biochem.* 309, 109–116. doi: 10.1016/S0003-2697(02)00410-4

Park, J., Joo, J., Kwon, S. G., Jang, Y., and Hyeon, T. (2007). Synthesis of monodisperse spherical nanocrystals. *Angew. Chem. Int. Ed.* 46, 4630–4660. doi: 10.1002/anie.200603148

Patolsky, F., Weizmann, Y., and Willner, I. (2004). Long-range electrical contacting of redox enzymes by SWCNT connectors. *Angew. Chem. Int. Ed.* 43, 2113–2117. doi: 10.1002/anie.200353275

Paton, K. R., Varrla, E., Backes, C., Smith, R. J., Khan, U., O'Neill, A., et al. (2014). Scalable production of large quantities of defect-free few-layer graphene by shear exfoliation in liquids. *Nat. Mater.* 13, 624–630. doi: 10.1038/nmat3944

Pedersen, D. B, and Duncan, E. J. S. (2005). "Surface plasmon resonance spectroscopy of gold nanoparticle-coated substrates," in *Technical Report, DRDC Suffield TR 2005-109* (Suffield, CT Defence RandD Canada).

Petryayeva, E., and Algar, W. R. (2014). Multiplexed homogeneous assays of proteolytic activity using a smartphone and quantum dots. *Anal. Chem.* 86, 3195–3202. doi: 10.1021/ac500131r

Petryayeva, E., Algar, W. R., and Medintz, I. L (2013). Quantum dots in bioanalysis: a review of applications across various platforms for fluorescence spectroscopy and imaging. *Appl. Spectrosc.* 67, 215–252. doi: 10.1366/12-06948

Pingarrón, J. M., Yáñez-Sedeño, P., and González-Cortés, A. (2008). Gold nanoparticle-based electrochemical biosensors *Electrochim. Acta.* 53, 5848–5866. doi: 10.1016/j.electacta.2008.03.005

Putzbach, W., and Ronkainen, N. (2013). Immobilization techniques in the fabrication of nanomaterial-based electrochemical biosensors: a review. *Sensors* 13, 4811–4840. doi: 10.3390/s130404811.

Ratinac, K. R., Yang, W., Gooding, J. J., Thordarson, P., and Braet, F. (2011). Graphene and related materials in electrochemical sensing. *Electroanalysis* 23, 803–826. doi: 10.1002/elan.201000545

Reiss, P., Protière, M., and Li, L. (2009). Core/Shell semiconductor nanocrystals. *Small* 5,154–168. doi: 10.1002/smll.200800841

Resch-Genger, U., Grabolle, M., Cavaliere-Jaricot, S., Nitschke, R., and Nann, T. (2008). Quantum dots versus organic dyes as fluorescent labels. *Nat. Meth.* 5, 763–775. doi: 10.1038/nmeth.1248

Reynolds, R. A., Mirkin, C. A., and Letsinger, R. L (2000). Homogeneous, nanoparticle-based quantitative colorimetric detection of oligonucleotides. *J. Am. Chem. Soc.* 122, 3795–3796. doi: 10.1021/ja000133k

Saha, K., Agasti, S. S., Kim, C., Li, X., and Rotello, V. M. (2012). Gold nanoparticles in chemical and biological sensing. *Chem. Rev.* 112, 2739–2779. doi: 10.1021/cr2001178

Shao, Y., Wang, J., Wu, H., Liu, J., Aksay, I. A., and Lin, Y. (2010). Graphene based electrochemical sensors and biosensors: a review. *Electroanalysis* 22, 1027–1036. doi: 10.1002/elan.200900571

Silveira, C. M., Baur, J., Holzinger, M., Moura, J. J. G., Cosnier, S., and Almeida, M. G. (2010b). Enhanced direct electron transfer of a multihemic nitrite reductase on single-walled carbon nanotube modified electrodes. *Electroanalysis* 22, 2973–2978. doi: 10.1002/elan.201000363

Silveira, C. M., Gomes, S. P., Araújo, A. N., Montenegro, M. C. B. S. M., Todorovic, S., Viana, A. S., et al. (2010a). An efficient non-mediated amperometric biosensor for nitrite determination. *Biosens. Bioelectron.* 25, 2026–2032. doi: 10.1016/j.bios.2010.01.031

So, M.-K., Xu, C., Loening, A. M., Gambhir, S. S., and Rao, J. (2006). Self-illuminating quantum dot conjugates for *in vivo* imaging. Nat. Biotechnol. 24, 339–343. doi: 10.1038/nbt1188

Sun, C., Lee, J. S. H., and Zhang, M. (2008). Magnetic nanoparticles in MR imaging and drug delivery. *Adv. Drug Deliv. Rev.* 60, 1252–1265. doi: 10.1016/j.addr.2008.03.018

Tamanaha, C. R., Mulvaney, S. P., Rife, J. C., and Whitman, L. J (2008). Magnetic labeling, detection, and system integration. *Biosens. Bioelectron.* 24, 1–13. doi: 10.1016/j.bios.2008.02.009

Tang, X., Bansaruntip, S., Nakayama, N., Yenilmez, E., Chang, Y. l., and Wang, Q. (2006). Carbon nanotube DNA sensor and sensing mechanism. *Nano Lett.* 6, 1632–1636. doi: 10.1021/nl060613v

Valentini, F., Carbone, M., and Palleschi, G. (2013). Carbon nanostructured materials for applications in nano-medicine, cultural heritage, and electrochemical biosensors. *Anal. Bioanal. Chem.* 405, 451–465. doi: 10.1007/s00216-012-6351-6

Vamvakaki, V., and Chaniotakis, N. A (2007). Carbon nanostructures as transducers in biosensors. *Sens. Actuators B Chem.* 126, 193–197. doi: 10.1016/j.snb.2006.11.042

Veitch, N. C (2004). Horseradish peroxidase: a modern view of a classic enzyme. *Phytochemistry* 65, 249–259. doi: 10.1016/j.phytochem.2003.10.022

Wang, J. (2005). Carbon-nanotube based electrochemical biosensors: a review. *Electroanalysis* 17, 7–14. doi: 10.1002/elan.200403113

Wang, S. X., and Li, G. (2008). Advances in giant magnetoresistance biosensors with magnetic nanoparticle tags: review and outlook. *Magn. IEEE Trans.* 44, 1687–1702. doi: 10.1109/TMAG.2008.920962

Wei, H., Ratchford, D., Li, X., Xu, H., and Shih, C.-K. (2009). Propagating surface plasmon induced photon emission from quantum dots. *Nano Lett.* 9, 4168–4171. doi: 10.1021/nl9023897

Weller, H. (1993). Colloidal semiconductor q-particles: chemistry in the transition region between solid state and molecules. *Angew. Chem. Int. Ed. Engl.* 32, 41–53. doi: 10.1002/anie.199300411

Wijaya, E., Lenaerts, C., Maricot, S., Hastanin, J., Habraken, S., Vilcot, J.-P., et al. (2011). Surface plasmon resonance-based biosensors: from the development of different SPR structures to novel surface functionalization strategies. *Curr. Opin. Solid State Mater. Sci.* 15, 208–224. doi: 10.1016/j.cossms.2011.05.001

Wilchek, M., and Bayer, E. (1988). The avidin-biotin complex in bioanalytical applications. *Anal. Biochem.* 171, 1–32. doi: 10.1016/0003-2697(88)90120-0.

Willner, B., Katz, E., and Willner, I. (2006). Electrical contacting of redox proteins by nanotechnological means. *Curr. Opin. Biotechnol.* 17, 589–596. doi: 10.1016/j.copbio.2006.10.008

Xiao, Y., Patolsky, F., Katz, E., Hainfeld, J. F., and Willner, I. (2003). "Plugging into Enzymes": nanowiring of redox enzymes by a gold nanoparticle. *Science* 299, 1877–1881. doi: 10.1126/science.1080664

Xu, Q., Mao, C., Liu, N.-N., Zhu, J.-J., and Sheng, J. (2006). Direct electrochemistry of horseradish peroxidase based on biocompatible carboxymethyl chitosan–gold nanoparticle nanocomposite. *Biosens. Bioelectron.* 22, 768–773. doi: 10.1016/j.bios.2006.02.010

Xu, W., Xue, X., Li, T., Zeng, H., and Liu, X. (2009). Ultrasensitive and selective colorimetric DNA detection by nicking endonuclease assisted nanoparticle amplification. *Angew. Chem. Int. Ed.* 48, 6849–6852. doi: 10.1002/anie.200901772

Yang, W., Ratinac, K. R., Ringer, S. P., Thordarson, P., Gooding, J. J., and Braet, F. (2010). Carbon nanomaterials in biosensors: should you use nanotubes or graphene? *Angew. Chem. Int. Ed.* 49, 2114–2138. doi: 10.1002/anie.200903463

Zeng, S., Yu, X., Law, W.-C., Zhang, Y., Hu, R., Dinh, X.-Q., et al. (2013). Size dependence of Au NP-enhanced surface plasmon resonance based on differential phase measurement. *Sens. Actuators B Chem.* 176, 1128–1133. doi: 10.1016/j.snb.2012.09.073

Zhang, C.-Y., Yeh, H.-C., Kuroki, M. T., and Wang, T.-H. (2005). Single-quantum-dot-based DNA nanosensor. *Nat. Mater.* 4, 826–831. doi: 10.1038/nmat1508

Zhou, M., Zhai, Y., and Dong, S. (2009). Electrochemical sensing and biosensing platform based on chemically reduced graphene oxide. *Anal. Chem.* 81, 5603–5613. doi: 10.1021/ac900136z

Conflict of Interest Statement: The authors declare that the research was conducted in the absence of any commercial or financial relationships that could be construed as a potential conflict of interest.

11

Life without double-headed non-muscle myosin II motor proteins

Venkaiah Betapudi[1,2]*

[1] Department of Cellular and Molecular Medicine, Lerner Research Institute, Cleveland Clinic, Cleveland, OH, USA
[2] Department of Physiology and Biophysics, Case Western Reserve University, Cleveland, OH, USA

Edited by:
Jianjun Chen, University of Chicago, USA

Reviewed by:
Christian Stock, University of Muenster, Germany
Jianfeng Chen, Chinese Academy of Sciences, China

*Correspondence:
Venkaiah Betapudi, Department of Cellular and Molecular Medicine (NC10), Cleveland Clinic, Lerner Research Institute, 9500 Euclid Avenue, Cleveland, OH 44195, USA
e-mail: betapuv@ccf.org;
vxb19@case.edu

Non-muscle myosin II motor proteins (myosin IIA, myosin IIB, and myosin IIC) belong to a class of molecular motor proteins that are known to transduce cellular free-energy into biological work more efficiently than man-made combustion engines. Nature has given a single myosin II motor protein for lower eukaryotes and multiple for mammals but none for plants in order to provide impetus for their life. These specialized nanomachines drive cellular activities necessary for embryogenesis, organogenesis, and immunity. However, these multifunctional myosin II motor proteins are believed to go awry due to unknown reasons and contribute for the onset and progression of many autosomal-dominant disorders, cataract, deafness, infertility, cancer, kidney, neuronal, and inflammatory diseases. Many pathogens like HIV, Dengue, hepatitis C, and Lymphoma viruses as well as *Salmonella* and *Mycobacteria* are now known to take hostage of these dedicated myosin II motor proteins for their efficient pathogenesis. Even after four decades since their discovery, we still have a limited knowledge of how these motor proteins drive cell migration and cytokinesis. We need to enrich our current knowledge on these fundamental cellular processes and develop novel therapeutic strategies to fix mutated myosin II motor proteins in pathological conditions. This is the time to think how to relieve the hijacked myosins from pathogens in order to provide a renewed impetus for patients' life. Understanding how to steer these molecular motors in proliferating and differentiating stem cells will improve stem cell based-therapeutics development. Given the plethora of cellular activities non-muscle myosin motor proteins are involved in, their importance is apparent for human life.

Keywords: myosin II, motor proteins, molecular machines, cell migration, cytokinesis, cancer, pathogenesis, microparticles

INTRODUCTION

Machines are involved in driving virtually every aspect of modern human life, and so are myosin motor proteins in driving cellular life. Myosin motor proteins are specialized molecular machines that convert cellular free-energy into mechanical work (Bustamante et al., 2004). It is largely believed that the myosin-performed mechanical work intersects with almost every facet of cell biology. In fact, myosins play a central role in driving cellular activities that are necessary for singing a courtship song in flies, reproduction, childbirth, growth, development, and immunity as well as predisposing humans to a certain degree of risk for diseases (Stedman et al., 2004; Maravillas-Montero and Santos-Argumedo, 2012; Slonska et al., 2012; Chakravorty et al., 2014; Min et al., 2014; Pecci et al., 2014).

The biological cell is equipped with a wide variety of motor proteins that are divided into cytoskeletal (myosin, kinesin, dynein), polymerization (actin, microtubule, dynamin), rotary (F_0F_1-ATP synthase), and nucleic acid (RNA and DNA polymerases, Helicase, Topoisomerases, RSC, SW1/SNF complex, SMC, viral DNA packaging protein) motor proteins to perform specific and dedicated cellular functions (Kolomeisky,

2013; Howard, 2014). Interestingly, these specialized molecular machines not only operate in a world where Brownian motion and viscous forces dominate but also work more efficiently than man-made combustion engines (van den Heuvel et al., 2007; Kabir et al., 2011). No biological cell can operate in the absence of these molecular machines. Most of these motor proteins are ubiquitously expressed but the expression of some of these motor proteins depends on cell and tissue type. The present review is about myosin motor protein, an essential component of the cytoskeletal system that is made up of proteins encoded by 441 genes in human. The human genome contains 40 genes that encode myosin motor proteins.

The term "myosin" (myo- + -ose + -in) means within muscle and was used to describe proteins with ATPase activity found originally in striated and smooth muscle cells (Pollard and Korn, 1973). The term "myo" was originated from "mys" to denote muscle in Greek. More than 140 myosins are reported in eukaryotes except in red algae and diplomonad protists (Vale, 2003). The majority of myosins have distinct head, neck, and tail domains and they are categorized into 35 different classes based on phylogenic analysis of their conserved heads, domain architectures, specific amino acid polymorphisms, and organismal distributions (Richards and Cavalier-Smith, 2005; Foth et al., 2006; Odronitz et al., 2007). Each class of myosins received a roman numeral. If more than one myosin of the same class is expressed in an organism, they are named in an alphabetical order according to their discovery. The present review is focused on current understanding and recent advances in various aspects of selected class II myosins as well as their regulation and relevance to human life and diseases.

CLASS II MYOSINS (MYOSIN II)

More than seven decades ago, an unknown myosin with ATPase activity was reported in the extracts of muscles (Engelhardt and Liubimova, 1994). Later, that unknown muscle myosin was identified as a class II myosin and then called conventional myosin and or the founding member of myosin super family. Class II myosins are expressed in all eukaryotes except plants. More than 34 class II myosins are reported in different organisms to date (Bagshaw, 1993). At least one myosin II is believed to be expressed in all eukaryotic cells. Based on motor or tail domain sequences and cell type expressions, class II myosins are further divided into four different sub-classes or groups. They are (1) *Acanthamoeba* or *Dictyostelium* myosins, (2) yeast myosins, (3) skeletal or cardiac or sarcomeric myosins, and (4) vertebrate smooth muscle or non-muscle myosins. Class II myosins are believed to be originated in unikonts that are ancestral eukaryotes with or without a single flagellum, including amoebozoans, fungi, and holozoans (Richards and Cavalier-Smith, 2005). While simple unicellular organisms like amoeba adopted a single myosin II gene, complex multicellular organisms except *Drosophila* acquired multiples of them during evolution. The human genome has over 40 myosin genes, and 15 of them are class II myosin genes (*MYH1*, *MYH2*, *MYH3*, *MYH4*, *MYH6*, *MYH7*, *MYH7B*, *MYH8*, *MYH9*, *MYH10*, *MYH11*, *MYH13*, *MYH14*, *MYH15*, *MYH16*) but not all of them are active (Berg et al., 2001). *MYH11* encodes myosin II in smooth muscles but its splice variants result in four distinct isoforms (Matsuoka et al., 1993). *MYH9*, *MYH10*, and *MYH14* located on different chromosomes encode myosin IIA, myosin IIB, and myosin IIC, respectively (**Figure 1**). These myosin II motor proteins are expressed exclusively in non-muscle cells, therefore called non-muscle myosin II motor proteins (Simons et al., 1991; Toothaker et al., 1991; Leal et al., 2003; Golomb et al.,

2004). Myosin IIA, myosin IIB, and myosin IIC are expressed in every human non-muscle cell with a few exceptions; however, their expressions depend on cell and tissue types (Kawamoto and Adelstein, 1991; Golomb et al., 2004). No tissue or cell type appears to express all three non-muscle myosin II motor proteins but many cell types express at least one or two of them under normal physiological conditions. Myosin IIA and myosin IIB are expressed in endothelial and epithelial cells at similar levels. However, myosin IIB and myosin IIC are expressed abundantly in nervous and lung tissue, respectively. Myosin IIA is the only conventional myosin II motor protein expressed in the circulating platelets. Thus, preferential expression of myosin II motor proteins in different cell types reflects their specialization in mediating separate, dedicated, and probably non-redundant cellular functions. Why doesn't a single cell or tissue type express all three myosin II motor proteins is yet to be clearly understood. Perhaps, the cell specific expression of myosin II paralogs is critical for maintaining different cell and tissue types.

Myosin II motor proteins are mostly found in the cytoplasm of quiescent cells except in the nuclei of proliferating myoblasts (Rodgers, 2005). The cytosolic myosin II motor proteins undergo transient localization to contractile ring during cytokinesis. Myosin II motor protein using ATP as a cytosolic fuel generates mechanical forces required for separation of daughter cells during cytokinesis. However, the specific roles and underlying mechanisms of myosin II paralogs during cytokinesis are not clearly understood. The functional and mechanical roles of non-muscle myosin II motor proteins are extensively investigated in migrating cells for the past two decades. Many laboratories reported myosin IIA and myosin IIB with specific roles in mediating cell shape changes and interaction with matrix during migration. Cells prefer to make periodic extension and retraction of their lamellipodia during migration by unknown mechanisms. Interestingly, myosin IIA and myosin IIB motor proteins localize distinctly in the lamellipodia of migrating cells. On one hand, myosin IIB promotes lamellipodia and growth cone extensions and on the other, myosin IIA drives retraction of cell membrane during cell migration (Rochlin et al., 1995; Brown and Bridgman, 2003; Betapudi, 2010). The specific roles of myosin IIC motor protein in driving cell migration are not clearly understood. Myosin II activity is necessary for keratinocytes' migration, a critical step in the re-epithelialization of human skin wound (Betapudi et al., 2010). Myosin II motor proteins are also required for internalization of the cell surface receptors including EGFR and CXCR4 (Rey et al., 2007; Kim et al., 2012). Myosin II-mediated mechanical forces have been implicated in operating the activity of contractile vacuoles to expel additional water and toxic materials from the soil-living amoeba in hypo-osmotic conditions (Betapudi and Egelhoff, 2009). Myosin II motor proteins have also been implicated in the mediation of viral infection (van et al., 2002; Arii et al., 2010), microparticle secretion (Betapudi et al., 2010), and cell death (Solinet and Vitale, 2008; Flynn and Helfman, 2010; Tang et al., 2011), however, their specific roles and underlying mechanisms remain unclear. Lower eukaryotes, such as amoeba can survive with certain developmental defects in the absence of myosin II (Xu et al., 1996) but the expression of all three myosin II motor proteins are necessary for

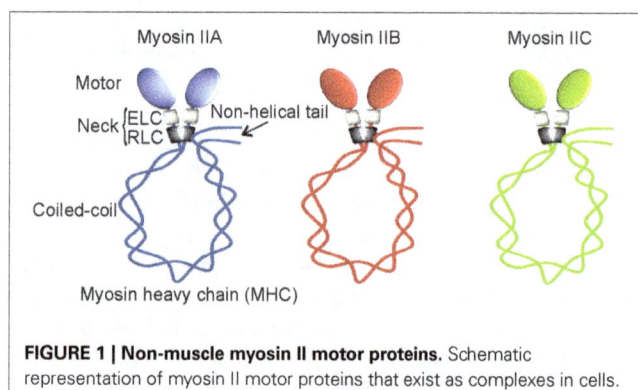

FIGURE 1 | Non-muscle myosin II motor proteins. Schematic representation of myosin II motor proteins that exist as complexes in cells.

mouse embryo growth and development (Conti and Adelstein, 2008).

NON-MUSCLE MYOSIN II COMPLEX

In line with multiple components involved in the assembly of man-made machines, biological cells also build their molecular machines using multiple polypeptides that are encoded by different genes. For instance, myosin II motor protein exists as a complex consisting of six non-covalently associated polypeptides that are encoded by a single myosin II and two different non-myosin genes. Each myosin II complex with 525 kDa molecular weight is composed of a myosin II heavy chain (MHC) homodimer, two essential light chains (ELC), and two regulatory light chains (RLC). Based on their extraction methods, ELC and RLC are also called alkali and 5,5′-dithiobis/2-nitrobenzoate (DTNB) light chains, respectively. While MHC with 226 kDa molecular weight is encoded by a myosin II gene, both ELC with 16 kDa and RLC with 22 kDa molecular weights are considered as non-myosin proteins of myosin II complex. Both ELC and RLC are commonly found in all myosin II complexes. Alternatively spliced MHC, ELC, and RLC are known to be expressed in certain tissue but our current knowledge on their specificities is still limited. Both heavy and light chain peptides undergo the UCS (UNC-45/Cro1/She4) chaperone-mediated proper folding and assembly regulation in order to form a functional myosin II complex in the Golgi apparatus (Gazda et al., 2013; Hellerschmied and Clausen, 2013). This understudied complicated assembly process is common for all three myosin II motor proteins remains elusive. Transcriptional regulations of ELC and RLC are not clearly understood; however, MHC expressions of all three myosin II motor proteins are under the control of house-keeping promoters having no TATA elements (Kawamoto, 1994; Weir and Chen, 1996). However, differential expressions of MHCs were observed in response to serum and mitotic stimulants (Kawamoto and Adelstein, 1991; Toothaker et al., 1991). Elevated levels of MHCs were found in many types of tumor tissues (our unpublished results) but their underlying mechanisms are not clearly understood.

The MHC of class II myosins can be subdivided into distinct head, neck, and tail functional domains. Except C-terminal tail pieces, the MHCs of myosin IIA, myosin IIB, and myosin IIC share a significant protein sequence similarity in their motor domains. The N-terminal catalytic globular head or motor domain has binding sites for actin and ATP. Motor domain is also called the functional engine of myosin II motor protein. Myosin II motor domain undergoes an ATP-dependent conformational change in order to control its interaction with actin filaments, a key element of the cell strategy to convert cellular free-energy into protein motion or mechanical work. Despite having a significant sequence similarity, myosin II motor domains carry different binding affinities for actin filaments. Thus, myosin IIA, IIB, and IIC are believed to perform mechanical work with different energetic efficiencies in cells. Myosin II motor domain is followed by a neck region consisting of two conserved IQ motifs (IQxxxRGxxxR); however, myosins of other classes may have more or less than two IQ motifs (Cheney and Mooseker, 1992). IQ motifs form an amphiphilic uninterrupted seven-turn

α-helix with binding affinity for either light chains or calmodulin in Ca^{+2}-independent manner. ELC and RLC occupy the first and second IQ motifs of the neck region, respectively. ELC binds IQ motif to give stability for MHC; however, RLC offers both stability and functional regulation to MHC. IQ domain allows light chains to acquire either compact or extended conformation. Thus, neck region with light chains attached acts as a linker and lever arm for myosin II motor domain to amplify energy conversion into mechanical work. The length of the neck region is believed to have direct impact on myosin II motor speed and energy transduction into mechanical work (Uyeda et al., 1996). The neck region of all myosins have IQ motifs except class XIV Toxoplasma myosin A (Heintzelman and Schwartzman, 1997). IQ motif with approximately 25 amino acids in length is widely distributed in nature, thus, ELC also binds other myosins of class V, VI, and VII as well as non-myosin proteins carrying IQ motifs, but RLC exclusively binds to myosins of class II and XVIII (Chen et al., 2007; Tan et al., 2008). Myosin II neck region is followed by a tail domain with variable amino acid sequences. The tail domain with coiled-coil α-helices terminates into a short non-helical tailpiece. The coiled-coil tail domain undergoes homodimerization to form a single rod-like structure. Thus, myosin II complex has two globular heads or motor domains with a single coiled-coil rod-like structure hence called double-headed myosin II motor protein. Myosin II complex attains a compact folded conformation due to a "proline-kink" at the junction of head and rod domains, and attachment of its C-terminal tail domain to RLC as depicted in **Figure 1** (Onishi and Wakabayashi, 1982; Trybus et al., 1982; Craig et al., 1983). Thus, the myosin II complex with compact folded structure sediments at 10 S (Svedberg) and therefore called 10S form. The myosin II complex in 10S form shows high binding affinity for ADP and inorganic phosphate (Pi), and virtually no enzyme activity (Cross et al., 1986, 1988). However, the activated myosin II complex exists in an elongated conformation due to its C-terminal tail detachment from RLC. The activated myosin II complex in an elongated form sediments at 6 S and therefore called 6S form (Trybus and Lowey, 1984). Myosin II motor proteins with elongated conformation tend to assemble into highly ordered parallel and anti-parallel thick filaments due to intermolecular interactions between coiled-coil tail domains. Interestingly, myosin II tail domains form large aggregates without proper filamentation in the absence of RLC (Pastra-Landis and Lowey, 1986; Rottbauer et al., 2006). Thus, RLC-controlled tail-domain filamentation and motor domain interaction with actin filaments are the most important aspects of cell strategy for converting ATP released free-energy into force and mechanical work using myosin II motor proteins.

RLC PHOSPHORYLATION IN REGULATING MYOSIN II ACTIVITY

Myosin IIA, myosin IIB, and myosin IIC paralogs with 60–80% sequence similarity at the amino acid level and same quaternary structure appear to be diverged from a common ancestor more than 600 million years ago, however, they display different regulatory mechanisms under normal physiological conditions (Jung et al., 2008). Role of RLC phosphorylation in regulating myosin II activity in many cell and tissue types is extensively investigated since its discovery in rabbit skeletal muscle myosins more than three decades ago (Casadei et al., 1984). RLC perhaps does not exist alone but when remains associated with the neck region of MHC undergoes reversible phosphorylation on its S1, S2, T9, T18, and S19 amino acids in order to turn-on and turn-off myosin II motor complexes in cells (**Figure 2**). RLC phosphorylation on S19 alone or on both T18 and S19 amino acids turns-on myosin II motor complex by increasing its ATPase activity and extended 6S conformation that allows simultaneous assembly into thick filaments (Wendt et al., 2001; Somlyo and Somlyo, 2003; Betapudi et al., 2006, 2010). However, RLC phosphorylation does not affect myosin II motor domain affinity for actin filaments (Sellers et al., 1982). RLC phosphorylation on S1, S2, and S9 or dephosphorylation on T18 and S19 amino acids turns-off myosin II complex by allowing acquisition of monomeric 10S compact conformation and no filamentation.

RLC reversible phosphorylation is tightly regulated by both myosin specific phosphatase and a wide variety of kinases including myosin light chain kinase (MLCK/MYLK), Rho-associated coiled-coil-containing kinase (ROCK), leucine zipper interacting protein kinase (ZIPK) or death associated protein kinase 3 (DAPK3), citron kinase or citron rho-interactive kinase (CRIK) or Serine/threonine-protein kinase 21 (STK21), myotonic dystrophy kinase-related CDC42-binding kinase (MRCK/CDC42BP). These kinases are known to phosphorylate RLC on T18 and S19 amino acids to activate myosin II complexes in different cell types. Protein kinase C (PKC) phosphorylates S1, S2, and S3 amino acids to inactivate myosin II in dividing cells (Nishikawa et al., 1984; Varlamova et al., 2001; Beach et al., 2011). Interestingly, all these kinases display specific intracellular localizations and respond to a wide variety of signal transduction pathways in order to phosphorylate RLC and activate myosin II motor proteins in many cell types. MLCK in response to Ca^{+2}-calmodulin activates myosin II that is localized next to cell membrane (Totsukawa et al., 2004). The site-specific intracellular localization and activity of MLCK are regulated by several kinases including p21 activated kinase 1 (PAK1), Abl tyrosine kinase, Src, and arrest defective 1 in many cell types (Sanders et al., 1999; Dudek et al., 2004; Shin et al., 2008). RhoA, a small GTP-binding protein activates both ROCK and citron kinase in the central part of cell. The actin binding protein, Shroom3 directs ROCK intracellular

localization, and RLC phosphorylation in neuroepithelial cells (Haigo et al., 2003; Hildebrand, 2005). DAPK3 predominantly displays nuclear localization and phosphorylates RLC in the cells that are undergoing apoptosis in a Ca^{2+}/calmodulin-independent manner (Murata-Hori et al., 1999). PKC phosphorylates RLC in the presence of Ca^{+2} and DAG (diacylglycerol) and or phorbol esters in mitotic cells (Varlamova et al., 2001). Both intracellular site-specific RLC reversible phosphorylation and myosin II activation are tightly controlled by protein phosphatase 1 (PP1), a ubiquitously expressed myosin specific phosphatase (Xia et al., 2005; Matsumura and Hartshorne, 2008; Rai and Egelhoff, 2011). All the regulators of RLC phosphorylation are also known to phosphorylate other substrates in cells. For instance, MLCK is implicated in phosphorylating a proline-rich protein tyrosine kinase 2 (PYK2/PTK2B) or focal adhesion kinase 2 (FAK2) that is involved in promoting lung vascular endothelial cell permeability during sepsis (Xu et al., 2008). ROCK also directly phosphorylates LIM kinase and MYPT1, a regulatory subunit of PP1 in many types of cells and tissues (Kimura et al., 1996; Leung et al., 1996). MYPT1 phosphorylation inactivates PP1 and this leads to a marked increase in RLC phosphorylation and myosin II activation. MYPT1 phosphorylation is also regulated by ZIPK, MRCK, and PKC in many cell and tissue types. PKC also phosphorylates MHC to regulate myosin II activity in cells under normal physiological conditions. MLCK-A is the only RLC phosphorylating kinase identified in *Dictyostelium* to date (Tan and Spudich, 1990). Unlike MLCK in mammalian cells, MLCK-A phosphorylates S13 of RLC in the absence of Ca^{+2}-calmodulin (Tan and Spudich, 1990). The RLC phosphorylation on S13 amino acid increases myosin II motor activity and regulates cell morphological changes without affecting normal growth and development of amoeba (Griffith et al., 1987; Chen et al., 1994; Uyeda et al., 1996; Liu et al., 1998; Matsumura, 2005). Except reversible phosphorylation, no other posttranslational modification of RLC that has a role in regulating myosin II activity is known to date.

MHC PHOSPHORYLATION IN REGULATING MYOSIN II ACTIVITY

MHC phosphorylation was first reported in macrophages in the early 1980s and after nearly a decade its role in regulating myosin II filamentation and localization was documented in lower eukaryotes like *Acanthamoeba* and *Dictyostelium disoideum* (Collins and Korn, 1980; Kuczmarski and Spudich, 1980; Trotter, 1982; Kuznicki et al., 1983; Trotter et al., 1985; Barylko et al., 1986; Pasternak et al., 1989; Egelhoff et al., 1993). According to computational prediction of phosphorylation sites, the heavy chains of myosin IIA, IIB, and IIC appear to undergo phosphorylation on multiple residues in the head, neck, and tail domains; however, only a few sites in the coiled-coil and non-helical tail regions of their C-terminal ends are reported to date. The MHC of myosin IIA undergoes phosphorylation on T1800, S1803, and S1808 in the coiled-coil and on S1943 residues in the non-helical tail regions (**Figure 3**). Myosin IIB and myosin IIC heavy chains also undergo phosphorylation on multiple sites in the coiled-coil and non-helical tail regions of their C-terminal ends (Dulyaninova and Bresnick, 2013). Many kinases including

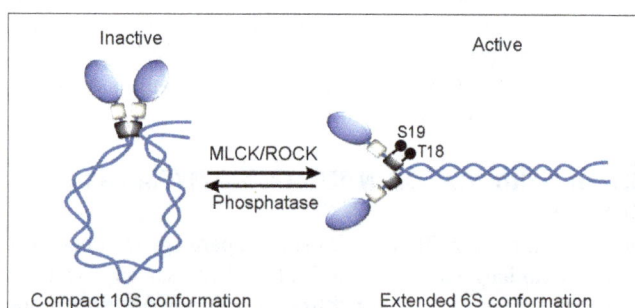

FIGURE 2 | Mechanism of the activation of myosin II motor proteins.
RLC phosphorylation by MLCK and ROCK or other kinases turns on myosin II motor protein *in vivo*.

Mechanotransduction

Mts1

S1916 → S1943
← S1943

S1927

MHCKA
MHCKB
MHCKC
TRPM6
TRPM7
PKC
CK2

Phosphatases

ATP

Actin filaments

FIGURE 3 | Myosin II motor proteins-mediated mechanotransduction in cells. Several myosin II heavy chain specific protein kinases activate myosin II motor proteins. The activated myosin II associates with actin filaments to generate contractile forces using cellular ATP.

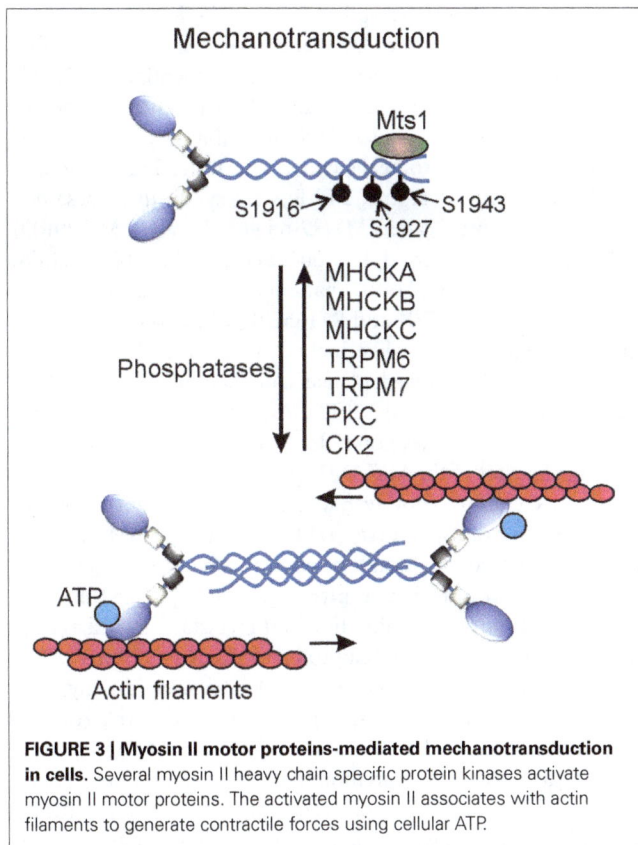

casein kinase 2 (CK2), the members of PKC as well as alpha-kinase family are involved in phosphorylating C-terminal ends of all three MHCs in normal physiological and pathological conditions (Murakami et al., 1998; Dulyaninova et al., 2005; Clark et al., 2008a,b; Ronen and Ravid, 2009). PKC members are involved in phosphorylating S1916 and S1937 residues of myosin IIA and myosin IIB, respectively (Conti et al., 1991; Even-Faitelson and Ravid, 2006). PKC is also involved in phosphorylating other multiple serine residues in myosin IIB and threonine residues in myosin IIC coiled-coil regions (Murakami et al., 1998; Ronen and Ravid, 2009). CK2 is known to phosphorylate S1943 residue in the non-helical tail region of myosin IIA *in vitro*. CK2 was implicated in the regulation of myosin II assembly and localization especially in pathological conditions. However, neither chemical inhibition nor siRNA-mediated depletion of CK2 showed any effect on S1943 phosphorylation or breast cancer cell migration on fibronectin coated surfaces (Betapudi et al., 2011). CK2 is also involved in phosphorylating multiple residues in the coiled-coil and non-helical tail regions of myosin IIB and myosin IIC (Murakami et al., 1998; Ronen and Ravid, 2009; Rosenberg et al., 2013). Thus, CK2 clearly plays a critical role in regulating myosin II-mediated cellular functions in other pathological conditions.

In addition to PKC and CK2, several members of the alpha-kinase family are involved in phosphorylating myosin II heavy chains in mammals and *Dictyostelium discoideum*. Alpha kinases belong to a small and unique group of protein kinases with catalytic domains having a little or no similarity at amino acid level with the catalytic domains of conventional protein kinases

(Ryazanov et al., 1999; De la Roche et al., 2002; Drennan and Ryazanov, 2004; Scheeff and Bourne, 2005; Middelbeek et al., 2010). Conventional protein kinases usually find their phosphorylating sites in β-turns, loops, and irregular structures of their substrates; however, the first member of the alpha-kinase family prefers to phosphorylate amino acids located in the α-turns of their cellular targets hence called α-kinases (Vaillancourt et al., 1988; Luck-Vielmetter et al., 1990). But recent *in vitro* phosphorylation studies showed that alpha-kinases also target residues present in the non-alpha helical structures of their cellular substrates (Jorgensen et al., 2003; Clark et al., 2008a). Members of the alpha-kinase family are identified only in eukaryotes to date (Ryazanov et al., 1999; Scheeff and Bourne, 2005). Transient receptor potential melastatin 6 (TRPM6) and Transient receptor potential melastatin 7 (TRPM7) kinases are among the total six alpha-kinases identified in human to date. TRPM6 and TRPM7 kinases belong to a large protein family of transient receptor potential cation channels that are involved in sensing mechanical stress, pain, temperature, taste, touch, and osmolarity (Ramsey et al., 2006; Middelbeek et al., 2010; Su et al., 2010; Runnels, 2011; Mene et al., 2013). Both TRPM6 and TRPM7 kinases phosphorylate T1800, S1803, and S1808 residues in the coiled-coil region of MHC to control myosin IIA filamentation and association with actin filaments (Clark et al., 2008a,b). These multifunctional kinases also phosphorylate several residues in the non-helical tail regions of myosin IIB and myosin IIC to control myosin II filamentation. MHC undergoes phosphorylation on T1823, T1833, and T2029 residues in the tail region of myosin II in *Dictyostelium* (De la Roche et al., 2002). Phosphorylation of these sites controls myosin II filamentation and plays critical roles in regulating growth and development of *Dictyostelium*. Except vWKa kinase, all other identified alpha-kinase family members including MHCK-A, MHCK-B, MHCK-C, and MHCK-D are involved in phosphorylating these sites in *Dictyostelium* (Egelhoff et al., 2005; Yumura et al., 2005; Underwood et al., 2010). Although vWKa does not directly phosphorylate MHC *in vitro* but regulates myosin II expression and filamentation in cells by unknown mechanism (Betapudi et al., 2005). Unlike other alpha kinases involved in regulating myosin II, vWKa displays specific sub-cellular localization to contractile vacuoles that are known to expel toxic metals and excess water from the cytoplasm of amoeba. Though the underlying mechanisms are yet to be uncovered, the myosin II-mediated mechanical work has been implicated in regulating the dynamics of contractile vacuoles and survival of *Dictyostelium discoideum* in abnormal osmotic conditions (Betapudi and Egelhoff, 2009). vWKa regulates myosin II expression and filament disassembly by unknown mechanisms to protect amoeba from osmotic shock death (Betapudi and Egelhoff, 2009). Phosphatases specific to the heavy chains of myosin II motor proteins are yet to be identified in mammals.

Many proteins including S100A4, lethal giant larvae (Lgl), myosin binding protein H, and S100P bind MHCs to control phosphorylation and filament assembly of myosin II in flies and mammals (Kriajevska et al., 1994; Ford et al., 1997; Vasioukhin, 2006; Du et al., 2012; Hosono et al., 2012). Lgl is a tumor suppressor protein and forms a complex with C-terminal ends of the MHC of myosin II to control cell proliferation. However, the

Lgl-myosin II complex dissociates when myosin II heavy chain is phosphorylated by PKC (Strand et al., 1994; Kalmes et al., 1996; Plant et al., 2003; Betschinger et al., 2005). Lgl binds coiled-coil regions of the MHC to control myosin II filamentation and localization (De et al., 1999; Dahan et al., 2012). Deletion of the Lgl located specific region in the human chromosome 17 has been implicated in the development of Smith-Magenis Syndrome, a developmental disorder that affects many body parts, intellectual disability, and sleep disturbances (Smith et al., 1986; Koyama et al., 1996; De et al., 2001). However, role of mutated Lgl in controlling myosin II phosphorylation and cellular functions remains elusive. The metastasis factor mts1 also called S100A4 or calvasculin, a member of the S100 family of calcium-binding proteins, binds C-terminal ends of the MHC of myosin II. Binding of S100A4 to C-terminal ends of the MHC promotes phosphorylation on S1943 and disassembly of myosin II filamentation; however, the underlying mechanisms remain unknown to date (Li et al., 2003; Badyal et al., 2011; Mitsuhashi et al., 2011; Kiss et al., 2012). S100-P, another member of S100 family of calcium-binding proteins and a novel therapeutic target for cancer, interacts with myosin II in cells. S100-P has been implicated in controlling myosin II filamentation and cell migration (Du et al., 2012). Myosin binding protein H (MYBPH) binds ROCK1 to control RLC phosphorylation and cell migration. MYBPH also binds MHC to control myosin II filamentation and cell migration; however, the underlying mechanisms are not clearly understood (Hosono et al., 2012). Recent studies suggest that the unassembled myosin II with phosphorylated RLC plays a role in the initiation of focal adhesion complexes formation and cell membrane extension (Shutova et al., 2012). It would be interesting to understand the coordinated regulation of RLC and MCH phosphorylation adopted by cell to regulate myosin II filamentation and cellular functions. Though the underlying mechanisms are not clearly understood, Tropomyosin, an integral part of the actin cytoskeleton system in cells has been implicated in regulating myosin II localization to plasma membrane and stress fiber formation (Bryce et al., 2003). Myosin II activity is also controlled by Supervillin, an actin filament binding and cell membrane associated scaffolding protein. Supervillin binds MLCK to control RLC phosphorylation and myosin II activity (Takizawa et al., 2007). Thus, non-muscle myosin II motor proteins are regulated by several proteins at multiple levels to perform dedicated cellular functions.

MYOSIN II MOTOR PROTEINS IN PREDISPOSING HUMANS TO DISEASES

Plants live normal life without class II myosins but mammals require these multifunctional molecular machines for survival and growth. Because the MYH9 germline-ablated mice without myosin IIA die on 6.5 embryonic day (E) due to defective cell-cell interaction and lack of polarized visceral endoderm (Conti et al., 2004). The MYH10 germline-ablated mice with no myosin IIB survive till E14.5 and then die due to brain and cardiac developmental defects (Tullio et al., 1997, 2001). However, the MYH14-ablated mice in the absence of myosin IIC can survive with no obvious defects till adulthood but require the expression of myosin IIB (Ma et al., 2010). Misregulation,

mutations, and alternative splicing of MYH9, MY10, and MY14 predispose humans to the onset and progression of many diseases (**Table 1**). More than 45 mutations are identified in MYH9 to date and some of them are linked to a large number of autosomal-dominant disorders including May-Hegglin anomaly, Sebastian platelet syndrome, Fetchner syndrome, Bernard-Soulier syndrome, Alport syndrome, and Epstein syndrome. These diseases are collectively called MYH9-related diseases (MYH9RD) (Kelley et al., 2000; Burt et al., 2008; Pecci et al., 2008; Balduini et al., 2011). The MYH9RD patients with mutations in the motor domain (R702C/H and R1165C/L) of myosin IIA develop deafness, cataract, Döhle-like inclusions, nephritis, and thrombocytopenia with enlarged platelets in their middle age (Pecci et al., 2008, 2014; De et al., 2013). An estimated 30–70 percent of MYH9RD patients develop kidney disease in their early adulthood. Leukocytes of the MYH9RD patients carry non-functional myosin IIA clumps. However, patients carrying mutations in the tail domain of myosin IIA (D1424H/N/Y, V1516M, E1841K, R1933X) show no symptoms of clinical relevance (Pecci et al., 2010). The overexpression of myosin IIA is implicated in causing enhanced cancer cell migration and metastasis as well as lung and kidney tumor invasion (Gupton and Waterman-Storer, 2006; Derycke et al., 2011; Xia et al., 2012). However, this hypothesis is downplayed by recent reports on myosin IIA roles in the posttranscriptional stabilization of p53 activity and repression of squamous cell carcinoma in mice (Schramek et al., 2014). A chimeric MYH9-Alk transcript formed by the fusion of MYH9 and ALK (anaplastic lymphoma kinase) was observed in anaplastic large cell lymphoma but its disease relevance is yet to be established (Lamant et al., 2003). No mutation in MYH10 that is relevant to a disease with any clinical symptom is reported to date; however, recently an E908X de novo mutation is reported in patients with microcephaly, hydrocephalus, cerebral, and cerebellar atrophy. An indirect link in between the expression of myosin IIB and progression of several diseases including, megakaryopoiesis, myocardial infarction, scar tissue formation, demyelination, and juvenile-onset neuronal ceroid lipofuscinosis (JCNL) or Batten disease is established (Antony-Debre et al., 2012). Batten disease, a lysosomal storage disorder is caused by mutations in CLN3 that encodes a lysosomal membrane binding chaperone known to interact directly with myosin IIB. Mutations in CLN3 are believed to affect interaction with myosin IIB as well as retrograde and anterograde trafficking in the Golgi complexes (Getty et al., 2011). Patients carrying CLN3 mutations show symptoms of seizures, psychomotor disturbances, dementia, and loss of vision (Cotman and Staropoli, 2012). Patients carrying mutations in MYH14 are also linked to many diseases including hereditary blindness (DFNA4), hoarseness, peripheral neuropathy, and myopathy (Donaudy et al., 2004; Choi et al., 2011). In addition, patients expressing aberrant splicing products of MYH14 develop myotonic dystrophy type 1 (DM1), a progressive multisystem genetic disorder that affects 1 in 8000 people worldwide (Rinaldi et al., 2012; Kumar et al., 2013).

In addition, the overexpression of myosin II upstream regulators ROCK and Mts1 is implicated in spreading cancer (Sandquist et al., 2006; Boye and Maelandsmo, 2010; Kim and Adelstein,

Table 1 | Defects and associated diseases of myosin II motor proteins and their regulators.

Gene	Mutation/Defect	Disease	Common symptoms
MYH9	R702C/H R1165C/L and many*	MYH9RD (May-Hegglin anomaly, Sebastian platelet syndrome, Fetchner, Bernard-Soulier syndrome, Alport syndrome, Epstein syndrome)	Thrombocytopenia, enlarged platelets, deafness, cataract, nephritis, and Döhle-like inclusions.
	MYH9-Alk chimeric Transcript**	Anaplastic large cell lymphoma	Blood cancer, painless swelling of lymph nodes, and rapid weight loss.
	Overexpression$	Cancer metastasis	–
MYH10	E908X (de novo)	Microcephaly, hydrocephalus, cerebral and cerebellar atrophy	Small head, dwarfism or short stature, delayed motor, and speech functions.
	Downregulation	Megakaryopoiesis, myocardial infarction, demyelination, Batten disease	Chest pain, dizziness, nausea, ocular paralysis, speech problem, and impaired vision.
MYH14	S7X, S120L, G376C, R726S, L976F	Hereditary blindness, hearing impairment (DFNA4), peripheral neuropathy, myopathy, hoarseness	Deafness, loss of vision, burning pain, numbness, changes in skin, hair or nail, dizziness, and paralysis.
	Aberrant splicing	Myotonic dystrophy type 1 (DM1) or Steinert disease	Weakness.
ROCK	Overexpression$	Cancer metastasis	–
Mts1	Overexpression$	Cancer metastasis	–
Dmlc2#	Δ2–46, S66A, S67A	Impairment of courtship	Inability of a fly to sing a courtship song.
MYLK	G601E	Cancer	–
	P147S	Asthma	–
	SNP	Asthma, acute lung injury, sepsis	–
CLN3@	L101P, L170P, Y199X, Q295K	Seizures, dementia, and psychomotor disturbances	Loss of vision and memory, mood swings, poor judgment.
TRPM6	R56X, S141L, R484X, S590X, Δ427–429, Δ736–737, Δ1260–1280	Seizures, Hypocalcemia, tetany, hypomagnesemia	Abnormal eye movement, convulsions, fatigue, numbness, anxiety, depression, dementia.
LLgl	Δ17p11.2^	Smith-Magenis Syndrome	Intellectual disability, sleep disturbances, behavior problems, defects in many body parts.

*Refer Burt et al. (2008), **Found in the lymphocytes of lymphoma patients, $Implicated, #Encodes RLC in Drosophila, @Interacts directly with myosin IIB, Δ deletion, SNP, single nucleotide polymorphism; ^LLgl, located region in chromosome 17.

2011). Mutations in RLC were shown to affect singing male courtship song in flies (Chakravorty et al., 2014). Mutations in RLC phosphorylating *MYLK* are linked to cancer (Greenman et al., 2007) and familial aortic dissections that may cause sudden death (Wang et al., 2010). A few race-specific single nucleotide polymorphism variants of *MYLK* are linked to asthma, acute lung injury and sepsis (Gao et al., 2006, 2007; Flores et al., 2007). Hypomagnesemia patients with secondary hypocalcemia carry mutations in *TRPM6* that is known to regulate MHC phosphorylation (Schlingmann et al., 2002; Walder et al., 2002). Though the underlying mechanisms are not clearly understood, myosin II motor proteins are believed to be hijacked by many pathogens such as herpes simplex virus type 1 for egression (van et al., 2002; Arii et al., 2010), murine leukemia virus for infection (Lehmann et al., 2005), and *Salmonella* bacteria for growth in macrophages (Wasylnka et al., 2008). Kaposi's sarcoma herpes simplex virus that is known to cause AIDS related neoplasm manipulates c-Cbl and myosin II-mediated signaling pathway to induce macropinocytosis in order to infect blood vessels (Sharma-Walia et al., 2010). Some pathogens like HIV-1 selectively down-regulates myosin IIA in kidney and cause renal disease probably to escape clearance through urine (Hays et al., 2012). Dengue virus type 2 activates Rac1 and Cdc42-mediated signaling pathway to regulate myosin II for successful infection of host cells (Zamudio-Meza et al., 2009). Respiratory Syncytial Virus (RSV) that is known to cause severe respiratory tract infections is believed to activate actomyosin system for improved pathogenesis (Krzyzaniak et al., 2013). Pathogens like hepatitis C virus induce development of autoantibodies having binding affinity for myosin IIA perhaps as a part of escape strategy from host defense network (von Muhlen et al., 1995).

CONCLUSION AND PERSPECTIVES

Molecular motor proteins are largely accepted as the most efficient transducers of cellular free energy into biological work that is critical for the sustenance of life. Class II myosins especially non-muscle myosin IIA, myosin IIB, and myosin IIC motor proteins are emerged as the main mechanotransducers of cellular-free energy that is necessary for driving multiple biological processes ranging from birth to death in mammals' life. During the past two decades, research on myosin II motor proteins was focused on understanding the underlying mechanisms of myosin II-mediated mechanotransduction in many biological systems. It is also proven beyond reasonable doubt that murine life does not exist without the expression of non-muscle myosin II motor proteins (Conti and Adelstein, 2008). Interestingly, many patients with mutated myosin IIA, myosin IIB, and myosin IIC paralogs are reported but none without these biological nanomachines to date. The extrapolation of these findings with caution may suggest that life in mammals does not exist without the expression of non-muscle myosin II motor proteins. Therefore, the emergence of genes that encode non-muscle myosin II motor proteins perhaps is a turning point in the evolution of mammals. During this process, humans acquired three different genes *Myh10*, *Myh11*, and *Myh14* with a significant homology in nucleotide sequence. It is generally believed that humans do require the expression of all three functional non-muscle myosin II motor proteins to maintain normal growth, development, and disease resistance. But why human cell and tissue types display differential expression of myosin II paralogs still remains unanswered. Part of the reasons could be due to their specialization in mediating dedicated functions that are specific to each cell and tissue type. However, this concept will benefit from further understanding of structural and posttranslational modifications of all three different myosin II complexes. Although we made progress in identifying several mutations in myosin II motors proteins and their regulating proteins, very little is known about the disease-relevant mutations in myosin II motor proteins. Novel strategies for management and diagnosis of MYH9RD patients are required (Althaus and Greinacher, 2010). This area of research requires additional attention to gain more insights for the development of myosin II-based novel therapeutic approaches in future. Many modern cell biologists recognize myosin II motor proteins as key drivers of cell migration and cytokinesis that are known to go awry in cancer and other pathological conditions. Although overexpression of myosin II motor proteins has been implicated in driving cancer progression and metastasis, further understanding of their specific expression profiles in every cancer type will help designing therapeutic developments. Also, expanding our limited knowledge on the expression of chimeric as well as alternate splicing products of non-muscle myosin II motor proteins in pathological conditions will allow development of treatment options. During the past two decades, we made a very limited progress on understanding how pathogens hijack non-muscle myosin II motor proteins for their efficient infection and propagation. Understanding what made these dedicated molecular machines to work for the interests of pathogens is no less than a challenge to cell biologists in future. We are yet to understand how myosin II motor proteins mediate release of microvesicles that are known to make inter cellular communications and promote progression of many human diseases. Myosin II-mediated mechanotransduction has been implicated in the regulation of stem cell proliferation and differentiation (Chen et al., 2014). Additional efforts to understand the mechanical roles of myosin IIA, IIB, and IIC motor proteins will have a significant impact on stem cells-based tissue engineering, synthetic bioengineering, and therapeutic development.

ACKNOWLEDGMENTS

The present review article was written with a particular theme on non-muscle myosin II motor proteins. The author apologizes for the omission of any study that has direct relevance with the main theme of the present article. The present work would not have been possible without the mentorship of author's previous mentors Professors Seyed E. Hasnain and Thomas T. Egelhoff, and critical reading by Dr. Saurabh Chattopadhyay as well as private and public funding agencies including the U.S. National Institute of Health.

REFERENCES

Althaus, K., and Greinacher, A. (2010). MYH-9 related platelet disorders: strategies for management and diagnosis. *Transfus. Med. Hemother.* 37, 260–267. doi: 10.1159/000320335

Antony-Debre, I., Bluteau, D., Itzykson, R., Baccini, V., Renneville, A., Boehlen, F., et al. (2012). MYH10 protein expression in platelets as a biomarker of RUNX1 and FLI1 alterations. *Blood* 120, 2719–2722. doi: 10.1182/blood-2012-04-422352

Arii, J., Goto, H., Suenega, T., Oyama, M., Kozuka-Hata, H., Imai, T., et al. (2010). Non-muscle myosin IIA is a functional entry receptor for herpes simplex virus-1. *Nature* 467, 859–862. doi: 10.1038/nature09420

Badyal, S. K., Basran, J., Bhanji, N., Kim, J. H., Chavda, A. P., Jung, H. S., et al. (2011). Mechanism of the Ca(2)+-dependent interaction between S100A4 and tail fragments of nonmuscle myosin heavy chain IIA. *J. Mol. Biol.* 405, 1004–1026. doi: 10.1016/j.jmb.2010.11.036

Bagshaw, C. R. (1993). *Muscle Contraction.* 2nd Edn. London: Chapman and Hill.

Balduini, C. L., Pecci, A., and Savoia, A. (2011). Recent advances in the understanding and management of MYH9-related inherited thrombocytopenias. *Br. J. Haematol.* 154, 161–174. doi: 10.1111/j.1365-2141.2011.08716.x

Barylko, B., Tooth, P., and Kendrick-Jones, J. (1986). Proteolytic fragmentation of brain myosin and localisation of the heavy-chain phosphorylation site. *Eur. J. Biochem.* 158, 271–282. doi: 10.1111/j.1432-1033.1986.tb09747.x

Beach, R. J., Licate, S. L., Crish, F. J., and Egelhoff, T. T. (2011). Analysis of the role of Ser1/Ser2/Thr9 phosphorylation on myosin II assembly and function in live cells. *BMC Cell Biol.* 12:52. doi: 10.1186/1471-2121-12-52

Berg, J. S., Powell, B. C., and Cheney, R. E. (2001). A millennial myosin census. *Mol. Biol. Cell* 12, 780–794. doi: 10.1091/mbc.12.4.780

Betapudi, V. (2010). Myosin II motor proteins with different functions determine the fate of lamellipodia extension during cell spreading. *PLoS ONE* 5:e8560. doi: 10.1371/journal.pone.0008560

Betapudi, V., and Egelhoff, T. T. (2009). Roles of an unconventional protein kinase and myosin II in amoeba osmotic shock responses. *Traffic* 10, 1773–1784. doi: 10.1111/j.1600-0854.2009.00992.x

Betapudi, V., Gokulrangan, G., Chance, M. R., and Egelhoff, T. T. (2011). A proteomic study of myosin II motor proteins during tumor cell migration. *J. Mol. Biol.* 407, 673–686. doi: 10.1016/j.jmb.2011.02.010

Betapudi, V., Licate, L. S., and Egelhoff, T. T. (2006). Distinct roles of non-muscle myosin II isoforms in the regulation of MDA-MB-231 breast cancer cell spreading and migration. *Cancer Res.* 66, 4725–4733. doi: 10.1158/0008-5472.CAN-05-4236

Betapudi, V., Mason, C., Licate, L., and Egelhoff, T. T. (2005). Identification and characterization of a novel alpha-kinase with a von Willebrand factor A-like motif localized to the contractile vacuole and Golgi complex in *Dictyostelium discoideum*. *Mol. Biol. Cell* 16, 2248–2262. doi: 10.1091/mbc.E04-07-0639

Betapudi, V., Rai, V., Beach, J. R., and Egelhoff, T. (2010). Novel regulation and dynamics of myosin II activation during epidermal wound responses. *Exp. Cell Res.* 316, 980–991. doi: 10.1016/j.yexcr.2010.01.024

Betschinger, J., Eisenhaber, F., and Knoblich, J. A. (2005). Phosphorylation-induced autoinhibition regulates the cytoskeletal protein Lethal (2) giant larvae. *Curr. Biol.* 15, 276–282. doi: 10.1016/j.cub.2005.01.012

Boye, K., and Maelandsmo, G. M. (2010). S100A4 and metastasis: a small actor playing many roles. *Am. J. Pathol.* 176, 528–535. doi: 10.2353/ajpath.2010.090526

Brown, M. E., and Bridgman, P. C. (2003). Retrograde flow rate is increased in growth cones from myosin IIB knockout mice. *J. Cell Sci.* 116, 1087–1094. doi: 10.1242/jcs.00335

Bryce, N. S., Schevzov, G., Ferguson, V., Percival, J. M., Lin, J. J. Matsumura, F., et al. (2003). Specification of actin filament function and molecular composition by tropomyosin isoforms. *Mol. Biol. Cell* 14, 1002–1016. doi: 10.1091/mbc.E02-04-0244

Burt, R. A., Joseph, J. E., Milliken, S., Collinge, J. E., and Kile, B. T. (2008). Description of a novel mutation leading to MYH9-related disease. *Thromb. Res.* 122, 861–863. doi: 10.1016/j.thromres.2008.06.011

Bustamante, C., Chemla, Y. R., Forde, N. R., and Izhaky, D. (2004). Mechanical processes in biochemistry. *Annu. Rev. Biochem.* 73, 705–748. doi: 10.1146/annurev.biochem.72.121801.161542

Casadei, J. M., Gordon, R. D., Lampson, L. A., Schotland, D. L., and Barchi, R. L. (1984). Monoclonal antibodies against the voltage-sensitive Na^+ channel from mammalian skeletal muscle. *Proc. Natl. Acad. Sci. U.S.A.* 81, 6227–6231. doi: 10.1073/pnas.81.19.6227

Chakravorty, S., Vu, H., Foelber, V., and Vigoreaux, J. O. (2014). Mutations of the *Drosophila* myosin regulatory light chain affect courtship song and reduce reproductive success. *PLoS ONE* 9:e90077. doi: 10.1371/journal.pone.0090077

Chen, A. K., Chen, X., Lim, Y. M., Reuveny, S., and Oh, S. K. (2014). Inhibition of ROCK-myosin II signaling pathway enables culturing of human pluripotent stem cells on microcarriers without extracellular matrix coating. *Tissue Eng. Part C Methods* 20, 227–238. doi: 10.1089/ten.tec.2013.0191

Chen, P., Ostrow, B. D., Tafuri, S. R., and Chisholm, R. L. (1994). Targeted disruption of the *Dictyostelium* RMLC gene produces cells defective in cytokinesis and development. *J. Cell Biol.* 127, 1933–1944. doi: 10.1083/jcb.127.6.1933

Chen, Z., Naveiras, O., Balduini, A., Mammoto, A., Conti, M. A., Adelstein, R. S., et al. (2007). The May-Hegglin anomaly gene MYH9 is a negative regulator of platelet biogenesis modulated by the Rho-ROCK pathway. *Blood* 110, 171–179. doi: 10.1182/blood-2007-02-071589

Cheney, R. E., and Mooseker, M. S. (1992). Unconventional myosins. *Curr. Opin. Cell Biol.* 4, 27–35. doi: 10.1016/0955-0674(92)90055-H

Choi, B. O., Kang, S. H., Hyun, Y. S., Kanwal, S., Park, S. W., Koo, H., et al. (2011). A complex phenotype of peripheral neuropathy, myopathy, hoarseness, and hearing loss is linked to an autosomal dominant mutation in MYH14. *Hum. Mutat.* 32, 669–677. doi: 10.1002/humu.21488

Clark, K., Middelbeek, J., Dorovkov, M. V., Figdor, C. G., Ryazanov, A. G., Lasonder, E., et al. (2008a). The alpha-kinases TRPM6 and TRPM7, but not eEF-2 kinase, phosphorylate the assembly domain of myosin IIA, IIB and IIC. *FEBS Lett.* 582, 2993–2997. doi: 10.1016/j.febslet.2008.07.043

Clark, K., Middelbeek, J., Lasonder, E., Dulyaninova, N. G., Morrice, N. A., Ryazanov, A. G., et al. (2008b). TRPM7 regulates myosin IIA filament stability and protein localization by heavy chain phosphorylation. *J. Mol. Biol.* 378, 790–803. doi: 10.1016/j.jmb.2008.02.057

Collins, J. H., and Korn, E. D. (1980). Actin activation of Ca^{2+}-sensitive Mg^{2+}-ATPase activity of *Acanthamoeba* myosin II is enhanced by dephosphorylation of its heavy chains. *J. Biol. Chem.* 255, 8011–8014.

Conti, M. A., and Adelstein, R. S. (2008). Nonmuscle myosin II moves in new directions. *J. Cell Sci.* 121, 11–18. doi: 10.1242/jcs.007112

Conti, M. A., Even-Ram, S., Liu, C., Yamada, K. M., and Adelstein, R. S. (2004). Defects in cell adhesion and the visceral endoderm following ablation of non-muscle myosin heavy chain II-A in mice. *J. Biol. Chem.* 279, 41263–41266. doi: 10.1074/jbc.C400352200

Conti, M. A., Sellers, J. R., Adelstein, R. S., and Elzinga, M. (1991). Identification of the serine residue phosphorylated by protein kinase C in vertebrate nonmuscle myosin heavy chains. *Biochemistry* 30, 966–970. doi: 10.1021/bi00218a012

Cotman, S. L., and Staropoli, J. F. (2012). The juvenile Batten disease protein, CLN3, and its role in regulating anterograde and retrograde post-Golgi trafficking. *Clin. Lipidol.* 7, 79–91. doi: 10.2217/clp.11.70

Craig, R., Smith, R., and Kendrick-Jones, J. (1983). Light-chain phosphorylation controls the conformation of vertebrate non-muscle and smooth muscle myosin molecules. *Nature* 302, 436–439. doi: 10.1038/302436a0

Cross, R. A., Cross, K. E., and Sobieszek, A. (1986). ATP-linked monomer-polymer equilibrium of smooth muscle myosin: the free folded monomer traps ADP.Pi. *EMBO J.* 5, 2637–2641.

Cross, R. A., Jackson, A. P., Citi, S., Kendrick-Jones, J., and Bagshaw, C. R. (1988). Active site trapping of nucleotide by smooth and non-muscle myosins. *J. Mol. Biol.* 203, 173–181. doi: 10.1016/0022-2836(88)90100-3

Dahan, I., Yearim, A., Touboul, Y., and Ravid, S. (2012). The tumor suppressor Lgl1 regulates NMII-A cellular distribution and focal adhesion morphology to optimize cell migration. *Mol. Biol. Cell* 23, 591–601. doi: 10.1091/mbc.E11-01-0015

De, L. C., Mechler, B. M., and Bryant, P. J. (1999). What is *Drosophila* telling us about cancer? *Cancer Metastasis Rev.* 18, 295–311.

De, L. H., de Blois, M. C., Vekemans, M., Sidi, D., Villain, E., Kindermans, C., et al. (2001). beta(1)-adrenergic antagonists improve sleep and behavioural disturbances in a circadian disorder, Smith-Magenis syndrome. *J. Med. Genet.* 38, 586–590. doi: 10.1136/jmg.38.9.586

De, R. D., Zieger, B., Platokouki, H., Heller, P. G., Pastore, A., Bottega, R., et al. (2013). MYH9-related disease: five novel mutations expanding the spectrum of causative mutations and confirming genotype/phenotype correlations. *Eur. J. Med. Genet.* 56, 7–12. doi: 10.1016/j.ejmg.2012.10.009

De la Roche, M. A., Smith, J. L., Betapudi, V., Egelhoff, T. T., and Cote, G. P. (2002). Signaling pathways regulating *Dictyostelium* myosin II. *J. Muscle Res. Cell Motil.* 23, 703–718. doi: 10.1023/A:1024467426244

Derycke, L., Stove, C., Vercoutter-Edouart, A. S., De, W. O., Dolle, L., Colpaert, N., et al. (2011). The role of non-muscle myosin IIA in aggregation and invasion of human MCF-7 breast cancer cells. *Int. J. Dev. Biol.* 55, 835–840. doi: 10.1387/ijdb.113336ld

Donaudy, F., Snoeckx, R., Pfister, M., Zenner, H. P., Blin, N., Di, S. M., et al. (2004). Nonmuscle myosin heavy-chain gene MYH14 is expressed in cochlea and mutated in patients affected by autosomal dominant hearing impairment (DFNA4). *Am. J. Hum. Genet.* 74, 770–776. doi: 10.1086/383285

Drennan, D., and Ryazanov, A. G. (2004). Alpha-kinases: analysis of the family and comparison with conventional protein kinases. *Prog. Biophys. Mol. Biol.* 85, 1–32. doi: 10.1016/S0079-6107(03)00060-9

Du, M., Wang, G., Ismail, T. M., Gross, S., Fernig, D. G., Barraclough, R., et al. (2012). S100P dissociates myosin IIA filaments and focal adhesion sites to reduce cell adhesion and enhance cell migration. *J. Biol. Chem.* 287, 15330–15344. doi: 10.1074/jbc.M112.349787

Dudek, S. M., Jacobson, J. R., Chiang, E. T., Birukov, K. G., Wang, P., Zhan, X., et al. (2004). Pulmonary endothelial cell barrier enhancement by sphingosine 1-phosphate: roles for cortactin and myosin light chain kinase. *J. Biol. Chem.* 279, 24692–24700. doi: 10.1074/jbc.M313969200

Dulyaninova, N. G., and Bresnick, A. R. (2013). The heavy chain has its day: regulation of myosin-II assembly. *Bioarchitecture* 3, 77–85. doi: 10.4161/bioa.26133

Dulyaninova, N. G., Malashkevich, V. N., Almo, S. C., and Bresnick, A. R. (2005). Regulation of myosin-IIA assembly and Mts1 binding by heavy chain phosphorylation. *Biochemistry* 44, 6867–6876. doi: 10.1021/bi0500776

Egelhoff, T. T., Croft, D., and Steimle, P. A. (2005). Actin activation of myosin heavy chain kinase A in *Dictyostelium*: a biochemical mechanism for the spatial regulation of myosin II filament disassembly. *J. Biol. Chem.* 280, 2879–2887. doi: 10.1074/jbc.M410803200

Egelhoff, T. T., Lee, R. J., and Spudich, J. A. (1993). *Dictyostelium* myosin heavy chain phosphorylation sites regulate myosin filament assembly and localization *in vivo*. *Cell* 75, 363–371. doi: 10.1016/0092-8674(93)80077-R

Engelhardt, W. A., and Liubimova, M. N. (1994). [Myosin and adenosine triphosphatase (Nature, 144, 688, Oct. 14, 1939)]. *Mol. Biol. (Mosk.)* 28, 1229–1230.

Even-Faitelson, L., and Ravid, S. (2006). PAK1 and aPKCzeta regulate myosin II-B phosphorylation: a novel signaling pathway regulating filament assembly. *Mol. Biol. Cell* 17, 2869–2881. doi: 10.1091/mbc.E05-11-1001

Flores, C., Ma, S. F., Maresso, K., Ober, C., and Garcia, J. G. (2007). A variant of the myosin light chain kinase gene is associated with severe asthma in African Americans. *Genet. Epidemiol.* 31, 296–305. doi: 10.1002/gepi.20210

Flynn, P. G., and Helfman, D. M. (2010). Non-muscle myosin IIB helps mediate TNF cell death signaling independent of actomyosin contractility (AMC). *J. Cell Biochem.* 110, 1365–1375. doi: 10.1002/jcb.22653

Ford, H. L., Silver, D. L., Kachar, B., Sellers, J. R., and Zain, S. B. (1997). Effect of Mts1 on the structure and activity of nonmuscle myosin II. *Biochemistry* 36, 16321–16327. doi: 10.1021/bi971182l

Foth, B. J., Goedecke, M. C., and Soldati, D. (2006). New insights into myosin evolution and classification. *Proc. Natl. Acad. Sci. U.S.A.* 103, 3681–3686. doi: 10.1073/pnas.0506307103

Gao, L., Flores, C., Fan-Ma, S., Miller, E. J., Moitra, J., Moreno, L., et al. (2007). Macrophage migration inhibitory factor in acute lung injury: expression, biomarker, and associations. *Transl. Res.* 150, 18–29. doi: 10.1016/j.trsl.2007.02.007

Gao, L., Grant, A., Halder, I., Brower, R., Sevransky, J., Maloney, J. P., et al. (2006). Novel polymorphisms in the myosin light chain kinase gene confer risk for acute lung injury. *Am. J. Respir. Cell Mol. Biol.* 34, 487–495. doi: 10.1165/rcmb.2005-0404OC

Gazda, L., Pokrzywa, W., Hellerschmied, D., Lowe, T., Forne, I., Mueller-Planitz, F., et al. (2013). The myosin chaperone UNC-45 is organized in tandem modules to support myofilament formation in *C. elegans*. *Cell* 152, 183–195. doi: 10.1016/j.cell.2012.12.025

Getty, A. L., Benedict, J. W., and Pearce, D. A. (2011). A novel interaction of CLN3 with nonmuscle myosin-IIB and defects in cell motility of Cln3(-/-) cells. *Exp. Cell Res.* 317, 51–69. doi: 10.1016/j.yexcr.2010.09.007

Golomb, E., Ma, X., Jana, S. S., Preston, Y. A., Kawamoto, S., Shoham, N. G., et al. (2004). Identification and characterization of nonmuscle myosin II-C, a new member of the myosin II family. *J. Biol. Chem.* 279, 2800–2808. doi: 10.1074/jbc.M309981200

Greenman, C., Stephens, P., Smith, R., Dalgliesh, G. L., Hunter, C., Bignell, G., et al. (2007). Patterns of somatic mutation in human cancer genomes. *Nature* 446, 153–158. doi: 10.1038/nature05610

Griffith, L. M., Downs, S. M., and Spudich, J. A. (1987). Myosin light chain kinase and myosin light chain phosphatase from *Dictyostelium*: effects of reversible phosphorylation on myosin structure and function. *J. Cell Biol.* 104, 1309–1323. doi: 10.1083/jcb.104.5.1309

Gupton, S. L., and Waterman-Storer, C. M. (2006). Spatiotemporal feedback between actomyosin and focal-adhesion systems optimizes rapid cell migration. *Cell* 125, 1361–1374. doi: 10.1016/j.cell.2006.05.029

Haigo, S. L., Hildebrand, J. D., Harland, R. M., and Wallingford, J. B. (2003). Shroom induces apical constriction and is required for hinge-point formation during neural tube closure. *Curr. Biol.* 13, 2125–2137. doi: 10.1016/j.cub.2003.11.054

Hays, T., D'Agati, V. D., Garellek, J. A., Warren, T., Trubin, M. E., Hyink, D. P., et al. (2012). Glomerular MYH9 expression is reduced by HIV-1. *AIDS* 26, 797–803. doi: 10.1097/QAD.0b013e328351f6cf

Heintzelman, M. B., and Schwartzman, J. D. (1997). A novel class of unconventional myosins from *Toxoplasma gondii*. *J. Mol. Biol.* 271, 139–146. doi: 10.1006/jmbi.1997.1167

Hellerschmied, D., and Clausen, T. (2013). Myosin chaperones. *Curr. Opin. Struct. Biol.* 25C, 9–15. doi: 10.1016/j.sbi.2013.11.002

Hildebrand, J. D. (2005). Shroom regulates epithelial cell shape via the apical positioning of an actomyosin network. *J. Cell Sci.* 118, 5191–5203. doi: 10.1242/jcs.02626

Hosono, Y., Usukura, J., Yamaguchi, T., Yanagisawa, K., Suzuki, M., and Takahashi, T. (2012). MYBPH inhibits NM IIA assembly via direct interaction with NMHC IIA and reduces cell motility. *Biochem. Biophys. Res. Commun.* 428, 173–178. doi: 10.1016/j.bbrc.2012.10.036

Howard, J. (2014). Jonathon Howard: motor proteins go walkabout. Interview by Caitlin Sedwick. *J. Cell Biol.* 204, 150–151. doi: 10.1083/jcb.2042pi

Jorgensen, R., Ortiz, P. A., Carr-Schmid, A., Nissen, P., Kinzy, T. G., and Andersen, G. R. (2003). Two crystal structures demonstrate large conformational changes in the eukaryotic ribosomal translocase. *Nat. Struct. Biol.* 10, 379–385. doi: 10.1038/nsb923

Jung, H. S., Burgess, S. A., Billington, N., Colegrave, M., Patel, H., Chalovich, J. M., et al. (2008). Conservation of the regulated structure of folded myosin 2 in species separated by at least 600 million years of independent evolution. *Proc. Natl. Acad. Sci. U.S.A.* 105, 6022–6026. doi: 10.1073/pnas.0707846105

Kabir, A. M., Kakugo, A., Gong, J. P., and Osada, Y. (2011). How to integrate biological motors towards bio-actuators fueled by ATP. *Macromol. Biosci.* 11, 1314–1324. doi: 10.1002/mabi.201100060

Kalmes, A., Merdes, G., Neumann, B., Strand, D., and Mechler, B. M. (1996). A serine-kinase associated with the p127-l(2)gl tumour suppressor of *Drosophila* may regulate the binding of p127 to nonmuscle myosin II heavy chain and the attachment of p127 to the plasma membrane. *J. Cell Sci.* 109(Pt 6), 1359–1368.

Kawamoto, S. (1994). Evidence for an internal regulatory region in a human nonmuscle myosin heavy chain gene. *J. Biol. Chem.* 269, 15101–15110.

Kawamoto, S., and Adelstein, R. S. (1991). Chicken nonmuscle myosin heavy chains: differential expression of two mRNAs and evidence for two different polypeptides. *J. Cell Biol.* 112, 915–924. doi: 10.1083/jcb.112.5.915

Kelley, M. J., Jawien, W., Ortel, T. L., and Korczak, J. F. (2000). Mutation of MYH9, encoding non-muscle myosin heavy chain A, in May-Hegglin anomaly. *Nat. Genet.* 26, 106–108. doi: 10.1038/79069

Kim, J. H., and Adelstein, R. S. (2011). LPA(1) -induced migration requires non-muscle myosin II light chain phosphorylation in breast cancer cells. *J. Cell. Physiol.* 226, 2881–2893. doi: 10.1002/jcp.22631

Kim, J. H., Wang, A., Conti, M. A., and Adelstein, R. S. (2012). Nonmuscle myosin II is required for internalization of the epidermal growth factor receptor and modulation of downstream signaling. *J. Biol. Chem.* 287, 27345–27358. doi: 10.1074/jbc.M111.304824

Kimura, K., Ito, M., Amano, M., Chihara, K., Fukata, Y., Nakafuku, M., et al. (1996). Regulation of myosin phosphatase by Rho and Rho-associated kinase (Rho-kinase). *Science* 273, 245–248. doi: 10.1126/science.273.5272.245

Kiss, B., Duelli, A., Radnai, L., Kekesi, K. A., Katona, G., and Nyitray, L. (2012). Crystal structure of the S100A4-nonmuscle myosin IIA tail fragment complex reveals an asymmetric target binding mechanism. *Proc. Natl. Acad. Sci. U.S.A.* 109, 6048–6053. doi: 10.1073/pnas.1114732109

Kolomeisky, A. B. (2013). Motor proteins and molecular motors: how to operate machines at the nanoscale. *J. Phys. Condens. Matter* 25:463101. doi: 10.1088/0953-8984/25/46/463101

Koyama, K., Fukushima, Y., Inazawa, J., Tomotsune, D., Takahashi, N., Nakamura, Y., et al. (1996). The human homologue of the murine Llglh gene (LLGL) maps within the Smith-Magenis syndrome region in 17p11.2. *Cytogenet. Cell Genet.* 72, 78–82.

Kriajevska, M. V., Cardenas, M. N., Grigorian, M. S., Ambartsumian, N. S., Georgiev, G. P., and Lukanidin, E. M. (1994). Non-muscle myosin heavy chain as a possible target for protein encoded by metastasis-related mts-1 gene. *J. Biol. Chem.* 269, 19679–19682.

Krzyzaniak, M. A., Zumstein, M. T., Gerez, J. A., Picotti, P., and Helenius, A. (2013). Host cell entry of respiratory syncytial virus involves macropinocytosis followed by proteolytic activation of the F protein. *PLoS. Pathog.* 9:e1003309. doi: 10.1371/journal.ppat.1003309

Kuczmarski, E. R., and Spudich, J. A. (1980). Regulation of myosin self-assembly: phosphorylation of *Dictyostelium* heavy chain inhibits formation of thick filaments. *Proc. Natl. Acad. Sci. U.S.A.* 77, 7292–7296. doi: 10.1073/pnas.77.12.7292

Kumar, A., Agarwal, S., Agarwal, D., and Phadke, S. R. (2013). Myotonic dystrophy type 1 (DM1): a triplet repeat expansion disorder. *Gene* 522, 226–230. doi: 10.1016/j.gene.2013.03.059

Kuznicki, J., Albanesi, J. P., Cote, G. P., and Korn, E. D. (1983). Supramolecular regulation of the actin-activated ATPase activity of filaments of *Acanthamoeba* Myosin II. *J. Biol. Chem.* 258, 6011–6014.

Lamant, L., Gascoyne, R. D., Duplantier, M. M., Armstrong, F., Raghab, A., Chhanabhai, M., et al. (2003). Non-muscle myosin heavy chain (MYH9): a new partner fused to ALK in anaplastic large cell lymphoma. *Genes Chromosomes Cancer* 37, 427–432. doi: 10.1002/gcc.10232

Leal, A., Endele, S., Stengel, C., Huehne, K., Loetterle, J., Barrantes, R., et al. (2003). A novel myosin heavy chain gene in human chromosome 19q13.3. *Gene* 312, 165–171. doi: 10.1016/S0378-1119(03)00613-9

Lehmann, M. J., Sherer, N. M., Marks, C. B., Pypaert, M., and Mothes, W. (2005). Actin- and myosin-driven movement of viruses along filopodia precedes their entry into cells. *J. Cell Biol.* 170, 317–325. doi: 10.1083/jcb.200503059

Leung, T., Chen, X. Q., Manser, E., and Lim, L. (1996). The p160 RhoA-binding kinase ROK alpha is a member of a kinase family and is involved in the reorganization of the cytoskeleton. *Mol. Cell Biol.* 16, 5313–5327.

Li, Z. H., Spektor, A., Varlamova, O., and Bresnick, A. R. (2003). Mts1 regulates the assembly of nonmuscle myosin-IIA. *Biochemistry* 42, 14258–14266. doi: 10.1021/bi0354379

Liu, X., Ito, K., Morimoto, S., Hikkoshi-Iwane, A., Yanagida, T., and Uyeda, T. Q. (1998). Filament structure as an essential factor for regulation of *Dictyostelium* myosin by regulatory light chain phosphorylation. *Proc. Natl. Acad. Sci. U.S.A.* 95, 14124–14129. doi: 10.1073/pnas.95.24.14124

Luck-Vielmetter, D., Schleicher, M., Grabatin, B., Wippler, J., and Gerisch, G. (1990). Replacement of threonine residues by serine and alanine in a phosphorylatable heavy chain fragment of *Dictyostelium* myosin II. *FEBS Lett.* 269, 239–243. doi: 10.1016/0014-5793(90)81163-I

Ma, X., Jana, S. S., Conti, M. A., Kawamoto, S., Claycomb, W. C., and Adelstein, R. S. (2010). Ablation of nonmuscle myosin II-B and II-C reveals a role for nonmuscle myosin II in cardiac myocyte karyokinesis. *Mol. Biol. Cell* 21, 3952–3962. doi: 10.1091/mbc.E10-04-0293

Maravillas-Montero, J. L., and Santos-Argumedo, L. (2012). The myosin family: unconventional roles of actin-dependent molecular motors in immune cells. *J. Leukoc. Biol.* 91, 35–46. doi: 10.1189/jlb.0711335

Matsumura, F. (2005). Regulation of myosin II during cytokinesis in higher eukaryotes. *Trends Cell Biol.* 15, 371–377. doi: 10.1016/j.tcb.2005.05.004

Matsumura, F., and Hartshorne, D. J. (2008). Myosin phosphatase target subunit: many roles in cell function. *Biochem. Biophys. Res. Commun.* 369, 149–156. doi: 10.1016/j.bbrc.2007.12.090

Matsuoka, R., Yoshida, M. C., Furutani, Y., Imamura, S., Kanda, N., Yanagisawa, M., et al. (1993). Human smooth muscle myosin heavy chain gene mapped to chromosomal region 16q12. *Am. J. Med. Genet.* 46, 61–67. doi: 10.1002/ajmg.1320460110

Mene, P., Punzo, G., and Pirozzi, N. (2013). TRP channels as therapeutic targets in kidney disease and hypertension. *Curr. Top. Med. Chem.* 13, 386–397. doi: 10.2174/1568026611313030013

Middelbeek, J., Clark, K., Venselaar, H., Huynen, M. A., and van Leeuwen, F. N. (2010). The alpha-kinase family: an exceptional branch on the protein kinase tree. *Cell Mol. Life Sci.* 67, 875–890. doi: 10.1007/s00018-009-0215-z

Min, S. Y., Ahn, H. J., Park, W. S., and Kim, J. W. (2014). Successful renal transplantation in MYH9-related disorder with severe macrothrombocytopenia: first report in Korea. *Transplant. Proc.* 46, 654–656. doi: 10.1016/j.transproceed.2013.11.144

Mitsuhashi, M., Sakata, H., Kinjo, M., Yazawa, M., and Takahashi, M. (2011). Dynamic assembly properties of nonmuscle myosin II isoforms revealed by combination of fluorescence correlation spectroscopy and fluorescence cross-correlation spectroscopy. *J. Biochem.* 149, 253–263. doi: 10.1093/jb/mvq134

Murakami, N., Chauhan, V. P., and Elzinga, M. (1998). Two nonmuscle myosin II heavy chain isoforms expressed in rabbit brains: filament forming properties, the effects of phosphorylation by protein kinase C and casein kinase II, and location of the phosphorylation sites. *Biochemistry* 37, 1989–2003. doi: 10.1021/bi971959a

Murata-Hori, M., Suizu, F., Iwasaki, T., Kikuchi, A., and Hosoya, H. (1999). ZIP kinase identified as a novel myosin regulatory light chain kinase in HeLa cells. *FEBS Lett.* 451, 81–84. doi: 10.1016/S0014-5793(99)00550-5

Nishikawa, M., Sellers, J. R., Adelstein, R. S., and Hidaka, H. (1984). Protein kinase C modulates *in vitro* phosphorylation of the smooth muscle heavy meromyosin by myosin light chain kinase. *J. Biol. Chem.* 259, 8808–8814.

Odronitz, F., Hellkamp, M., and Kollmar, M. (2007). diArk–a resource for eukaryotic genome research. *BMC Genomics* 8:103. doi: 10.1186/1471-2164-8-103

Onishi, H., and Wakabayashi, T. (1982). Electron microscopic studies of myosin molecules from chicken gizzard muscle I: the formation of the intramolecular loop in the myosin tail. *J. Biochem.* 92, 871–879.

Pasternak, C., Flicker, P. F., Ravid, S., and Spudich, J. A. (1989). Intermolecular versus intramolecular interactions of *Dictyostelium* myosin: possible regulation by heavy chain phosphorylation. *J. Cell Biol.* 109, 203–210. doi: 10.1083/jcb.109.1.203

Pastra-Landis, S. C., and Lowey, S. (1986). Myosin subunit interactions. Properties of the 19,000-dalton light chain-deficient myosin. *J. Biol. Chem.* 261, 14811–14816.

Pecci, A., Klersy, C., Gresele, P., Lee, K. J., De Rocco, D., Bozzi, V., et al. (2014). MYH9-related disease: a novel prognostic model to predict the clinical evolution of the disease based on genotype-phenotype correlations. *Hum. Mutat.* 35, 236–247. doi: 10.1002/humu.22476

Pecci, A., Panza, E., De Rocco, D., Pujol-Moix, N., Girotto, G., Podda, L., et al. (2010). MYH9 related disease: four novel mutations of the tail domain of myosin-9 correlating with a mild clinical phenotype. *Eur. J. Haematol.* 84, 291–297. doi: 10.1111/j.1600-0609.2009.01398.x

Pecci, A., Panza, E., Pujol-Moix, N., Klersy, C., Di Bari, F., Bozzi, V., et al. (2008). Position of nonmuscle myosin heavy chain IIA (NMMHC-IIA) mutations predicts the natural history of MYH9-related disease. *Hum. Mutat.* 29, 409–417. doi: 10.1002/humu.20661

Plant, P. J., Fawcett, J. P., Lin, D. C., Holdorf, A. D., Binns, K., Kulkarni, S., et al. (2003). A polarity complex of mPar-6 and atypical PKC binds, phosphorylates and regulates mammalian Lgl. *Nat. Cell Biol.* 5, 301–308. doi: 10.1038/ncb948

Pollard, T. D., and Korn, E. D. (1973). *Acanthamoeba* myosin. II. Interaction with actin and with a new cofactor protein required for actin activation of Mg^{2+} adenosine triphosphatase activity. *J. Biol. Chem.* 248, 4691–4697.

Rai, V., and Egelhoff, T. T. (2011). Role of B regulatory subunits of protein phosphatase type 2A in myosin II assembly control in *Dictyostelium discoideum*. *Eukaryot. Cell* 10, 604–610. doi: 10.1128/EC.00296-10

Ramsey, I. S., Delling, M., and Clapham, D. E. (2006). An introduction to TRP channels. *Annu. Rev. Physiol.* 68, 619–647. doi: 10.1146/annurev.physiol.68.040204.100431

Rey, M., Valenzuela-Fernandez, A., Urzainqui, A., Yanez-Mo, M., Perez-Martinez, M., Penela, P., et al. (2007). Myosin IIA is involved in the endocytosis of CXCR4 induced by SDF-1alpha. *J. Cell Sci.* 120, 1126–1133. doi: 10.1242/jcs.03415

Richards, T. A., and Cavalier-Smith, T. (2005). Myosin domain evolution and the primary divergence of eukaryotes. *Nature* 436, 1113–1118. doi: 10.1038/nature03949

Rinaldi, F., Terracciano, C., Pisani, V., Massa, R., Loro, E., Vergani, L., et al. (2012). Aberrant splicing and expression of the non muscle myosin heavy-chain gene MYH14 in DM1 muscle tissues. *Neurobiol. Dis.* 45, 264–271. doi: 10.1016/j.nbd.2011.08.010

Rochlin, M. W., Itoh, K., Adelstein, R. S., and Bridgman, P. C. (1995). Localization of myosin II A and B isoforms in cultured neurons. *J. Cell Sci.* 108(Pt 12), 3661–3670.

Rodgers, B. D. (2005). Insulin-like growth factor-I downregulates embryonic myosin heavy chain (eMyHC) in myoblast nuclei. *Growth Horm. IGF. Res.* 15, 377–383. doi: 10.1016/j.ghir.2005.08.001

Ronen, D., and Ravid, S. (2009). Myosin II tailpiece determines its paracrystal structure, filament assembly properties, and cellular localization. *J. Biol. Chem.* 284, 24948–24957. doi: 10.1074/jbc.M109.023754

Rosenberg, M. M., Ronen, D., Lahav, N., Nazirov, E., Ravid, S., and Friedler, A. (2013). High resolution characterization of myosin IIC protein tailpiece and its effect on filament assembly. *J. Biol. Chem.* 288, 9779–9789. doi: 10.1074/jbc.M112.430173

Rottbauer, W., Wessels, G., Dahme, T., Just, S., Trano, N., Hassel, D., et al. (2006). Cardiac myosin light chain-2: a novel essential component of thick-myofilament assembly and contractility of the heart. *Circ. Res.* 99, 323–331. doi: 10.1161/01.RES.0000234807.16034.fe

Runnels, L. W. (2011). TRPM6 and TRPM7: a Mul-TRP-PLIK-cation of channel functions. *Curr. Pharm. Biotechnol.* 12, 42–53. doi: 10.2174/138920111793937880

Ryazanov, A. G., Pavur, K. S., and Dorovkov, M. V. (1999). Alpha-kinases: a new class of protein kinases with a novel catalytic domain. *Curr. Biol.* 9, R43–R45. doi: 10.1016/S0960-9822(99)80006-2

Sanders, L. C., Matsumura, F., Bokoch, G. M., and de, L. P. (1999). Inhibition of myosin light chain kinase by p21-activated kinase. *Science* 283, 2083–2085. doi: 10.1126/science.283.5410.2083

Sandquist, J. C., Swenson, K. I., Demali, K. A., Burridge, K., and Means, A. R. (2006). Rho kinase differentially regulates phosphorylation of nonmuscle myosin II isoforms A and B during cell rounding and migration. *J. Biol. Chem.* 281, 35873–35883. doi: 10.1074/jbc.M605343200

Scheeff, E. D., and Bourne, P. E. (2005). Structural evolution of the protein kinase-like superfamily. *PLoS Comput. Biol.* 1:e49. doi: 10.1371/journal.pcbi.0010049

Schlingmann, K. P., Weber, S., Peters, M., Niemann Nejsum, L., Vitzthum, H., Klingel, K., et al. (2002). Hypomagnesemia with secondary hypocalcemia is caused by mutations in TRPM6, a new member of the TRPM gene family. *Nat Genet.* 31, 166–170. doi: 10.1038/ng889

Schramek, D., Sendoel, A., Segal, J. P., Beronja, S., Heller, E., Oristian, D., et al. (2014). Direct *in vivo* RNAi screen unveils myosin IIa as a tumor suppressor of squamous cell carcinomas. *Science* 343, 309–313. doi: 10.1126/science.1248627

Sellers, J. R., Eisenberg, E., and Adelstein, R. S. (1982). The binding of smooth muscle heavy meromyosin to actin in the presence of ATP. Effect of phosphorylation. *J. Biol. Chem.* 257, 13880–13883.

Sharma-Walia, N., Paul, A. G., Bottero, V., Sadagopan, S., Veettil, M. V., Kerur, N., et al. (2010). Kaposi's sarcoma associated herpes virus (KSHV) induced COX-2: a key factor in latency, inflammation, angiogenesis, cell survival and invasion. *PLoS Pathog.* 6:e1000777. doi: 10.1371/journal.ppat.1000777

Shin, D. H., Chun, Y. S., Lee, D. S., Huang, L. E., and Park, J. W. (2008). Bortezomib inhibits tumor adaptation to hypoxia by stimulating the FIH-mediated repression of hypoxia-inducible factor-1. *Blood* 111, 3131–3136. doi: 10.1182/blood-2007-11-120576

Shutova, M., Yang, C., Vasiliev, J. M., and Svitkina, T. (2012). Functions of nonmuscle myosin II in assembly of the cellular contractile system. *PLoS ONE* 7:e40814. doi: 10.1371/journal.pone.0040814

Simons, M., Wang, M., McBride, O. W., Kawamoto, S., Yamakawa, K., Gdula, D., et al. (1991). Human nonmuscle myosin heavy chains are encoded by two genes located on different chromosomes. *Circ. Res.* 69, 530–539. doi: 10.1161/01.RES.69.2.530

Slonska, A., Polowy, R., Golke, A., and Cymerys, J. (2012). Role of cytoskeletal motor proteins in viral infection. *Postepy Hig. Med. Dosw. (Online)* 66, 810–817. doi: 10.5604/17322693.1016360

Smith, A. C., McGavran, L., Robinson, J., Waldstein, G., MacFarlane, J., Zonona, J., et al. (1986). Interstitial deletion of (17)(p11.2p11.2) in nine patients. *Am. J. Med. Genet.* 24, 393–414. doi: 10.1002/ajmg.1320240303

Solinet, S., and Vitale, M. L. (2008). Isoform B of myosin II heavy chain mediates actomyosin contractility during TNFalpha-induced apoptosis. *J. Cell Sci.* 121, 1681–1692. doi: 10.1242/jcs.022640

Somlyo, A. P., and Somlyo, A. V. (2003). Ca²⁺ sensitivity of smooth muscle and nonmuscle myosin II: modulated by G proteins, kinases, and myosin phosphatase. *Physiol. Rev.* 83, 1325–1358. doi: 10.1152/physrev.00023.2003

Stedman, H. H., Kozyak, B. W., Nelson, A., Thesier, D. M., Su, L. T., Low, D. W., et al. (2004). Myosin gene mutation correlates with anatomical changes in the human lineage. *Nature* 428, 415–418. doi: 10.1038/nature02358

Strand, D., Jakobs, R., Merdes, G., Neumann, B., Kalmes, A., Heid, H. W., et al. (1994). The *Drosophila* lethal(2)giant larvae tumor suppressor protein forms homo-oligomers and is associated with nonmuscle myosin II heavy chain. *J. Cell Biol.* 127, 1361–1373. doi: 10.1083/jcb.127.5.1361

Su, L. T., Chen, H. C., Gonzalez-Pagan, O., Overton, J. D., Xie, J., Yue, L., et al. (2010). TRPM7 activates m-calpain by stress-dependent stimulation of p38 MAPK and c-Jun N-terminal kinase. *J. Mol. Biol.* 396, 858–869. doi: 10.1016/j.jmb.2010.01.014

Takizawa, N., Ikebe, R., Ikebe, M., and Luna, E. J. (2007). Supervillin slows cell spreading by facilitating myosin II activation at the cell periphery. *J. Cell Sci.* 120, 3792–3803. doi: 10.1242/jcs.008219

Tan, I., Yong, J., Dong, J. M., Lim, L., and Leung, T. (2008). A tripartite complex containing MRCK modulates lamellar actomyosin retrograde flow. *Cell* 135, 123–136. doi: 10.1016/j.cell.2008.09.018

Tan, J. L., and Spudich, J. A. (1990). *Dictyostelium* myosin light chain kinase. Purification and characterization. *J. Biol. Chem.* 265, 13818–13824.

Tang, H. W., Wang, Y. B., Wang, S. L., Wu, M. H., Lin, S. Y., and Chen, G. C. (2011). Atg1-mediated myosin II activation regulates autophagosome formation during starvation-induced autophagy. *EMBO J.* 30, 636–651. doi: 10.1038/emboj.2010.338

Toothaker, L. E., Gonzalez, D. A., Tung, N., Lemons, R. S., Le Beau, M. M., Arnaout, M. A., et al. (1991). Cellular myosin heavy chain in human leukocytes: isolation of 5′ cDNA clones, characterization of the protein, chromosomal localization, and upregulation during myeloid differentiation. *Blood* 78, 1826–1833.

Totsukawa, G., Wu, Y., Sasaki, Y., Hartshorne, D. J., Yamakita, Y., Yamashiro, S., et al. (2004). Distinct roles of MLCK and ROCK in the regulation of membrane protrusions and focal adhesion dynamics during cell migration of fibroblasts. *J. Cell Biol.* 164, 427–439. doi: 10.1083/jcb.200306172

Trotter, J. A. (1982). Living macrophages phosphorylate the 20,000 Dalton light chains and heavy chains of myosin. *Biochem. Biophys. Res. Commun.* 106, 1071–1077. doi: 10.1016/0006-291X(82)91820-4

Trotter, J. A., Nixon, C. S., and Johnson, M. A. (1985). The heavy chain of macrophage myosin is phosphorylated at the tip of the tail. *J. Biol. Chem.* 260, 14374–14378.

Trybus, K. M., Huiatt, T. W., and Lowey, S. (1982). A bent monomeric conformation of myosin from smooth muscle. *Proc. Natl. Acad. Sci. U.S.A.* 79, 6151–6155. doi: 10.1073/pnas.79.20.6151

Trybus, K. M., and Lowey, S. (1984). Conformational states of smooth muscle myosin. Effects of light chain phosphorylation and ionic strength. *J. Biol. Chem.* 259, 8564–8571.

Tullio, A. N., Accili, D., Ferrans, V. J., Yu, Z. X., Takeda, K., Grinberg, A., et al. (1997). Nonmuscle myosin II-B is required for normal development of the mouse heart. *Proc. Natl. Acad. Sci. U.S.A.* 94, 12407–12412. doi: 10.1073/pnas.94.23.12407

Tullio, A. N., Bridgman, P. C., Tresser, N. J., Chan, C. C., Conti, M. A., Adelstein, R. S., et al. (2001). Structural abnormalities develop in the brain after ablation of the gene encoding nonmuscle myosin II-B heavy chain. *J. Comp Neurol.* 433, 62–74. doi: 10.1002/cne.1125

Underwood, J., Greene, J., and Steimle, P. A. (2010). Identification of a new mechanism for targeting myosin II heavy chain phosphorylation by *Dictyostelium* myosin heavy chain kinase B. *BMC Res. Notes* 3:56. doi: 10.1186/1756-0500-3-56

Uyeda, T. Q., Abramson, P. D., and Spudich, J. A. (1996). The neck region of the myosin motor domain acts as a lever arm to generate movement. *Proc. Natl. Acad. Sci. U.S.A.* 93, 4459–4464. doi: 10.1073/pnas.93.9.4459

Vaillancourt, J. P., Lyons, C., and Cote, G. P. (1988). Identification of two phosphorylated threonines in the tail region of *Dictyostelium* myosin II. *J. Biol. Chem.* 263, 10082–10087.

Vale, R. D. (2003). The molecular motor toolbox for intracellular transport. *Cell* 112, 467–480. doi: 10.1016/S0092-8674(03)00111-9

van, L. H., Elliott, G., and O'Hare, P. (2002). Evidence of a role for nonmuscle myosin II in herpes simplex virus type 1 egress. *J. Virol.* 76, 3471–3481. doi: 10.1128/JVI.76.7.3471-3481.2002

van den Heuvel, M. G., Bolhuis, S., and Dekker, C. (2007). Persistence length measurements from stochastic single-microtubule trajectories. *Nano Lett.* 7, 3138–3144. doi: 10.1021/nl071696y

Varlamova, O., Spektor, A., and Bresnick, A. R. (2001). Protein kinase C mediates phosphorylation of the regulatory light chain of myosin-II during mitosis. *J. Muscle Res. Cell Motil.* 22, 243–250. doi: 10.1023/A:1012289905754

Vasioukhin, V. (2006). Lethal giant puzzle of Lgl. *Dev. Neurosci.* 28, 13–24. doi: 10.1159/000090749

von Muhlen, C. A., Chan, E. K., Peebles, C. L., Imai, H., Kiyosawa, K., and Tan, E. M. (1995). Non-muscle myosin as target antigen for human autoantibodies in patients with hepatitis C virus-associated chronic liver diseases. *Clin. Exp. Immunol.* 100, 67–74. doi: 10.1111/j.1365-2249.1995.tb03605.x

Walder, R. Y., Landau, D., Meyer, P., Shalev, H., Tsolia, M., Borochowitz, Z., et al. (2002). Mutation of TRPM6 causes familial hypomagnesemia with secondary hypocalcemia. *Nat. Genet.* 3, 171–174. doi: 10.1038/ng901

Wang, L., Guo, D. C., Cao, J., Gong, L., Kamm, K. E., Regalado, E., et al. (2010). Mutations in myosin light chain kinase cause familial aortic dissections. *Am. J. Hum. Genet.* 87, 701–707. doi: 10.1016/j.ajhg.2010.10.006

Wasylnka, J. A., Bakowski, M. A., Szeto, J., Ohlson, M. B., Trimble, W. S., Miller, S. I., et al. (2008). Role for myosin II in regulating positioning of *Salmonella*-containing vacuoles and intracellular replication. *Infect. Immun.* 76, 2722–2735. doi: 10.1128/IAI.00152-08

Weir, L., and Chen, D. (1996). Characterization of the nonmuscle myosin heavy chain IIB promoter: regulation by E2F. *Gene Expr.* 6, 45–57.

Wendt, T., Taylor, D., Trybus, K. M., and Taylor, K. (2001). Three-dimensional image reconstruction of dephosphorylated smooth muscle heavy meromyosin reveals asymmetry in the interaction between myosin heads and placement of subfragment 2. *Proc. Natl. Acad. Sci. U.S.A.* 98, 4361–4366. doi: 10.1073/pnas.071051098

Xia, D., Stull, J. T., and Kamm, K. E. (2005). Myosin phosphatase targeting subunit 1 affects cell migration by regulating myosin phosphorylation and actin assembly. *Exp. Cell Res.* 304, 506–517. doi: 10.1016/j.yexcr.2004.11.025

Xia, Z. K., Yuan, Y. C., Yin, N., Yin, B. L., Tan, Z. P., and Hu, Y. R. (2012). Nonmuscle myosin IIA is associated with poor prognosis of esophageal squamous cancer. *Dis. Esophagus* 25, 427–436. doi: 10.1111/j.1442-2050.2011.01261.x

Xu, J., Gao, X. P., Ramchandran, R., Zhao, Y. Y., Vogel, S. M., and Malik, A. B. (2008). Nonmuscle myosin light-chain kinase mediates neutrophil transmigration in sepsis-induced lung inflammation by activating beta2 integrins. *Nat. Immunol.* 9, 880–886. doi: 10.1038/ni.1628

Xu, X. S., Kuspa, A., Fuller, D., Loomis, W. F., and Knecht, D. A. (1996). Cell-cell adhesion prevents mutant cells lacking myosin II from penetrating aggregation streams of *Dictyostelium*. *Dev. Biol.* 175, 218–226. doi: 10.1006/dbio.1996.0109

Yumura, S., Yoshida, M., Betapudi, V., Licate, L. S., Iwadate, Y., Nagasaki, A., et al. (2005). Multiple myosin II heavy chain kinases: roles in filament assembly control and proper cytokinesis in *Dictyostelium*. *Mol. Biol. Cell* 16, 4256–4266. doi: 10.1091/mbc.E05-03-0219

Zamudio-Meza, H., Castillo-Alvarez, A., Gonzalez-Bonilla, C., and Meza, I. (2009). Cross-talk between Rac1 and Cdc42 GTPases regulates formation of filopodia required for dengue virus type-2 entry into HMEC-1 cells. *J. Gen. Virol.* 90, 2902–2911. doi: 10.1099/vir.0.014159-0

Conflict of Interest Statement: The author declares that the research was conducted in the absence of any commercial or financial relationships that could be construed as a potential conflict of interest.

12

The dehydrogenase region of the NADPH oxidase component Nox2 acts as a protein disulfide isomerase (PDI) resembling PDIA3 with a role in the binding of the activator protein p67^{phox}

Edna Bechor, Iris Dahan, Tanya Fradin, Yevgeny Berdichevsky, Anat Zahavi, Aya Federman Gross, Meirav Rafalowski and Edgar Pick*

The Julius Friedrich Cohnheim Laboratory of Phagocyte Research, Department of Clinical Microbiology and Immunology, Sackler School of Medicine, Tel Aviv University, Tel Aviv, Israel

Edited by:
Bulent Mutus, University of Windsor, Canada

Reviewed by:
Elisa Pagnin, University of Padova, Italy
Lei Wang, Chinese Academy of Sciences, China

*Correspondence:
Edgar Pick, Department of Clinical Microbiology and Immunology, Sackler School of Medicine, Rooms 832/833, Sackler School of Medicine Building, Ramat Aviv University Campus, Tel Aviv University, Tel Aviv 69978, Israel
e-mail: epick@post.tau.ac.il

The superoxide (O_2^{-})-generating NADPH oxidase of phagocytes consists of a membrane component, cytochrome b_{558} (a heterodimer of Nox2 and p22^{phox}), and four cytosolic components, p47^{phox}, p67^{phox}, p40^{phox}, and Rac. The catalytic component, responsible for O_2^{-} generation, is Nox2. It is activated by the interaction of the dehydrogenase region (DHR) of Nox2 with the cytosolic components, principally with p67^{phox}. Using a peptide-protein binding assay, we found that Nox2 peptides containing a ^{369}CysGlyCys^{371} triad (CGC) bound p67^{phox} with high affinity, dependent upon the establishment of a disulfide bond between the two cysteines. Serially truncated recombinant Nox2 DHR proteins bound p67^{phox} only when they comprised the CGC triad. CGC resembles the catalytic motif (CGHC) of protein disulfide isomerases (PDIs). This led to the hypothesis that Nox2 establishes disulfide bonds with p67^{phox} via a thiol-dilsulfide exchange reaction and, thus, functions as a PDI. Evidence for this was provided by the following: (1) Recombinant Nox2 protein, which contained the CGC triad, exhibited PDI-like disulfide reductase activity; (2) Truncation of Nox2 C-terminal to the CGC triad or mutating C369 and C371 to R, resulted in loss of PDI activity; (3) Comparison of the sequence of the DHR of Nox2 with PDI family members revealed three small regions of homology with PDIA3; (4) Two monoclonal anti-Nox2 antibodies, with epitopes corresponding to regions of Nox2/PDIA3 homology, reacted with PDIA3 but not with PDIA1; (5) A polyclonal anti-PDIA3 (but not an anti-PDIA1) antibody reacted with Nox2; (6) p67^{phox}, in which all cysteines were mutated to serines, lost its ability to bind to a Nox2 peptide containing the CGC triad and had an impaired capacity to support oxidase activity *in vitro*. We propose a model of oxidase assembly in which binding of p67^{phox} to Nox2 via disulfide bonds, by virtue of the intrinsic PDI activity of Nox2, stabilizes the primary interaction between the two components.

Keywords: protein disulfide isomerase, superoxide, NADPH oxidase, Nox2, p67^{phox}, synthetic peptides, Cys-Gly-Cys triad, disulfide bonds

INTRODUCTION

Oxygen-derived radicals are the principal mediators responsible for the killing of pathogenic microorganisms by phagocytes. All oxygen radicals produced by phagocytes are derived from the superoxide anion (O_2^{-}), generated by the NADPH-derived one-electron reduction of molecular oxygen. The process is catalyzed by a membrane-imbedded 91 kDa flavoprotein, known as Nox2, which is associated with a second protein of 22 kDa (p22^{phox}), to form the flavocytochrome b_{558} heterodimer. Nox2 is 570 residues-long and comprises six transmembrane α-helices linked by three outside-facing loops and two cytosol-facing loops, and a cytosolic segment, extending from residue 288 to 570. Nox2 contains all redox stations supporting the flow of electrons from NADPH to oxygen. These are an NADPH-binding site and non-covalently bound FAD, present in the cytosolic segment, and two non-identical hemes present in the third and fifth membrane helices (reviewed in Quinn and Gauss, 2004).

Abbreviations: CGC, ^{357}CysGlyCys^{371}; CGD, chronic granulomatous disease; DE-GSSG, dieosin glutathione disulfide; DTT, dithiothreitol; DHR, dehydrogenase region; EC$_{50}$, half maximal effective concentration; FPLC, fast protein liquid chromatography; GSG, glutathione disulfide; HMBA, 4-(hydroxymercuri)benzoic acid; IC$_{50}$, half maximal inhibitory concentration; LiDS, lithium dodecyl sulfate; mBBr, monobromobimane; NEM, *N*-ethylmaleimide; O_2^{-}, superoxide; PAO, phenylarsine oxide; PBS, phosphate buffered saline pH 7.3; PCMB, *p*-chloromercuribenzoate; PDI, protein disulfide isomerase; ROS, reactive oxygen species; V_{max}, maximal velocity.

In the resting phagocyte, no electron flow occurs along the redox centers in Nox2. The initiation of the electron flow is mediated by a conformational change in Nox2, resulting from the interaction with regulatory proteins present in the cytosol. The regulatory cytosolic components are p47^{phox}, p67^{phox}, p40^{phox}, and the small GTPase Rac (1 or 2) and they translocate to the membrane environment of Nox2, to generate the NADPH oxidase complex (briefly, oxidase), a process known as oxidase assembly (reviewed in Groemping and Rittinger, 2005). In the intact cell, p47^{phox}, p67^{phox}, and Rac are all required for the induction of O_2^- production but it is yet unsettled whether direct interaction of all components with Nox2 is required. There is evidence for p67^{phox} being the key component responsible for the causation of a conformational remodeling of Nox2 (Kreck et al., 1996; Gorzalczany et al., 2000). Major unsolved issues are the identities of region(s) in Nox2 and p67^{phox} participating in the interaction among the two. It has been found that an "activation domain" comprising residues 199–210 (Han et al., 1998) or a wider region, extending from residue 190 to 208 (Sumimoto, 2008) in p67^{phox} is essential for oxidase activation but not for the actual p67^{phox}–Nox2 interaction. Experimental evidence related to the latter suggests that the binding site for Nox2 is located N-terminal to residue 198 (Dang et al., 2002) and, most likely, between residues 187 and 198 or 194 and 198 (Federman Gross et al., 2012).

The cytosolic segment of Nox2 is also known as the dehydrogenase region (DHR) by virtue of the fact that it contains the NADPH- and FAD-binding sites and is homologous to the prokaryotic protein ferredoxin-NADP$^+$ reductase. The DHR comprises binding sites for p47^{phox} and Rac but, so far, there is no solid evidence for the identity of the binding site(s) for p67^{phox}. We approached this question by applying the "peptide walking" methodology (Joseph and Pick, 1995; Dahan et al., 2012; Dahan and Pick, 2012). Overlapping 15-mer peptides, corresponding to the DHR of Nox2 (residues 288–570) were attached to 96-well plates and reacted with recombinant 6His-p67^{phox} in the fluid phase; peptide-bound p67^{phox} was detected by peroxidase-conjugated anti-polyHis antibody. It was found that p67^{phox} binds preferentially to two peptides, corresponding to residues 357–371 (termed Nox2 peptide 24) and 369–383 (termed Nox2 peptide 28) (Dahan and Pick, manuscript in preparation). The peptides share a ^{369}CysGlyCys^{371} (CGC) triad, located at the C-terminus of peptide 24 and the N-terminus of peptide 28. The CGC triad is present in the DHR of Nox2 of all species, down to amphibians, and is absent in Nox1, 3, 4, and 5 (Kawahara et al., 2007). Peptides derived from Nox4, corresponding to Nox2 peptides 24 and 28 by sequence alignment but lacking the CGC triad, did not bind p67^{phox}. It is of interest that Nox4 generates oxygen radicals constitutively, without a requirement for cytosolic activators, such as p67^{phox} (Bedard and Krause, 2007).

Replacing C369 or C371 with Arg or Ser abolished binding of p67^{phox} to peptides 24 and 28. A 369Cys to Arg mutation in Nox2 causes chronic granulomatous disease (CGD) of the X91$^+$ form, with normal expression of Nox2 but impaired production of O_2^-, impaired translocation of cytosolic components, and low FAD binding (Leusen et al., 2000; Debeurme et al., 2010).

We next found that the introduction of an intramolecular disulfide bond between C369 and C371 in Nox2 peptides 24 and 28 resulted in a marked increase in the binding of p67^{phox}. Reduction of the disulfide bond and alkylation of the reduced thiols totally abolished binding of p67^{phox} (Fradin et al., 2011, 2012; Pick, 2012; Fradin and Pick, manuscript in preparation). An important observation was that enhanced binding of p67^{phox} was evident only when the disulfide bond was established between two non-adjacent cysteines and between cysteines present in the same peptide; when the CGC triad was replaced by CCG and a disulfide bond established between the adjacent cysteines or the disulfide bond linked C369 or C371 on two peptides, forming a dimer, no enhanced binding of p67^{phox} was found.

These observations are to be related to a large body of early work by several groups showing that thiol alkylating agents interfere with oxidase activation in intact phagocytes and in in vitro systems. Thus, N-ethylmaleimide (NEM) inhibited oxidase activation in intact phagocytes, if added before activation (Cohen and Chovaniec, 1978; Yamashita et al., 1984) and in vitro (Shpungin et al., 1989) and was shown to act on a membrane component (Shpungin et al., 1989). Similar results were obtained with 4-(hydroxymercuri)benzoic acid [HMBA, known in the past as p-chloromercuribenzoate (PCMB)]. A solubilized membrane-associated fraction, prepared from stimulated neutrophils, was inactivated by HMBA (Bellavite et al., 1983; Gabig and Lefker, 1984) and macrophage membrane liposomes treated with HMBA were inactive in a cell-free oxidase activation system (Pick et al., 1987). Less defined evidence exists for cysteines being the target for the oxidase inhibitors apocynin, VAS2870, and ebselen, and the specific components affected were not rigorously identified.

More recent testimony for the involvement of Nox2-localized cysteines in oxidase function was provided by the discovery of the marked inhibitory effect of phenylarsine oxide (PAO), an agent which reacts with vicinal dithiols to form a disulfide-like complex (van Iwaarden et al., 1992). PAO was found to inhibit oxidase activation in neutrophils and in vitro, when preincubated with the membrane fraction, a reaction that was reversed by reducing agents (Le Cabec and Maridonneau-Parini, 1995). PAO was proposed to act on the pair of cysteines 369 and 371 in Nox2 (Doussiere et al., 1998). More support for the role of vicinal cysteines in Nox2 in oxidase activation was offered by the inhibitory effect of the fungal toxin gliotoxin (Waring and Beaver, 1996), which contains a bridged disulfide ring capable of forming mixed disulfides with thiol-containing proteins. Gliotoxin was reported to inhibit oxidase activation in neutrophils (Tsunawaki et al., 2004) and in vitro, with the vicinal cysteines 369 and 371 in Nox2, considered as likely targets (Nishida et al., 2005). Gliotoxin did not affect Nox4 (Serrander et al., 2007), in good agreement with the facts that Nox4 does not contain the CGC triad and does not require cytosolic assistance for generation of reactive oxygen species (ROS).

On considering our results on the binding of p67^{phox}–Nox2 peptides containing the CGC triad, in the disulfide form, and the accumulated evidence in support of a role for these cysteines in oxidase assembly, we reasoned that, at a certain stage in the process of oxidase assembly, the intramolecular disulfide bond in Nox2 might be converted to an intermolecular disulfide bond between Nox2 and cysteines in p67^{phox} by a thiol—disulfide exchange reaction. It is likely that the primary interaction between the Nox2 DHR and p67^{phox} is based on specific binding sites in the two partners and does not involve disulfide bonds. The establishment of disulfide bonds between cysteines in the Nox2 CGC triad and cysteines in p67^{phox} is a secondary event with a stabilizing role.

It is our hypothesis that Nox2 serves as an endogenous protein disulfide isomerase (PDI), when the cysteines in the CGC triad are in the disulfide form. PDIs are multi-domain proteins belonging to the thioredoxin superfamily (reviewed in Collet and Messens, 2010) and to the PDI gene family, which comprises 21 members, varying in size, domain composition and tissue expression (reviewed in Ellgaard and Ruddock, 2005; Appenzeller-Herzog and Ellgaard, 2008; Galligan and Petersen, 2012; Ali Khan and Mutus, 2014). PDIs can catalyze thiol—disulfide oxidation and reduction and disulfide rearrangement (isomerization) and also function as chaperones. PDIs contain two thioredoxin-like catalytic domains, with a characteristic CXXC active site motif. This is CGHC, in most PDIs, as opposed to the CGPC sequence, typical of thioredoxin.

The proposal that Nox2 acts as a PDI is backed by the following body of evidence: (a) The CGC triad closely mimics the two CGHC catalytic motifs of PDI; thus CGC has a disulfide reduction potential (E°′) of −167 mV, which is very close to that of the CGHC motif of PDI (−180 mV) (Woycechowsky and Raines, 2003); (b) The disulfide bond in CGC is relatively unstable with high loop opening (k_o) and closing (k_c) constants, a property favorable to PDI activity (Zhang and Snyder, 1989; Kersteen and Raines, 2003); (c) Small peptides CGC and RKCGC, in the disulfide form, can mimic the actions of PDI protein (Kersteen and Raines, 2003; Wang et al., 2011), and (d) The CGC triad resembles similar motifs appearing in a large family of eukaryotic and prokaryotic enzymes known as thiol-disulfide oxidoreductases, which contain one or more di-cysteine motifs and, sometimes, a flavin cofactor (Sevier and Kaiser, 2002).

In the present work we present the experimental evidence in support of the hypothesis that the DHR of Nox2 functions as a PDI, leading to establishment of disulfide bonds with p67^{phox} and assuring the stabilization of the NADPH oxidase complex. This conclusion is based on the use of a wide range of methodologies, comprising the generation of recombinant DHR of Nox2 and p67^{phox} and mutants of these, protein - protein and peptide—protein binding assays, an enzymatic PDI reductase assay, immunologic characterization, the use of PDI inhibitors, and bioinformatics.

MATERIALS AND METHODS
SYNTHETIC PEPTIDES
Synthetic peptides corresponding to selected sequence segments of Nox2 were made by two companies: Mimotopes (Clayton, Victoria, Australia), and Bachem (Bubendorf, Switzerland). Peptides were 15-residues-long, with a biotin tag at the N-terminus (attached by a Ser-Gly-Ser-Gly spacer) and a C-terminal amide. Peptides were of a purity of at least 70%, documented by reversed phase chromatography, and the size, confirmed by MALDI-mass spectroscopy. Peptides were dissolved in a mixture

of 75 parts 1-methyl-2-pyrrolidone and 25 parts water (v/v), to a concentration of 1.5 mM, to serve as stock solutions for further dilution, and kept frozen at −75°C in small aliquots. Working solutions were freshly prepared on the day of performing the experiments.

CHEMICALS
Common laboratory chemicals, at the highest purity available, were purchased from Merck KGaA (Darmstadt, Germany), or Sigma-Aldrich (St. Louis, MO, USA). The following potential PDI inhibitors were used: phenylarsine oxide (Sigma-Aldrich); bacitracin (mixture of 9 bacitracins, mainly type A; Sigma-Aldrich); gliotoxin (from *Gliocladium fimbriatum*, Fermentek, Jerusalem, Israel); PDI inhibitor 16F16 (Sigma-Aldrich); ribonuclease A, with scrambled disulfide bonds from bovine pancreas (Sigma-Aldrich), rutin hydrate (Sigma-Aldrich), and PDI inhibitor III, PACMA 31 (Merck). Dieosin glutathione disulfide (DE-GGSG), used in the PDI reductase assay, was a kind gift of Dr. Bulent Mutus (University of Windsor, Ontario, Canada). Nickel Sepharose 6 Fast Flow was obtained from GE Healthcare Bio-Sciences AB, Uppsala, Sweden. Monobromobimane Fluoropure grade (mBBr) was obtained from Molecular Probes, Life Technologies, Thermo Fisher Scientific, Waltham, MA, USA.

RECOMBINANT PDI
Recombinant PDI A1 (human; >95% pure; Cat. No. enz-262), made in *E. coli*, comprising a 12-His tag at the N-terminus, and recombinant PDI A3 (human; >95% pure; Cat. No. enz-474), made in *E. coli*, comprising residues 25–505 and a 37-His tag at the N-terminus, were obtained from ProSpec-Tany TechnoGene (Ness Ziona, Israel).

RECOMBINANT CYTOSOLIC NADPH OXIDASE COMPONENTS
Recombinant p67^{phox}, p47^{phox}, and Rac1 (Q61L mutant), with an N-terminal 6His tag, were expressed in *E. coli* and purified on Nickel Sepharose, as described before (Mizrahi et al., 2010). Recombinant non-prenylated Rac1 was prenylated *in vitro* by recombinant mammalian geranylgeranyltransferase type I (a gift of Dr. Carolyn Weinbaum, Duke University), as described before (Gorzalczany et al., 2002).

GENERATION OF RECOMBINANT p67^{phox} WITH 4 Cys TO Ser MUTATIONS
We generated recombinant p67^{phox}(1–212), in which cysteines 40, 45, 121, and 165 were mutated to serine. To construct a 6His-tagged protein in which the four cysteines were concurrently changed to serine, we used a synthetic gene encoding p67^{phox} amino acids 1–212, comprising the four mutations in plasmid pIDTSmart AMP (Integrated DNA Technologies, Coralville, IA, USA). This was subcloned into the *Bam*HI-*Eco*RI restriction sites of plasmid pET-30a-His$_6$, as described for wild-type p67^{phox} (Mizrahi et al., 2010). The mutant protein was expressed in *E. coli* and purified on Nickel Sepharose, essentially as described for wild-type p67^{phox}. The only deviation from the procedure applied to wild-type p67^{phox} was the supplementation of all buffers used for washing the Nickel Sepharose beads, and for binding and eluting the protein, with 0.2% (v/v) Triton X-100 (Sigma-Aldrich),

with the purpose of preventing hydrophobic interaction with Nickel Sepharose. Such interaction was more pronounced with the mutant protein.

PHAGOCYTE MEMBRANES

Phagocyte membranes were prepared from guinea pig peritoneal macrophages elicited by the injection of mineral oil (Bromberg and Pick, 1984). The membranes were solubilized in 40 mM n-octyl ß-D-glucopyranoside and reconstituted into liposomes by dialysis against detergent-free buffer as described previously (Shpungin et al., 1989). The specific cytochrome b_{558} heme content of membrane vesicles was measured by the difference spectrum of sodium dithionite-reduced *minus* oxidized samples (Pick et al., 1987).

PREPARATION OF NusA-Nox2 FUSION PROTEINS

Recombinant fusion proteins linking parts of the DHR of human Nox2 (residues 288–570) with the *E. coli* protein NusA were constructed in order to obtain soluble Nox2 preparations. Recombinant NusA was also made, to serve as a negative control. The method was based on the ability of NusA to markedly improve the solubility of the fusion partner while conserving biological activity (Davis et al., 1999). Details of the construction of the fusion proteins are described under "Construction of NusA-Nox2 fusion proteins" in Supplementary Material. Fusion proteins were constructed to contain C-terminal to NusA the following Nox2 truncations: residues 328–570, 357–570, 372–570, 387–570, 408–570, 444–570, and 462–570. In addition, the fusion protein NusA-Nox2(357–570) was subjected to mutagenesis, in which Cys^{369} and Cys^{371} were mutated to Arg. The mutants were constructed using the "QuickChange Site-Directed Mutagenesis Kit" (Agilent Technologies, Santa Clara, CA, USA). Plasmid pET-43a-Nox2(357–570) was used as a template. The integrity of the mutant genes was confirmed by DNA sequencing. The fusion proteins and NusA were expressed in *E. coli* and purified on Nickel Sepharose, as described for the recombinant cytosolic NADPH oxidase components (Mizrahi et al., 2010). The NusA-Nox2 fusion proteins and NusA were further purified by fast protein liquid chromatography (FPLC) gel filtration on a HiLoad 16/60 Superdex 200 prep grade column (GE Healthcare Bio-Sciences AB), as described before (Mizrahi et al., 2010). This procedure separated a polymeric fraction from a homodimer fraction (197 kDa in size for NusA-Nox2(357–570), and 145 kDa, for NusA). The purified homodimers were used in enzymatic and immunologic assays.

DETERMINATION OF PROTEIN CONCENTRATION AND PURITY

The protein concentration of the recombinant proteins was measured by the method of Bradford (1976), modified for use with 96-wells microplates, using Bio-Rad protein assay dye reagent concentrate (Bio-Rad Laboratories, Hercules, CA, USA) and bovine γ-globulin as a standard. The level of purity of the recombinant proteins was assessed by SDS-PAGE analysis.

SDS-PAGE

SDS-PAGE was performed in a XCell *SureLock* Mini-Cell, using precast 12%, 1-mm thick, 10 well, NuPAGE Bis-Tris gels, and NuPAGE MOPS SDS running buffer. Run was at a constant voltage of 200 V for 50 min. All items were obtained from Novex, Life Technologies, Carlsbad, CA, USA, and the manufacturer's instructions were followed. Molecular weight standards (Precision Plus, All Blue, range 250–10 kDa, prestained) were purchased from Bio-Rad. The gels were stained by Instant Blue (Expedeon, Harston, UK).

ANTIBODIES

The following anti-human Nox2 antibodies were obtained from Santa Cruz Biotechnology (Santa Cruz, CA, USA): anti-Nox2 54.1 (mouse monoclonal, recognizing epitope 381–390 in the cytosolic segment of Nox2; Burritt et al., 1995; Baniulis et al., 2005); anti-Nox2 NL7 (mouse monoclonal, recognizing epitope 498–506 in the cytosolic segment of Nox2; Burritt et al., 2003), and anti-Nox2 C-15 (goat polyclonal raised against a peptide in the C-terminal region of Nox2; sequence is unknown). The following anti-human PDIA3 antibodies were used in this study: anti-PDIA3 H-220 (rabbit polyclonal, raised against residues 108–207; Santa Cruz); anti-PDIA3 HPA003230 (rabbit polyclonal, raised against residues 101–218; Sigma-Aldrich, Prestige Antibodies); anti-PDIA3 NBP-84797 (rabbit polyclonal, raised against residues 246–377; Novus Biologicals, Littleton, CO, USA), and anti-PDIA3 ABE1032 (rabbit polyclonal, raised against whole human recombinant PDIA3; Merck). Anti-human PDA1 antibody H-160 (rabbit polyclonal, raised against residues 211–370) was obtained from Santa Cruz. Anti-polyhistidine antibody A7058 (mouse monoclonal, peroxidase conjugate) was obtained from Sigma-Aldrich. Anti-Nus-Tag antibody (mouse monoclonal, with high affinity for NusA protein for detection of fusion proteins containing the Nus-Tag expressed with the pET-41.1 vector) was obtained from Novagen, EMD Chemicals, Merck. Second anti-mouse IgG (A3562), anti-rabbit IgG (A3687), and anti-goat IgG (A7650) antibodies, all conjugated with alkaline phosphatase, were obtained from Sigma-Aldrich. Rabbit anti-mouse IgG, conjugated with peroxidase (Cat. No. 315-035-003) was obtained from Jackson ImmunoResearch, West Grove, PA, USA.

IMMUNOBLOTTING

Immunoblotting was performed essentially as described before (Knoller et al., 1991), with a number of modifications. The blocking buffer consisted of NaCl (136 mM), Tris-HCl (25 mM), and KCl (2.68 mM), pH 7.5, supplemented 3 g% of bovine serum albumin (A4053, 96%, Sigma-Aldrich) and 0.1% (v/v) Triton X-100. A Pierce G2 Fast Blotter semi-dry transfer apparatus and 1-Step Transfer Buffer (Prod. No. 84731) were used, following the manufacturer's instructions (Thermo Fisher Scientific). Second antibodies were alkaline phosphatase conjugates and bands were detected by direct color development, using SIGMAFAST BCIP-NBT reagent (B5655, Sigma-Aldrich).

PROTEIN—PROTEIN BINDING ASSAY

This assay was used to assess the binding of NusA-Nox2 fusion proteins to p67^{phox}. p67^{phox} was diluted in phosphate buffered saline (PBS) (consisting of 137 mM NaCl, 2.7 mM KCl, 4.3 mM Na_2PO_4, and 1.4 mM KH_2PO_4, pH 7.3) to a concentration of

0.5 μM. 200 μl amounts were added to the wells of a 96-well plate (Immulon 4 HBX ultra-high binding flat bottom polystyrene, Cat. No. 3855, Thermo Labsystems, Helsinki, Finland). p67^{phox} was allowed to attach to the wells for 16–18 h at 4°C, the well contents being mixed on an orbital shaker. Unattached protein was removed by repeated washing with PBS supplemented with 0.1% v/v Tween 20 (Sigma-Aldrich), using a Wellwash Versa microplate washer (Thermo Fisher Scientific). Next, the surface-attached p67^{phox} was exposed to 200 μl/well of NusA-Nox2 or NusA, diluted to a concentration of 0.5 μM in PBS supplemented with 0.1% Tween 20 and 1% w/v casein, sodium salt (Sigma-Aldrich), for 1 h at room temperature, the well contents being mixed on an orbital shaker. After repeated washing, we added 200 μl/well anti-NusA antibody, diluted 1/2500 in in PBS supplemented with 0.1% Tween 20 and 1% w/v casein, for 1 h at room temperature with shaking. This was followed, after further washing, by 200 μl/well of peroxidase conjugated anti-mouse IgG, diluted 1/2500 in PBS supplemented with 0.1% Tween 20 and 1% w/v casein, and incubation for 1 h at room temperature, with shaking. After washing off the second antibody, bound peroxidase was quantified by adding 200 μl/well of tetramethyl benzidine + substrate chromogen reagent (DakoCytomation, Glostrup, Denmark) and measuring the increase in absorbance at 650 nm over time, for 10 min (Dahan and Pick, manuscript in preparation). Absorbance was measured in a Spectramax 340 microplate reader, in the kinetic mode (Molecular Devices, Sunnyvale, CA, USA) fitted with SoftMax Pro 5.2 software. Results were expressed as mAbs at 650 nm per min.

PEPTIDE—PROTEIN BINDING ASSAY

This assay was used to assess the binding of p67^{phox}, in solution, to surface-attached synthetic Nox2 peptides. The procedure is a modified version of that described before (Morozov et al., 1998; Dahan et al., 2002). A brief description of the procedure is presented here. Synthetic Nox2 peptides, with a biotin tag at either the N- or C-terminus, were diluted to a concentration of 1 μM in PBS supplemented with 1% casein. 200 μl/well volumes were added to streptavidin-coated 96 well plates (BioBind Assembly, streptavidin-coated, Cat. No. 95029263, Thermo Fisher Scientific) and the plates incubated for 1 h at room temperature, the well contents being mixed on an orbital shaker. After removing the unattached peptide by repeated washing with PBS supplemented with 1% v/v Tween 20, the wells were filled with 200 μl volumes of p67^{phox} at a concentration of 1.5 μM in PBS, supplemented with 1% casein, and the plates kept for 16–18 h at at 4°C with shaking on an orbital shaker. After removal of unattached p67^{phox} by repeated washing, 200 μl/well of a 1/3000 dilution of peroxidase-conjugated anti-polyhistidine antibody in PBS, supplemented with 0.1% Tween 20 and 1% w/v casein, were added and the plate incubated for 1 h at room temperature, with shaking. Bound peroxidase was quantified as described for the protein—protein binding assay.

MEASURING PDI REDUCTASE ACTIVITY

The various methods of assaying PDI activity were recently reviewed (Watanabe et al., 2014). Preliminary experiments were performed using the turbidimetric assay based on the reduction

of disulfides in oxidized insulin (Holmgren, 1979; Martinez-Galisteo et al., 1993) but this was abandoned when it was found not to be sufficiently sensitive for assessing the PDI activity of NusA-Nox2 fusion proteins. Consequently, we opted for a highly sensitive fluorescent method of measuring the disulfide reductase activity of PDI (Raturi and Mutus, 2007). The principle of this method is the use of DE-GSSG, in which two eosin molecules, covalently attached to glutathione disulfide (GSSG), exhibit fluorescent self quenching due to their proximity. On addition of PDI, in the presence of a minimal concentration of dithiothreitol (DTT), the disulfide bond in GSSG is reduced and the separation of the eosin molecules results in a ~70-fold increase in fluorescence. We first tested the feasibility of using this assay by measuring the activity of recombinant PDIA1 and PDIA3, following the assay conditions of Raturi and Mutus (2007). The assay mixture consisted of various concentrations of PDI in 1 ml volumes of 0.1 M potassium phosphate buffer, pH 7.0 supplemented with 2 mM EDTA, 150 nM DE-GSSG, and 5 μM DTT. Control mixtures consisted of PDI, in the absence of DTT, and of DTT, in the absence of PDI. The kinetics of fluorescence was initially measured in an FP-750 spectrofluorometer (JASCO Corporation, Tokyo, Japan), with excitation at 519 nm and emission, at 539 nm, in a stirred cuvette, for 15–30 min in the time course mode. The necessity to perform numerous measurements simultaneously and repeated measurements on the same sample led us to switch to the use of 96 well plates and a Gemini XPS microplate spectrofluorometer (Molecular Devices) fitted with SoftMax Pro 5.2 software. The assays were performed in Microfluor 1 Black flat bottom low background plates (Part 7605, Thermo Labsystems, Franklin, MA, USA). To each well were added, in a 200 μl volume, an assay mixture consisting of various concentrations of PDI in 0.1 M potassium phosphate buffer, pH 7.0 supplemented with 2 mM EDTA, 0.8 μM DE-GSSG, and 12.5 μM DTT. The kinetics of fluorescence was measured, with excitation at 519 nm and emission, at 545 nm, for 30 min, at 24°C, with six readings on the same well and mixing of the wells between readings. Results were expressed as V_{max} (milli relative fluorescence units per min), following selection of the linear segment of the increase in fluorescence curve.

ASCERTAINING THE ABSENCE OF CYSTEINES IN THE p67^{phox} MUTANT PROTEIN

The conversion of cysteines to serines in the p67^{phox} mutant was confirmed at the stage of constructing the mutant by DNA sequencing. We, nevertheless, ascertained the presence of cysteines in wild-type p67^{phox}(1–212) and their absence in the mutant, by the binding or absence of binding of the thiol reagent monobromobimane (mBBR). mBBR is non-fluorescent when free and becomes fluorescent upon binding to thiol groups in proteins (Kosower and Kosower, 1995).

CELL-FREE NADPH OXIDASE ACTIVATION ASSAYS

Two variations of the cell-free NADPH oxidase assay were used. The canonical assay, known as the amphiphile- and p47^{phox}-dependent assay, involves the participation of phagocyte membranes (as a source of cytochrome b_{558}), p47^{phox}, p67^{phox}, Rac1 in the GTP-bound form, and an activating anionic amphiphile, such as arachidonate or lithium dodecyl sulfate (LiDS) (Bromberg and Pick, 1984, 1985). The second variation, known as the amphiphile- and p47^{phox}-independent assay, involves the participation of phagocyte membranes, p67^{phox}, and Rac1 and does not require an amphiphilic activator (Gorzalczany et al., 2000). Detailed descriptions of both methodologies have been published (Pick, 2014). In the vast majority of assays, NADPH oxidase activity was quantified by the production of superoxide (O$_2^{\cdot-}$), measured by the reduction of oxidized cytochrome c. In some cases, when compounds, such as certain PDI inhibitors, interfered with cytochrome c reduction, we used NADPH consumption, as an alternative (Sha'ag, 1989). Results were expressed as turnover values (mol O$_2^{\cdot-}$/s/mol cytochrome b_{558} heme).

EFFECT OF PDI INHIBITORS ON CELL-FREE NADPH OXIDASE AVTIVATION

A number of agents described as PDI inhibitors were tested for an effect on cell-free NADPH oxidase activation. Based on the hypothesis that the target of such inhibitors is located in the phagocyte membrane and, specifically, in Nox2, the cell-free assays were designed in a manner to allow pre-incubation of the inhibitors with the phagocyte membranes for 5 min, at room temperature. This was followed by the addition of p47^{phox}, p67^{phox}, Rac1 and LiDS (in the canonical assay), or p67^{phox} and prenylated Rac1 (in the amphiphile- and p47^{phox}-independent assay) and further incubation for 90 s or 5 min, respectively.

GRAPH PLOTTING

Plotting of graphs and kinetic analyses were executed by using GraphPad Prism Version 6.05 (GraphPad Software, San Diego, CA, USA).

RESULTS

EXCHANGING THE CGC TRIAD IN Nox2 PEPTIDE 24 WITH CGHC OR CGPC CONSERVES BUT DOES NOT AUGMENT THE ENHANCEMENT OF p67^{phox} BINDING

We found that two synthetic 15-mer peptides, derived from the DHR of Nox2, which share a CGC triad at either the C- or N-terminus (designated peptides 24 and 28, respectively) bind full-length (1–526) and truncated (1–212) p67^{phox} with low affinity and the introduction of an intramolecular disulfide bond linking cysteines 369 and 371 leads to a marked increase in the binding of p67^{phox}. Based on the resemblance of the CXC motif with the CXXC motif characteristic of the catalytic motifs of PDI and thioredoxin, we synthesized analogs of Nox2 peptide 24 (IVGDWTEGLFNACGC) in which the CGC sequence was replaced by either CGHC (the catalytic motif of PDI) or CGPC (the catalytic motif of thioredoxin). Both peptide analogs were prepared in the reduced and the disulfide forms and examined for the ability to bind p67^{phox}, in comparison with the unmodified peptide 24. As apparent in **Figure 1**, the introduction of a disulfide bond in both modified peptides (CGHC and CGPC) caused a significant increase in p67^{phox} binding but the levels of binding were inferior to those measured with the disulfide form of the

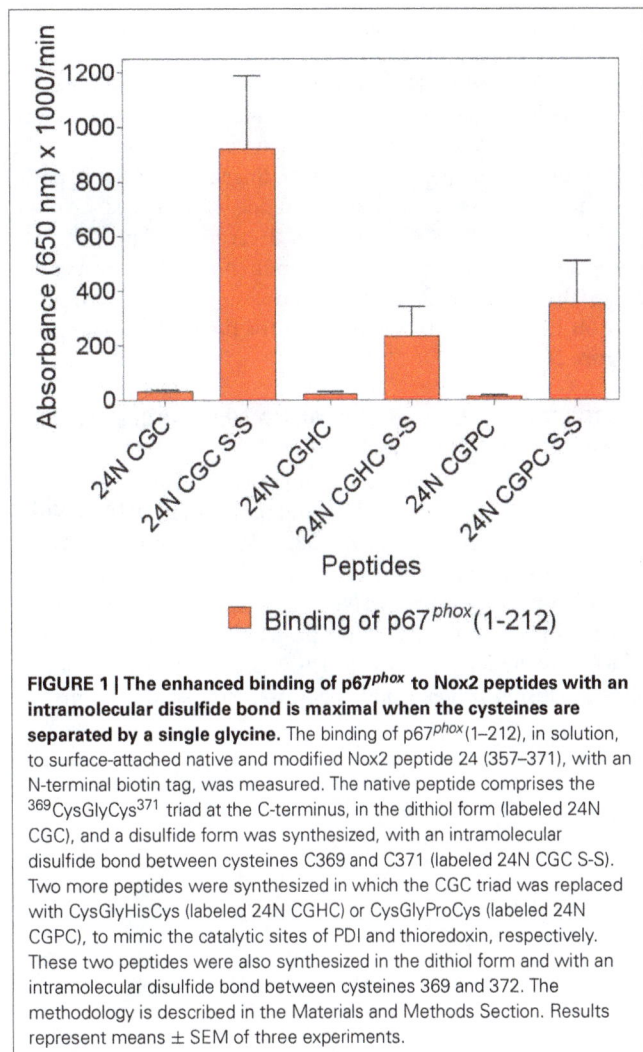

FIGURE 1 | The enhanced binding of p67^{phox} to Nox2 peptides with an intramolecular disulfide bond is maximal when the cysteines are separated by a single glycine. The binding of p67^{phox}(1–212), in solution, to surface-attached native and modified Nox2 peptide 24 (357–371), with an N-terminal biotin tag, was measured. The native peptide comprises the ^{369}CysGlyCys^{371} triad at the C-terminus, in the dithiol form (labeled 24N CGC), and a disulfide form was synthesized, with an intramolecular disulfide bond between cysteines C369 and C371 (labeled 24N CGC S-S). Two more peptides were synthesized in which the CGC triad was replaced with CysGlyHisCys (labeled 24N CGHC) or CysGlyProCys (labeled 24N CGPC), to mimic the catalytic sites of PDI and thioredoxin, respectively. These two peptides were also synthesized in the dithiol form and with an intramolecular disulfide bond between cysteines 369 and 372. The methodology is described in the Materials and Methods Section. Results represent means ± SEM of three experiments.

original peptide 24. Identical results were obtained when peptide 24 and the modified peptides, all in the reduced form, were subjected to oxidation by 1 mM H$_2$O$_2$, a procedure found to lead to the formation of intramolecular disulfide bonds between vicinal cysteines. These results are in good agreement with the findings of Woycechowsky and Raines (2003), showing that a CGC tripeptide, with an amidated C-terminus, is a good functional mimic of PDI.

ALKYLATION OR OXIDATION OF THIOLS IN p67^{phox} INTERFERES WITH BINDING TO Nox2 PEPTIDES CONTAINING THE CGC TRIAD

The findings that binding of p67^{phox} to Nox2 peptides containing the CGC triad was abolished when either C369 or C371 were exchanged to arginine or serine, and that Nox4 peptides, corresponding by alignment to Nox2 peptides 24 and 28, did not bind p67^{phox}, as well as the clinical correlate showing that a C369 to R mutation in Nox2 causes the X91$^+$ form of CGD, with impaired translocation of cytosolic components, represent strong indicators for disulfide bonds being established between Nox2 and p67^{phox} cysteines. To test this hypothesis directly, we treated p67^{phox} with the thiol alkylating agent NEM or the thiol oxidants diamide and H_2O_2 and measured binding to peptide 24 in the native and disulfide form. We found that pretreatment of p67^{phox} with either NEM (**Figure 2A**) or diamide (**Figure 2B**) totally abolished binding to Nox2 peptide 24 in the disulfide form. Both reagents also suppressed the low affinity binding to the native form of the peptide. These findings support the hypothesis that some form of thiol—disulfide exchange is involved in the binding of p67^{phox} to Nox2. Treatment of p67^{phox} with H_2O_2 also markedly reduced the binding of p67^{phox} to peptide 24 in the disulfide form but a surprisingly high concentration oh H_2O_2(10 mM) was required for the effect to take place (**Figure 2C**).

CONSTRUCTION OF RECOMBINANT FUSION PROTEINS COMPRISING SEGMENTS OF THE DHR OF Nox2

We have generated a number of recombinant fusion proteins consisting of the *E. coli* protein NusA and parts of the DHR of Nox2, starting at residues 328, 357, 372, 387, 408, 444, and 462, all ending at the C-terminus of Nox2 (residue 570) (**Figure 3**). As a control protein, we expressed NusA not fused to Nox2. Most work was performed with the fusion protein NusA-Nox2(357–570), which was the longest form which was predominantly soluble. Since all fusion proteins and NusA, too, consisted of a mixture

of polymers and homodimers, we purified the proteins by gel filtration and used the homodimer fraction in all experiments. As seen in **Figure 3**, fusion protein NusA-Nox2(357–570) comprised residues 357–383, which correspond to the overlapping peptides 24 and 28, and, thus, contained the CGC triad. It also contained all parts of the NADPH-binding site and part of the FAD-ribityl binding site. The five shorter truncated proteins lacked the CGC triad.

For the immunologic characterization of the fusion proteins, we used two mouse monoclonal anti-human Nox2 antibodies. Antibody 54.1 reacted with Nox2 epitope 381-390 and antibody NL7 reacted with epitope 498–506. The locations of the two epitopes on the serially truncated Nox2 proteins are indicated in **Figure 3**. All truncated Nox2 proteins were subjected to immunoblot analysis. As apparent in **Figure 4A**, antibody 54.1 reacted with proteins NusA-Nox2(357–570) and (372–570) and weakly, with (387–570), in perfect correlation with the presence and absence of the relevant epitopes. Antibody NL7 reacted, as expected, with all the truncated proteins, in accordance with the presence of the relevant epitope on all truncations (**Figure 4B**). Thus, the structure of all NusA-Nox2 fusion proteins to be used in the present work, predicted by DNA sequencing, was confirmed by their immunologic characteristics.

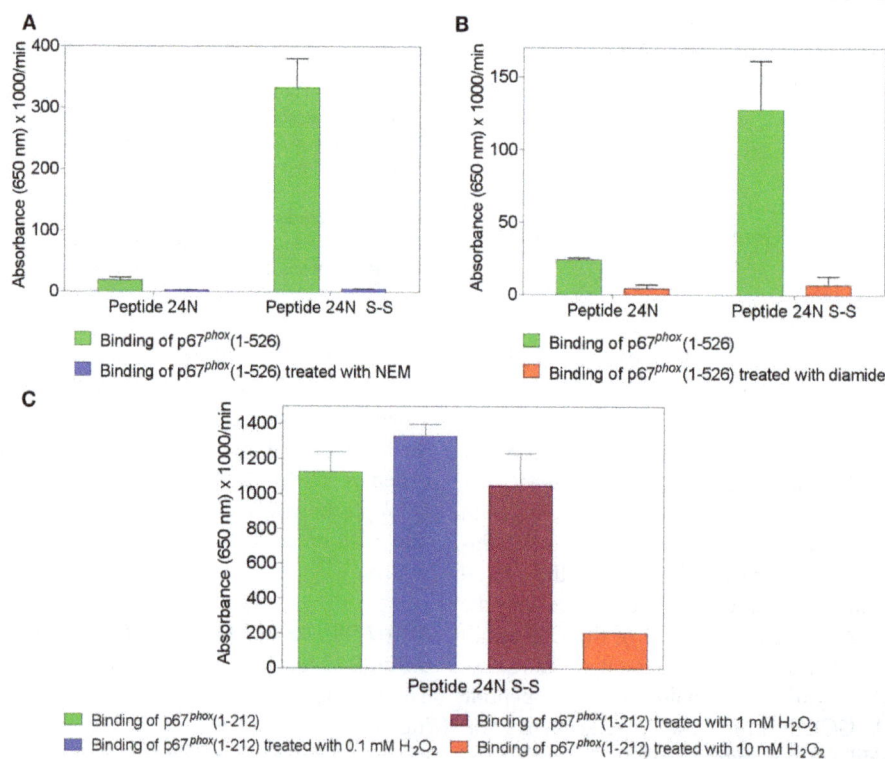

FIGURE 2 | Treatment of p67^{phox} with *N*-ethylmaleimide (NEM), diamide, or H_2O_2 reverses the enhanced binding to Nox2 peptides containing an intramolecular disulfide bond. p67^{phox}(1–526) (1.5 μM) was incubated with NEM (0.1 mM) (**A**) or diamide (5 mM) (**B**), for 1 h at room temperature. p67^{phox}(1–212) (1.5 μM) was exposed to H_2O_2 (0.1, 1, or 10 mM) for 2 h at 4°C (**C**). Control preparations were supplemented with equal volumes of PBS for the same time interval. The binding of treated and untreated p67^{phox}, in solution, to surface-attached Nox2 peptide 24 (357–371), with an N-terminal biotin tag, was measured. Binding to peptide 24 in the dithiol form (labeled 24N) and/or only (in the case of exposure to H_2O_2) to the peptide with an intramolecular disulfide bond between cysteines C369 and C371 (labeled 24N S-S) was assessed. Results shown in all panels represent means ± SEM of three experiments.

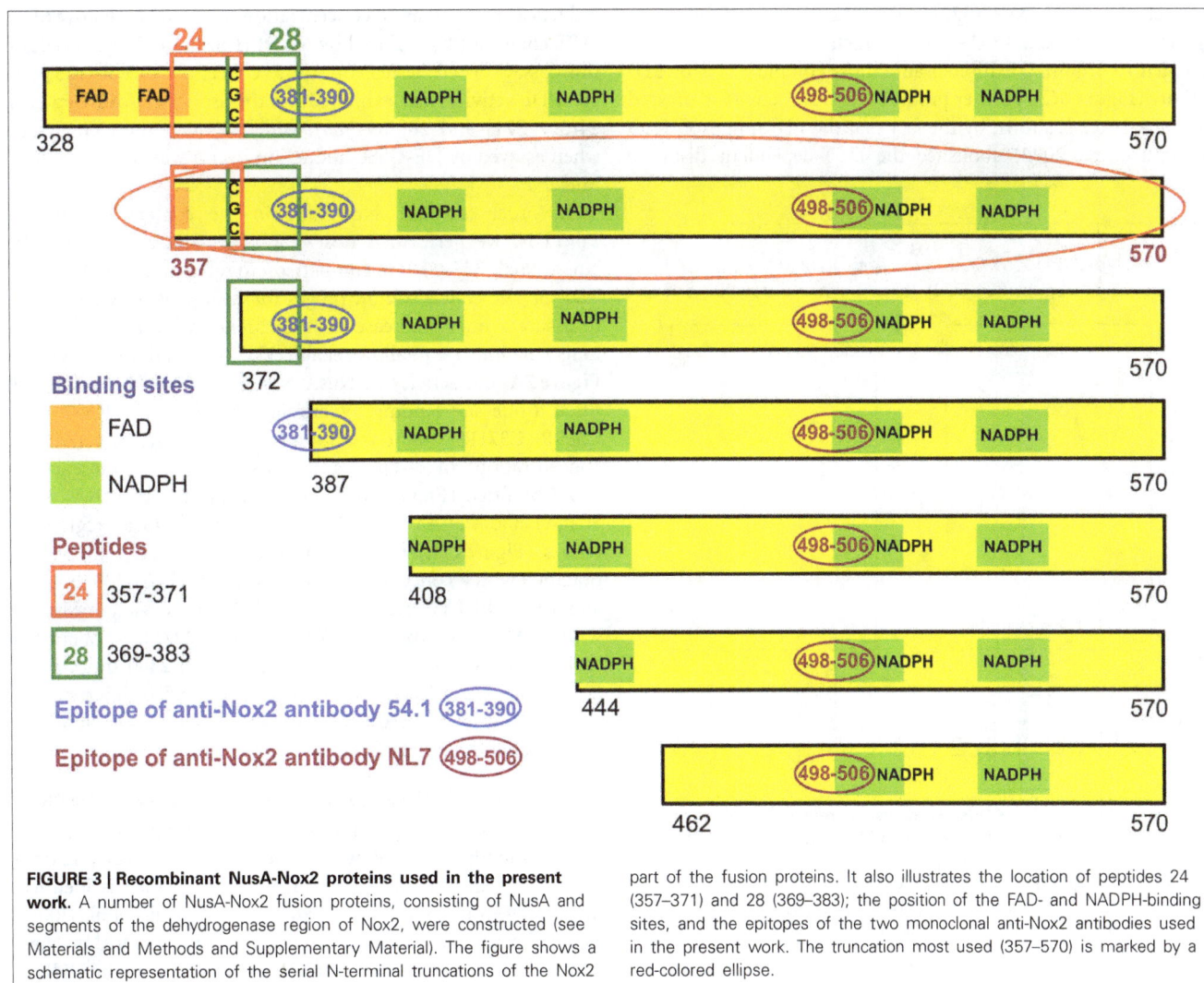

FIGURE 3 | Recombinant NusA-Nox2 proteins used in the present work. A number of NusA-Nox2 fusion proteins, consisting of NusA and segments of the dehydrogenase region of Nox2, were constructed (see Materials and Methods and Supplementary Material). The figure shows a schematic representation of the serial N-terminal truncations of the Nox2 part of the fusion proteins. It also illustrates the location of peptides 24 (357–371) and 28 (369–383); the position of the FAD- and NADPH-binding sites, and the epitopes of the two monoclonal anti-Nox2 antibodies used in the present work. The truncation most used (357–570) is marked by a red-colored ellipse.

BINDING OF NusA-Nox2 FUSION PROTEINS TO p67^phox IS DEPENDENT ON THE PRESENCE OF THE CGC TRIAD IN Nox2

We used a protein—protein binding assay to measure the ability of the various NusA-Nox2 truncations to bind to p67^{phox}. NusA not fused to Nox2 segments was used as a negative control. In these experiments, the homodimeric forms of NusA-Nox2 and NusA, derived by purification by gel filtration, were used. As seen in **Figures 5A,B**, only NusA-Nox2(328–570) and (357–570) exhibited significant binding to p67^{phox}. The shorter truncations bound to p67^{phox} to an extent similar to that of unfused NusA. These results demonstrate the need for the presence of the CGC triad in Nox2 for binding to p67^{phox} since only NusA-Nox2(328–570) and (357–570) comprise the CGC triad (see **Figure 3**). Binding to p67^{phox}(1–212) exceeded that to p67^{phox}(1–526), in accordance with data indicating that a conformational change in p67^{phox}(1–526) is required for optimal binding (Dang et al., 2002; Federman Gross et al., 2012).

THE NusA-Nox2 FUSION PROTEIN EXHIBITS PDI REDUCTASE ACTIVITY

The availability of a soluble and well-characterized fusion protein comprising a major part of the DHR of Nox2 offered the opportunity to test it for the presence of enzymatic PDI activity. We first used the insulin reduction assay (Holmgren, 1979), with both the rate of aggregation and the lag time serving as kinetic parameters (Martinez-Galisteo et al., 1993). For standardizing the method, we assayed recombinant PDIA1 and PDIA3 and its sensitivity was found sufficient for measuring the activity of both PDIs, down to a concentration of 10 nM. We, thus, attempted to apply the insulin method for assessing the reductase activity of NusA-Nox2(357–570), up to a concentration of 5 μM, but no activity could be detected.

We next turned to using a higher sensitivity assay, in which the DTT-dependent disulfide reductase activity is measured by the relief of fluorescent self quenching of the PDI pseudo-substrate DE-GSSG, as described by Raturi and Mutus (2007). As apparent in **Figure 6A**, NusA-Nox(357–570), at a concentration of 2 μM, exhibited PDI activity in marked excess over that measured with DTT alone. The reductase activity showed clear dose dependency (**Figure 6B**). Since the only catalytic motif present in Nox2 that resembles the canonical motif of PDI (CGHC) is the CGC triad, we looked for a correlation between the presence of the CGC triad and activity. Such correlation was indeed found, as shown by the facts that NusA-Nox2(372–570), which lacks the CGC triad, as well as NusA-Nox2(357–570) in which C369 and C371 were mutated to arginine, exhibited an almost complete loss of PDI activity (**Figure 6C**). Further proof for the PDI activity expressed by NusA-Nox2 is shown by the fact that the PDI inhibitor, PAO (Gallina et al., 2002) abolished the DTT-dependent disulfide

reductase activity at a concentration of $50\,\mu M$ (**Figure 6D**). DTT could not be replaced by NADPH in the DE-GSSG reduction assay, when tested up to a concentration of $125\,\mu M$. No PDI activity was exhibited by the isolated Nox2 peptides 24(357–371) and 28(369–383), which contain the CGC triad, when assayed by DE-GSSG reduction over a wide concentration range.

In order to place Nus-Nox2 in the context of canonical PDIs, we performed dose-response reductase assays with NusA-Nox2(357–570) in comparison to recombinant PDIA1 and PDIA3. The concentration ranges for each protein were chosen to assure a linear increase in fluorescence within a sufficiently long time interval to allow reliable V_{max} calculations. As seen in **Figure 7A**, the activity of NusA-Nox2(357–570) $(1–4\,\mu M)$ best fitted a one site binding hyperbola. The NusA-Nox2(357–570) C369R, C371R mutant was found to lack reductase activity at concentrations of $1–4\,\mu M$ and, thus, no dose-response curve could be fitted (**Figure 7B**). The activities of both PDIA1 (10–80 nM) and PDIA3 (1.25–10 nM) generated linear regression curves (**Figures 7C,D**). By using extrapolation of V_{max} values obtained with a concentration of 10 nM PDIA1 (37,303 ± 1279) and 10 nM PDIA3 (1,206,911 ± 81,022) to the V_{max} measured with 1 μM NusA-Nox2(357–570) (49,164 ± 2271), we calculated that PDIA1 was about 75 times and PDIA3, 2400 times more active than NusA-Nox2(357–570). It is of possible interest that, using this assay, the reductase activity of PDIA3 was about 30 times higher than that of PDIA1.

Nox2 AND PDIA3 SHARE SMALL REGIONS OF SEQUENCE HOMOLOGY

In the course of the generation of a mouse monoclonal anti-Nox2 antibody (54.1), it was found that the antibody reacted with a protein known as GRp58 or ERp57, which is identical to the PDIA3 member of the PDI family (Baniulis et al., 2005). The authors noted that the cross-reaction was explained by Nox2 and PDIA3 sharing a five-residues motif (AVDGP, in the DHR of

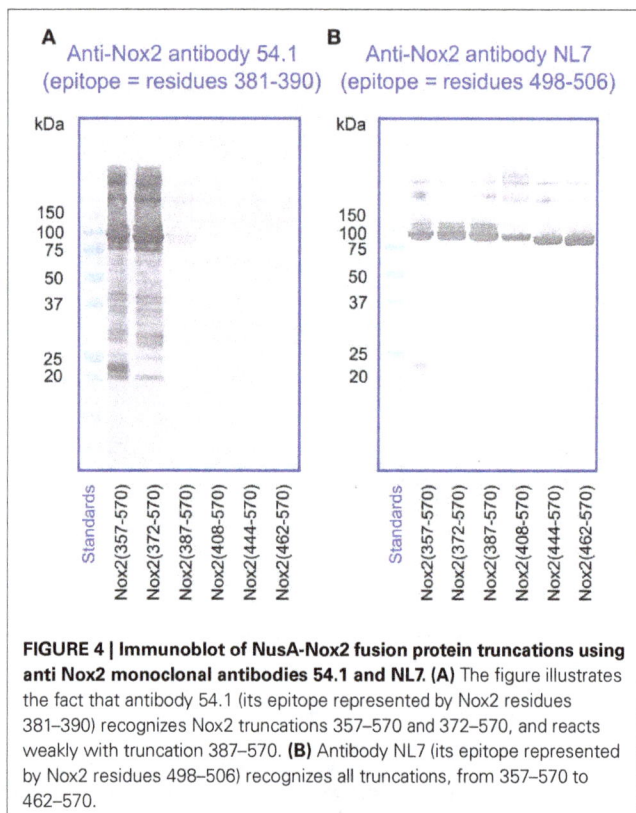

FIGURE 4 | Immunoblot of NusA-Nox2 fusion protein truncations using anti Nox2 monoclonal antibodies 54.1 and NL7. (A) The figure illustrates the fact that antibody 54.1 (its epitope represented by Nox2 residues 381–390) recognizes Nox2 truncations 357–570 and 372–570, and reacts weakly with truncation 387–570. **(B)** Antibody NL7 (its epitope represented by Nox2 residues 498–506) recognizes all truncations, from 357–570 to 462–570.

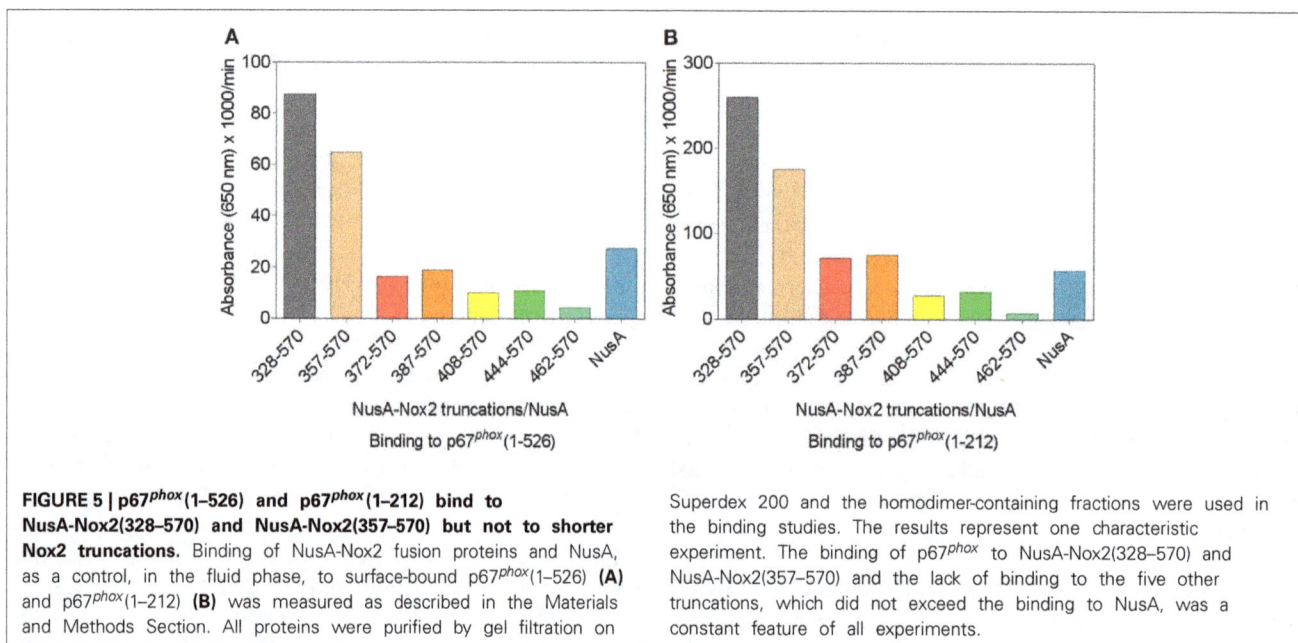

FIGURE 5 | p67phox(1–526) and p67phox(1–212) bind to NusA-Nox2(328–570) and NusA-Nox2(357–570) but not to shorter Nox2 truncations. Binding of NusA-Nox2 fusion proteins and NusA, as a control, in the fluid phase, to surface-bound p67phox(1–526) **(A)** and p67phox(1–212) **(B)** was measured as described in the Materials and Methods Section. All proteins were purified by gel filtration on

Superdex 200 and the homodimer-containing fractions were used in the binding studies. The results represent one characteristic experiment. The binding of p67phox to NusA-Nox2(328–570) and NusA-Nox2(357–570) and the lack of binding to the five other truncations, which did not exceed the binding to NusA, was a constant feature of all experiments.

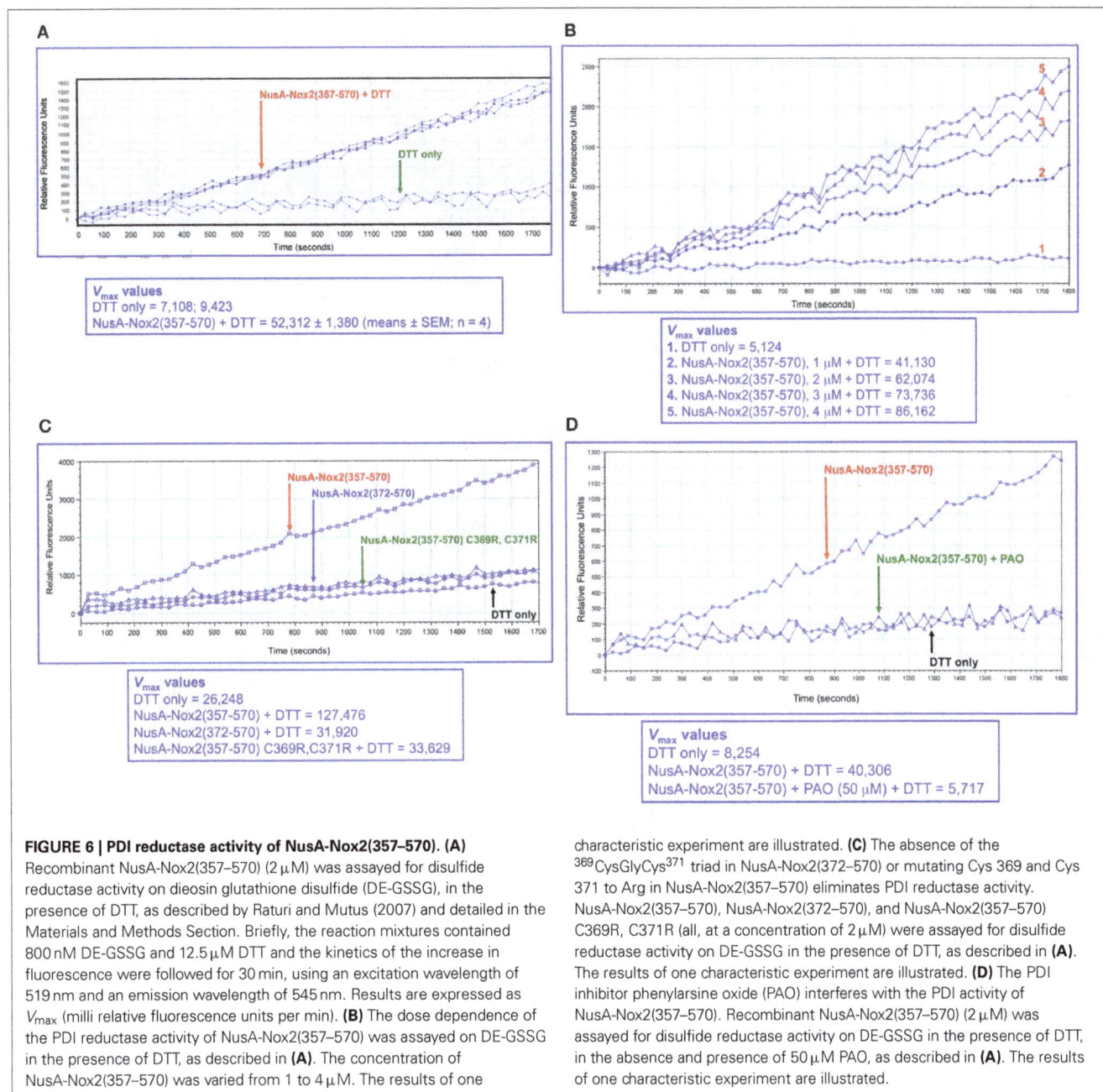

FIGURE 6 | PDI reductase activity of NusA-Nox2(357–570). (A) Recombinant NusA-Nox2(357–570) (2 µM) was assayed for disulfide reductase activity on dieosin glutathione disulfide (DE-GSSG), in the presence of DTT, as described by Raturi and Mutus (2007) and detailed in the Materials and Methods Section. Briefly, the reaction mixtures contained 800 nM DE-GSSG and 12.5 µM DTT and the kinetics of the increase in fluorescence were followed for 30 min, using an excitation wavelength of 519 nm and an emission wavelength of 545 nm. Results are expressed as V_{max} (milli relative fluorescence units per min). **(B)** The dose dependence of the PDI reductase activity of NusA-Nox2(357–570) was assayed on DE-GSSG in the presence of DTT, as described in **(A)**. The concentration of NusA-Nox2(357–570) was varied from 1 to 4 µM. The results of one characteristic experiment are illustrated. **(C)** The absence of the ^{369}CysGlyCys^{371} triad in NusA-Nox2(372–570) or mutating Cys 369 and Cys 371 to Arg in NusA-Nox2(357–570) eliminates PDI reductase activity. NusA-Nox2(357–570), NusA-Nox2(372–570), and NusA-Nox2(357–570) C369R, C371R (all, at a concentration of 2 µM) were assayed for disulfide reductase activity on DE-GSSG in the presence of DTT, as described in **(A)**. The results of one characteristic experiment are illustrated. **(D)** The PDI inhibitor phenylarsine oxide (PAO) interferes with the PDI activity of NusA-Nox2(357–570). Recombinant NusA-Nox2(357–570) (2 µM) was assayed for disulfide reductase activity on DE-GSSG in the presence of DTT, in the absence and presence of 50 µM PAO, as described in **(A)**. The results of one characteristic experiment are illustrated.

Nox2, and AYDGP, in PDIA3). We noticed that yet another mouse monoclonal antibody to Nox2 (NL7) reacted with an epitope in the DHR of Nox2 (KDVITG) which shares four residues with sequence KDLIQG, in PDIA3. Finally, yet another minor, three-residue identity between the DHR of Nox2 and PDIA3 (IVG), was noticed (for all similarities, see **Figure 8**). The above Nox2 sequences exhibit no similarity with PDIA1.

The residues in the Nox2 sequence AVDGP shared with PDIA3 are also present in Nox3; the shared residues in Nox2 sequence KDVITG are also present in Nox5, and shared residues IVG are also present in Nox4. However, as noted in the Introduction, the CGC triad is specific for Nox2. The AYDGP sequence in PDIA3 is located in the catalytic domain *a*, C-terminal to the CGHC motif;

and the IVG and KDLIQG sequences are located in domains *b* and *b′*, respectively (Coe and Michalak, 2010; Kozlov et al., 2010).

IMMUNOBLOT ANALYSIS CONFIRMS Nox2/PDIA3 SEQUENCE SIMILARITIES

We subjected preparations of NusA-Nox2(357–570) and (372–570), NusA, recombinant PDI3 and PDIA1, and macrophage membranes to immunoblotting with the two anti-Nox2 antibodies 54.1 and NL7. As apparent in **Figures 9A,B**, anti-Nox2 antibodies 54.1 and NL7 reacted with PDIA3 but not with PDIA1. The reactivity of antibody 54.1 with PDIA3 was more pronounced than that of antibody NL7. As expected, and as shown in **Figure 4**, both antibodies reacted strongly with

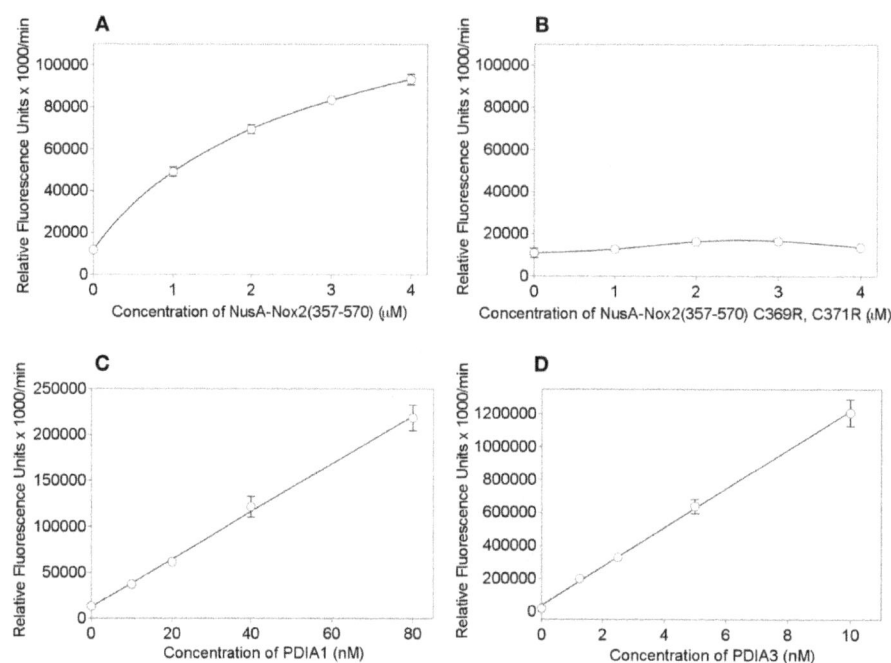

FIGURE 7 | PDI reductase activities of NusA-Nox2(357–570) compared to those of the NusA-Nox2(357–570) C369R, C371R mutant and to PDIA1 and PDIA3. (A) PDI reductase activity of recombinant NusA-Nox2(357–570) was assayed in a concentration range of 1–4 μM and plotted as non-linear regression (one site binding equation). (B) PDI reductase activity of recombinant NusA-Nox2(357–570) C369R, C371R mutant was assayed in a concentration range of 1–4 μM and plotted as a cubic spline curve. (C) PDI reductase activity of recombinant PDIA1 was assayed in a concentration range of 10–80 nM and plotted as linear regression. (D) PDI reductase activity of recombinant PDIA3 was assayed in a concentration range of 1–10 nM and plotted as linear regression. The assays were performed as described by Raturi and Mutus (2007) and detailed in the Materials and Methods Section. Results represent means ± SEM of 9 (NusA-Nox2(357–570), 4 (NusA-Npx2(357–570) C369R, C371R), 4 (PDIA1), and 3 (PDIA3) experiments.

NusA-Nox2(357–570) and (372–570) and detected a broader and more diffuse band of about 54–58 kDa in macrophage membranes, corresponding to the characteristics of guinea pig Nox2 (Knoller et al., 1991). A polyclonal goat anti-Nox2 antibody (C-15), raised against a peptide in the C-terminal region of Nox2, which reacted with NusA-Nox2(357-570), did not react with PDIA3 (result not shown, see **Table 1**). We suggest that the ability of the two anti-Nox2 antibodies to react with PDIA3 is due to the similarity of the Nox2 epitopes recognized by antibodies 54.1 and NL7 with sequences AYDGP and KDLIQG in PDIA3, respectively (see **Figure 8**).

In a reciprocal series of experiments, we found that the polyclonal anti-PDIA3 antibody H-220 reacts strongly with NusA-Nox2(357–570) but not with Nusa-Nox2(372–570) and NusA (**Figure 10A**). Its specificity is proven by its ability to recognize recombinant PDIA3 but not PDIA1. The antibody does not seem to recognize Nox2 in macrophage membranes in spite of the high level of homology of human and guinea pig Nox2. It, however, detects a sharp band in the macrophage membrane, of a size somewhat smaller than recombinant PDIA3 (see Discussion of this finding in the paragraph below). In contrast to anti-PDIA3, polyclonal anti-PDIA1 antibody does not react with NusA-Nox2(357–570) (**Figure 10B**). Its specificity is shown by it recognizing recombinant PDIA1, though there is some cross-reaction with PDIA3. Anti-PDIA1 also recognizes an antigen in the macrophage membrane (double band), possibly PDIA1.

These findings are best explained by the likelihood that polyclonal anti-PDIA3 antibody H-220, raised against residues 108–207 of PDIA3, comprises antibodies against sequence AYDGP (residues 114–118), similar to sequence AVDGP (residues 386–390) in Nox2, and, possibly IVG (residues 156–158), identical to sequence IVG (residues 357–359) in Nox2. We have no explanation for the lack of reaction of anti-PDIA3 H-220 with truncation NusA-Nox2(372–570), unless the participation of epitope IVG [absent in NusA-Nox2(372–570)] is required.

We next investigated the identity of the sharp band detected by anti-PDIA3 H-220 in macrophage membranes. For this purpose, we immunoblotted four individual batches of macrophage membranes with monoclonal anti-Nox2 antibody 54.1, in parallel with four polyclonal anti-PDIA3 antibodies (H-220, HPA003230, NBP-84797, and ABE1032; the characteristics of the antibodies are described in Materials and Methods). With the exception of H-220, none of the anti-PDIA3 antibodies recognized recombinant NusA-Nox2(357–570) and none of the antibodies recognized Nox2 in macrophage membranes. However, all four anti-PDIA3 antibodies detected a protein of about 58 kDa in all batches of macrophage membranes in the shape of a sharply defined narrow band, indicating that PDIA3 is present in the membrane (see **Figure 11**, illustrating results obtained with anti-Nox2 antibody 54.1 and anti-PDIA3 antibody HPA003230). Monoclonal anti-Nox2 antibody 54.1 reacted with Nox2 in membranes (a diffuse band) but also with what appears to

FIGURE 8 | Amino acid sequence similarities between the dehydrogenase region of Nox2 and PDIA3. The figure illustrates three small regions (3–6 residues) exhibiting partial sequence identities between Nox2 and PDIA3. The representation of the motifs in non-phagocytic Noxes (Noxes 1, 3, 4, and 5) is also shown. Monoclonal anti-Nox2 antibody 54.1 recognizes Nox2 residues ^{381}KLPKIAVDGP^{390} and might recognize PDIA3 residues ^{114}AYDGP^{118} (see **Figure 9A**). Monoclonal anti-Nox2 antibody NL7 recognizes Nox2 residues ^{498}EKDVITGLK^{506} and might recognize PDIA3 residues ^{252}KDLIQG^{257} (see **Figure 9B**). Polyclonal anti-PDIA3 antibody H-220 was raised against PDIA3 residues 108–207 and could potentially recognize PDIA3 ^{114}AYDGP^{118} and ^{156}IVG^{158} and Nox2 residues ^{386}AVDGP^{390} and ^{357}IVG^{359}, respectively (see **Figure 10A**). The regions of Nox2/PDIA3 homology do not exist in the sequence of PDIA1, in good agreement with the fact that the two anti-Nox2 antibodies do not recognize PDIA1 (see **Figures 9A,B**) and a polyclonal anti-PDIA1 antibody does not recognize Nox2 (see **Figure 10B**). Identical residues in different proteins are in red font; the CGC triad in Nox2 is in orange font.

be PDIA3 present in the membrane, appearing as a narrow band on the background of the more diffuse Nox2, corresponding in size to the band detected by anti-PDIA3 antibody HPA003230. The presence of PDIA3 in macrophage membranes might offer the explanation for the inability of anti-PDIA3 antibody H-220 to recognize Nox2 in the membrane; this could be due to competition by membrane PDIA3, to which the antibody is likely to bind with higher affinity. All results obtained by immunoblotting are summarized in **Table 1**. The presence of PDIA3 (and PDIA1) in macrophage membranes, although expected in light of the ubiquity of PDIs even outside the endoplasmic reticulum (Turano et al., 2002), poses a methodological challenge when investigating the PDI-like function of Nox2 and emphasizes the advantage of working with recombinant Nox2.

MUTATING CYSTEINES IN p67^{phox}(1–212) TO SERINE PREVENTS BINDING TO A Nox2 DHR PEPTIDE CONTAINING THE CGC TRIAD

The data presented so far suggest that the CGC triad in the DHR of Nox2 serves as a pseudo-PDI catalytic motif leading to the binding of p67^{phox} to Nox2 via a thiol-disulfide exchange reaction. Since such a reaction must involve cysteine(s) in p67^{phox} and since oxidation or alkylation of cysteines in p67^{phox} abolished binding to CGC-containing Nox2 peptides (see **Figure 2**), we reasoned that mutating the cysteines in p67^{phox} should prevent binding to these peptides and also affect oxidase activation.

Full-length p67^{phox} contains nine cysteines and p67^{phox}(1–212) contains four. Based on the knowledge that p67^{phox}(1–212) binds to NusA-Nox2 and to CGC-containing Nox2 peptides and supports oxidase activation *in vitro* with an efficiency identical

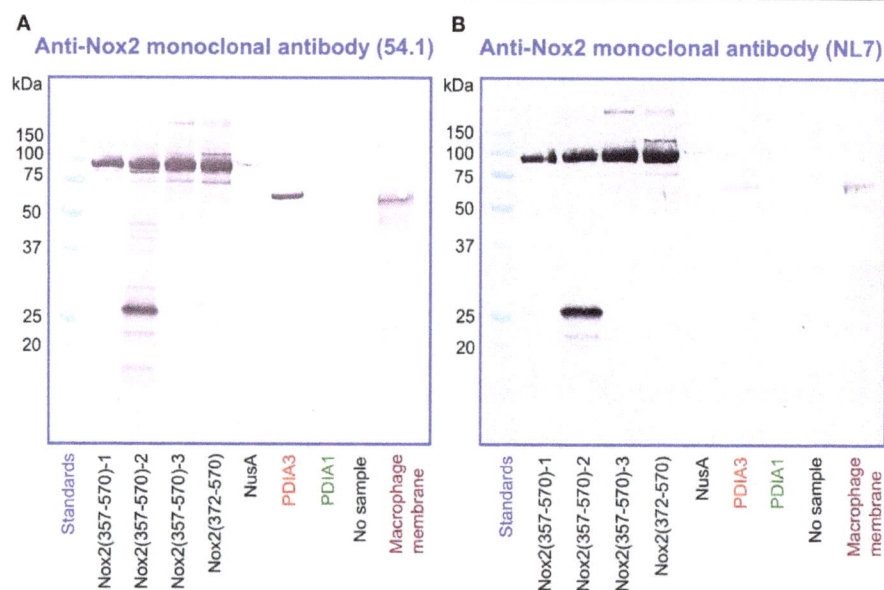

A Anti-Nox2 monoclonal antibody (54.1)

B Anti-Nox2 monoclonal antibody (NL7)

FIGURE 9 | Two monoclonal anti-Nox2 antibodies react with PDIA3 but not with PDIA1 by immunoblotting. (A) Monoclonal anti-Nox2 antibody 54.1 (epitope, residues 381–390) reacts with recombinant NusA-Nox2(357–570) and NusA-Nox2(372–570) but not with NusA. It reacts strongly with recombinant PDIA3 but not with recombinant PDIA1. It recognizes a protein of about 54–58 kDa (diffuse band) in the guinea pig macrophage membrane, corresponding to Nox2 (Knoller et al., 1991). **(B)**

Monoclonal anti-Nox2 antibody NL7 (epitope, residues 498–506) reacts with recombinant NusA-Nox2(357–570) and NusA-Nox2(372–570) but not with NusA. It reacts moderately with recombinant PDIA3 but not with recombinant PDIA1. It recognizes a protein of about 54–58 kDa (diffuse band) in the guinea pig macrophage membrane, corresponding to Nox2. Nox2(357–570) numbered 1, 2, and 3 represent three batches of NusA-Nox2(357–570).

Table 1 | Cross-reactivity of some anti-Nox2 antibodies with PDIA3 and of some anti-PDIA3 antibodies with Nox2.

Protein	Antibodies							
	Anti-Nox2 54.1	Anti-Nox2 NL7	Anti-Nox2 C-15	Anti-PDIA3 H-220	Anti-PDIA3 HPA003230	Anti-PDIA3 NBP1-84797	Anti-PDIA3 ABE1032	Anti-PDIA1 H-160
	Mouse monoclonal	Mouse monoclonal	Goat polyclonal	Rabbit Polyclonal	Rabbit Polyclonal	Rabbit polyclonal	Rabbit polyclonal	Rabbit polyclonal
NusA-Nox2 (357–570)	+	+	+	+	−	−	−	−
NusA-Nox2 (372–570)	+	+	ND[a]	−	−	ND[a]	−	ND[a]
PDIA3 (recombinant)	+	+	−	+	+	+	+	+/−[b]
PDIA3 (in macrophage membrane)	+	+	ND[a]	+	+	+	+	+[c]
PDIA1 (recombinant)	−	−	−	−	−	−	+/−[b]	+

[a] ND, not determined.

[b] Weak reaction in comparison to that with the specific antigen.

[c] PDIA1 was detected; detection of PDIA3 could not be established with certainty because of similarity in size with PDIA1.

to that of the full-length protein, we mutated the four cysteines present in p67^{phox}(1–212) (cysteines 40, 45, 121, and 165) to serines. The wild-type protein and the mutated protein were found by gel filtration to have a native molecular mass of 28.4 and 27.4 kDa, respectively, in good agreement with the theoretical

value of 25.5 kDa. The only difference between the two proteins was an apparent increase in the hydrophobic character of the mutant protein (see Materials and Methods). We confirmed the absence of cysteines in the mutant protein by comparing the binding of the thiol probe mBBr (Kosower and Kosower, 1995)

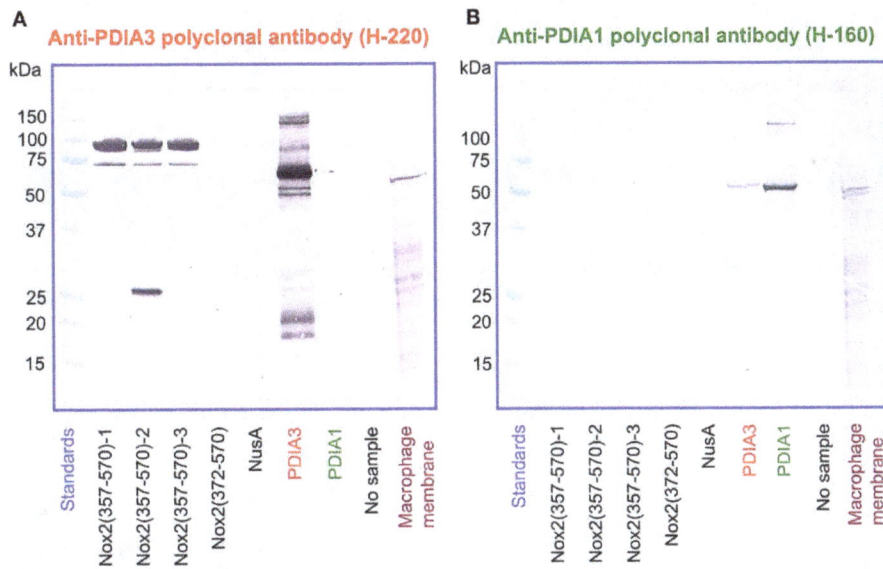

FIGURE 10 | A polyclonal anti-PDIA3 antibody but not a polyclonal anti-PDIA1 antibody reacts with Nox2 by immunoblotting. (A) Polyclonal anti-PDIA3 antibody H-220 reacts with recombinant NusA-Nox2(357–570) but not with NusA-Nox2(372–570) and NusA. It reacts strongly with recombinant PDIA3 but not with recombinant PDIA1. It also recognizes a protein in the guinea pig macrophage membrane (sharp single band). **(B)** Polyclonal anti-PDIA1 antibody H-160 does not react with recombinant NusA-Nox2(357–570) and NusA-Nox2(372–570). It reacts with recombinant PDIA1 and weakly, with recombinant PDIA3. It also recognizes a protein in the guinea pig macrophage membrane (sharp double band). Nox2(357–570) numbered 1, 2, and 3 represent three batches of NusA-Nox2(357–570).

to the wild-type and mutant proteins. Wild-type protein indeed bound mBBr, whereas the mutant protein did not (results not shown).

The binding of wild type and mutant p67^{phox} to Nox2 DHR peptide 24 comprising the CGC triad, in either the dithiol or disulfide form, was assessed by the peptide—protein binding assay. As apparent in **Figures 12A,B**, the wild type protein exhibited moderate binding to peptide 24 in the reduced form and much enhanced binding to the disulfide form. The mutant protein lacked binding ability to both forms of the peptide. The total absence of binding of the mutant protein to the disulfide form of peptide 24 is especially noteworthy.

MUTATING CYSTEINES IN p67^{phox}(1–212) TO SERINE INTERFERES WITH NADPH OXIDASE ACTIVATION

We next assayed wild-type and mutant p67^{phox}(1–212) for their ability to support oxidase activation in the cell-free system. We compared the two proteins in the amphiphile- and p47^{phox}-dependent and in the amphiphile- and p47^{phox}-independent systems. As seen in **Figures 13A,B**, there was a two-fold increase in the EC$_{50}$ of the mutant p67^{phox} in both assays, indicating that that the mutant protein has a significantly impaired ability to support oxidase activation. In accordance with our hypothesis, the mutant protein did not loose its activity completely, as expected from an effect of the stability of the assembled oxidase complex, as opposed to an effect on the primary interaction, which would result in a total loss of activity. Cysteine mutagenesis was applied in the past to p47^{phox} and shown to result in a paradoxical increase in the ability to support oxidase activation by an unknown mechanism (Babior, 2002).

THE MAJORITY OF PDI INHIBITORS DO NOT AFFECT NADPH OXIDASE ACTIVATION *IN VITRO*

It has been reported that two PDI inhibitors, bacitracin and scrambled RNAs, inhibit oxidase activation in an amphiphile-dependent cell-free system, consisting of neutrophil membranes and recombinant cytosolic components (Paes et al., 2011). As referred to in the Introduction, PAO and gliotoxin, two compounds binding to vicinal cysteines and, most likely, acting on C369 and C371 in Nox2, were reported to inhibit oxidase activation. PAO is considered to act as a PDI inhibitor (Gallina et al., 2002) and direct evidence for this was presented by Raturi et al. (2005) and also appears in **Figure 6D**. Direct proof for gliotoxin acting as a PDI inhibitor is lacking.

We, thus, engaged in a systematic investigation of the effect of a number of compounds, reported to act as PDI inhibitors on a variety of targets and in various situations, on cell-free oxidase activation. The amphiphile- and p47^{phox}-independent assay was used, in order to reduce the number of potential targets for the inhibitors. Since p47^{phox} was described to associate with exogenous PDI (Paes et al., 2011), we preferred to work in a system free of p47^{phox}. In addition to PAO and gliotoxin, the following PDI inhibitors were tested: bacitracin (Dickerhof et al., 2011), scrambled RNAse (Essex and Li, 1999), 16F16 (Hoffstrom et al., 2010), rutin (Jasuja et al., 2012), and PACMA 31 (Xu et al., 2012). In all assays, the potential inhibitor was preincubated with the membrane component before the induction of oxidase complex assembly. As shown in **Figure 14**, significant inhibition of oxidase activation was caused only by PAO (**Figure 14A**), in accordance with earlier work (Le Cabec and Maridonneau-Parini, 1995; Doussiere et al., 1998). However, in preliminary experiments

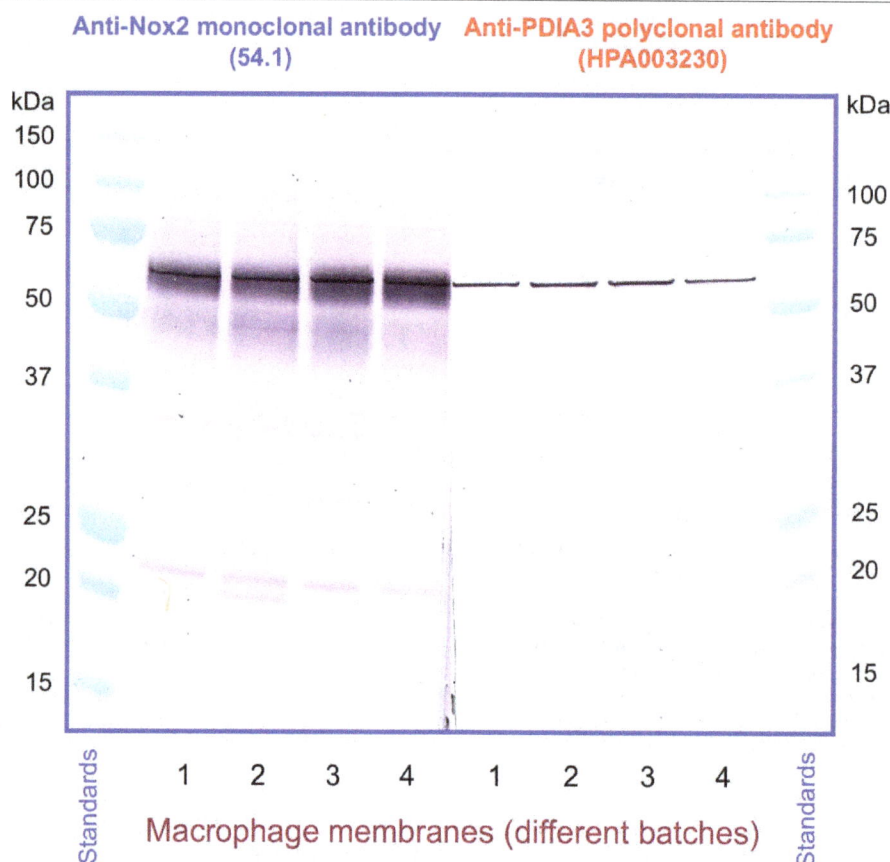

FIGURE 11 | Detection of PDIA3 in macrophage membranes by both anti-Nox2 and anti-PDIA3 antibodies. Monoclonal anti-Nox2 antibody 54.1, which recognizes recombinant PDIA3 (see **Figure 9A**), is shown to detect in macrophage membranes an antigen most likely to be PDIA3. This appears in four distinct membrane batches as a sharply defined band on the background of the more diffuse Nox2; the overlap is caused by the almost identical size of guinea pig Nox2 and PDIA3. A polyclonal anti-PDIA3 antibody (HPA003230), found not to cross-react with Nox2 (see **Table 1**), detects PDIA3 in macrophage membranes. The position and character of the bands detected by anti-PDIA3 strengthens the proposal that the sharp band detected by anti-Nox2 represents PDIA3.

(results not shown), we found that PAO also inhibited oxidase activation supported by p67^{phox} with cysteines mutated to serines, a finding which raises questions about its mechanism of action. It was, indeed, reported that PAO also had an effect on the heme of Nox2 (Doussiere et al., 1999). Gliotoxin was found to be a poor inhibitor, when tested up to a concentration of 200 μM, and generated an atypical dose-response curve, a result which differed from the marked inhibitory effect reported in the past (Nishida et al., 2005). Bacitracin was also found to be a poor inhibitor, with an IC$_{50}$ of 602 μM and a maximal inhibition not exceeding 60%, and scrambled RNAse was ineffective, up to a concentration of 14.6 μM. Lack of an inhibitory effect was also found for 16F16 (up to 14.6 μM), rutin (up to 400 μM), and PACMA 31 (up to 500 μM).

These results do not counter our hypothesis that the DHR of Nox2 functions as a PDI. Nox2 possesses only one copy of a catalytic site resembling CGHC and lacks the equivalent of a substrate binding domain (b'). Bacitracin is unlikely to be an effective inhibitor, first because its target in PDI is the b' domain and its linker to domain a', both absent in Nox2, and second because commercial bacitracin consists predominantly of

bacitracin A, of low inhibitory potency (Dickerhof et al., 2011). Isomerization by PDI also requires the b' domain (Ellgaard and Ruddock, 2005) and, thus, an interaction between Nox2 and scrambled RNAse is unlikely. The mechanism of action of 16F16 and rutin are not fully understood but it is of interest that rutin was found to lack an inhibitory effect on PDIA3 (Jasuja et al., 2012). The lack of effect of PACMA 31 is also expected, considering the exquisite specificity of the compound for PDIA1 and, more specifically for cysteines in the catalytic motif in domain a', in conjunction with additional non-cysteine residues (Xu et al., 2012).

DISCUSSION

Our results support a model of NADPH oxidase assembly and consequent activation in which the cytosolic component p67^{phox}, most directly responsible for activation, establishes a disulfide bond with a cysteine belonging to the ^{369}CysGlyCys^{371} triad in the DHR of Nox2. This interaction serves as a stabilizer of the oxidase complex and is, most likely, secondary to the primary binding of p67^{phox} to Nox2, which involves specific binding sites on both components.

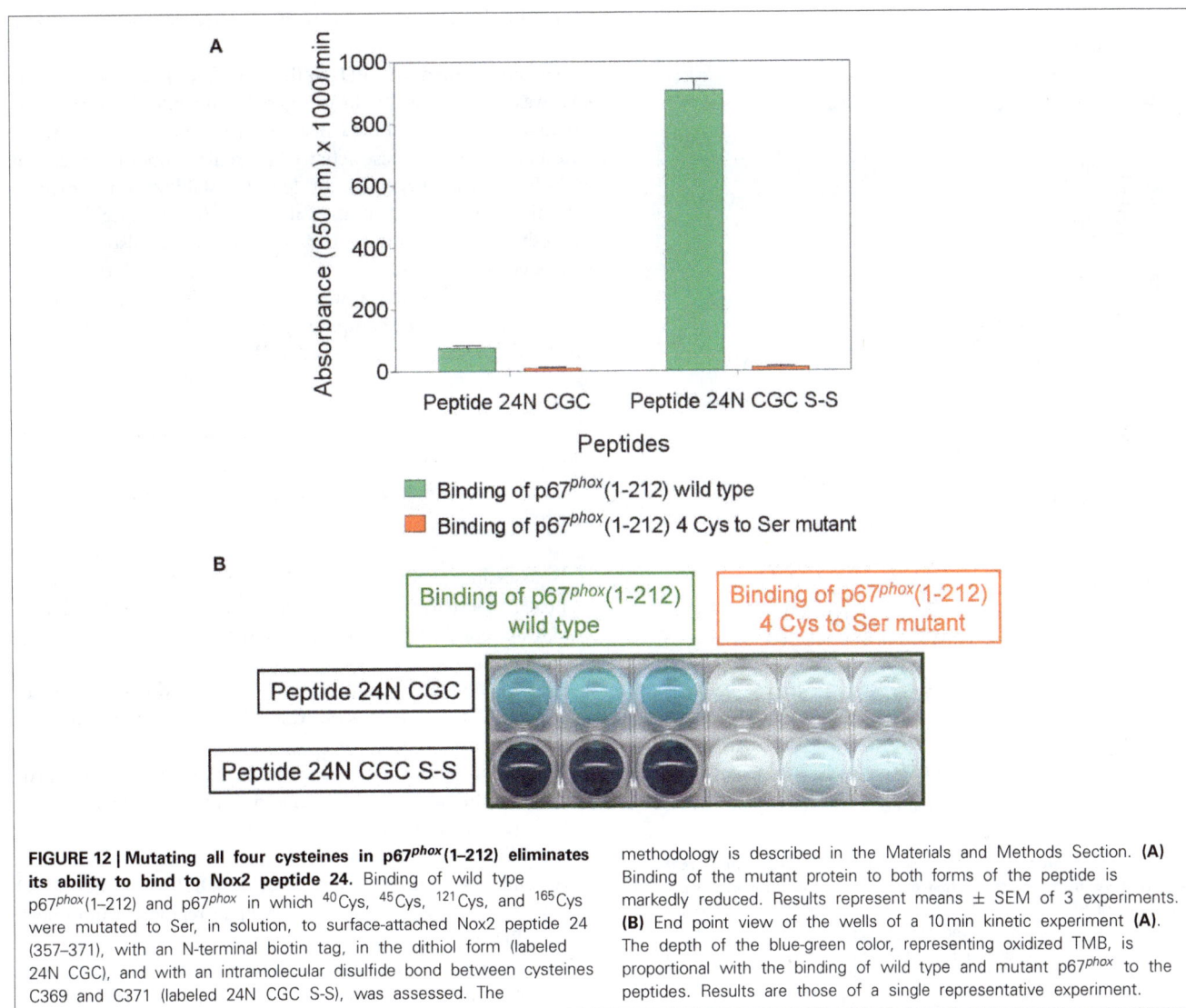

FIGURE 12 | Mutating all four cysteines in p67^{phox}(1–212) eliminates its ability to bind to Nox2 peptide 24. Binding of wild type p67^{phox}(1–212) and p67^{phox} in which ^{40}Cys, ^{45}Cys, ^{121}Cys, and ^{165}Cys were mutated to Ser, in solution, to surface-attached Nox2 peptide 24 (357–371), with an N-terminal biotin tag, in the dithiol form (labeled 24N CGC), and with an intramolecular disulfide bond between cysteines C369 and C371 (labeled 24N CGC S-S), was assessed. The methodology is described in the Materials and Methods Section. **(A)** Binding of the mutant protein to both forms of the peptide is markedly reduced. Results represent means ± SEM of 3 experiments. **(B)** End point view of the wells of a 10 min kinetic experiment **(A)**. The depth of the blue-green color, representing oxidized TMB, is proportional with the binding of wild type and mutant p67^{phox} to the peptides. Results are those of a single representative experiment.

Earlier findings that were suggestive of such a mechanism were: (a) p67^{phox} binds to Nox2 peptides sharing the CGC triad at either the C- or N-termini; (b) p67^{phox} does not bind to peptides when either C369 or C371 are replaced by arginine or serine; (c) The CGC triad is present only in Nox2, the most potent generator of superoxide among all members of the Nox family, and the one most dependent on regulation by cytosolic components, and (d) The introduction of a disulfide bond between C369 and C371 in Nox2 peptides greatly enhances binding of p67^{phox} and its reduction to the dithiol form abolishes the binding.

Work by other groups also supports our hypothesis. Thus, patients with a C369 to R mutation in Nox2 suffer from the X91$^+$ form of CGD, with normal expression of Nox2 but impaired translocation of cytosolic components (Leusen et al., 2000; Debeurme et al., 2010). It was also reported that exposure of phagocyte membranes to H$_2$O$_2$ or of p67^{phox} to irradiation generating oxygen radicals, with a presumed effect on cysteines, inhibits oxidase assembly but was without effect on the assembled complex (Ostuni et al., 2010).

An important prerequisite for binding of p67^{phox} to Nox2 was the presence of a disulfide bond between cysteines 369 and 371. The characteristics of this bond are critical for binding, as shown by the requirement for cysteines to be separated by one non-cysteine residue and for their presence in the same molecule. Contrary to expectations, replacing the native CGC triad in the Nox2 peptides with the PDI (CGHC) or thioredoxin (CGPC) catalytic motifs did not result in enhanced binding. The requirement for a disulfide bond in Nox2 peptides for binding of p67^{phox} was one of the first indicators for a PDI-like function of Nox2, in accordance with the thiol—disulfide exchange reactions catalyzed by canonical PDIs (Ellgaard and Ruddock, 2005; Appenzeller-Herzog and Ellgaard, 2008). This raises the issue of the physiological mechanism to take place in the intact phagocyte responsible for the generation of the disulfide bond, when taking into account the fact that the cytosolic environment of the DHR of Nox2 is a reducing one. Also, such bond has to return to the dithiol state in order to assure the reversibility of the Nox2—p67^{phox} link. The stability of this link is a yet

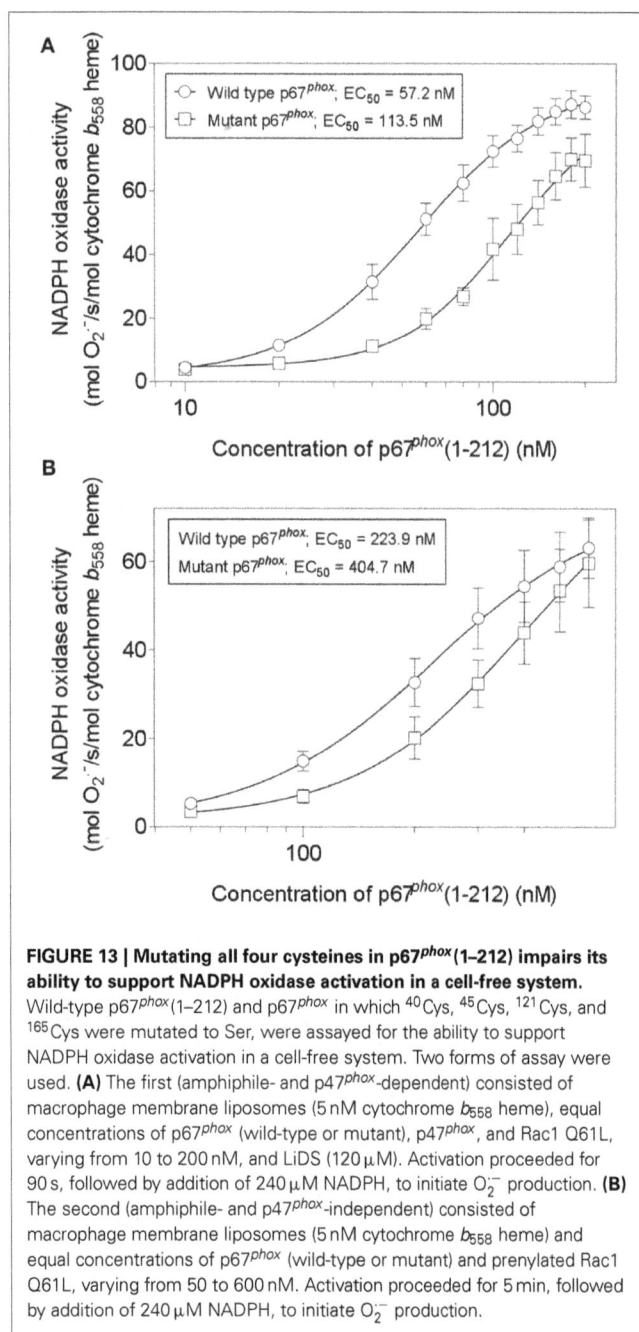

FIGURE 13 | Mutating all four cysteines in p67^{phox}(1–212) impairs its ability to support NADPH oxidase activation in a cell-free system. Wild-type p67^{phox}(1–212) and p67^{phox} in which ^{40}Cys, ^{45}Cys, ^{121}Cys, and ^{165}Cys were mutated to Ser, were assayed for the ability to support NADPH oxidase activation in a cell-free system. Two forms of assay were used. **(A)** The first (amphiphile- and p47^{phox}-dependent) consisted of macrophage membrane liposomes (5 nM cytochrome b_{558} heme), equal concentrations of p67^{phox} (wild-type or mutant), p47^{phox}, and Rac1 Q61L, varying from 10 to 200 nM, and LiDS (120 μM). Activation proceeded for 90 s, followed by addition of 240 μM NADPH, to initiate $O_2^{\cdot-}$ production. **(B)** The second (amphiphile- and p47^{phox}-independent) consisted of macrophage membrane liposomes (5 nM cytochrome b_{558} heme) and equal concentrations of p67^{phox} (wild-type or mutant) and prenylated Rac1 Q61L, varying from 50 to 600 nM. Activation proceeded for 5 min, followed by addition of 240 μM NADPH, to initiate $O_2^{\cdot-}$ production.

of oxidized PDI enhanced oxidase activation in a cell-free system.

The involvement of a PDI in the assembly of the NADPH oxidase was championed by the group of F. Laurindo. They provided extensive experimental evidence for a role of a PDI in the regulation of NADPH oxidase activity in vascular smooth muscle and endothelial cells, suggesting a role in the stabilization of oxidase subunits assembly (Laurindo et al., 2008). A more specific role in the activation of phagocyte oxidase activation was also described, expressed in the ability of exogenous oxidized PDI to enhance and of reduced PDI to diminish oxidase activation *in vitro* (Paes et al., 2011). A peptide comprising the CGHC PDI catalytic motif inhibited oxidase activation and PDI–p47^{phox} and PDI–p22^{phox} associations were demonstrated (Santos et al., 2009; Paes et al., 2011).

We propose that, independently of the regulation of Nox2-dependent oxidase activity by an exogenous PDI, the DHR of Nox2 itself functions as an intrinsic PDI. The evidence for this is as follows:

1. By assessing the binding of p67^{phox} to seven NusA-Nox2 fusion proteins, serially truncated at the N-terminus of the Nox2 moiety, we found that binding required the presence of the CGC triad.
2. Fusion protein NusA-Nox2(357–570), which comprises the CGC triad, expressed dose-dependent PDI reductase activity.
3. Truncation of the Nox2 moiety in the NusA-Nox2 fusion protein C-terminal to the CGC triad or mutating C369 and C371 to R, resulted in the loss of PDI activity.
4. The PDI activity of NusA-Nox2 was suppressed by PAO.
5. A comparison of the sequence of the DHR of human Nox2 with several PDIs, revealed three small regions of homology with PDIA3.
6. Two monoclonal anti-Nox2 antibodies, with epitopes corresponding to two regions of Nox2/PDIA3 similarity, reacted with PDIA3 but not with PDIA1.
7. One polyclonal anti-PDIA3 antibody (but not an anti-PDIA1 antibody) reacted strongly with recombinant NusA-Nox2.
8. p67^{phox}(1–212) in which all four cysteines were mutated to serines lost the ability to bind to a Nox2 peptide comprising the CGC triad and its capacity to activate the oxidase was impaired.

unsolved issue and depends on the conditions leading to the activation of the oxidase (van Bruggen et al., 2004; Tlili et al., 2012).

We can envisage several mechanisms for the oxidation of cysteines in Nox2. The oxidizing agent might be H_2O_2 derived by dismutation of $O_2^{\cdot-}$ generated by the oxidase itself, leaking into the cytosolic milieu of the DHR of Nox2 (Enyedi et al., 2013). Another possibility is the intercalation of the CGC triad into the redox cascade of Nox2, from NADPH, via FAD/FADH$_2$, to heme (Cross and Segal, 2004). An additional hypothetical mechanism is oxidation by an exogenous PDI in the oxidized form or by another thiol oxidase. Paes et al. (2011) indeed showed that addition

PDIA3 (also known as ERp57 or GRp58) is a 505-residues long protein, with a molecular mass of 56.78 kDa, found in the endoplasmic reticulum but also on the cell surface, plasma membrane, nucleus, cytosol, and in the secreted form (Frickel et al., 2004; Coe and Michalak, 2010). Its structure is similar to other PDIs (33% overall identity), with two catalytic domains (*a* and *a'*) comprising the CGHC motif and disulfide reduction potentials of −167 and −156 mV, respectively. It catalyzes disulfide reduction, dithiol oxidation, and disulfide isomerization. Its reductase activity was reported to be 20 times less efficient than that of PDIA1, using the insulin assay (Frickel et al., 2004); this is in disagreement with our finding, based on the DE-GSSG reduction assay, of a 30-fold higher activity of PDIA3 compared to PDIA1. PDIA3 promotes

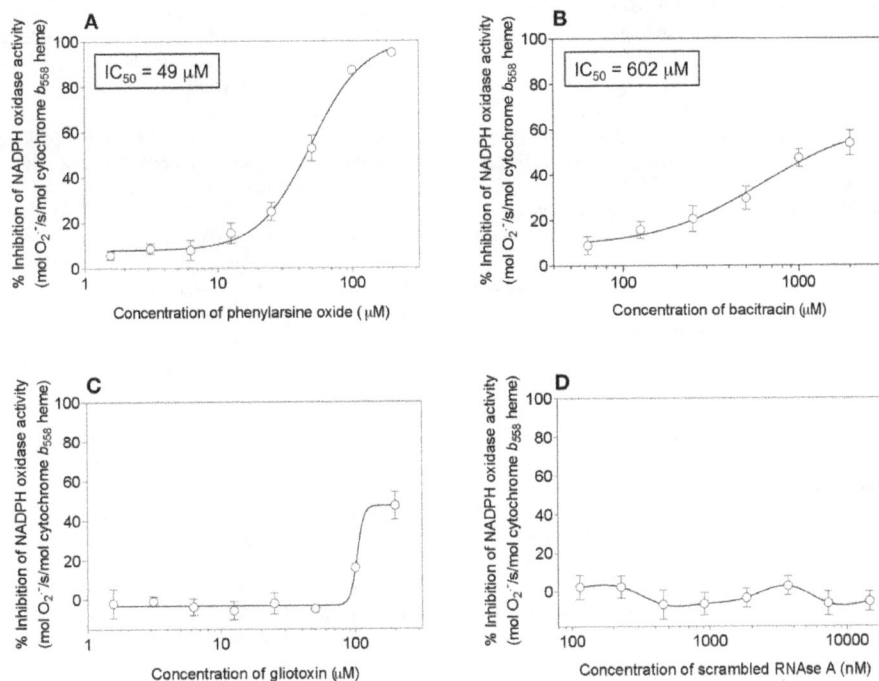

FIGURE 14 | Effect of PDI inhibitors on NADPH oxidase activation in a cell-free system. Selected PDI inhibitors were tested for their ability to interfere with NADPH oxidase activation in an amphiphile- and p47^{phox}-independent system. The reaction mixtures contained macrophage membrane liposomes (5 nM cytochrome b_{558} heme), p67^{phox}(1–212) (300 nM), and prenylated Rac1 Q61L (300 nM). The membrane liposomes were preincubated with the PDI inhibitors for 5 min, followed by the addition of p67^{phox} and Rac1, incubation for 5 min in the absence of an amphiphilic activator, and the addition of 240 μM NADPH, to initiate O_2^{-} production. The effect of the following inhibitors is described in the figure: **(A)** phenylarsine oxide (1.56–200 μM); **(B)** bacitracin (62.5–2000 μM); **(C)** gliotoxin (1.56–200 μM), and **(D)** scrambled RNAse (0.114–14.6 μM). Results represent means ± SEM of 3 experiments, for each of the inhibitors tested. IC$_{50}$ values are indicated for phenylarsine oxide and bacitracin but could not be determined for gliotoxin and scrambled RNAse, due to the absence of curves amenable to kinetic analysis.

folding of glycoproteins in the endoplasmic reticulum and is a key component of the MHC-I peptide loading complex (Purcell and Elliot, 2008). A newly discovered role is that of a GDP dissociation inhibitor for the Ras family small GTPase Ra1A (Brymora et al., 2012).

The PDI reductase activity of NusA-Nox2(357–570) is much lower than that of recombinant PDIA1 and PDIA3, when compared by the DE-GSSG reduction assay. The reason for this might be the difference between the catalytic motifs (CGC, in Nox2, and CGHC, in most PDIs), the presence of two catalytic motifs in most PDIs, or the lesser affinity of Nox2 for the DE-GSSG substrate. Our results do not allow the allocation of functional roles to regions of homology between Nox2 and PDIA3. An evolutionary link between the two proteins is also unlikely. The most conservative interpretation is to look upon Nox2 as a relative of the large family of thiol-disulfide oxidoreductases, characterized by a CXXC motif (reviewed in Sevier and Kaiser, 2002).

In conclusion, we have described a rather unique situation, in which one protein (Nox2) mimics the enzymatic function of a seemingly unrelated protein (PDIA3), with which it shares a similar catalytic motif and minor sequence homology. We plan to expand this study by identifying the individual cysteines in p67^{phox} involved in binding to Nox2 and by designing NADPH oxidase inhibitors focused on the Nox2 CGC - p67^{phox} interaction.

AUTHORS CONTRIBUTIONS

Edna Bechor, Iris Dahan, Tanya Fradin, Yevgeny Berdichevsky, Anat Zahavi, Aya Federman Gross, Meirav Rafalowski and Edgar Pick were involved in the design and performance of the experiments and in the analysis of the results. Edgar Pick planed the conceptual framework and wrote the manuscript, which was approved by all the authors.

ACKNOWLEDGMENTS

This work was supported by grants No. 49/09 and 300/13 from the Israel Science Foundation, the Roberts Fund, and the Joseph and Shulamit Salomon Fund. We thank Adva Yeheskel for assistance with bioinformatics.

REFERENCES

Ali Khan, H., and Mutus, B. (2014). Protein disulfide isomerase a multifunctional protein with multiple physiological roles. *Front. Chem.* 2:70. doi: 10.3389/fchem.2014.00070

Appenzeller-Herzog, C., and Ellgaard, L. (2008). The human PDI family: versatility packedinto a single fold. *Biochim. Biophys. Acta* 1783, 535–548. doi: 10.1016/j.bbamcr.2007.11.010

Babior, B. M. (2002). The activity of leukocyte NADPH oxidase: regulation by p47^{phox} cysteine and serine residues. *Antiox. Redox Signal.* 4, 35–38. doi: 10.1089/152308602753625834

Baniulis, D., Nakano, Y., Nauseef, W. M., Banfi, B., Cheng, G., Lambeth, D. J., et al. (2005). Evaluation of two anti-gp91^{phox} antibodies as immunoprobes for Nox family proteins: mAb 54.1 recognizes recombinant full-length Nox2, Nox3 and the C-terminal domains of Nox1-4 and reacts with GRP 58. *Biochim. Biophys. Acta* 1752, 186–196. doi: 10.1016/j.bbapap.2005.07.018

Bedard, K., and Krause, K.-H. (2007). The Nox family of ROS-generating NADPH oxidases: physiology and pathophysiology. *Physiol. Rev.* 87, 245–313. doi: 10.1152/physrev.00044.2005

Bellavite, P., Cross, A. R., Serra, M. C., Davoli, A., Jones, O. T. G., and Rossi, F. (1983). The cytochrome b and flavin content and properties of the O_2^-–forming NADPH oxidase solubilized from activated neutrophils. *Biochim. Biophys. Acta* 746, 40–47. doi: 10.1016/0167-4838(83)90008-0

Bradford, M. M. (1976). A rapid and sensitive method for the quantitation of microgram quantities of protein utilizing the principle of protein-dye binding. *Anal. Biochem.* 72, 248–254. doi: 10.1016/0003-2697(76)90527-3

Bromberg, Y., and Pick, E. (1984). Unsaturated fatty acids stimulate NADPH-dependent superoxide production by cell-free system derived from macrophages. *Cell. Immunol.* 88, 213–221. doi: 10.1016/0008-8749(84)90066-2

Bromberg, Y., and Pick, E. (1985). Activation of NADPH-dependent superoxide production in a cell-free system by sodium dodecyl sulfate. *J. Biol. Chem.* 260, 13539–13545.

Brymora, A., Duggin, I. G., Berven, L. A., van Dam, E. M., Roufogalis, B. D., and Robinson, P. J. (2012). Identification and characterization of the RalA-ERp57 interaction: evidence for GDI activity of ERp57. *PLoS ONE* 7:e50879. doi: 10.1371/journal.pone.0050879

Burritt, J. B., Foubert, T. R., Baniulis, D., Lord, C. I., Taylor, R. M., Mills, J. S., et al. (2003). Functional epitope on human neutrophil flavocytochrome b_{558}. *J. Immunol.* 170, 6082–6089. doi: 10.4049/jimmunol.170.12.6082

Burritt, J. B., Quinn, M. T., Jutila, M. A., Bond, C. W., and Jesaitis, A. J. (1995). Topological mapping of neutrophil cytochrome *b* epitopes with phage-display libraries. *J. Biol. Chem.* 270, 16974–16980. doi: 10.1074/jbc.270.28.16974

Coe, H., and Michalak, M. (2010). ERp57, a multifunctional endoplasmic reticulum resident oxidoreductase. *Int. J. Biochem. Cell Biol.* 42, 796–799. doi: 10.1016/j.biocel.2010.01.009

Cohen, H., and Chovaniec, M. E. (1978). Superoxide production by digitonin-stimulated guinea pig granulocytes. The effects of N-ethyl maleimide, divalent cations, and glycolytic mitochondrial inhibitors on the activation of the super-oxide generating system. *J. Clin. Invest.* 61, 1088–1096. doi: 10.1172/JCI109008

Collet, J.-F., and Messens, J. (2010). Structure, function, and mechanism of thioredoxin proteins. *Antioxid. Redox Signal.* 13, 1205–1216. doi: 10.1089/ars.2010.3114

Cross, A. R., and Segal, A. W. (2004). The NADPH oxidase of professional phagocytes –prototype of the NOX electron transport chain. *Biochim. Biophys. Acta* 1657, 1–22. doi: 10.1016/j.bbabio.2004.03.008

Dahan, I., Issaeva, I., Sigal, N., Gorzalczany, Y., and Pick, E. (2002). Mapping of functional domains in the p22^{phox} subunit of flavocytochrome b_{559} participating in the assembly of the NADPH oxidase complex by "peptide walking." *J. Biol. Chem.* 277, 8421–8432. doi: 10.1074/jbc.M109778200

Dahan, I., Molshanski-Mor, S., and Pick, E. (2012). Inhibition of NADPH oxidase activation by peptides mapping within the dehydrogenase region of Nox2 - A "peptide walking" study. *J. Leukoc. Biol.* 91, 501–515. doi: 10.1189/jlb.1011507

Dahan, I., and Pick, E. (2012). Strategies for identifying synthetic peptides to act as inhibitors of NADPH oxidases, or "All that you did and did not want to know about Nox inhibitory peptides." *Cell. Mol. Life Sci.* 69, 2283–2305. doi: 10.1007/s00018-012-1007-4

Dang, P. M.-C., Cross, A. R., Quinn, M. T., and Babior, B. M. (2002). Assembly of the neutrophil respiratory burst oxidase: a direct interaction between p67^{phox} and cytochrome b_{558} II. *Proc. Natl. Acad. Sci. U.S.A.* 99, 4262–4265. doi: 10.1073/pnas.072345299

Davis, G. D., Elisee, C., Newham, D. M., and Harrison, R. G. (1999). New fusion protein systems designed to give soluble expression in *Escherichia coli*. *Biotechnol. Bioeng.* 65, 382–388.

Debeurme, F., Picciocchi, A., Dagher, M.-C., Grunwald, D., Beaumel, S., Fieschi, F., et al. (2010). Regulation of NADPH oxidase activity in phagocytes. Relationship between FAD/NADPH binding and oxidase complex assembly. *J. Biol. Chem.* 285, 33197–33208. doi: 10.1074/jbc.M110.151555

Dickerhof, N., Kleffmann, T., Jack, R., and McCormick, S. (2011). Bacitracin inhibits the reductive activity of protein disulfide isomerase by disulfide bond formation with free cysteines in the substrate-binding domain. *FEBS J.* 278, 2034–2043. doi: 10.1111/j.1742-4658.2011.08119.x

Doussiere, J., Bouzidi, F., Poinas, A., Gaillard, J., and Vignais, P. V. (1999). Kinetic study of the activaton of the neutrophil oxidase by arachidonic acid. Antagonistic effects of arachidonic acid and phenylarsine oxide. *Biochemistry* 38, 16394–16406. doi: 10.1021/bi991502w

Doussiere, J., Poinas, A., Blais, C., and Vignais, P. V. (1998). Phenylarsine oxide as an inhibitor of the activation of the neutrophil NADPH oxidase. Identification of the β subunit of the flavocytochrome b component of the NADPH oxidase as a target site for phenylarsine oxide by photoaffinity labeling and pho-toinactivation. *Eur. J. Biochem.* 251, 649–658. doi: 10.1046/j.1432-1327.1998.2510649.x

Ellgaard, L., and Ruddock, L. W. (2005). The human protein disulfide isomerase family: sustrate interactions and functional properties. *EMBO Rep.* 6, 28–32. doi: 10.1038/sj.embor.7400311

Enyedi, B., Zana, M., Donkó, A., and Geiszt, M. (2013). Spatial and temporal analysis of NADPH oxidase-generated hydrogen peroxide signals by novel fluorescent reporter proteins. *Antiox. Redox Signal.* 19, 523–534. doi: 10.1089/ars.2012.4594

Essex, D. W., and Li, M. (1999). Protein disulfide isomerase mediates platelet aggregation and secretion. *Br. J. Haematol.* 104, 448–454. doi: 10.1046/j.1365-2141.1999.01197.x

Federman Gross, A., Zahavi, A., Berdichevsky, Y., Mizrahi, A., and Pick, E. (2012). Binding of p67^{phox} to the dehydrogenase region of Nox2 – Protein–protein interaction study reveals a requirement for a conformational change in p67^{phox}. *Eur. J. Clin. Invest.* 42, 25.

Fradin, T., Dahan, I., Molshanski-Mor, S., Mizrahi, A., Berdichevsky, Y., and Pick, E. (2011). A dithiol - disulfide switch in the cytosolic part of Nox2 controls NADPH oxidase assembly. *Eur. J. Clin. Invest.* 41, 29.

Fradin, T., Mizrahi, A., and Pick, E. (2012). From complex biology to elementary chemistry- A redox switch in Nox2 regulates p67^{phox} - Nox2 interaction. *Eur. J. Clin. Invest.* 42, 36.

Frickel, E.-M., Frei, P., Bouvier, M., Stafford, W. F., Helenius, A., Glockshuber, R., et al. (2004). Erp57 is a multifunctional thiol-disulfide oxidorerductase. *J. Biol. Chem.* 279, 18277–18287. doi: 10.1074/jbc.M314089200

Gabig, T. G., and Lefker, B. A. (1984). Catalytic properties of the resolved flavo-protein and cytochrome b components of the NADPH-dependent O_2^- - generating oxidase from human neutrophils. *Biochem. Biophys. Res. Commun.* 118, 430–436. doi: 10.1016/0006-291X(84)91321-4

Gallina, A., Hanley, T. M., Mandel, R., Trahey, M., Broder, C. C., Viglianti, G. A., et al. (2002). Inhibitors of protein disulfide isomerase prevent cleavage of disul-fide bonds in receptor-bound glycoprotein 120 and prevent HIV-1 entry. *J. Biol. Chem.* 277, 50579–50588. doi: 10.1074/jbc.M204547200

Galligan, J. J., and Petersen, D. R. (2012). The human protein disulfide isomerase gene family. *BMC Hum. Genomics* 6:6. doi: 10.1186/1479-7364-6-6

Gorzalczany, Y., Alloul, N., Sigal, N., Weinbaum, C., and Pick, E. (2002). A preny-lated p67phox-Rac1 chimera elicits NADPH-dependent superoxide production by phagocyte membranes in the absence of an activator and of p47phox. Conversion of a pagan NADPH oxidase to monotheism. *J. Biol. Chem.* 277, 18605–18610. doi: 10.1074/jbc.M202114200

Gorzalczany, Y., Sigal, N., Itan, M., Lotan, O., and Pick, E. (2000). Targeting of Rac1 to the phagocyte membrane is sufficient for the induction of NADPH oxidase assembly. *J. Biol. Chem.* 275, 40073–40081. doi: 10.1074/jbc.M0060 13200

Groemping, Y., and Rittinger, K. (2005). Activation and assembly of the NADPH oxidase: a structural perspective. *Biochem. J.* 386, 401–416. doi: 10.1042/BJ20041835

Han, C.-H., Freeman, J. L. R., Lee, T. H., Motalebi, S. A., and Lambeth, J. D. (1998). Regulation of the neutrophil respiratory burst oxidase – identification of an activation domain in p67phox. *J. Biol. Chem.* 273, 16663–16668.

Hoffstrom, B. G., Kaplan, A., Letso, R., Schmid, R. S., Turmel, G. J., Lo, D. C., et al. (2010). Inhibitors of protein disulfide isomerase suppress apoptosis induced by misfolded proteins. *Nature Chem. Biol.* 6, 900–906. doi: 10.1038/nchembio.467

Holmgren, A. (1979). Thioredoxin catalyzes the reduction of insulin disulfides by dithiothreitol and dihydrolipoamide. *J. Biol. Chem.* 254, 9627–9632.

Joseph, G., and Pick, E. (1995). "Peptide walking" is a novel method of map-ping functional domains in proteins. Its application to the Rac1-dependent activation of NADPH oxidase. *J. Biol. Chem.* 270, 29079–29082.

Kawahara, T., Quinn, M. T., and Lambeth, J. D. (2007). Molecular evolution of the reactive oxygen-generating NADPH oxidase (Nox/Duox) family of enzymes. *BMC Evol. Biol.* 7:109. doi: 10.1186/1471-2148-7-109

Kersteen, E. A., and Raines, R. T. (2003). Catalysis of protein folding disulfide isomerase and small-molecule mimics. *Antioxid. Redox Signal.* 5, 413–424. doi: 10.1089/152308603768295159

Knoller, S., Shpungin, S., and Pick, E. (1991). The membrane-associated component of the amphiphile-activated cytosol-dependent superoxide forming NADPH oxidase of macrophages is identical to cytochrome b_{559}. *J. Biol. Chem.* 266, 2795–2804.

Kosower, E. M., and Kosower, N. S. (1995). Bromobimane probes for thiols. *Methods Enzymol.* 251, 133–148.

Kozlov, G., Määttänen, P., Thomas, D. Y., and Gehring, K. (2010). A structural view of the PDI family of proteins. *FEBS J.* 277, 3924–3936. doi: 10.1111/j.1742-4658.2010.07793.x

Kreck, M. L., Freeman, J. L., Abo, A., and Lambeth, J. D. (1996). Membrane association of Rac is required for high activity of the respiratory burst oxidase. *Biochemistry* 35, 15683–15692. doi: 10.1021/bi962064l

Laurindo, F. R. M., Fernandes, D. C., Amanso, A. M., Lopes, L. R., and Santos, C. X. C. (2008). Novel role of protein disulfide isomerase in the regulation of NADPH oxidase activity; Pathophysiological implications in vascular diseases. *Antioxid. Redox Signal.* 10, 1101–1113. doi: 10.1089/ars.2007.2011

Le Cabec, V., and Maridonneau-Parini, I. (1995). Complete and reversible inhibition of NADPH oxidase in human neutrophils by phenyarsine oxide at a step distal to membrane translocation of the enzyme subunits. *J. Biol. Chem.* 270, 2067–2073.

Leusen, J. H. W., Meischl, C., Eppink, M. H. M., Hilarius, P. M., de Boer, M., Weening, R. S., et al. (2000). Four novel mutations in the gene encoding gp91^{phox} of human NADPH oxidase: consequences for oxidase activity. *Blood* 95, 666–673.

Martinez-Galisteo, E., Padilla, C. A., Garcia-Alfonso, C., Lopez-Barea, J., and Barcena, J. A. (1993). Purification and properties of bovine thioredoxin system. *Biochimie* 75, 803–809. doi: 10.1016/0300-9084(93)90131-B

Mizrahi, A., Berdichevsky, Y., Casey, P. J., and Pick, E. (2010). A prenylated p47^{phox}-p67^{phox}-Rac1 chimera is a quintessential NADPH oxidase activator. Membrane association and functional capacity. *J. Biol. Chem.* 285, 25485–25499. doi: 10.1074/jbc.M110.113779

Morozov, I., Lotan, O., Joseph, G., Gorzalczany, Y., and Pick, E. (1998). Mapping of functional domains in p47^{phox} involved in the activation of NADPH oxidase by "peptide walking." *J. Biol. Chem.* 273, 15435–15444. doi: 10.1074/jbc.273.25.15435

Nishida, S., Yoshida, L. S., Shimoyama, T., Nunoi, H., Kobayashi, T., and Tsunawaki, S. (2005). Fungal metabolite gliotoxin targets flavocytochrome b_{558} in the activation of the human neutrophil NADPH oxidase. *Infect. Immun.* 73, 235–244. doi: 10.1128/IAI.73.1.235-244.2005

Ostuni, M. A., Gelinotte, M., Bizouarn, T., Baciou, L., and Houée-Levin, C. (2010). Targeting NADPH oxidase by reactive oxygen species reveals an initial sensitive step in the assembly process. *Free Rad. Biol. Med.* 49, 900–907. doi: 10.1016/j.freeradbiomed.2010.06.021

Paes, A. M. de A., Verissimo-Filho, S., Lopes Guimaraes, L., Silva, A. C. B., Takiuti, J. T., Santos, C. X., et al. (2011). Protein disulfide isomerase redox-dependent association with p47^{phox}: evidence for an organizer role in leukocyte NADPH oxidase activation. *J. Leukoc. Biol.* 90, 799–810. doi: 10.1189/jlb.0610324

Pick, E. (2012). A dithiol/disulfide redox switch in the dehydrogenase region in Nox2 regulates the assembly of the superoxide-generating NADPH oxidase of phagocytes. *Free Rad. Biol. Med.* 53, S57. doi: 10.1016/j.freeradbiomed.2012.08.540

Pick, E. (2014). "Cell-free NADPH oxidase activation assays: *in vitro* veritas," in *Neutrophil Methods and Protocols, 2nd Edn*, eds M. T. Quinn and F. R. DeLeo (New York, NY; Heidelberg; Dordrecht; London: Humana Press), 339–403. doi: 10.1007/978-1-62703-845-4_22

Pick, E., Bromberg, Y., Shpungin, S., and Gadba, R. (1987). Activation of the superoxide forming NADPH oxidase in a cell-free system by sodium dodecyl sulfate. Characterization of the membrane-associated component. *J. Biol. Chem.* 262, 16476–16483.

Purcell, A. W., and Elliot, T. (2008). Molecular machinations of the MHC-I peptide loading complex. *Curr. Opin. Immunol.* 20, 75–81. doi: 10.1016/j.coi.2007.12.005

Quinn, M. T., and Gauss, K. A. (2004). Structure and regulation of the neutrophil respiratory burst oxidase: comparison with nonphagocyte oxidases. *J. Leukoc. Biol.* 76, 760–781. doi: 10.1189/jlb.0404216

Raturi, A., and Mutus, B. (2007). Characterization of redox state and reductase activity of protein disulfide isomerase under different redox environments using a sensitive fluorescent assay. *Free Rad. Biol. Med.* 43, 62–70. doi: 10.1016/j.freeradbiomed.2007.03.025

Raturi, A., Vacratsis, P. O., Seslija, D., Lee, L., and Mutus, B. (2005). A direct, continuous, sensitive assay for protein disulfide isomerase based on fluorescence sef-quenching. *Biochem. J.* 391, 351–357. doi: 10.1042/BJ20050770

Santos, C. X. C., Stolf, B. S., Takemoto, P. V. A., Amanso, A. M., Lopes, L. R., Souza, E. B., et al. (2009). Protein disulfide isomerase (PDI) associates with NADPH oxidase and is required for phagocytosis of *Leishmania chagasi* promastigotes by macrophages. *J. Leukoc. Biol.* 86, 989–998. doi: 10.1189/jlb.0608354

Serrander, L., Cartier, L., Bedard, K., Banfi, B., Lardy, B., Plastre, O., et al. (2007). Nox4 activity is determined by mRNA levels and reveals unique pattern of ROS generation. *Biochem. J.* 406, 105–114. doi: 10.1042/BJ20061903

Sevier, C. S., and Kaiser, C. A. (2002). Formation and transfer of disulphide bonds in living cells. *Nature Rev. Mol. Cell. Biol.* 3, 836–847. doi: 10.1038/nrm954

Sha'ag, D. (1989). Sodium dodecyl sulfate dependent NADPH oxidation: an alternative method for assaying NADPH oxidase in a cell-free system. *J. Biochem. Biophys. Methods* 19, 121–128. doi: 10.1016/0165-022X(89)90056-0

Shpungin, S., Dotan, I., Abo, A., and Pick, E. (1989). Activation of the superoxide forming NADPH oxidase in a cell-free system by sodium dodecyl sulfate. Absolute lipid dependence of the solubilized enzyme. *J. Biol. Chem.* 264, 9195–9203.

Jasuja, R., Passam, F., Kennedy, D. R., Kim, S. J., van Hessem, L., Lin, L., et al. (2012). Protein disulfide isomerase inhibitors constitute a new class of antithrombotic agents. *J. Clin. Invest.* 122, 2104–2113. doi: 10.1172/JCI61228

Sumimoto, H. (2008). Structure, regulation and evolution of Nox-family NADPH oxidases that produce reactive oxygen species. *FEBS J.* 275, 3249–3277. doi: 10.1111/j.1742-4658.2008.06488.x

Tlili, A., Erard, M., Faure, M. -C., Baudin, X., Piolot, T., Dupré-Crochet, S., et al. (2012). Stable accumulation of p67^{phox} at the phagosomal membrane and ROS production within the phagosome. *J. Leukoc. Biol.* 91, 83–95. doi: 10.1189/jlb.1210701

Tsunawaki, S., Yoshida, L., Nishida, S., Kobayashi, T., and Shimoyama, T. (2004). Fungal metabolite gliotoxin inhibits assembly of the human respiratory burst NADPH oxidase. *Infect. Immun.* 72, 3373–3382. doi: 10.1128/IAI.72.6.3373-3382.2004

Turano, C., Coppari, S., Altieri, F., and Ferraro, A. (2002). Proteins of the PDI family: unpredicted non-ER locations and functions. *J. Cell. Physiol.* 193, 154–163. doi: 10.1002/jcp.10172

van Bruggen, R., Anthony, E., Fernandez-Borja, M., and Roos, D. (2004). Continuous translocation of Rac2 and the NADPH oxidase component p67^{phox} during phagocytosis. *J. Biol. Chem.* 279, 9097–9102. doi: 10.1074/jbc.M309284200

van Iwaarden, P. R., Driessen, A. J. M., and Konings, W. N. (1992). What we can learn from the effects of thiol reagents on transport proteins. *Biochim. Biophys. Acta* 1113, 161–170. doi: 10.1016/0304-4157(92)90037-B

Wang, G.-Z., Dong, X.-Y., and Sun, Y. (2011). Peptide disulfides CGC and RKCGC facilitate oxidative protein refolding. *Biochem. Eng.* 55, 169–175. doi: 10.1016/j.bej.2011.04.002

Waring, P., and Beaver, J. (1996). Gliotoxin and related epipolythiodioxopiperazines. *Gen. Pharmac.* 27, 1311–1316. doi: 10.1016/S0306-3623(96)00083-3

Watanabe, M. M., Laurindo, F. R. M., and Fernandes, D. C. (2014). Methods of measuring protein disulfide isomerase activity: a critical overview. *Front. Chem.* 2:73. doi: 10.3389/fchem.2014.00073

Woycechowsky, K. J., and Raines, R. T. (2003). The CXC motif: a functional mimic of protein disulfide isomerase. *Biochemistry* 42, 5387–5394. doi: 10.1021/bi026993q

Xu, S., Butkevich, A. N., Yamada, R., Zhou, Y., Debnath, B., Duncan, R., et al. (2012). Discovery of an orally active small-molecule irreversible inhibitor of protein disulfide isomerase for ovarian cancer treatment. *Proc. Natl. Acad. Sci. U.S.A.* 109, 16348–16353. doi: 10.1073/pnas.1205226109

Yamashita, T., Someya, A., and Tzuzawa-Kido, Y. (1984). Effect of maleimide derivatives on superoxide-generating system of guinea pig neutrophils stimulated by different soluble stimuli. *Eur. J. Biochem.* 145, 71–76. doi: 10.1111/j.1432-1033.1984.tb08523.x

Zhang, R., and Snyder, G. H. (1989). Dependence of formation of small disulfide loops in two-cysteine peptides on the number and types of intervening amino acids. *J. Biol. Chem.* 264, 18472–18479.

Conflict of Interest Statement: The authors declare that the research was conducted in the absence of any commercial or financial relationships that could be construed as a potential conflict of interest.

13

Estimation of kinetic parameters related to biochemical interactions between hydrogen peroxide and signal transduction proteins

Paula M. Brito[1,2,3] *and Fernando Antunes*[4]*

[1] URIA-Centro de Patogénese Molecular, Faculdade de Farmácia, Universidade de Lisboa, Lisboa, Portugal
[2] Instituto de Medicina Molecular, Faculdade de Medicina da Universidade de Lisboa, Lisboa, Portugal
[3] Faculdade de Ciências da Saúde, Universidade da Beira Interior, Covilhã, Portugal
[4] Departamento de Química e Bioquímica and Centro de Química e Bioquímica, Faculdade de Ciências, Universidade de Lisboa, Lisboa, Portugal

Edited by:
Bulent Mutus, University of Windsor, Canada

Reviewed by:
Peizhong Mao, Oregon Health and Science University, USA
Saptarshi Kar, The University of Western Australia, Australia

*Correspondence:
Fernando Antunes, Departamento de Química e Bioquímica and Centro de Química e Bioquímica, Faculdade de Ciências, Universidade de Lisboa, Campo Grande, P-1749-016 Lisboa, Portugal
e-mail: fantunes@fc.ul.pt

The lack of kinetic data concerning the biological effects of reactive oxygen species is slowing down the development of the field of redox signaling. Herein, we deduced and applied equations to estimate kinetic parameters from typical redox signaling experiments. H_2O_2-sensing mediated by the oxidation of a protein target and the switch-off of this sensor, by being converted back to its reduced form, are the two processes for which kinetic parameters are determined. The experimental data required to apply the equations deduced is the fraction of the H_2O_2 sensor protein in the reduced or in the oxidized state measured in intact cells or living tissues after exposure to either endogenous or added H_2O_2. Either non-linear fittings that do not need transformation of the experimental data or linearized plots in which deviations from the equations are easily observed can be used. The equations were shown to be valid by fitting to them virtual time courses simulated with a kinetic model. The good agreement between the kinetic parameters estimated in these fittings and those used to simulate the virtual time courses supported the accuracy of the kinetic equations deduced. Finally, equations were successfully tested with real data taken from published experiments that describe redox signaling mediated by the oxidation of two protein tyrosine phosphatases, PTP1B and SHP-2, which are two of the few H_2O_2-sensing proteins with known kinetic parameters. Whereas for PTP1B estimated kinetic parameters fitted in general the present knowledge, for SHP-2 results obtained suggest that reactivity toward H_2O_2 as well as the rate of SHP-2 regeneration back to its reduced form are higher than previously thought. In conclusion, valuable quantitative kinetic data can be estimated from typical redox signaling experiments, thus improving our understanding about the complex processes that underlie the interplay between oxidative stress and redox signaling responses.

Keywords: redox regulation, redox signaling, kinetics, rate constant, PTP1B, SHP-2, protein tyrosine phosphatases

INTRODUCTION

Being higher reductions states of molecular dioxygen, reactive oxygen species are present in all aerobic organisms. Initially, these species were seen as harmful species that caused or participated in the etiology of many diseases through oxidative damage, but more recently physiological roles mediated by the modulation of the redox state of biomolecules were attributed to reactive oxygen species (Sies, 2014). Today redox biology is an established field. As Berzelius put it the venom is in the dose, and reactive oxygen species have different roles depending on their concentration. This work is centered on hydrogen peroxide (H_2O_2), a reactive oxygen species that has the properties of a second messenger (Forman et al., 2010) and participates in many pathways, including insulin (Mahadev et al., 2001; Haque et al., 2011), mitogenic (Irani et al., 1997), inflammatory, and apoptotic signaling (Oakley et al., 2009; Tschopp and Schroder, 2010). Having a relative low chemical reactivity, H_2O_2 reacts mainly with metal centers and with thiol compounds, such as cysteine residues in proteins (Marinho et al., 2014). Examples of H_2O_2 targets are PerR, a metal-dependent transcription factor that is inhibited by H_2O_2 in a Fenton-like reaction, and protein tyrosine phosphatases (PTPs), which are inhibited upon oxidation of cysteine residues in their active center (Tanner et al., 2011; Marinho et al., 2014; Sies, 2014). The list of proteins containing cysteine residues that were observed to be oxidized by H_2O_2 is vast, near 200 (Le Moan et al., 2006; Martínez-Acedo et al., 2012), and continues to increase as investigators find new targets for H_2O_2. In contrast, kinetic parameters concerning oxidation by H_2O_2 have been

Abbreviations: PTP1B, Protein tyrosine phosphatase 1; PTPs, protein tryrosine phosphatases; the Src homology 2 (SH2) domain containing phosphotyrosine phosphatase 2 (Shp-2).

measured only for a few of these proteins (Ferrer-Sueta et al., 2011; Tanner et al., 2011), leading several researchers to point out the lack of proper quantitative data as a barrier to the development of the field (Brigelius-Flohé and Flohé, 2011; Buettner et al., 2013). Importantly, the triggering of biphasic responses by H_2O_2 in a narrow concentration range has important biological implications. For example, in H4IIEC hepatocytes H_2O_2 can either enhance or impair insulin signaling depending on its concentration (Iwakami et al., 2011). This dual role was attributed to the different sensitivity of PTP1B inhibition and JNK activation, two kinases that stimulate and inhibit insulin signaling, respectively. Thus, while H_2O_2 is an essential component of the insulin signaling pathway, it may also mediate the etiology of insulin resistance (Fisher-Wellman and Neufer, 2012). Although the underlying data is known for some time, such picture only emerged recently, probably because in absence of a quantitative framework, these biphasic responses were often interpreted as contradictory findings that were dependent of the biological model used or simply reflected non-reproducible experimental results. To study such complex responses it is advantageous to apply a quantitative and integrative approach typical of systems biology (Buettner et al., 2013), where the reactivity of targets toward H_2O_2 is determined to undercover which pathways operate *in vivo* under different conditions. In this work, we address how kinetic parameters can be determined from typical experiments performed in redox signaling.

Based on a simple reaction scheme representing H_2O_2 signaling, we started by deducing kinetic equations that are tailored to estimate kinetic parameters from experimental data. Next, to test the validity of the deduced equations, virtual experiments carried out under different conditions of H_2O_2 exposure were simulated with a kinetic model, and the results were fitted to the equations deduced. The agreement between the kinetic parameters obtained in these fittings and those used to obtain the virtual time courses was used as a criterion to decide on the accuracy of the kinetic equations deduced. Finally, to evaluate the applicability of kinetic equations to real data, experimental results described in the literature focusing on PTP-dependent signaling were fitted to the equations deduced. Two PTPs, PTP1B, and SHP-2, for which kinetic rate constants are known, were chosen as test cases. Our study demonstrates that insightful kinetic parameters related to biochemical interactions between H_2O_2 and signal transduction proteins can be estimated from typical H_2O_2-signaling experiments by applying the equations deduced here.

THEORY AND METHODS
MASTER EQUATION
A minimal mathematical analytical model was set up to describe a signaling event triggered by H_2O_2, according to the following two reactions:

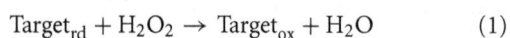

$$\text{Target}_{rd} + H_2O_2 \rightarrow \text{Target}_{ox} + H_2O \qquad (1)$$

$$\text{Target}_{ox} \rightarrow \text{Target}_{rd} \qquad (2)$$

In the first reaction, the reduced form of a sensor protein target (Target_{rd}) is oxidized by H_2O_2, modifying its activity, which

results in the modulation of a signaling pathway. In the second reaction, the oxidized target (Target_{ox}) is switched-off by being regenerated back to the reduced form. A specific example of these two reactions is the inhibition of PTPs by oxidation of cysteine residues in their active center, which are reactivated upon reduction of this site; the temporary inhibition of these phosphatases increases the level of phosphorylation of their targets, thus promoting the signaling process. For these two reactions rate laws were defined as follows:

- For the H_2O_2-dependent oxidation step (1) $v_1 = k_{activation} \times [\text{Target}_{rd}]$, where $k_{activation} = k_{target+H2O2} \times [H_2O_2]$. $k_{target+H2O2}$ is the rate constant for the direct oxidation of the target protein by H_2O_2.
- For the switch-off step (2), $v_2 = k_{switchoff} \times [\text{Target}_{ox}]$. The total concentration of the target protein is assumed to be constant within the duration of the experiment ($[\text{Target}]_{total} = [\text{Target}_{ox}] + [\text{Target}_{rd}]$), and so $v_2 = k_{switchoff} \times ([\text{Target}]_{total} - [\text{Target}_{rd}])$.

Based on these two chemical reactions, the following differential equation was set up, where Target_{rd} is the fraction of the target protein in the reduced state, t is time, and d/dt stands for the differential operator:

$$\frac{d\text{Target}_{rd}}{dt} = k_{switchoff}\left(1 - \text{Target}_{rd}\right) - k_{activation}\text{Target}_{rd} \qquad (3)$$

The master equation describing the time course of Target_{rd} is given by the analytical solution of Equation (3):

$$\text{Target}_{rd}\big|_t = \frac{k_{switchoff}}{k_{switchoff} + k_{activation}} + e^{-(k_{switchoff}+k_{activation})\times t}$$
$$\times \left(\text{Target}_{rd}\big|_0 - \frac{k_{switchoff}}{k_{switchoff} + k_{activation}}\right) \qquad (4)$$

With

$$\text{Target}_{ox}\big|_t = 1 - \text{Target}_{rd}\big|_t \qquad (5)$$

$\text{Target}_{rd}\big|_t$ and $\text{Target}_{ox}\big|_t$ are the fractions of the target protein in the reduced and oxidized state at time t, respectively. Once the experimental variation of these fractions with time is known and the fraction of reduced target at time 0 ($\text{Target}_{ox}\big|_0$) is measured, a non-linear fit to Equation (4) can be applied to estimate the kinetic parameters $k_{activation}$ and $k_{switchoff}$. One possibility is to estimate these two unknown parameters from a two-parameter non-linear fitting. Alternatively, if one of these parameters is already known, only the remaining unknown parameter is estimated from a one-parameter non-linear fitting.

Next, we linearized Equation (4) so that kinetic parameters can be determined from linear plots, in which deviations from the master Equation (4) are easier to observe. To linearize Equation (4), the steady-state (ss) fraction of protein present in the reduced form, $\text{Target}_{ox}\big|_{ss}$, was obtained by letting t to tend to infinite, resulting in Equation (6):

$$\text{Target}_{\text{rd}}\big|_{ss} = \frac{k_{switchoff}}{k_{switchoff} + k_{activation}} \qquad (6)$$

Equation (6) was used to rewrite and linearize Equation (4) as Equations (7) and (8). If in the experimental time course this steady-state is not observed, Equation (4) cannot be linearized according to this procedure.

$$\frac{\text{Target}_{\text{rd}}\big|_t - \text{Target}_{\text{rd}}\big|_{ss}}{\text{Target}_{\text{rd}}\big|_0 - \text{Target}_{\text{rd}}\big|_{ss}} = e^{-\left(k_{switchoff} + k_{activation}\right)\times t} \qquad (7)$$

$$\ln\left(\frac{\text{Target}_{\text{rd}}\big|_t - \text{Target}_{\text{rd}}\big|_{ss}}{\text{Target}_{\text{rd}}\big|_0 - \text{Target}_{\text{rd}}\big|_{ss}}\right) = -\left(k_{switchoff} + k_{activation}\right) \times t \qquad (8)$$

A plot of $\ln\left(\frac{\text{Target}_{\text{rd}}|t - \text{Target}_{\text{rd}}|ss}{\text{Target}_{\text{rd}}|0 - \text{Target}_{\text{rd}}|ss}\right)$ vs. time gives a linear relationship with slope $= -\left(k_{switchoff} + k_{activation}\right)$.

Finally, combining this slope with Equation (6), kinetic parameters are estimated as:

$$k_{switchoff} = -\text{slope} \times \text{Target}_{reduced}\big|_{ss} \qquad (9a)$$

$$k_{switchoff} = -\text{slope} - k_{switchoff} \qquad (9b)$$

SIMPLIFICATION OF THE MASTER EQUATION
Simplified forms of Equation (4) that apply to specific experimental conditions may constitute a useful alternative to estimate kinetic parameters.

Absence of H₂O₂
On the assumption that $k_{activation} = 0$, i.e., H_2O_2 is absent in the system, Equation (4) was simplified as Equation (10).

$$\text{Target}_{\text{rd}}\big|_t - 1 = \left(\text{Target}_{\text{rd}}\big|_0 - 1\right) \times e^{-k_{switchoff}\times t} \qquad (10a)$$

Or

$$\text{Target}_{\text{ox}}\big|_t = \text{Target}_{\text{ox}}\big|_0 \times e^{-k_{switchoff}\times t} \qquad (10b)$$

This equation is applied to determine $k_{switchoff}$ from time courses that follow the return of the sensor protein to its reduced form. Taking the logarithmic of both sides of Equation (10b):

$$\ln\left(\text{Target}_{\text{ox}}\big|_t\right) = \ln\left(\text{Target}_{\text{ox}}\big|_0\right) - k_{switchoff} \times t \qquad (11)$$

A plot of $\ln\left(\text{Target}_{\text{ox}}\big|_t\right)$ vs. time produces a straight line with $k_{switchoff} = -\text{slope}$.

No target reduction
Equation (12), another simplified form of Equation (4), was obtained by ignoring target reduction, i.e., $k_{switchoff} = 0$.

$$\text{Target}_{\text{rd}}\big|_t = \text{Target}_{\text{rd}}\big|_0 \times e^{-k_{activation}\times t} \qquad (12)$$

This equation is used to estimate $k_{activation}$ from short time courses when target reduction is still negligible. Taking the logarithmic of both sides of Equation (12):

$$\ln\left(\text{Target}_{\text{rd}}\big|_t\right) = \ln\left(\text{Target}_{\text{rd}}\big|_0\right) - k_{activation} \times t \qquad (13)$$

A plot of $\ln\left(\text{Target}_{\text{rd}}\big|_t\right)$ vs. time produces a straight line with $k_{activation} = -\text{slope}$.

CONCENTRATION STUDIES
In all previous equations, H_2O_2 is a hidden variable that influences $k_{activation}$ and kinetic parameters are estimated from experiments in which the time course of the oxidation state of the target protein is followed. If cells are exposed to various concentrations of H_2O_2, kinetic parameters may also be estimated by following the variation of the oxidation state of the target protein as a function of the H_2O_2 concentration at a given time point. To this end, $k_{activation}$ was replaced by $k_{target + H2O2} \times [H_2O_2]$ in Equation (4), forming Equation (14):

$$\text{Target}_{\text{rd}}\big|_t = \frac{k_{switchoff}}{k_{switchoff} + k_{target + H_2O_2} \times [H_2O_2]}$$
$$+ e^{-\left(k_{switch} + k_{target + H_2O_2}\times [H_2O_2]\right)\times t}\left(\text{Target}_{\text{rd}}\big|_0\right.$$
$$\left. - \frac{k_{switchoff}}{k_{switchoff} + k_{target + H_2O_2} \times [H_2O_2]}\right) \qquad (14)$$

For a known fixed t, a non-linear two-parameter fitting of $\text{Target}_{\text{rd}}|t$ vs. $[H_2O_2]$ allows to estimate $k_{switchoff}$ and $k_{target + H2O2}$. As before, if one of the two parameters is already known a one-parameter non-linear fitting may be used to determine the other parameter.

Concerning Equation (8), after specifying the H_2O_2 concentration explicitly this equation became:

$$\ln\left(\frac{\text{Target}_{\text{rd}}\big|_t - \text{Target}_{\text{rd}}\big|_{ss}}{\text{Target}_{\text{rd}}\big|_0 - \text{Target}_{\text{rd}}\big|_{ss}}\right) = -k_{switchoff} \times t - k_{target + H_2O_2}$$
$$\times t \times [H_2O_2] \qquad (15)$$

A plot of $\ln\left(\frac{\text{Target}_{\text{rd}}|t - \text{Target}_{\text{rd}}|ss}{\text{Target}_{\text{rd}}|0 - \text{Target}_{\text{rd}}|ss}\right)$ vs. $[H_2O_2]$ gives a linear relationship with slope $= -k_{target + H_2O_2} \times t$ and intercept $= -k_{switchoff} \times t$. In order to use this equation, the fraction of reduced target reached at steady-state ($\text{Target}_{\text{rd}}|ss$) must be known previously. Therefore, time courses are needed for each H_2O_2 concentration in order to obtain this value, which lessens the applicability of this equation.

In absence of target reduction, the equivalent of Equation (13) was deduced as:

$$\ln\left(\text{Target}_{\text{rd}}\big|_t\right) = \ln\left(\text{Target}_{\text{rd}}\big|_0\right)$$
$$- k_{target + H_2O_2} \times t \times [H_2O_2] \qquad (16)$$

A plot of $\ln\left(\text{Target}_{\text{rd}}\big|_t\right)$ vs. $[H_2O_2]$ produces a straight line with slope $= -k_{target + H_2O_2} \times t$.

Importantly, the $[H_2O_2]$ in these equations refers to the intracellular $[H_2O_2]$ that reacts with the target. Therefore, in order to estimate $k_{target + H2O2}$, this concentration must be known. Intracellular $[H_2O_2]$ attained when cells are exposed to extracellular H_2O_2 can be estimated from the gradient between extracellular and intracellular H_2O_2. If this gradient is unknown, then these equations may be applied with the extracellular H_2O_2 concentrations, but the value of $k_{target+H2O2}$ obtained is referred to extracellular H_2O_2 concentrations, with the true value being higher. If instead $k_{target+H2O2}$ is known *a priori*, equations may be used to estimate the gradient between extracellular and intracellular H_2O_2.

VALIDATION OF EQUATIONS

To test the validity of the equations, the following mathematical kinetic model was set up. This model simulates ideal experiments in which cells are exposed to extracellular H_2O_2 or are stimulated to produce endogenous H_2O_2 in a receptor-mediated process. A key characteristic of these virtual experiments is that the kinetic parameters concerning H_2O_2 signaling are known a priori, corresponding to the kinetic parameters introduced in the model. Thus, by fitting the virtual time courses to the equations deduced previously, the validity of the equations can be tested objectively. If the equations are valid, kinetic parameters obtained in these fittings should be similar to those used in the kinetic model. In addition, by varying several parameters of the model, namely those concerning the experimental set up describing H_2O_2 exposure, experimental conditions in which the equations are valid may be defined.

The model is described by the following differential equations, which take into account two compartments, one referring to the extracellular space (V_{out}) and the other to the cell volume (V_{in}). Multicompartmentation was implemented as described previously (Alves et al., 2006).

$$\frac{d\,[H_2O_2]_{out}}{dt} = H_2O_2\_production\_out$$
$$+H_2O_2\_export \times V_{in}/V_{out} - H_2O_2\_import$$

$$\frac{d\,[H_2O_2]_{in}}{dt} = H_2O_2\_production\_in + H_2O_2\_import$$
$$\times V_{out}/V_{in} - H_2O_2\_export - v\_GPx - v\_Target$$

$$\frac{d\,[target_{rd}]}{dt} = v\_switch\_off - v\_Target$$

Reactions considered in the model (**Table 1**) were: extracellular production of H_2O_2 ($H_2O_2\_production\_out$), which simulates, for example, production of H_2O_2 by glucose oxidase added to the incubation medium; intracellular production of H_2O_2 ($H_2O_2\_production\_in$), which simulates the endogenous production triggered by a receptor-mediated process; permeation of H_2O_2 across the plasma membrane into ($H_2O_2\_import$) and out of the cell ($H_2O_2\_export$); consumption of H_2O_2 by an antioxidant enzyme (v_Gpx) and by a sensor protein target (v_Target); and finally, the switch-off mechanism of the target protein (v_switch_off). **Table 2** shows the parameters used. Although kinetics and respective rate constants are based on published values, this model does not intend to model a particular cell or a specific signaling pathway. Reactivities of the target and the antioxidant enzyme toward H_2O_2 were based on that of PTP1B (Barrett et al., 1999) and glutathione peroxidase (GPx) (Flohe, 1979; Forstrom and Tappel, 1979), respectively. Levels of GPx, permeability constant for H_2O_2 across the plasma membrane,

Table 1 | Reactions and respective rate laws included in the kinetic model.

Reaction	Name	Rate law
$\rightarrow H_2O_{2out}$	$H_2O_2\_production\_out$	$v\_\,H_2O_{2out}$
$\rightarrow H_2O_{2in}$	$H_2O_2\_production\_in$	$k\_H_2O_{2in}$ or $k\_H_2O_{2in} \times$ sine(time/1200 $\times$ 3.14)
$H_2O_{2out} \rightarrow H_2O_{2in}$	$H_2O_2\_import$	$Ps \times A/V_{out} \times [H_2O_{2out}]$
$H_2O_{2in} \rightarrow H_2O_{2out}$	$H_2O_2\_export$	$Ps \times A/V_{in} \times [H_2O_{2in}]$
$H_2O_{2in} \xrightarrow{GPx} H_2O$	v_GPx	$k_{GPx} \times [GPx] \times [H_2O_{2in}]$
$H_2O_{2in} + Target_{rd} \rightarrow Target_{ox} + H_2O$	v_Target	$k_{Target+H2O2} \times [Target_{rd}] \times [H_2O_{2in}]$
$Target_{ox} \rightarrow Target_{rd}$	v_switch_off	$k_{switchoff} \times ([Target_{tot}] - [Target_{rd}])$

Table 2 | Parameter values used in the kinetic model.

Parameter	Value	Parameter	Value
Ps	$2.0\,\mu m\ s^{-1}$	k_{GPx}	$6 \times 10^7\ M^{-1}\ s^{-1}$
A	$627\,\mu m^2$	GPx	$2 \times 10^{-7}\ M$
V_{in}	$1472\,\mu m^3$	$k_{Target+H2O2}$	$40\ M^{-1}\ s^{-1}$
V_{out}	$679 \times V_{in}$	$k_{switchoff}$	$1 \times 10^{-3}\ s^{-1}$
$v\_H_2O_{2out}$	$(0\text{–}23.4) \times 10^{-7}\ M\ s^{-1}$	$Target_{tot}$	$8.3 \times 10^{-9}\ M$
kH_2O_{2in}	$(0\text{–}5) \times 10^{-3}\ M\ s^{-1}$		

All simulations used these parameters values; for cases where a range of values is indicated, the actual value used in the simulation is indicated in the respective figure legend.

V_{out} and V_{in} were taken from Antunes and Cadenas (2000). $k_{switchoff}$ was obtained from the lower range of values estimated in this work, based on previously published experiments. The resulting differential equations were solved numerically with PLAS (Voit, 1991). In the kinetic model, concentrations of $Target_{rd}$ and $Target_{ox}$ were used. The respective fractions were calculated subsequently so that simulation data could be analyzed with the equations deduced here. The parameter $k_{activation}$ is not a rate constant in the numerical model, but it was calculated as $k_{target+H2O2} \times [H_2O_{2in}]$; when $[H_2O_{2in}]$ was not constant, for example when a bolus addition of H_2O_2 or the endogenous nonconstant production of H_2O_2 was simulated, an average $[H_2O_{2in}]$ was used.

RESULTS

As a first step to test the equations derived here, redox signaling experiments were simulated to generate data that was introduced into the equations in order to determine kinetic parameters. The validity of the equations was checked by comparing the kinetic parameters obtained with those used in the simulations.

VALIDATION OF EQUATIONS WITH SIMULATED EXPERIMENTS

The exposure of cell cultures to extracellular H_2O_2 initiates cellular responses that differ from those caused by the intracellular release of H_2O_2 triggered by receptor-mediated mechanisms (Forman, 2007), being the main difference the additional signal transduction pathways initiated in the first case. Nevertheless, the control of H_2O_2 delivery achieved by the extracellular exposure makes this approach more suitable for the purpose of estimating kinetic parameters. We simulated both approaches with the kinetic model described in Theory and Methods, starting with the extracellular exposure to H_2O_2.

Extracellular addition of H_2O_2

Cells may be exposed to extracellular H_2O_2 either by bolus additions or by incubation with steady-state concentrations of H_2O_2. In the bolus addition, a single dose of H_2O_2 is added to cells, constituting the most common method of exposing cell cultures to extracellular H_2O_2. It has the advantage of simplicity, but the results obtained are strongly dependent on the specific assay conditions (Marinho et al., 2013a). Among other factors, cell density and, for adherent-growing cells, the volume of incubation media dramatically affect the results.

In the steady-state methodology, exposure to H_2O_2 is calibrated so that cells are exposed for a known concentration of H_2O_2 that remains constant during the assay. Although more complex, this approach has much better experimental reproducibility with the actual H_2O_2 concentration in the assay being independent of experimental conditions. The implementation of this methodology is described in detailed in (Covas et al., 2013; Cyrne et al., 2013; Marinho et al., 2013a).

Steady-state. In the deduction of the kinetic equations, $k_{activation}$ ($k_{target + H2O2} \times [H_2O_2]$) was considered to be a constant parameter, i.e., H_2O_2 was assumed to be constant with time. So, as a positive control we started by analyzing the results simulated with a steady-state incubation, a case in which the equations tested should be valid.

In the first condition analyzed, cellular exposure to H_2O_2 (**Figure 1A**, curve 2) was long enough so that a balance between oxidation of the target that senses H_2O_2 and its regeneration was achieved. As observed in **Figure 1A**, curve 1, initially the target was oxidized until its oxidation state reached a near steady-state value as given by Equation (6). Simulated results were transformed according to Equation (8) (**Figure 1B**, blue line), with $k_{activation}$ and $k_{switchoff}$ being estimated from Equations (9A) and (9B). The estimations obtained, $k_{activation} = 8.1 \times 10^{-4}\,s^{-1}$ and $k_{switchoff} = 1.0 \times 10^{-3}\,s^{-1}$, matched closely the respective expected values of $7.9 \times 10^{-4}\,s^{-1}$ and $1.0 \times 10^{-3}\,s^{-1}$, which were used in the simulations to draw curve 1 in **Figure 1A**. Note that the fitting to Equation (8) departed from linearity for longer time points (**Figure 1B**), when $Target_{rd}$ approached its steady-state value (**Figure 1B**). This behavior was caused by small uncertainties in this value, which must be known in order to plot data according to Equation (8). As an alternative, the two parameters were also obtained from a non-linear fitting to

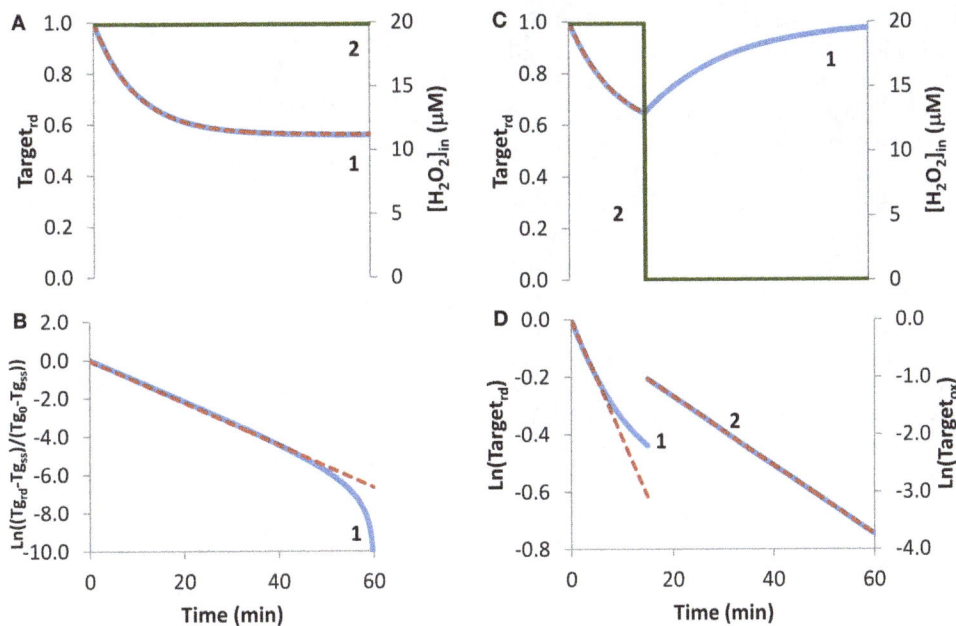

FIGURE 1 | Determination of kinetic parameters from simulated experiments when extracellular H_2O_2 was delivered as a steady-state. H_2O_2 concentration (curve 2, green line) was maintained during the duration of the experiment **(A)** or was stopped at 15 min **(C)**. **(A)** The profile of the fraction of the H_2O_2 target in the reduced form ($Target_{rd}$, blue line) was obtained by simulation of the kinetic model described in the Theory and Methods Section with $v\_H_2O_{2out} = 3.51 \times 10^{-7}\,M\,s^{-1}$ and $[H_2O_{2out}] = 3 \times 10^{-4}\,M$ at time = 0, (intracellular H_2O_2 production was absent, $k\_\,H_2O_{2in} = 0$); red dashed line is the two-parameter non-linear fitting to Equation (4) used to estimate the kinetic parameters $k_{activation}$ and $k_{switchoff}$. **(B)** Fitting of the profile of $Target_{rd}$ obtained in **(A)** to Equation

(8) (curve 1, blue line) using $Target_{rd\_0} = 1$ and $Target_{rd\_ss} = 0.56$, (Tg in y-axis title means Target). $k_{activation}$ and $k_{switchoff}$ were estimated from the slope of the straight line by applying Equations (9A) and (9B). **(C)** The profile of the fraction of $Target_{rd}$ (curve 1, blue line) was obtained as in **(A)** but $v\_H_2O_{2out}$ and $[H_2O_2]_{out}$ were set to zero at 15 min; red dashed line is the one-parameter non-linear fitting to Equation (4) in which $k_{switchoff}$ obtained in **(D)** curve 2 was used as an input to estimate $k_{activation}$. **(D)** $Target_{rd}$ was fitted to Equation (13) (curve 1, left y-axis) while H_2O_2 was present, afterwards $Target_{ox}$ was fitted to Equation (11) (curve 2, right y-axis); $k_{activation}$ was estimated from the linear part of the fitting to Equation (13), $k_{switchoff}$ from the fitting to Equation (11).

Equation (4) (**Figure 1A**, red dashed line), which neither requires data transformation nor knowing the value of the fraction of reduced target at steady-state. In this case, the estimated parameters, $k_{activation} = 7.9 \times 10^{-4}\,s^{-1}$ and $k_{switchoff} = 1.0 \times 10^{-3}\,s^{-1}$, matched exactly the expected values.

If the fraction of the reduced target does not reach a steady-state because, for example, a balance between its oxidation and regeneration is not attained before the exposure to H_2O_2 is terminated, Equation (8) cannot be applied. To illustrate this situation, a simulation was done under the exact same conditions as before with the exception that H_2O_2 exposure lasted only for 15 min (**Figure 1C**). Experimentally, this is equivalent to either replacing the external media to remove the H_2O_2 generating system, such as glucose oxidase, or by adding external catalase to the incubation media. This simulation was analyzed with Equation (13), an equation deduced ignoring the regeneration of the reduced form of the target. Good enough estimations were obtained by using only the first stage of the time course, when the degree of target oxidation was still low, and therefore the contribution of its regeneration to the time course of $Target_{rd}$, could be ignored. From the slope of the initial linear part of the curve (**Figure 1D**, curve 1), a $k_{activation} = 6.8 \times 10^{-4}\,s^{-1}$ was obtained, which was close to the expected value of $7.9 \times 10^{-4}\,s^{-1}$. Concerning $k_{switchoff}$, this kinetic parameter was estimated by fitting to Equation (11) the part of the curve starting after removal of external H_2O_2. Note that Equation (11) was deduced assuming reduction of the oxidized target when H_2O_2 was absent. An excellent linear plot was observed (**Figure 1D**, curve 2) with the estimated $k_{switchoff}$ of $1.0 \times 10^{-3}\,s^{-1}$ matching the expected value. Kinetic parameters were also obtained by non-linear fittings of Equation (4) to the first part of the curve when H_2O_2 was still present, either as a two-parameter non-linear fitting in which the two parameters – $k_{activation}$ and $k_{switchoff}$ – were determined, or as a one-parameter non-linear fitting, in which only one of the parameters was estimated, with the other being obtained from the linear plots of **Figure 1B**. If the conditions of the experimental assay fulfill all the assumptions applied to deduce Equation (4), a two-parameter non-linear fitting is the best choice, since both parameters are obtained without transformation of the experimental data. However, if the assumptions are not all fulfilled, which is the most common situation, we advise to apply a one-parameter non-linear fitting to Equation (4), inputting as a known parameter $k_{switchoff}$ estimated from Equation (11), being $k_{activation}$ the unknown parameter. As always whatever the option taken, the goodness of the fitting should be inspected. The dashed line in **Figure 1C** was obtained as a one-parameter non-linear fitting using $k_{switchoff} = 1.0 \times 10^{-3}\,s^{-1}$ with the estimated $k_{activation}$ value of $7.9 \times 10^{-4}\,s^{-1}$ matching the expected value.

Overall, kinetic parameters estimated from simulated experiments in which H_2O_2 was delivered as a steady-state matched the expected values, validating the equations applied. This could be anticipated because Equation (4) relies on the key assumption that H_2O_2 concentration is constant during the experiment.

Bolus addition. The bolus addition set up, the most common experimental approach to expose cells to H_2O_2, was simulated in **Figure 2**. Upon incubation with a 1 mM bolus addition, the H_2O_2 sensor was oxidized, the $Target_{rd}$ fraction reached a minimum at approximately 12 min, then regeneration became more important than oxidation, and $Target_{rd}$ increased (**Figure 2A**, curve 1). Kinetics of H_2O_2 consumption depends on the experimental set up, and under the conditions of this simulation, H_2O_2 was fully consumed after 60 min (**Figure 2A**, curve 2). Nevertheless, the general pattern observed in this simulation served as a test case to check how the non-constant H_2O_2 concentration impacts the estimation of kinetic parameters.

Because the fraction of reduced target never reached a constant value, Equation (8) was not applied, and instead Equation (13) was used to estimate $k_{activation}$. Only the first part of the curve was considered (**Figure 2B**, curve 1), because shorter time courses minimize target regeneration, a process ignored by Equation (13). The $k_{activation}$ estimation of $1.8 \times 10^{-3}\,s^{-1}$ was close to the expected value of $1.7 \times 10^{-3}\,s^{-1}$. Concerning $k_{switchoff}$, this parameter was estimated by fitting to Equation (11) the part of the curve after $Target_{rd}$ reached its minimum (**Figure 2B**, curve 2). The presence of H_2O_2 in this part of the experiment promoted target oxidation, violating a key assumption behind Equation (11), and consequently deviations from linearity were observed. Even by using only the linear part of the curve, the $k_{switchoff}$ estimation of $6.7 \times 10^{-4}\,s^{-1}$

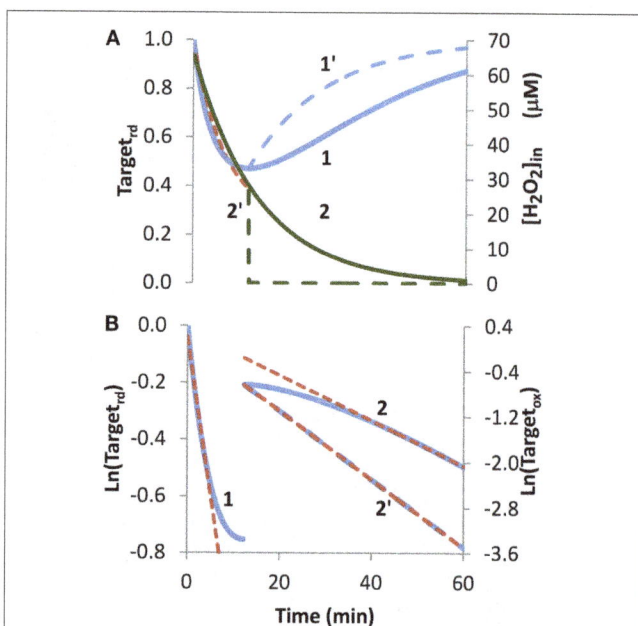

FIGURE 2 | Determination of kinetic parameters from simulated experiments when H_2O_2 was delivered as an extracellular bolus addition. (A) Curve 1 is the profile of $Target_{rd}$ fraction (blue line) obtained by simulation with v_H_2O_{2out} = 0 M s^{-1} and [H_2O_{2out}] = 1 × 10^{-3} M at time = 0, (intracellular H_2O_2 production was absent, k_ H_2O_{2in} = 0). Curve 1' is the profile of $Target_{rd}$ (blue line) obtained as in curve 1 but [H_2O_{2out}] was set to zero at 12 min; red dashed line is the one-parameter non-linear fitting to Equation (4) in which $k_{switchoff}$ obtained in **(B)** was used as an input to estimate $k_{activation}$. **(B)** Until 12 min results were fitted to Equation (13) (curve 1, left y-axis), afterwards were fitted to Equation (11) (curve 2, right y-axis); $k_{activation}$ and $k_{switchoff}$ were estimated from the linear part of the fittings to Equations (13) and (11), respectively.

underestimated the expected value of $1.0 \times 10^{-3}\,\mathrm{s}^{-1}$. Removal of external H_2O_2 at 12 min, when $Target_{rd}$ reached its minimum (**Figure 2A**, curve 2′), changed the regeneration profile of the H_2O_2 target (**Figure 2A**, curve 1′) and vastly improved the fitting to Equation (11) (**Figure 2B**, curve 2′) with the estimated $k_{switchoff}$ of $1.0 \times 10^{-3}\,\mathrm{s}^{-1}$ matching exactly the expected value.

Concerning the non-linear fitting to Equation (4), a $k_{activation} = 1.8 \times 10^{-3}\,\mathrm{s}^{-1}$ was obtained when $k_{switchoff} = 6.7 \times 10^{-4}\,\mathrm{s}^{-1}$ was used as an input. Alternatively, by inputting a $k_{switchoff} = 1.0 \times 10^{-3}\,\mathrm{s}^{-1}$ a $k_{activation} = 2.0 \times 10^{-3}\,\mathrm{s}^{-1}$ was obtained.

Overall, these results indicate that the proposed equations can be applied to experiments in which H_2O_2 is delivered as a bolus addition. The accuracy of parameter estimation improves if H_2O_2 is removed at the time when the reduced form of the target reaches its minimum.

Concentration studies.

Besides time courses, studies often evaluate how the concentration of H_2O_2 affects the oxidation state of the sensor target. We started by simulating the dependency of $Target_{rd}$ on external H_2O_2 concentration, delivered as a steady-state during 10 min (**Figure 3A**, curve 1). From the non-linear fitting to Equation (14), kinetic parameters that matched exactly the expected values were obtained ($k_{target+H2O2} = 2.6\,\mathrm{M}^{-1}\,\mathrm{s}^{-1}$ and $k_{switchoff} = 1.0 \times 10^{-3}\,\mathrm{s}^{-1}$). Note that $k_{target+H2O2}$ estimated from the fitting was based on external H_2O_2 concentrations. By considering the gradient between these and the intracellular H_2O_2 concentrations—15 in the present simulation—the estimated value of the rate constant of $40\,\mathrm{M}^{-1}\,\mathrm{s}^{-1}$ for the reaction between the target and H_2O_2 matched the value used in the simulation. Results were also linearized and fitted to Equation (16) (curve 2 in **Figure 3A**), which was deduced assuming absence of target reduction, i.e., $k_{switchoff} = 0\,\mathrm{s}^{-1}$. In this case, the estimated value of $1.7\,\mathrm{M}^{-1}\,\mathrm{s}^{-1}$ for $k_{target+H2O2}$, which was converted to $26\,\mathrm{M}^{-1}\,\mathrm{s}^{-1}$ when intracellular H_2O_2 concentrations were considered, underestimated the expected value.

To test how Equations (14) and (16) behave with data generated with bolus additions, the study of **Figure 3A** was repeated but now the H_2O_2 concentrations introduced in the equations were the initial bolus additions (**Figure 3B**, curve 1). Kinetic parameters obtained with the non-linear fitting to Equation (14) were $k_{target+H2O2} = 2.0\,\mathrm{M}^{-1}\,\mathrm{s}^{-1}$ (or $30\,\mathrm{M}^{-1}\,\mathrm{s}^{-1}$ if referred to intracellular H_2O_2), and $k_{switchoff} = 1.3 \times 10^{-3}\,\mathrm{s}^{-1}$. Linearization according to Equation (16) (curve 2 in **Figure 3B**) gave a $k_{target+H2O2}$ of $1.2\,\mathrm{M}^{-1}\,\mathrm{s}^{-1}$, (or $18\,\mathrm{M}^{-1}\,\mathrm{s}^{-1}$ if referred to intracellular H_2O_2). As expected, these estimations were less accurate than those obtained when H_2O_2 was delivered as a steady-state, but nevertheless they constitute satisfactory semi-quantitative estimations.

Receptor-mediated endogenous H_2O_2 production

The endogenous production of H_2O_2 upon cell stimulation by a ligand will give the best picture of the influence of H_2O_2 in a particular cell signaling pathway, as H_2O_2 production is both spatial and time restricted (Forman, 2007). Nevertheless, since the profile of H_2O_2 concentration generated is unknown this imposes

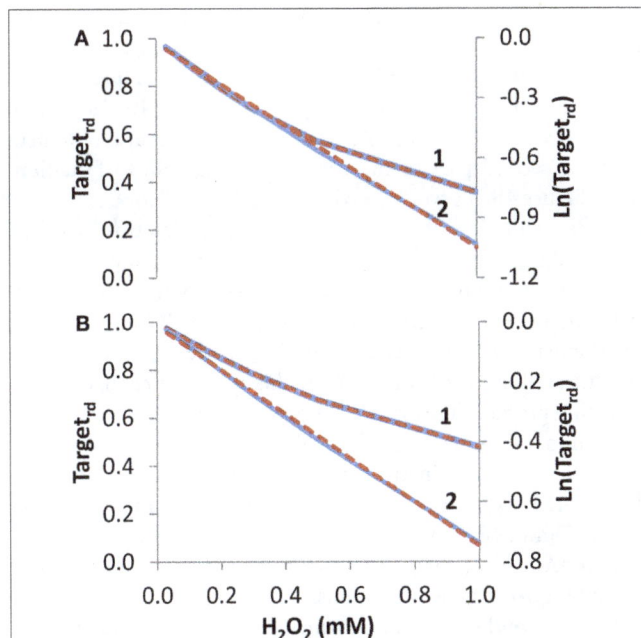

FIGURE 3 | Determination of kinetic parameters from simulated experiments when the concentration of extracellular H_2O_2 was changed. The reduced form of the target fraction ($Target_{rd}$) obtained at 10 min is plotted as a function of either H_2O_2 steady-state concentrations in **(A)** or initial bolus additions in **(B)** (curves 1, blue linen). Simulations were run with $v\_H_2O_{2out}$ varying in the range $(0.35–23.4) \times 10^{-7}\,\mathrm{M}\,\mathrm{s}^{-1}$ and $[H_2O_{2out}]$ in the range $(0.03–1) \times 10^{-3}\,\mathrm{M}$ at time $= 0$ in A, while in B $v\_H_2O_{2out}$ was set to zero and $[H_2O_{2out}]$ was changed in the range $(0.03–1) \times 10^{-3}\,\mathrm{M}$ at time $= 0$; intracellular H_2O_2 production was absent ($k\_H_2O_{2in} = 0$) in both cases. In **(A,B)**, results were analyzed with non-linear fits of $Target_{rd}$ to Equation (14) (curve 1, red dashed line) in order to estimate $k_{target+H2O2}$ and $k_{switchoff}$, or they were linearized according to Equation (16) (curve 2, blue line) with $k_{target+H2O2}$ being estimated from the slopes of the red dashed lines.

potential problems to the determination of kinetic parameters. To test how the kinetic equations behave under such circumstances, we started by simulating a case where H_2O_2 intracellular production was rapidly triggered and then set at a near constant value. This scenario worked as positive control and was analyzed as described previously for the extracellular addition of steady-state H_2O_2. The kinetic parameters obtained matched exactly the expected values or were very close to these values depending on the fittings applied (not shown). Nevertheless, this scenario is seldom achieved when H_2O_2 is produced endogenously, and next we tested the kinetic equations under non-constant H_2O_2 intracellular production.

H_2O_2 endogenous production was simulated with a sine-like function: there was an initial increase in the H_2O_2 concentration, reaching its maximum at 10 min, and then a decrease until H_2O_2 production stopped at 20 min (**Figures 4A,C**, curve 2). In this context, two scenarios were simulated. In the first, H_2O_2 production was high enough so that a near constant level of reduced target was observed (**Figure 4A**, curve 1), and accordingly results were fitted to Equation (8). The presence of a non-constant H_2O_2 production caused deviations from linearity (**Figure 4B**, curve

1). Nevertheless results obtained from the near-linear intermediate portion of the curve gave estimations, $k_{switchoff} = 7.8 \times 10^{-4}\,s^{-1}$ and $k_{activation} = 1.2 \times 10^{-2}\,s^{-1}$, that compared well with the expected parameters, $k_{switchoff} = 1.0 \times 10^{-3}\,s^{-1}$ and $k_{activation} = 1.0 \times 10^{-2}\,s^{-1}$. As before, $k_{switchoff}$ was also obtained from the second part of the curve by fitting data to Equation (11) (**Figure 4B**, curve 2), giving a $k_{switchoff} = 7.6 \times 10^{-4}\,s^{-1}$, an underestimation of the expected value. As described for the bolus addition, removal of H_2O_2 from the system after $Target_{rd}$ reached its minimum improved the estimations (not shown). In real experiments, the effect of this addition will be dependent on whether removal of extracellular H_2O_2 decreases the localized intracellular levels of H_2O_2. If this occurs, a change in the reduction profile of the oxidized target should be observed. In this simulation, the application of non-linear fittings did not improve the estimations of kinetic parameters: from a non-linear fitting where $k_{switchoff} = 7.6 \times 10^{-4}\,s^{-1}$ was used as input (dashed line in **Figure 4A**) a $k_{activation}$ of $0.51 \times 10^{-2}\,s^{-1}$ was obtained (**Figure 4A**, dashed line) and a two-parameter non-linear fitting did not improve these estimations.

In the second simulation in which endogenous production of H_2O_2 followed a sine-like function, a constant level of reduced target was not observed (**Figure 4C**, curve 1). The $k_{activation}$ estimation of $1.0 \times 10^{-3}\,s^{-1}$, obtained from the linear portion of the plot according to Equation (13) (**Figure 4D**, curve 1), matched the expected value. The $k_{switchoff}$ estimation of $9.3 \times 10^{-4}\,s^{-1}$, obtained from the second part of the curve after fitting

data to Equation (11) (**Figure 4D**, curve 2), was close to the expected value of $1.0 \times 10^{-3}\,s^{-1}$. A one-parameter non-linear fitting (**Figure 4C**, dashed line) gave a $k_{activation}$ of $1.0 \times 10^{-3}\,s^{-1}$, i.e., the expected value, when a $k_{switchoff}$ of 9.3×10^{-4} was used as input.

Overall, when H_2O_2 production is not constant deviations from the equations derived here are expected. Nevertheless, estimated kinetic parameters are still satisfactory at a semi-quantitative level, and the deviations from linearity in the plots proposed here may be used as a useful tool to diagnose a non-constant H_2O_2 production.

FITS TO EXPERIMENTAL DATA

The use of simulation data was useful to test the validity of the equations deduced and to figure out how deviations from the assumptions behind their deduction affected the estimation of kinetic parameters. Nevertheless, simulation data points are virtually infinite and devoid of experimental error; in contrast real experiments contain a finite number of measurements with associated experimental error. To test how equations cope with these issues, they were applied to data obtained from the literature for two PTPs, PTP1B and SHP-2.

H_2O_2-external delivery

Two experiments in which H_2O_2 was added externally as a bolus addition (**Figure 5**) were analyzed. In the first experiment (Rinna et al., 2006), the time course of PTP1B oxidation was followed in a

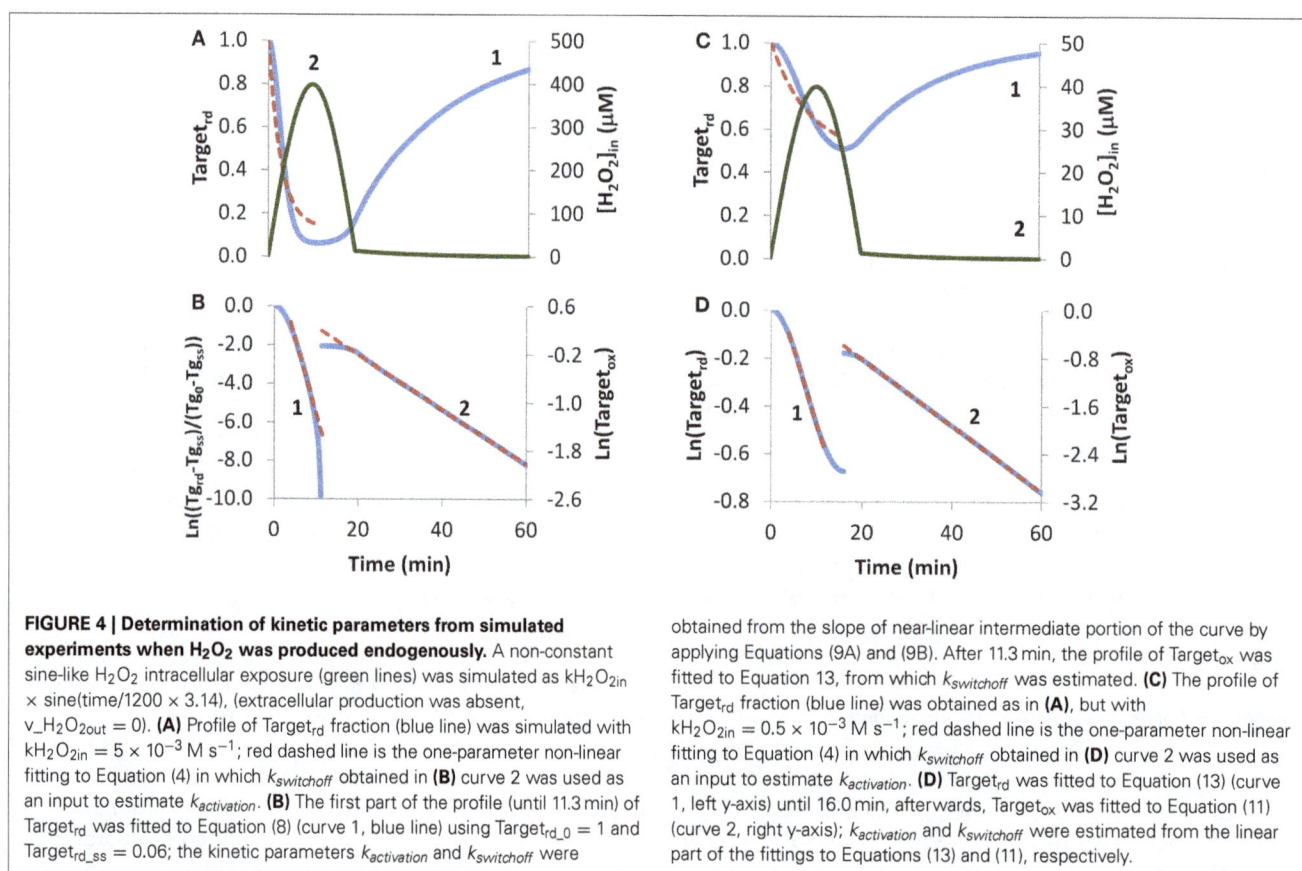

FIGURE 4 | Determination of kinetic parameters from simulated experiments when H_2O_2 was produced endogenously. A non-constant sine-like H_2O_2 intracellular exposure (green lines) was simulated as $kH_2O_{2in} \times sine(time/1200 \times 3.14)$, (extracellular production was absent, $v\_H_2O_{2out} = 0$). **(A)** Profile of $Target_{rd}$ fraction (blue line) was simulated with $kH_2O_{2in} = 5 \times 10^{-3}\,M\,s^{-1}$; red dashed line is the one-parameter non-linear fitting to Equation (4) in which $k_{switchoff}$ obtained in **(B)** curve 2 was used as an input to estimate $k_{activation}$. **(B)** The first part of the profile (until 11.3 min) of $Target_{rd}$ was fitted to Equation (8) (curve 1, blue line) using $Target_{rd\_0} = 1$ and $Target_{rd\_ss} = 0.06$; the kinetic parameters $k_{activation}$ and $k_{switchoff}$ were

obtained from the slope of near-linear intermediate portion of the curve by applying Equations (9A) and (9B). After 11.3 min, the profile of $Target_{ox}$ was fitted to Equation 13, from which $k_{switchoff}$ was estimated. **(C)** The profile of $Target_{rd}$ fraction (blue line) was obtained as in **(A)**, but with $kH_2O_{2in} = 0.5 \times 10^{-3}\,M\,s^{-1}$; red dashed line is the one-parameter non-linear fitting to Equation (4) in which $k_{switchoff}$ obtained in **(D)** curve 2 was used as an input to estimate $k_{activation}$. **(D)** $Target_{rd}$ was fitted to Equation (13) (curve 1, left y-axis) until 16.0 min, afterwards, $Target_{ox}$ was fitted to Equation (11) (curve 2, right y-axis); $k_{activation}$ and $k_{switchoff}$ were estimated from the linear part of the fittings to Equations (13) and (11), respectively.

rat alveolar macrophage cell line after addition of a 100 μM H_2O_2 bolus dose for 15 min (**Figure 5A**, curve 1). By fitting data to Equation (4) with a two parameter non-linear fitting, a $k_{activation}$ of $1.1 \times 10^{-3}\,s^{-1}$ and a $k_{switchoff}$ of $2.6 \times 10^{-3}\,s^{-1}$ were estimated. Alternatively, when results were linearized according to Equation (13) (**Figure 5A**, curve 2) a $k_{activation}$ of $0.59 \times 10^{-3}\,s^{-1}$ was obtained, after discarding the 15 min point. Note that also with simulation data a similar deviation from linearity at late time points was observed (**Figure 2B**, curve 1). By considering an external H_2O_2 concentration of 100 μM, the apparent first-order rate constant $k_{activation}$ in the range $(0.59–1.1) \times 10^{-3}\,s^{-1}$ was converted to a rate constant between the target and H_2O_2 ($k_{target+H2O2}$) of $5.9–11\,M^{-1}\,s^{-1}$. This value refers to extracellular H_2O_2, and so if the gradient between extracellular and intracellular H_2O_2 was considered the value of $k_{target+H2O2}$ for PTP1B would be higher.

In the second experiment, rat-1 fibroblasts were subjected to H_2O_2 bolus additions in the range 0–500 μM for 1 min, followed by the measurement of the oxidation level of SHP-2 (Meng et al.,

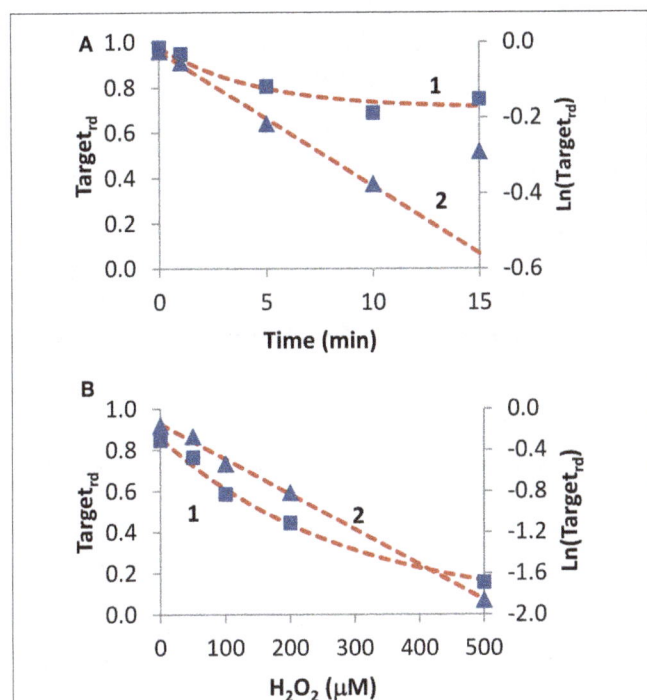

FIGURE 5 | Determination of kinetic parameters from experiments when H_2O_2 was delivered as an extracellular bolus addition. (A) Data (■) was taken from Figure 1 in Rinna et al. (2006). Gel was digitalized and analyzed with ImageJ (Rasband, 1997); fraction of oxidized target was calculated assuming a maximum level of oxidation of 31% (Rinna et al., 2006). Time course was fitted to Equation (4), with $k_{activation}$ and $k_{switchoff}$ as variables (curve 1, left y-axis), or was linearized (▲) and fitted to Equation (13) (curve 2, right y-axis). **(B)** Data (■) was taken from Figure 2A in Meng et al. (2002). Gel was digitalized and analyzed with ImageJ (Rasband, 1997); fraction of oxidized protein tyrosine phosphatase corresponding to 70 kDa (SHP-2) was calculated assuming that complete oxidation was achieved by 1 mM H_2O_2. Time course was fitted to Equation (14), with $k_{target+H2O2}$ and $k_{switchoff}$ as variables (curve 1, left y-axis), or was linearized (▲) and fitted to Equation (16) (curve 2, right y-axis).

2002) (**Figure 5B**, curve 1). After fitting data to Equation (14) (**Figure 5B**, curve 1) with a two parameter non-linear fitting, a $k_{target+H2O2}$ of $60\,M^{-1}\,s^{-1}$ and a $k_{switchoff}$ of $1.3 \times 10^{-3}\,s^{-1}$ were estimated. Linearization according to Equation (16) gave a $k_{target+H2O2}$ of $57\,M^{-1}\,s^{-1}$ (**Figure 5B**, curve 2). Again, $k_{target+H2O2}$ values refer to extracellular H_2O_2 concentrations. Even if a bolus addition was used, because short-term incubations of 1 min were done, the assumption of constant H_2O_2 behind the deduction of Equations (14) and (16) was verified.

Receptor-mediated signaling

To test how equations behave when analyzing receptor-mediated signaling, the following two experiments were considered. In the first, A431 human epidermoid carcinoma cells were stimulated by EGF, triggering H_2O_2 intracellular production that lead to PTP1B oxidation and inhibition (**Figures 6A,B**), while in the second experiment, rat-1 cells were stimulated with PDGF inducing SHP-2 oxidation (**Figures 6C,D**). In both cases, the profile of PTP oxidation did not reach a near steady-state, precluding the application of Equation (8). Concerning $k_{switchoff}$, estimations of $1.9 \times 10^{-3}\,s^{-1}$ and $8.7 \times 10^{-3}\,s^{-1}$ were obtained, respectively for PTP1B and SHP-2 reactivation, after fitting to Equation (11) the second part of the PTP oxidation curves (curves 2 in **Figures 6B,D**). For $k_{activation}$, estimations of $1.0 \times 10^{-3}\,s^{-1}$ and $9.3 \times 10^{-3}\,s^{-1}$ were obtained, respectively for PTP1B and SHP-2, after applying Equation (13) to linearize the first part of the PTP oxidation profile (curve 1 in **Figures 6B,D**). These $k_{activation}$ values were close to those obtained from non-linear fittings, $2.0 \times 10^{-3}\,s^{-1}$ and $9.7 \times 10^{-3}\,s^{-1}$ for PTP1B and SHP-2, respectively (dashed lines in **Figures 6A,C**).

Overall, data taken from literature fitted well to the equations deduced here, even if experiments analyzed were carried out without any special concern considering their application to estimate kinetic parameters.

DISCUSSION

Herein, we deduced equations to determine kinetic parameters from typical redox signaling experiments in which H_2O_2 is either added externally to cells or is endogenously produced following receptor activation by diverse cellular stimuli. The equations were shown to be accurate after fitting them to data generated by simulations. We also performed simulations in which the assumption that H_2O_2 is constant during the experiment was not fulfilled, that is, H_2O_2 was delivered as a bolus addition, or the endogenous production of H_2O_2 was not constant. Under these conditions, deviations from linearity were observed when simulation results were plotted according to the linear equations we deduced. Nevertheless, the estimated kinetic parameters were close to the parameters introduced in the simulations. Finally, we tested the application of the equations to real experiments with published experimental data concerning the H_2O_2 signaling mediated by inhibition of PTPs, namely PTP1B and SHP-2. While in general excellent fittings were obtained, in some cases deviations as those observed when H_2O_2 was added as a bolus addition were observed. In general, the estimated kinetic parameters (**Table 3**) are consistent with the published rate constants.

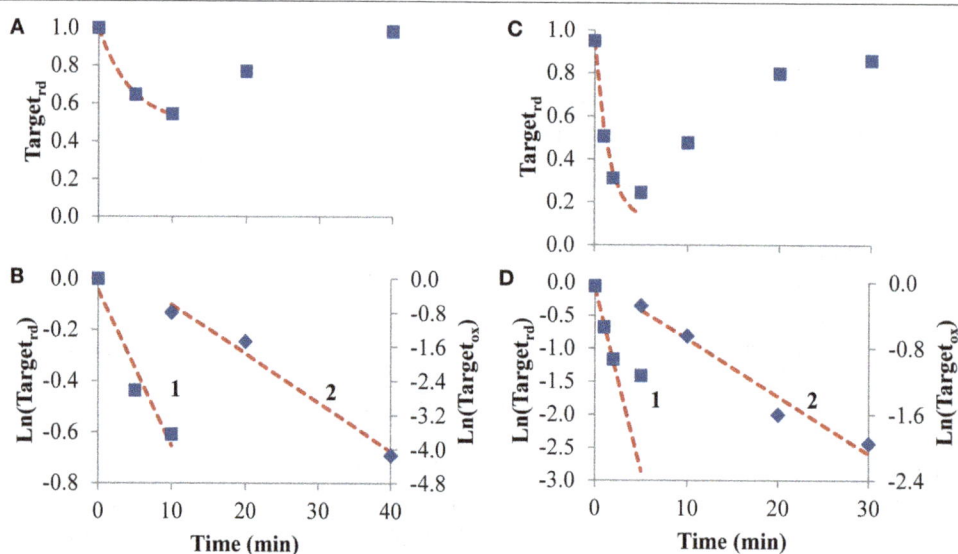

FIGURE 6 | Determination of kinetic parameters from experiments where intracellular H_2O_2 production was triggered by cell stimulation. (A) Data concerning PTP1B oxidation following stimulation of A431 cells with EGF (■) was taken from Figure 2 in Lee et al. (1998). (C) Data concerning SHP-2 oxidation following stimulation of rat-1 cells by PDGF (■) was taken from gel in Figure 4 in Meng et al. (2002) after digitalization and analysis with ImageJ (Rasband, 1997), and assuming that 100% of oxidation of SHP-2 was achieved in absence of iodoacetic acid as

described in Meng et al. (2002). The red dashed lines in (A,C) are one-parameter non-linear fittings to Equation (4) in which $k_{switchoff}$ values obtained from curves 2 in (B,D), respectively, were used as input to estimate $k_{activation}$. (B,D) Linearization of data shown in (A,C), respectively; results in the first part of the experiment were fitted to Equation (13) (curves 1, left y-axis), afterwards were fitted to Equation (11) (curves 2, right y-axis); $k_{activation}$ was estimated from the fitting to Equation (13), $k_{switchoff}$ from the fitting to Equation (11).

Table 3 | Kinetic parameters estimated in this work based on the analysis of published data.

PTP	$k_{switchoff}$		$k_{target+H2O2}$	$k_{activation}$
	External H_2O_2	Receptor-mediated H_2O_2 production	External H_2O_2	Receptor-mediated H_2O_2 production
PTP1B	2.6×10^{-3} s^{-1}	1.9×10^{-3} s^{-1}	5.9–11 M^{-1} s^{-1}	$(1–2) \times 10^{-3}$ s^{-1}
SHP-2	1.3×10^{-3} s^{-1}	8.7×10^{-3} s^{-1}	57–60 M^{-1} s^{-1}	$(9.3–9.7) \times 10^{-3}$ s^{-1}

Apparent first-order rate constant for the reactivation of PTP ($k_{switchoff}$), the rate constant for the inactivation of PTP by extracellular H_2O_2 ($k_{target+H2O2}$), and the apparent first-order rate constant for this activation ($k_{activation} = k_{target+H2O2} \times [H_2O_2]$) are shown.

Concerning the parameters describing redox signal switching-off ($k_{switchoff}$), which in the case of the PTPs analyzed here corresponds to their reactivation, the results summarized in Table 3 constitute, as far as we know, the first cell-based kinetic estimates for this process. This characterization is relevant because modulation of PTP reactivation regulates cell signaling (Dagnell et al., 2013). For PTP1B, $k_{switchoff}$ values in the range $(1.9–2.6) \times 10^{-3}$ s^{-1} were estimated, which are plausible taking into account the known data for PTP1B reactivation. In vitro, reduced thioredoxin (2 μM) reactivates oxidized PTP1B with an apparent rate constant of 1.4×10^{-3} s^{-1} (Parsons and Gates, 2013), which corresponds to a rate constant of 700 M^{-1} s^{-1} for this reaction. Thus, considering this rate constant and the $k_{switchoff}$ values determined here, we estimate the cellular concentration of reduced thioredoxin to be 2–3 μM. This range is close to the value observed experimentally in Jurkat T-cells, 0.43 μM (Adimora et al., 2010), with the difference observed being easily account for different cell

lines used or by the participation of alternative thioredoxin-related proteins, like the redoxin TRP14, in the reactivation of PTP1B (Dagnell et al., 2013). This agreement further strengths the validity of the approach we purpose here to reveal kinetic information hidden in typical redox signaling experiments.

For SHP-2, $k_{switchoff}$ was estimated in the range $(1.3–8.7) \times 10^{-3}$ s^{-1} (Table 3), which is similar or higher than the range estimated for PTP1B. This is unexpected because the reactivity of SHP-2 toward thioredoxin is about 20 times lower than PTP1B (Parsons and Gates, 2013). Either the cell line rat-1, where SHP-2 reactivation data was obtained (Meng et al., 2002), contains much higher levels of thioredoxin or an alternative system other than thioredoxin is reactivating SHP-2. The second alternative is supported by the observation that in cells lacking thioredoxin reductase TrxR1, a key partner of thioredoxin that keeps it in the reduced state, SHP-2 oxidation remains unchanged (Dagnell et al., 2013).

Concerning the oxidation of PTPs by H_2O_2, rate constants estimated from experiments in which extracellular H_2O_2 was added were $5.9–11\,M^{-1}\,s^{-1}$ for PTP1B and $57–60\,M^{-1}\,s^{-1}$ for SHP-2 (**Table 3**). These values were calculated based on the external H_2O_2 concentrations added to cells. The actual H_2O_2 concentration that oxidizes these targets is lower as H_2O_2 gradients across the plasma membrane are established when cells are incubated with extracellular H_2O_2 (Antunes and Cadenas, 2000; Marinho et al., 2013b). Thus, the value of these rate constants will be higher if they are based on the actual intracellular H_2O_2 concentrations that oxidize PTPs. For PTP1B, rate constants obtained in kinetic studies with purified PTP1B are in the range $9–43\,M^{-1}\,s^{-1}$ (Denu and Tanner, 1998; Barrett et al., 1999; Zhou et al., 2011; Marinho et al., 2014), and so a gradient between the extracellular and the intracellular concentration of H_2O_2 at the site of PTP1B oxidation is estimated to be in the range 2–7 for the experiments analyzed in this work, which matches the lower range of known gradients for human cell lines (Antunes and Cadenas, 2000; Makino et al., 2004; De Oliveira-Marques et al., 2007; Oliveira-Marques et al., 2013). However, gradients as high as 650 have been recently estimated taking into account the participation of peroxiredoxin (Huang and Sikes, 2014), whose role in the degradation of H_2O_2 is still an open issue (Benfeitas et al., 2014). Thus, the rate constants obtained for PTP1B fit the known quantitative data for the reactivity of this PTP with H_2O_2.

For SHP-2, the estimated rate constants of $57–60\,M^{-1}\,s^{-1}$ (**Table 3**) for its oxidation by H_2O_2 were higher than those determined *in vitro* with purified SHP-2, which are in the range $9–15\,M^{-1}\,s^{-1}$ (Chen et al., 2009; Zhou et al., 2011). Moreover, if the gradient of H_2O_2 across the plasma membrane is taken into account this difference will be even higher. Several possible explanations may account for this discrepancy. First, kinetic rate constants obtained *in vitro* with purified proteins may not reflect rate constants under *in vivo* conditions (Van Eunen et al., 2010, 2012). Second, peroxy-derivatives such as peroxymonocarbonate (Trindade et al., 2006; Zhou et al., 2011) and peroxymonophosphate (LaButti et al., 2007), which have higher reactivity with PTPs than H_2O_2, could be the actual species that oxidize SHP-2. Third, the primary sensor of H_2O_2 may not be SHP-2 but a high-reactive target that slowly relays the oxidation to SHP-2 (Winterbourn and Hampton, 2008; Forman et al., 2010; Brigelius-Flohé and Flohé, 2011; Ferrer-Sueta et al., 2011). Note that the models described here do not distinguish between a mechanism in which a low-reactive sensor is slowly oxidized by H_2O_2, from a mechanism in which a high-reactive sensor is rapidly oxidized by H_2O_2 and then, through a thiol-disulfide reshuffling transfer reaction, slowly oxidizes a low reactive sensor such as SHP-2. In general, known data about redox signaling pathways is consistent with either of these two scenarios (Marinho et al., 2014). Distinguishing between these possible alternative mechanisms will be possible after collecting rate constants in several cell lines upon the generalized application of the equations deduced here to redox signaling experiments.

The kinetic parameters estimated from experiments in which cells are activated by receptor-mediated pathways indicated that the apparent first-order rate constant for the oxidation of SHP-2 is about 5 times higher than that for PTP1B (**Table 3**). Because $k_{activation} = k_{target + H2O2} \times [H_2O_2]$, either the localized H_2O_2 intracellular concentration is higher in the experiment in which SHP-2 oxidation was observed, or $k_{target + H2O2}$ is higher for SHP-2 than for PTP1B, or both. In this regard, the EGF receptor, the H_2O_2 producing enzyme NOX2, and SHP-2 immunoprecipitated all together (Paulsen et al., 2012), supporting the possibility of a highly localized H_2O_2 signaling pool. For PTP1B, from the $k_{activation}$ estimation of $(1.0–2.0) \times 10^{-3}\,s^{-1}$ the local intracellular H_2O_2 concentration reached locally in A431 cells, when stimulated by EGF under the experimental conditions described in Lee et al. (1998), is estimated to be in the range $23–220\,\mu M$, assuming a $k_{target + H2O2}$ value in the range $9–43\,M^{-1}\,s^{-1}$. Such local concentrations, particularly those in the low range of these values, can potentially be reached upon the concerted action of local production of H_2O_2 by NADPH oxidases (Chen et al., 2008; Mishina et al., 2011; Paulsen et al., 2012) and localized inhibition of H_2O_2 removing enzymes (Woo et al., 2010; Rawat et al., 2013). In addition, it can also be suggested that H_2O_2 diffusion out of membrane-entrapped signaling microcompartments may be constrained, because biomembranes constitute a regulable barrier for H_2O_2 diffusion (Antunes and Cadenas, 2000; Branco et al., 2004; Bienert et al., 2007; Miller et al., 2010).

While the equations deduced here were applied successfully to typical signaling experiments, a few alterations in the way experiments are carried out will improve the accuracy of parameter estimation. When cells are exposed to extracellular H_2O_2, we suggest a steady-state delivery so that H_2O_2 is constant during the experiment (Marinho et al., 2013a), a key assumption considered in the deduction of the equations. If the use of a bolus addition is absolutely needed, we suggest short-term experiments so that the H_2O_2 decay caused by its cellular consumption is less significant. Finally, removal of H_2O_2 by adding catalase or replacing extracellular incubation media without H_2O_2, during the second part of the experiment when target reduction starts to predominate, improves the estimation of $k_{switchoff}$ values. This last suggestion may also be applied when H_2O_2 production is triggered by a receptor-mediated mechanism following cell stimulation.

In conclusion, the application of the equations deduced here to typical redox-signaling experiments reveals valuable quantitate kinetic information. Of note, the equations described require only measuring the relative levels of oxidation of a H_2O_2 sensor target and not absolute concentrations, thus facilitating their application to most experiments. While equations were tested with PTP signaling, they can be applied to other proteins that react with H_2O_2, such has thiol-proteins and those containing metal-centers. Being characterized by the presence of both multiple parallel pathways and biphasic effects, redox regulation is a field that will benefit from the widespread determination of kinetic parameters. Such knowledge is important to distinguish apparent contradictory biological effects of reactive oxygen species that are involved in pathological damaging pathways and, at the same time, are part of normal functional signaling pathways. In this way, the present knowledge on redox signaling and oxidative stress would be more efficiently translated into therapeutic applications.

ACKNOWLEDGMENTS

Supported by Fundação para a Ciência e a Tecnologia (FCT), Portugal (PEst-OE/QUI/UI0612/2013 and VIH/SAU/0020/2011).

REFERENCES

Adimora, N. J., Jones, D. P., and Kemp, M. L. (2010). A model of redox kinetics implicates the thiol proteome in cellular hydrogen peroxide responses. *Antioxid. Redox Signal.* 13, 731–743. doi: 10.1089/ars.2009.2968

Alves, R., Antunes, F., and Salvador, A. (2006). Tools for kinetic modeling of biochemical networks. *Nat. Biotechnol.* 24, 667–672. doi: 10.1038/nbt0606-667

Antunes, F., and Cadenas, E. (2000). Estimation of H2O2 gradients across biomembranes. *FEBS Lett.* 475, 121–126. doi: 10.1016/S0014-5793(00)01638-0

Barrett, W. C., DeGnore, J. P., Konig, S., Fales, H. M., Keng, Y. F., Zhang, Z. Y., et al. (1999). Regulation of PTP1B via glutathionylation of the active site cysteine 215. *Biochemistry (Mosc.)* 38, 6699–6705. doi: 10.1021/bi990240v

Benfeitas, R., Selvaggio, G., Antunes, F., Coelho, P. M. B. M., and Salvador, A. (2014). Hydrogen peroxide metabolism and sensing in human erythrocytes: a validated kinetic model and reappraisal of the role of peroxiredoxin II. *Free Radic. Biol. Med.* 74, 35–49. doi: 10.1016/j.freeradbiomed.2014.06.007

Bienert, G. P., Møller, A. L. B., Kristiansen, K. A., Schulz, A., Møller, I. M., Schjoerring, J. K., et al. (2007). Specific aquaporins facilitate the diffusion of hydrogen peroxide across membranes. *J. Biol. Chem.* 282, 1183–1192. doi: 10.1074/jbc.M603761200

Branco, M. R., Marinho, H. S., Cyrne, L., and Antunes, F. (2004). Decrease of H2O2 plasma membrane permeability during adaptation to H2O2 in *Saccharomyces cerevisiae*. *J. Biol. Chem.* 279, 6501–6506. doi: 10.1074/jbc.M311818200

Brigelius-Flohé, R., and Flohé, L. (2011). Basic principles and emerging concepts in the redox control of transcription factors. *Antioxid. Redox Signal.* 15, 2335–2381. doi: 10.1089/ars.2010.3534

Buettner, G. R., Wagner, B. A., and Rodgers, V. G. J. (2013). Quantitative redox biology: an approach to understanding the role of reactive species in defining the cellular redox environment. *Cell Biochem. Biophys.* 67, 477–483. doi: 10.1007/s12013-011-9320-3

Chen, C.-Y., Willard, D., and Rudolph, J. (2009). Redox regulation of SH2-domain-containing protein tyrosine phosphatases by two backdoor cysteines. *Biochemistry (Mosc.)* 48, 1399–1409. doi: 10.1021/bi801973z

Chen, K., Kirber, M. T., Xiao, H., Yang, Y., and Keaney, J. F. (2008). Regulation of ROS signal transduction by NADPH oxidase 4 localization. *J. Cell Biol.* 181, 1129–1139. doi: 10.1083/jcb.200709049

Covas, G., Marinho, H. S., Cyrne, L., and Antunes, F. (2013). Activation of Nrf2 by H2O2: *de novo* synthesis versus nuclear translocation. *Methods Enzymol.* 528, 157–171. doi: 10.1016/B978-0-12-405881-1.00009-4

Cyrne, L., Oliveira-Marques, V., Marinho, H. S., and Antunes, F. (2013). H2O2 in the induction of NF-κB-dependent selective gene expression. *Methods Enzymol.* 528, 173–188. doi: 10.1016/B978-0-12-405881-1.00010-0

Dagnell, M., Frijhoff, J., Pader, I., Augsten, M., Boivin, B., Xu, J., et al. (2013). Selective activation of oxidized PTP1B by the thioredoxin system modulates PDGF-? receptor tyrosine kinase signaling. *Proc. Natl. Acad. Sci. U.S.A.* 110, 13398–13403. doi: 10.1073/pnas.1302891110

Denu, J. M., and Tanner, K. G. (1998). Specific and reversible inactivation of protein tyrosine phosphatases by hydrogen peroxide: evidence for a sulfenic acid intermediate and implications for redox regulation. *Biochemistry (Mosc.)* 37, 5633–5642. doi: 10.1021/bi973035t

De Oliveira-Marques, V., Cyrne, L., Marinho, H., and Antunes, F. (2007). A quantitative study of NF-kappa B activation by H2O2: relevance in inflammation and synergy with TNF-alpha. *J. Immunol.* 178, 3893–3902. doi: 10.4049/jimmunol.178.6.3893

Ferrer-Sueta, G., Manta, B., Botti, H., Radi, R., Trujillo, M., and Denicola, A. (2011). Factors affecting protein thiol reactivity and specificity in peroxide reduction. *Chem. Res. Toxicol.* 24, 434–450. doi: 10.1021/tx100413v

Fisher-Wellman, K. H., and Neufer, P. D. (2012). Linking mitochondrial bioenergetics to insulin resistance via redox biology. *Trends Endocrinol. Metab.* 23, 142–153. doi: 10.1016/j.tem.2011.12.008

Flohe, L. (1979). Glutathione peroxidase: fact and fiction. *Ciba Found. Symp.* 65, 95–122.

Forman, H. J. (2007). Use and abuse of exogenous H2O2 in studies of signal transduction. *Free Radic. Biol. Med.* 42, 926–932. doi: 10.1016/j.freeradbiomed.2007.01.011

Forman, H. J., Maiorino, M., and Ursini, F. (2010). Signaling functions of reactive oxygen species. *Biochemistry (Mosc.)* 49, 835–842. doi: 10.1021/bi9020378

Forstrom, J. W., and Tappel, A. L. (1979). Donor substrate specificity and thiol reduction of glutathione disulfide peroxidase. *J. Biol. Chem.* 254, 2888–2891.

Haque, A., Andersen, J. N., Salmeen, A., Barford, D., and Tonks, N. K. (2011). Conformation-sensing antibodies stabilize the oxidized form of PTP1B and inhibit its phosphatase activity. *Cell* 147, 185–198. doi: 10.1016/j.cell.2011.08.036

Huang, B. K., and Sikes, H. D. (2014). Quantifying intracellular hydrogen peroxide perturbations in terms of concentration. *Redox Biol.* 2, 955–962. doi: 10.1016/j.redox.2014.08.001

Irani, K., Xia, Y., Zweier, J. L., Sollott, S. J., Der, C. J., Fearon, E. R., et al. (1997). Mitogenic signaling mediated by oxidatnts in ras-transformed fibroblasts. *Science* 275, 1649–1652. doi: 10.1126/science.275.5306.1649

Iwakami, S., Misu, H., Takeda, T., Sugimori, M., Matsugo, S., Kaneko, S., et al. (2011). Concentration-dependent dual effects of hydrogen peroxide on insulin signal transduction in H4IIEC hepatocytes. *PLoS ONE* 6:e27401. doi: 10.1371/journal.pone.0027401

LaButti, J., Chowdhury, G., Reilly, T. J., and Gates, K. S. (2007). Redox regulation of protein tyrosine phosphatase 1B (PTP1B) by peroxymonophosphate (= O3POOH). *J. Am. Chem. Soc.* 129:5320. doi: 10.1021/ja070194j

Lee, S. R., Kwon, K. S., Kim, S. R., and Rhee, S. G. (1998). Reversible inactivation of protein-tyrosine phosphatase 1B in A431 cells stimulated with epidermal growth factor. *J. Biol. Chem.* 273, 15366–15372. doi: 10.1074/jbc.273.25.15366

Le Moan, N., Clement, G., Le Maout, S., Tacnet, F., and Toledano, M. B. (2006). The *Saccharomyces cerevisiae* proteome of oxidized protein thiols: contrasted functions for the thioredoxin and glutathione pathways. *J. Biol. Chem.* 281, 10420–10430. doi: 10.1074/jbc.M513346200

Mahadev, K., Zilbering, A., Zhu, L., and Goldstein, B. J. (2001). Insulin-stimulated hydrogen peroxide reversibly inhibits protein-tyrosine phosphatase 1b *in vivo* and enhances the early insulin action cascade. *J. Biol. Chem.* 276, 21938–21942. doi: 10.1074/jbc.C100109200

Makino, N., Sasaki, K., Hashida, K., and Sakakura, Y. (2004). A metabolic model describing the H2O2 elimination by mammalian cells including H2O2 permeation through cytoplasmic and peroxisomal membranes: comparison with experimental data. *Biochim. Biophys. Acta* 1673, 149–159. doi: 10.1016/j.bbagen.2004.04.011

Marinho, H. S., Cyrne, L., Cadenas, E., and Antunes, F. (2013a). H2O2 delivery to cells: steady-state versus bolus addition. *Methods Enzymol.* 526, 159–173. doi: 10.1016/B978-0-12-405883-5.00010-7

Marinho, H. S., Cyrne, L., Cadenas, E., and Antunes, F. (2013b). The cellular steady-state of H2O2: latency concepts and gradients. *Methods Enzymol.* 527, 3–19. doi: 10.1016/B978-0-12-405882-8.00001-5

Marinho, H. S., Real, C., Cyrne, L., Soares, H., and Antunes, F. (2014). Hydrogen peroxide sensing, signaling and regulation of transcription factors. *Redox Biol.* 2, 535–562. doi: 10.1016/j.redox.2014.02.006

Martínez-Acedo, P., Núñez, E., Gómez, F. J. S., Moreno, M., Ramos, E., Izquierdo-Álvarez, A., et al. (2012). A novel strategy for global analysis of the dynamic thiol redox proteome. *Mol. Cell. Proteomics* 11, 800–813. doi: 10.1074/mcp.M111.016469

Meng, T.-C., Fukada, T., and Tonks, N. K. (2002). Reversible oxidation and inactivation of protein tyrosine phosphatases *in vivo*. *Mol. Cell* 9, 387–399. doi: 10.1016/S1097-2765(02)00445-8

Miller, E. W., Dickinson, B. C., and Chang, C. J. (2010). Aquaporin-3 mediates hydrogen peroxide uptake to regulate downstream intracellular signaling. *Proc. Natl. Acad. Sci. U.S.A.* 107, 15681–15686. doi: 10.1073/pnas.1005776107

Mishina, N. M., Tyurin-Kuzmin, P. A., Markvicheva, K. N., Vorotnikov, A. V., Tkachuk, V. A., Laketa, V., et al. (2011). Does cellular hydrogen peroxide diffuse or act locally? *Antioxid. Redox Signal.* 14, 1–7. doi: 10.1089/ars.2010.3539

Oakley, F. D., Abbott, D., Li, Q., and Engelhardt, J. F. (2009). Signaling components of redox active endosomes: the redoxosomes. *Antioxid. Redox Signal.* 11, 1313–1333. doi: 10.1089/ARS.2008.2363

Oliveira-Marques, V., Silva, T., Cunha, F., Covas, G., Marinho, H. S., Antunes, F., et al. (2013). A quantitative study of the cell-type specific modulation

Estimation of kinetic parameters related to biochemical interactions between hydrogen peroxide and signal transduction...

191

of c-Rel by hydrogen peroxide and TNF-? *Redox Biol.* 1, 347–352. doi: 10.1016/j.redox.2013.05.004

Parsons, Z. D., and Gates, K. S. (2013). Thiol-dependent recovery of catalytic activity from oxidized protein tyrosine phosphatases. *Biochemistry (Mosc.)* 52, 6412–6423. doi: 10.1021/bi400451m

Paulsen, C. E., Truong, T. H., Garcia, F. J., Homann, A., Gupta, V., Leonard, S. E., et al. (2012). Peroxide-dependent sulfenylation of the EGFR catalytic site enhances kinase activity. *Nat. Chem. Biol.* 8, 57–64. doi: 10.1038/nchembio.736

Rasband, W. (1997). *ImageJ, U. S. National Institutes of Health.* Bethesda, MD. Available online at: http://rsb.info.nih.gov/ij/, 1997–2006

Rawat, S. J., Creasy, C. L., Peterson, J. R., and Chernoff, J. (2013). The tumor suppressor Mst1 promotes changes in the cellular redox state by phosphorylation and inactivation of peroxiredoxin-1 protein. *J. Biol. Chem.* 288, 8762–8771. doi: 10.1074/jbc.M112.414524

Rinna, A., Torres, M., and Forman, H. J. (2006). Stimulation of the alveolar macrophage respiratory burst by ADP causes selective glutathionylation of protein tyrosine phosphatase 1B. *Free Radic. Biol. Med.* 41, 86–91. doi: 10.1016/j.freeradbiomed.2006.03.010

Sies, H. (2014). Role of metabolic H2O2 generation: redox signalling and oxidative stress. *J. Biol. Chem.* 289, 8735–8741. doi: 10.1074/jbc.R113.544635

Tanner, J. J., Parsons, Z. D., Cummings, A. H., Zhou, H., and Gates, K. S. (2011). Redox regulation of protein tyrosine phosphatases: structural and chemical aspects. *Antioxid. Redox Signal.* 15, 77–97. doi: 10.1089/ars.2010.3611

Trindade, D. F., Cerchiaro, G., and Augusto, O. (2006). A role for peroxymonocarbonate in the stimulation of biothiol peroxidation by the bicarbonate/carbon dioxide pair. *Chem. Res. Toxicol.* 19, 1475–1482. doi: 10.1021/tx060146x

Tschopp, J., and Schroder, K. (2010). NLRP3 inflammasome activation: the convergence of multiple signalling pathways on ROS production? *Nat. Rev. Immunol.* 10, 210–215. doi: 10.1038/nri2725

Van Eunen, K., Bouwman, J., Daran-Lapujade, P., Postmus, J., Canelas, A. B., Mensonides, F. I. C., et al. (2010). Measuring enzyme activities under standardized *in vivo*-like conditions for systems biology. *FEBS J.* 277, 749–760. doi: 10.1111/j.1742-4658.2009.07524.x

Van Eunen, K., Kiewiet, J. A. L., Westerhoff, H. V., and Bakker, B. M. (2012). Testing biochemistry revisited: how *in vivo* metabolism can be understood from *in vitro* enzyme kinetics. *PLoS Comput Biol* 8:e1002483. doi: 10.1371/journal.pcbi.1002483

Voit, E. (1991). *Canonical Nonlinear Modeling: S-System Approach to Understanding Complexity.* New York, NY: Van Nostrand Reinhold.

Winterbourn, C. C., and Hampton, M. B. (2008). Thiol chemistry and specificity in redox signaling. *Free Radic. Biol. Med.* 45, 549–561. doi: 10.1016/j.freeradbiomed.2008.05.004

Woo, H. A., Yim, S. H., Shin, D. H., Kang, D., Yu, D.-Y., and Rhee, S. G. (2010). Inactivation of peroxiredoxin i by phosphorylation allows localized H2O2 accumulation for cell signaling. *Cell* 140, 517–528. doi: 10.1016/j.cell.2010.01.009

Zhou, H., Singh, H., Parsons, Z. D., Lewis, S. M., Bhattacharya, S., Seiner, D. R., et al. (2011). The Biological Buffer Bicarbonate/CO_2 Potentiates H_2O_2-Mediated Inactivation of Protein Tyrosine Phosphatases. *J. Am. Chem. Soc.* 133, 15803–15805. doi: 10.1021/ja2077137

Conflict of Interest Statement: The authors declare that the research was conducted in the absence of any commercial or financial relationships that could be construed as a potential conflict of interest.

14

Methods of measuring protein disulfide isomerase activity: a critical overview

Monica M. Watanabe, Francisco R. M. Laurindo and Denise C. Fernandes *

Vascular Biology Laboratory, Heart Institute (InCor), University of São Paulo School of Medicine, São Paulo, Brazil

Edited by:
Bulent Mutus, University of
Windsor, Canada

Reviewed by:
Laura De Gara, Università Campus
Bio-Medico di Roma, Italy
Elia Ranzato, University of Piemonte
Orientale "Amedeo Avogadro", Italy

*Correspondence:
Denise C. Fernandes, Vascular
Biology Laboratory, Heart Institute
(InCor), University of São Paulo
School of Medicine, Av Eneas C
Aguiar, 44, 9th floor, CEP
05403-000, São Paulo, Brazil
e-mail: denisef@usp.br

Protein disulfide isomerase is an essential redox chaperone from the endoplasmic reticulum (ER) and is responsible for correct disulfide bond formation in nascent proteins. PDI is also found in other cellular locations in the cell, particularly the cell surface. Overall, PDI contributes to ER and global cell redox homeostasis and signaling. The knowledge about PDI structure and function progressed substantially based on *in vitro* studies using recombinant PDI and chimeric proteins. In these experimental scenarios, PDI reductase and chaperone activities are readily approachable. In contrast, assays to measure PDI isomerase activity, the hallmark of PDI family, are more complex. Assessment of PDI roles in cells and tissues mainly relies on gain- or loss-of-function studies. However, there is limited information regarding correlation of experimental readouts with the distinct types of PDI activities. In this mini-review, we evaluate the main methods described for measuring the different kinds of PDI activity: thiol reductase, thiol oxidase, thiol isomerase and chaperone. We emphasize the need to use appropriate controls and the role of critical interferents (e.g., detergent, presence of reducing agents). We also discuss the translation of results from *in vitro* studies with purified recombinant PDI to cellular and tissue samples, with critical comments on the interpretation of results.

Keywords: protein disulfide isomerase, chaperone, thiols, reduction, isomerization, oxidation, redox signaling

Protein disulfide isomerase is present in most eukaryotic organisms and is an essential redox chaperone that catalyzes the introduction of disulfide bonds and the rearrangement of incorrect ones in nascent proteins into the endoplasmic reticulum lumen. PDI was the first protein with folding activity described in the literature (Goldberger et al., 1964), and is the founder of the mammalian PDI family. Currently there are 20 or more other PDI members containing one or more thioredoxin-like domains (Lu and Holmgren, 2014). PDI is formed by four thioredoxin-like structural domains, two of them containing the CGHC (cysteine-glycine-histidine-cysteine) catalytic motif capable of catalyzing sulfydryl oxidation and reduction/isomerization of disulfide bonds (for a revision see Hatahet and Ruddock, 2009). The two other domains form a central hydrophobic core pocket involved in substrate binding. A mobile arm close to the C-terminal site regulates, on a redox-dependent way, substrate access to the hydrophobic core (Wang et al., 2013). In addition to its role in disulfide introduction in nascent proteins, PDI has a chaperone effect, preventing protein aggregation in a way not directly related to its redox thiols, but rather to the hydrophobic pocket region away from the redox domains. PDI binds *in vitro* to a variety of molecules, from small peptides to proteins, while *in vivo* only a few client proteins were identified (Hatahet and Ruddock, 2009). In plasma membrane and pericellular compartments, PDI is involved in important biological processes such as thrombus formation, tissue factor regulation, platelet aggregation, cell adhesion and virus internalization. The multiple PDI cellular redox effects and its versatility

in binding to several proteins implicate that PDI may act as an emerging redox cell signaling adaptor (Laurindo et al., 2012) and a promising therapeutic target of several diseases (Xu et al., 2014).

There are several assays to measure PDI activity. Some assays are more specific to one particular PDI activity (e.g., thiol reduction or oxidation), while others focus in the measurement of PDI isomerase activity. Methods for PDI activity in general are used for 3 main purposes: (a) the study of protein folding by PDI to identify substrate intermediates, which requires elaborated analysis and detection by mass spectroscopy, (b) screening of PDI substrates or inhibitors, which demands fast and low-cost assays to be preferentially adopted for high-throughput system (HTPS) platforms, (c) understanding PDI function in (patho)physiological contexts by comparison of PDI activities in different experimental conditions in biological samples. Proteins such as bovine pancreatic tripsin inhibitor (BPTI) and ribonuclease T1 (RNaseT1) fit the requirements for studies of PDI-mediated protein folding, while insulin has been chosen for HTPS automation. However, PDI assays in biological samples are a considerable challenge. Some substrates commonly used for purified PDI assays (insulin or fluorescent GSSG) were already used in cell homogenates, but the interpretation is still difficult due to intrinsic interferents such as the presence of other reductants in the assay. The purpose of this review is to critically discuss the most used methods of measuring the different types of PDI activities (e.g., isomerase, oxidative refolding, reductase, and chaperone), with emphasis given to PDI in biological samples.

AN OVERVIEW OF PDI ACTIVITY ASSAYS

Depending on the starting material, i.e., the "substrate" of PDI, one PDI activity will be preferentially measured over others. Therefore, PDI assays can be classified based on the initial redox state of the substrate. When the substrate of a protein contains scrambled disulfides and PDI catalyzes its conversion to native state (and thus the recovery of substrate activity), this assay is named "isomerase assay" (e.g., scrambled RNase isomerization). In the case of a totally reduced protein, PDI will promote "oxidative refolding" in a series of thiol oxidation/reduction cycles and possibly isomerization reactions to promote substrate gain-of-function (e.g., reduced RNase oxidative folding). PDI reductase activity assays are easier to perform and constitute the most popular in the literature. According to substrate, reductase activity is followed through increase in turbidity or fluorescence changes (e.g., insulin reduction). Finally, using proteins that do not contain disulfide bonds as substrates, PDI chaperone activity can be measured by recovery of substrate activity and/or changes in substrate protein aggregation (e.g., GAPDH aggregation).

When developing a PDI assay, it is important to keep in mind that PDI does not have a known preferential group of substrates (such as Erp57, that preferentially folds glycosylated proteins, Jessop et al., 2007) and substrates used in PDI assays were not so far proven to be physiological PDI substrates. Also, PDI concentration at ER lumen is estimated around 0.2–0.5 mM (Lyles and Gilbert, 1991), so PDI would be in excess over many substrates, a condition that is not generally mimicked in these assays. Many *in vivo* conditions are not considered in such PDI reductase activity assays: PDI cellular compartmentalization, molecular crowding inside cells (which affects protein folding

stability, Zhou, 2013), and PDI recycling after substrate folding—promoted by PDI partners (e.g., endoplasmic reticulum oxidase-Ero1, Rancy and Thorpe, 2008 or oxidized peroxiredoxin-Prx4, Zito et al., 2010). Finally, another important issue is that although *in vitro* PDI chaperone and isomerase activities can be measured separately, *in vivo* PDI redox folding will not discriminate between both activities and, contrarily, seems to require isomerase and chaperone activities acting together (Laurindo et al., 2012). Thus, results obtained from *in vitro* assays should also be interpreted taking into account their limitations due to a reductionist design. A way to partially overcome such limitations is to preferentially combine the measurement of PDI activity with *in vivo* data and discuss them in the context of the specific physiological milieu being investigated.

ISOMERASE ASSAY

Assays for isomerase activity are based on gain-of-function of an inactive protein substrate containing unfolded disulfides (or scrambled disulfide bonds), so it is inactive. In the presence of PDI, disulfides bonds will be folded back to a "correct" position, thus allowing the measurement of activity of the recovered substrate. Two enzymes are more commonly used, scrambled RNase (scRNase, 4 disulfides) and riboflavin-binding protein (RfBP, 9 disulfides).

For RNase, which is reduced in denaturing conditions and then allowed to oxidize under air at room temperature to acquire random disulfide bonds (**Figure 1**), aliquots are removed to measure substrate activity recovery during incubation with PDI. Partial folded RNase can be quenched with excess alkylating agent before measuring RNAse activity recovery (Rancy and

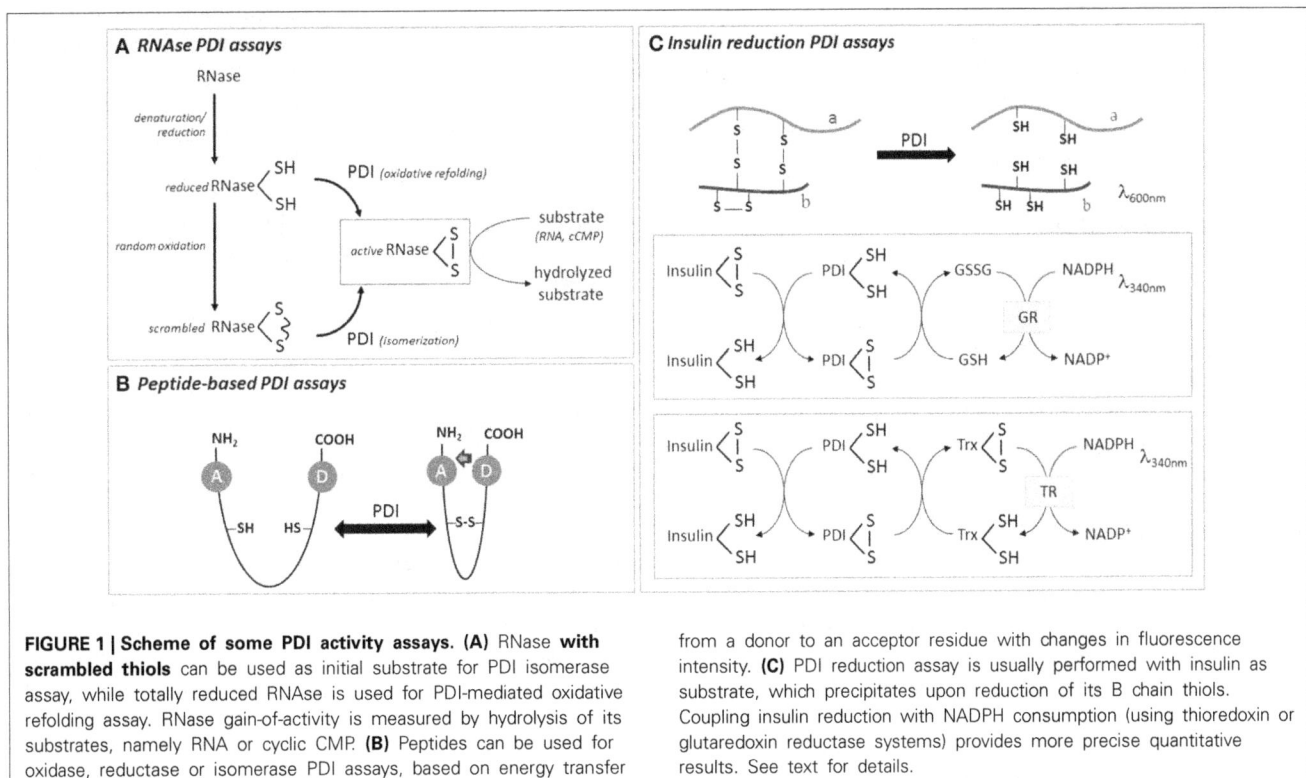

FIGURE 1 | **Scheme of some PDI activity assays. (A)** RNase **with scrambled thiols** can be used as initial substrate for PDI isomerase assay, while totally reduced RNAse is used for PDI-mediated oxidative refolding assay. RNase gain-of-activity is measured by hydrolysis of its substrates, namely RNA or cyclic CMP. **(B)** Peptides can be used for oxidase, reductase or isomerase PDI assays, based on energy transfer from a donor to an acceptor residue with changes in fluorescence intensity. **(C)** PDI reduction assay is usually performed with insulin as substrate, which precipitates upon reduction of its B chain thiols. Coupling insulin reduction with NADPH consumption (using thioredoxin or glutaredoxin reductase systems) provides more precise quantitative results. See text for details.

Thorpe, 2008), which is measured following the hydrolysis of 2 different RNAse substrates: high-molecular-weight RNA (Hillson et al., 1984) or cyclic cytidine monophosphate (cCMP) (El Hindy et al., 2014). The second provides higher increases in absorbance, but requires correction due to cCMP depletion over time and CMP-mediated RNase inhibition (Lyles and Gilbert, 1991). The assay buffer generally contains the redox pair GSH/GSSG, to drive PDI catalysis. One disadvantage of this method is that the most desirable is to obtain all experimental results from the same scRNase batch due to heterogeneity of scRNase among preparations. Otherwise, it is difficult to have reproducible results, independently if scRNase is home-made or commercial (Hatahet and Ruddock, 2009). This method has been extensively used in mechanistic PDI studies. It can also be applied in some particular cellular preparations, such as enriched-membrane fraction of vascular cells (Janiszewski et al., 2005) which does not contain soluble thioredoxin or glutaredoxin reductase systems, and thus may preferentially reflect PDI activity. Of note, it is important to confirm the presence of PDI in the analyzed specimen using western blotting (and even to use the amount of protein to normalize the activity).

The assay with RfBP is based on fluorescence quenching of free riboflavin due to its binding to apoRfBP (RfBP previously reduced under denaturing conditions to release riboflavin, followed by ferricyanide oxidation) (Rancy and Thorpe, 2008). ApoRfBP is mixed with riboflavin in the presence of PDI, and riboflavin fluorescence is measured directly. It is potentially a powerful assay, in which PDI is used in excess over substrate (30-fold), but is not a measurement of gain-of-function (Hatahet and Ruddock, 2009). This assay was not tested with biological samples yet. In homogenate samples the concentration of reagents would have to be optimized, in order to minimize the fluorescence interference of endogenous flavoenzymes, and secondarily, the interference of endogenous riboflavin binding proteins, such as riboflavin kinase or riboflavin transporters. This assay is not recommended for samples, such as plasma, with high levels of albumin or immunoglobulins (which also bind riboflavin).

OXIDATIVE REFOLDING ASSAY

Several fully reduced proteins can be used as substrates for PDI mediated-oxidative refolding. RNase and BPTI (3 disulfides) are considered the best model substrates of oxidative folding assay. In both cases, the intermediates formed during PDI-mediated oxidative folding are characterized by mass spectroscopy after thiol alkylation (Irvine et al., 2014). Although time-consuming, this experimental strategy is the best to estimate rate constants of PDI-dependent reactions during different steps of the folding process (Hatahet and Ruddock, 2009). Other frequently-used substrates include: lysozyme (4 disulfides, van den Berg et al., 1999) and 2 modified RNases, RNaseT1 (2 disulfides) and glutathionylated RNAseT1 (Ruoppolo and Freedman, 1995; Ruoppolo et al., 1996). Reduced RNaseT1 behaves in the assay as reduced RNase, while the glutathionylated substrate has to be deglutathionylated in the first step, a step that may also be mediated by PDI (Reinhardt et al., 2008; Townsend et al., 2009). Translation of oxidative refolding assays to biological samples has the same difficulties as those of isomerase activity discussed previously.

REDUCTION ASSAY

In this assay, PDI will reduce disulfides bonds present in the oxidized substrate. The most popular substrate, due to technical simplicity and low cost, is insulin. The reduction of insulin promotes the aggregation of insulin B chain and is followed by increase in turbidity. PDI is incubated with freshly prepared clear insulin solution in the presence of a reducing agent (DTT or GSH) (Holmgren, 1979). Better aggregation curves are obtained with B chain of insulin than A chain, which induces high background readouts (Karala and Ruddock, 2010). The kinetics of insulin reduction generates two parameters that can be used for relative quantification of different catalysts: the lag time (the time that takes to start precipitation, which can be set as an increase by 0.02 in absorbance over the baseline) and the rate of precipitation (the maximal absorbance increase per minute, Δabsorbance $\times$ min^{-1}). It is the method elected for screening PDI inhibitors in diverse experimental setups (Khan et al., 2011; Paes et al., 2011; Jasuja et al., 2012), and can be optimized for HTPS (Smith et al., 2004) or coupled to fluorogenic dye in commercial kits (ProteoStat™ PDI Assay Kit from Enzo Life Sciences and PDI Inhibitor Screening Assay Kit from AbCam). It was also used in cellular homogenates. To increase assay sensitivity, fluorescent insulin (FITC-insulin) can be used. Also, in order to improve assay quantification, it is also possible to couple insulin reduction with NADPH consumption by coupling with GSSG (or thioredoxin reductase, **Figure 1**). In this case, one enzyme unit is defined as the amount of PDI that catalyzes the formation of GSSG per min, and kinetic parameters can be quantified (Vmax, Kobs, Ki). The assay has to be optimized in a way that insulin aggregation does not interfere with NADPH absorbance (e.g., 340 nm) (Morjana and Gilbert, 1991; Lee et al., 2014).

Other compound used for PDI reduction assay is GSSG. Although GSSG is a poor substrate for PDI (Hatahet and Ruddock, 2009), GSSG covalently attached to eosin (Di-E-GSSG) provides a good sensitivity to the assay, able to detect around nM PDI (Raturi and Mutus, 2007). Di-E-GSSG is non-fluorescent due to self-quenching of two proximal eosin moieties, but after reduction the probe shows ~70-fold increase in fluorescence. This method has been already used for biological samples, such as endothelial cell (HMEC-1) homogenates (Muller et al., 2013) and plasma (Prado et al., 2013). Finally, there is a radioactive method in which the release of radioactive I^{125}-tyramine-SH is assessed, using I^{125}-tyramine-SS-poly-(D-lysine) as substrate, bound to anionic cell surface. This method was employed in intact cells (myeloid and erythrocytes) to measure cell surface PDI reductase activity (Gallina et al., 2002), although confounding effects can potentially occur due to activity of proteases able to cleave the substrate (Xu et al., 2014).

PEPTIDES AS PDI SUBSTRATES

Some peptides have been developed to measure PDI activities in vitro and might have promising results when tested in biological samples. Peptides linked to fluorescent moieties allow high sensitivity and direct measurement of fluorescence. One example is a peptide based on tachyplesin I (TI), a 17-residue antimicrobial peptide that crosses membranes (Kersteen et al., 2005). Scrambled TI (4 cysteines) is isomerized by PDI to its native conformation in which a tryptophan and a dansyl-linked lysine physically interact, allowing fluorescence resonance energy transference (FRET). Another peptide, which can be used for reduction or oxidation PDI assays, depending on initial peptide redox state, contains two cysteines separated by a non-aethyleneglycol spacer (Christiansen et al., 2004). When reduced, this peptide is fluorescent due to o-aminobenzoyl group in the N-terminal region. Upon oxidation, o-aminobenzoyl fluorescence is quenched due to proximity with nitrotyrosine in the C-terminal segment. As a control, a peptide in which a cysteine is changed to serine can be used. Although they are not commercially available, both peptides show the advantages of enhanced homogeneity as starting materials (compared to scrambled thiol-proteins) and direct measurement of peptide fluorescence (i.e., aminobenzoyl/dansyl groups) within samples.

CHAPERONE ASSAY

Assays for chaperone-like activity of PDI are based on the catalysis by PDI of the self-refolding process of completely denatured substrates, which does not require disulfide bonds for folding. The first method for chaperone-like PDI activity was proposed to distinguish PDI chaperone from disulfide isomerase activity, using D-glyceraldehyde-3-phosphate dehydrogenase as substrate (Cai et al., 1994). Other substrates that can be used are lactate dehydrogenase, chemically denatured-rhodanese or citrate synthase, and thermally denatured-alcohol dehydrogenase (Xu et al., 2014). Briefly, the denatured protein is diluted in refolding buffer, in the presence of excess PDI (5–10-fold) and changes in aggregation are followed by light scattering or turbidity. Since changes in aggregation may not correlate with efficient substrate gain-of-function, many authors also measure the reactivation of the substrate, which requires the dissociation of PDI: substrate

complex and optimization of substrate reactivation assay. One recently-reported assay uses green fluorescent protein as PDI substrate to increase sensitivity (Mares et al., 2011). However, it is important to note that oxidation can switch PDI conformation and interfere with its chaperone function (Wang et al., 2013).

CRITICAL COMMENTS ON THE MEASUREMENT OF PDI ACTIVITY IN BIOLOGICAL SAMPLES

All the assays discussed above can be, in theory, applied to biological samples since in general the latter contain high amounts of PDI. With the combination of specific PDI inhibitors, one may guarantee that the activity measured in the whole homogenates is in fact due to PDI and not to other reductase or reduction systems. However, in practice it not as simple as that. Indeed, so far only reductase assays were reportedly transposed to cellular samples. Measurement of reductase PDI activity in cell surface is less prone to endogenous interferents, and the assay can be combined with a neutralizing PDI antibody (Raturi and Mutus, 2007; Langer et al., 2013). In whole cell homogenates, reductase assays will be affected by other reductase systems such as thioredoxin or glutaredoxin. Indeed, the fluorescent Di-E-GSSG was recently employed to measure thioredoxin reductase activity in serum, plasma and lymphocyte lysate (Montano et al., 2014). Furthermore, biological samples might have endogenous unknown PDI inhibitors, as suggested by the observations from our laboratory indicating higher lag time for insulin reduction when exogenous PDI is incubated together with homogenates, as compared to exogenous PDI alone (**Figure 2A**). These observations reinforce the use of a panel of PDI inhibitors and/or the strategy of PDI gain/loss-of-function to provide reliable results for PDI activity assays. For example, in endothelial cells PDI

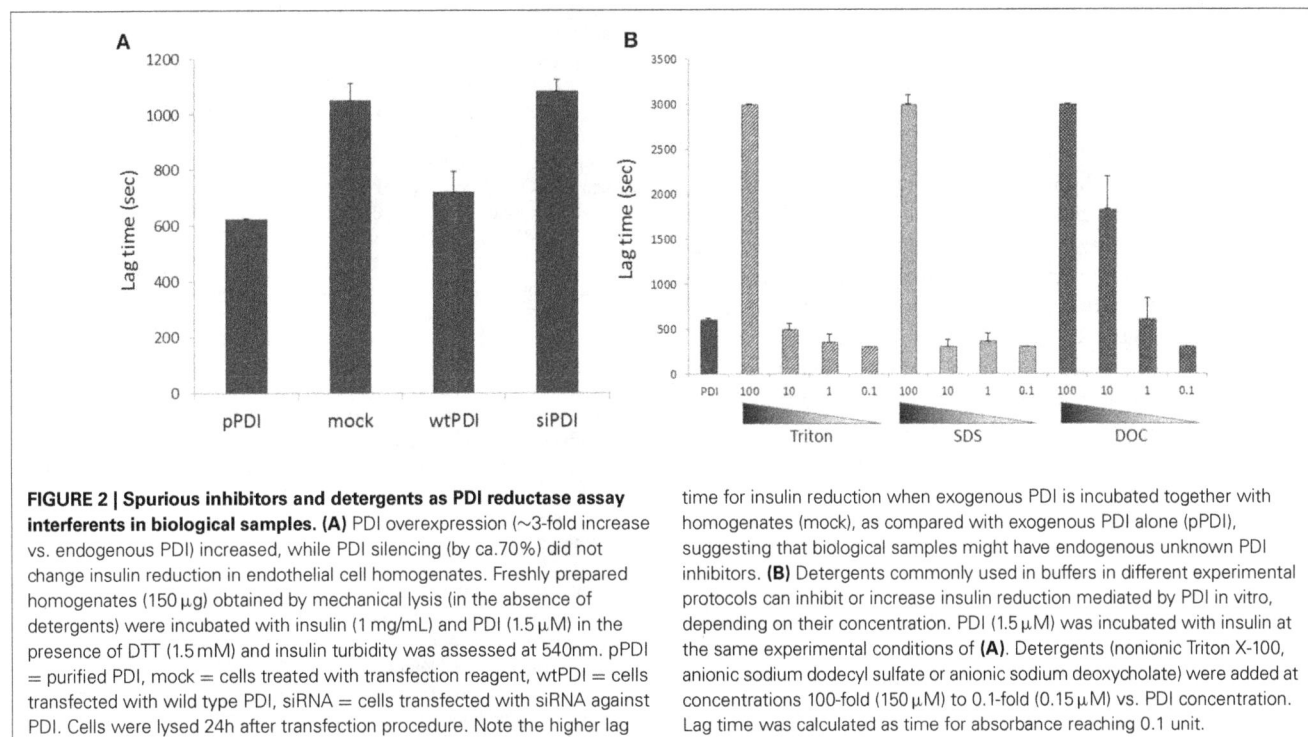

FIGURE 2 | Spurious inhibitors and detergents as PDI reductase assay interferents in biological samples. (A) PDI overexpression (~3-fold increase vs. endogenous PDI) increased, while PDI silencing (by ca.70%) did not change insulin reduction in endothelial cell homogenates. Freshly prepared homogenates (150 μg) obtained by mechanical lysis (in the absence of detergents) were incubated with insulin (1 mg/mL) and PDI (1.5 μM) in the presence of DTT (1.5 mM) and insulin turbidity was assessed at 540nm. pPDI = purified PDI, mock = cells treated with transfection reagent, wtPDI = cells transfected with wild type PDI, siRNA = cells transfected with siRNA against PDI. Cells were lysed 24h after transfection procedure. Note the higher lag

time for insulin reduction when exogenous PDI is incubated together with homogenates (mock), as compared with exogenous PDI alone (pPDI), suggesting that biological samples might have endogenous unknown PDI inhibitors. **(B)** Detergents commonly used in buffers in different experimental protocols can inhibit or increase insulin reduction mediated by PDI in vitro, depending on their concentration. PDI (1.5 μM) was incubated with insulin at the same experimental conditions of **(A)**. Detergents (nonionic Triton X-100, anionic sodium dodecyl sulfate or anionic sodium deoxycholate) were added at concentrations 100-fold (150 μM) to 0.1-fold (0.15 μM) vs. PDI concentration. Lag time was calculated as time for absorbance reaching 0.1 unit.

reductase activity was increased by 2-fold after overexpression of wild-type PDI, while no such increase was detected in cells overexpressing PDI mutated in all redox thiols (Muller et al., 2013). Using the coupled insulin reduction assay, Lee et al. (2014) showed that overexpression of both PDI and thioredoxin interaction protein (Txnip) increased NADPH consumption in cell homogenates compared to PDI transfected controls. In vascular cells, PDI overexpression decreased the lag time of insulin reduction kinetics, although PDI silencing had no effect (**Figure 2A**). These data also suggest that in complex samples, results obtained with PDI reduction assays might significantly reflect PDI expression levels rather than "intrinsic" PDI isomerase activity within cell. One possible alternative is to measure PDI activity after its immunoprecipitation, which reduces contaminants from the whole homogenate.

Several molecules are known to bind and inhibit PDI activity *in vitro* (e.g., nitric oxide, the hormones T3 and estrogen, the peptides somastotin and bacitracin, synthetic organic compounds as bisphenol A and phenylarsenic oxide). However, only few are able to inhibit PDI in some particular experimental *in vivo* conditions. Examples are the reversible PDI inhibitors derived from plants, juniferdin (Khan et al., 2011) and quercetin-3-rutinoside (Jasuja et al., 2012), as well as synthetic compounds that irreversibly inhibit PDI, 16F16, thiomuscimol (Hoffstrom et al., 2010), RB-11-ca (Banerjee et al., 2013), and PACMA31(Xu et al., 2012). These compounds may be used as additional controls during PDI activity assays, although none of these compounds to date have been shown to specifically inhibit PDI. The neutralizing anti-PDI (RL90) antibody is also a powerful control through its specificity, although recently Wu et al. (2012) showed some cross-reactivity with another member of PDI family, Erp57, in Di-E-SSG assay.

Another important experimental topic is the buffer used for cell and tissue lysis. In general they contain, in addition to reducing agents, one or more surfactants, necessary for cell membrane rupture. For example, the widely used RIPA lysis buffer contains NP-40 (1%, 16 mM), sodium deoxycholate (0.5%, 12 mM) and sodium dodecyl sulfate (0.1%, 3.5 mM) (Cold Spring Harbor Protocols). However, Triton was already shown to bind PDI b' domain and strongly inhibit enzyme activity (Klappa et al., 1998), and is typically used at 0.05% (0.77 mM) to inhibit purified PDI in reduction assay (Karala and Ruddock, 2010). Indeed, results from our laboratory indicate that surfactants with distinct physicochemical properties significantly alter PDI reductase activity *in vitro* (**Figure 2B**, same for NP40, CHAPS and saponin). Inhibition or even a small increase of PDI-mediated insulin reduction depends on the proportion between PDI and detergent molecules (**Figure 2B**). Thus, the best procedure for PDI activity assays is to obtain protein extracts by mechanical lysis in detergent-free buffers.

CONCLUSIONS AND PERSPECTIVES

Understanding the role of PDI in cells and tissues as a homeostatic redox signaling adaptor with chaperone properties will require different assays to assess PDI activity. Currently, a major challenge is to precisely measure overall PDI isomerase activity in cells and tissues, and not only one step of thiol rearrangement (oxidation or reduction). In addition, the assay should be performed preferentially *in situ*, since PDI activity may change depending on cell location or redox status. As long as the ideal PDI isomerase assay is yet unavailable for cells and tissues, PDI reductase activity has been measured instead in the cell surface and intact cells, as well as cell homogenates. Such results should be interpreted with care. Improvements in assay reliability will probably require not only advances into sensitivity and specificity, but optimization of adequate controls as well (neutralizing PDI antibodies or more specific emerging PDI inhibitors), or even cell fractionation to remove other reductase systems. Optimization of assays with fluorescent peptides for biological samples seems to be a forthcoming promising approach.

ACKNOWLEDGMENTS

Work supported by Fundação de Amparo à Pesquisa do Estado de São Paulo (FAPESP #09/54764-6), Centro de Pesquisa, Inovação e Difusão FAPESP (CEPID "Processos Redox em Biomedicina", Grant #13/07937-8) and Fundação Zerbini. We thank Dr. Celio X. Santos for critically revising this manuscript.

REFERENCES

Banerjee, R., Pace, N. J., Brown, D. R., and Weerapana, E. (2013). 1,3,5-Triazine as a modular scaffold for covalent inhibitors with streamlined target identification. *J. Am. Chem. Soc.* 135, 2497–2500. doi: 10.1021/ja400427e

Cai, H., Wang, C. C., and Tsou, C. L. (1994). Chaperone-like activity of protein disulfide isomerase in the refolding of a protein with no disulfide bonds. *J. Biol. Chem.* 269, 24550–24552.

Christiansen, C., St Hilaire, P. M., and Winther, J. R. (2004). Fluorometricpolyethyleneglycol-peptide hybrid substrates for quantitative assay of protein disulfide isomerase. *Anal. Biochem.* 333, 148–155. doi: 10.1016/j.ab.2004.06.027

El Hindy, M., Hezwani, M., Corry, D., Hull, J., El Amraoui, F., Harris, M., et al. (2014). The branched-chain aminotransferase proteins: novel redox chaperones for protein disulfide isomerase-implications in *Alzheimer's* disease. *Antioxid. Redox Signal* 20, 2497–2513. doi: 10.1089/ars.2012.4869

Gallina, A., Hanley, T. M., Mandel, R., Trahey, M., Broder, C. C., Viglianti, G. A., et al. (2002). Inhibitors of protein-disulfide isomerase prevent cleavage of disulfide bonds in receptor-bound glycoprotein 120 and prevent HIV-1 entry. *J. Biol. Chem.* 277, 50579–50588. doi: 10.1074/jbc.M204547200

Goldberger, R. F., Epstein, C. J., and Anfinsen, C. B. (1964). Purification and properties of a microsomal enzyme system catalyzing the reactivation of reduced ribonuclease and lysozyme. *J. Biol. Chem.* 239, 1406–1410.

Hatahet, F., and Ruddock, L. W. (2009). Protein disulfide isomerase: a critical evaluation of its function in disulfide bond formation. *Antioxid. Redox Signal* 11, 2807–2850. doi: 10.1089/ars.2009.2466

Hillson, D. A., Lambert, N., and Freedman, R. B. (1984). Formation and isomerization of disulfide bonds in proteins: protein-disulfide-isomerase. *Methods Enzymol.* 107, 281–294. doi: 10.1016/0076-6879(84)07018-X

Hoffstrom, B. G., Kaplan, A., Letso, R., Schmid, R. S., Turmel, G. J., Lo, D. C., et al. (2010). Inhibitors of protein disulfide isomerase suppress apoptosis induced by misfolded proteins. *Nat. Chem. Biol.* 6, 900–906. doi: 10.1038/nchembio.467

Holmgren, A. (1979). Thioredoxin catalyzes the reduction of insulin disulfides by dithiothreitol and dihydrolipoamide. *J. Biol. Chem.* 254, 9627–9632.

Irvine, A. G., Wallis, A. K., Sanghera, N., Rowe, M. L., Ruddock, L. W., Howard, M. J., et al. (2014). Protein disulfide-isomerase interacts with a substrate protein at all stages along its folding pathway. *PLoS ONE* 9:e82511. doi: 10.1371/journal.pone.0082511

Janiszewski, M., Lopes, L. R., Carmo, A. O., Pedro, M. A., Brandes, R. P., Santos, C. X., et al. (2005). Regulation of NAD(P)H oxidase by associated protein disulfide isomerase in vascular smooth muscle cells. *J. Biol. Chem.* 280, 40813–40819. doi: 10.1074/jbc.M509255200

Jasuja, R., Passam, F. H., Kennedy, D. R., Kim, S. H., van Hessem, L., Lin, L., et al. (2012). Protein disulfide isomerase inhibitors constitute a new class of antithrombotic agents. *J. Clin. Invest.* 122, 2104–2113. doi: 10.1172/JCI61228

Jessop, C. E., Chakravarthi, S., Garbi, N., Hämmerling, G. J., Lovell, S., and Bulleid, N. J. (2007). ERp57 is essential for efficient folding of glycoproteins sharing common structural domains. *EMBO J.* 26, 28–40. doi: 10.1038/sj.emboj.7601505

Karala, A. R., and Ruddock, L. W. (2010). Bacitracin is not a specific inhibitor of protein disulfide isomerase. *FEBS J.* 277, 2454–2462. doi: 10.1111/j.1742-4658.2010.07660.x

Kersteen, E. A., Barrows, S. R., and Raines, R. T. (2005). Catalysis of protein disulfide bond isomerization in a homogeneous substrate. *Biochemistry* 44, 12168–12178. doi: 10.1021/bi0507985

Khan, M. M. G., Simizu, S., Lai, N. S., Kawatani, M., Shimizu, T., and Osada, H. (2011). Discovery of a small molecule PDI inhibitor that inhibits reduction of HIV-1 envelope glycoprotein gp120. *ACS Chem. Biol.* 6, 245–251. doi: 10.1021/cb100387r

Klappa, P., Ruddock, L. W., Darby, N. J., and Freedman, R. B. (1998). The b'domain provides the principal peptide-binding site of protein disulfide isomerase but all domains contribute to binding of misfolded proteins. *EMBO J.* 17, 927–935. doi: 10.1093/emboj/17.4.927

Langer, F., Spath, B., Fischer, C., Stolz, M., Ayuk, F. A., Kröger, N., et al. (2013). Rapid activation of monocyte tissue factor by antithymocyte globulin is dependent on complement and protein disulfide isomerase. *Blood* 121, 2324–2335. doi: 10.1182/blood-2012-10-460493

Laurindo, F. R., Pescatore, L. A., and Fernandes, D. C. (2012). Protein disulfide isomerase in redox cell signaling and homeostasis. *Free Radic. Biol. Med.* 52, 1954–1969. doi: 10.1016/j.freeradbiomed.2012.02.037

Lee, S., Min Kim, S., Dotimas, J., Li, L., Feener, E. P., Baldus, S., et al.(2014). Thioredoxin-interacting protein regulates protein disulfide isomerases and endoplasmic reticulum stress. *EMBO Mol. Med.* 6, 732–743. doi: 10.15252/emmm.201302561

Lu, J., and Holmgren, A. (2014). The thioredoxin superfamily in oxidative protein folding. *Antioxid. Redox Signal* 21, 457–470. doi: 10.1089/ars.2014.5849

Lyles, M. M., and Gilbert, H. F. (1991). Catalysis of the oxidative folding of ribonuclease a by protein disulfide isomerase: dependence of the rate on the composition of the redox buffer. *Biochemistry* 30, 613–619. doi: 10.1021/bi00217a004

Mares, R. E., Meléndez-López, S. G., and Ramos, M. A. (2011). Acid-denatured green fluorescent protein (GFP) as model substrate to study the chaperone activity of protein disulfide isomerase. *Int. J. Mol. Sci.* 12, 4625–4636. doi: 10.3390/ijms12074625

Montano, S. J., Lu, J., Gustafsson, T. N., and Holmgren, A. (2014). Activity assays of mammalian thioredoxin and thioredoxin reductase: fluorescent disulfide substrates, mechanisms, and use with tissue samples. *Anal. Biochem.* 449, 139–146. doi: 10.1016/j.ab.2013.12.025

Morjana, N. A., and Gilbert, H. F. (1991). Effect of protein and peptide inhibitors on the activity of protein disulfide isomerase. *Biochemistry* 30, 4985–4990.

Muller, C., Bandemer, J., Vindis, C., Camaré, C., Mucher, E., Guéraud, F., et al. (2013). Protein disulfide isomerase modification and inhibition contribute to ER stress and apoptosis induced by oxidized low density lipoproteins. *Antioxid. Redox Signal* 18, 731–742. doi: 10.1089/ars.2012.4577

Paes, A. M., Veríssimo-Filho, S., Guimarães, L. L., Silva, A. C., Takiuti, J. T., Santos, C. X., et al. (2011). Protein disulfide isomerase redox-dependent association with p47(phox): evidence for an organizer role in leukocyte NADPH oxidase activation. *J. Leukoc. Biol.* 90, 799–810. doi: 10.1189/jlb.0610324

Prado, G. N., Romero, J. R., and Rivera, A. (2013). Endothelin-1 receptor antagonists regulate cell surface-associated protein disulfide isomerase in sickle cell disease. *FASEB J.* 27, 4619–4629. doi: 10.1096/fj.13-228577

Rancy, P. C., and Thorpe, C. (2008). Oxidative protein folding *in vitro*: a study of the cooperation between quiescin-sulfhydryl oxidase and protein disulfide isomerase. *Biochemistry* 47, 12047–12056. doi: 10.1021/bi801604x

Raturi, A., and Mutus, B. (2007). Characterization of redox state and reductase activity of protein disulfide isomerase under different redox environments using a sensitive fluorescent assay. *Free Radic. Biol. Med.* 43, 62–70. doi: 10.1016/j.freeradbiomed.2007.03.025

Reinhardt, C., von Brüh, M. L., Manukyan, D., Grahl, L., Lorenz, M., Altmann, B., et al. (2008). Protein disulfide isomerase acts as an injury response signal that enhances fibrin generation via tissue factor activation. *J. Clin. Invest.* 118, 1110–1122. doi: 10.1172/JCI32376

Ruoppolo, M., and Freedman, R. B. (1995). Refolding by disulfide isomerization: the mixed disulfide between ribonuclease T1 and glutathione as a model refolding substrate. *Biochemistry* 34, 9380–9388. doi: 10.1021/bi00029a014

Ruoppolo, M., Freedman, R. B., Pucci, P., and Marino, G. (1996). Glutathione-dependent pathways of refolding of RNase T1 by oxidation and disulfide isomerization: catalysis by protein disulfide isomerase. *Biochemistry* 35, 13636–13646. doi: 10.1021/bi960755b

Smith, A. M., Chan, J., Oksenberg, D., Urfer, R., Wexler, D. S., Ow, A., et al. (2004). A high-throughput turbidometric assay for screening inhibitors of protein disulfide isomerase activity. *J. Biomol. Screen.* 9, 614–620. doi: 10.1177/1087057104265292

Townsend, D. M., Manevich, Y., He, L., Xiong, Y., Bowers, R. R. Jr., Hutchens, S., et al. (2009). Nitrosative stress-induced s-glutathionylation of protein disulfide isomerase leads to activation of the unfolded protein response. *Cancer Res.* 69, 7626–7634. doi: 10.1158/0008-5472.CAN-09-0493

van den Berg, B., Chung, E. W., Robinson, C. V., Mateo, P. L., and Dobson, C. M. (1999). The oxidative refolding of hen lysozyme and its catalysis by protein disulfide isomerase. *EMBO J.* 18, 4794–4803. doi: 10.1093/emboj/18.17.4794

Wang, C., Li, W., Ren, J., Fang, J., Ke, H., Gong, W., et al. (2013). Structural insights into the redox-regulated dynamic conformations of human protein disulfide isomerase. *Antioxid. Redox Signal* 19, 36–45. doi: 10.1089/ars.2012.4630

Wu, Y., Ahmad, S. S., Zhou, J., Wang, L., Cully, M. P., and Essex, D. W. (2012). The disulfide isomerase ERp57 mediates platelet aggregation, hemostasis, and thrombosis. *Blood* 119, 1737–1746. doi: 10.1182/blood-2011-06-360685

Xu, S., Butkevich, A. N., Yamada, R., Zhou, Y., Debnath, B., Duncan, R., et al. (2012). Discovery of an orally active small-molecule irreversible inhibitor of protein disulfide isomerase for ovarian cancer treatment. *Proc. Natl. Acad. Sci. U.S.A.* 109, 16348–16353. doi: 10.1073/pnas.1205226109

Xu, S., Sankar, S., and Neamati, N. (2014). Protein disulfide isomerase: a promising target for cancer therapy. *Drug Discov. Today* 19, 222–240. doi: 10.1016/j.drudis.2013.10.017

Zhou, H. Z. (2013). Influence of crowded cellular environments on protein folding, binding, and oligomerization: biological consequences and potentials of atomistic modeling. *FEBS Lett.* 587, 1053–1061. doi: 10.1016/j.febslet.2013.01.064

Zito, E., Melo, E. P., Yang, Y., Wahlander, A., Neubert, T. A., and Ron, D. (2010). Oxidative protein folding by an endoplasmic reticulum-localized peroxiredoxin. *Mol. Cell* 40, 787–797. doi: 10.1016/j.molcel.2010.11.010

Conflict of Interest Statement: The authors declare that the research was conducted in the absence of any commercial or financial relationships that could be construed as a potential conflict of interest.

15

Copper mediated controlled radical copolymerization of styrene and 2-ethylhexyl acrylate and determination of their reactivity ratios

Bishnu P. Koiry and Nikhil K. Singha *

Rubber Technology Centre, Indian Institute of Technology Kharagpur, Kharagpur, India

Edited by:
Clemens Kilian Weiss, Universtiy of Applied Science Bingen, Germany

Reviewed by:
Frederik Roman Wurm, Max Planck Institute for Polymer Research, Germany
Yasuhiro Matsuda, Shizuoka University, Japan
Jerome Claverie, Université du Québec à Montréal, Canada

*Correspondence:
Nikhil K. Singha, Rubber Technology Centre, Indian Institute of Technology Kharagpur, Kharagpur 721302, West Bengal, India
e-mail: nks@rtc.iitkgp.ernet.in

Copolymerization is an important synthetic tool to prepare polymers with desirable combination of properties which are difficult to achieve from the different homopolymers concerned. This investigation reports the copolymerization of 2-ethylhexyl acrylate (EHA) and styrene using copper bromide (CuBr) as catalyst in combination with N,N,N′,N″,N″- pentamethyldiethylenetriamine (PMDETA) as ligand and 1-phenylethyl bromide (PEBr) as initiator. Linear kinetic plot and linear increase in molecular weights vs. conversion indicate that copolymerization reactions were controlled. The copolymer composition was calculated using ^{1}H NMR studies. The reactivity ratio of styrene and EHA (r_1 and r_2) were determined using the Finemann–Ross (FR), inverted Finemann–Ross (IFR), and Kelen–Tudos (KT) methods. Thermal properties of the copolymers were also studied by using TGA and DSC analysis.

Keywords: controlled radical polymerization, ATRP, polystyrene, poly(2-ethylhexyl acrylate), copolymers, reactivity ratio, thermal properties

INTRODUCTION

The homopolymer and copolymers of 2-ethylhexyl acrylate (EHA) have very good film formation characteristics and have very good low temperature flexibility, because of the presence of branched and longer alkyl pendant group in the EHA. They have also low volume shrinkage. So the copolymers of EHA are widely used in paints, coating and adhesive applications (Skeist, 1977; Plessis et al., 2001; Webster and Crain, 2002). Copolymerization is an important synthetic tool which can control the thermal and mechanical properties of the polymers (Kavousian et al., 2004). In the copolymers of styrene and EHA, the properties of the copolymers can be monitored by controlled incorporation of the respective comonomers styrene and EHA. Polystyrene possess high glass transition temperature, T_g (~100°C) where as poly(2-ethylhexyl acrylate) (PEHA) has low T_g of −60°C. Copolymers of styrene and EHA are important components for hard coating (Plessis et al., 2001) and also are used as blend compatibilizers (Haldankar, 2001). Conventional free radical polymerization (FRP) leads to uncontrolled molecular weight and broad dispersity (Đ). There is also gel formation tendency, because of the several side reactions during FRP. This makes them difficult to apply for paints and coating material application owing to high viscosity (Solomon and Moad, 1995; Odian, 2004). Since 1990s there have been spectacular advances in the field of controlled radical polymerization (CRP). There were several CRP techniques namely, atom transfer radical polymerization (ATRP) (Kamigaito et al., 2001; Matyjaszewski and Jia, 2001; Kavitha and Singha, 2008), nitroxide-mediated polymerization (NMP) (Harth et al., 2001; Hawker et al., 2001), and reversible addition–fragmentation chain transfer (RAFT) (Chiefari et al.,

1998; Moad et al., 2000; Moad, 2006; Barner-Kowollik, 2008). Among the different CRP techniques, ATRP is applicable to polymerize wide range of monomers and can be carried out at wide range of temperature (−20 to 200°C) (Haloi et al., 2009). ATRP has been successfully applied to synthesize a wide range of polymers with varied molecular weights, different architectures, functionalities etc. Transition metal catalyzed CRP known as ATRP is an important CRP method (Matyjaszewski, 2000) which is carried out in presence of an active alkyl halide using a transition metal halide as catalyst in combination with a suitable ligand (Matyjaszewski and Jia, 2001). A wide variety of copolymers can be prepared via ATRP with controlled molecular weight, functionality, and low dispersity. There are several reports on the copolymerization of different acrylate monomers using ATRP and determination of their reactivity ratios. The reactivity ratios are important parameters for a set of monomer. The reactivity ratios of monomers predict the copolymer composition as well as the sequence distribution of the comonomers. It also predicts the properties of the copolymer. ATRP provides random copolymer with similar chain compositions which is very much different from FRP (Matyjaszewski, 2002). During the polymerization reaction, the polymer chains grow simultaneously and thus all the polymer chains have same composition. However, in FRP macromolecular chains start growing at different times during the polymerization and the monomer composition continuously changes. As a result, in FRP the different chains will have different compositions in the end (Solomon and Moad, 1995). The reactivity ratios of co-monomers in a CRP are somewhat different from the same in FRP. This is because in CRP processes there is intermittent activation-deactivation of the active species which results in different rates of consumption of comonomers (Matyjaszewski, 2002; Braunecker and Matyjaszewski, 2007). For example, Mignard et al. reported the reactivity ratios of the copolymerization of styrene and butyl acrylate (BA) via NMP at 120°C in solution. They reported the reactivity ratios for styrene and BA within the range of 0.60–1.2 ($r_{styrene}$) and 0.16–0.29 (r_{BA}) respectively (Mignard et al., 2004). Arehart and Matyjaszewski reported the reactivity ratios of styrene and BA prepared via ATRP at 110°C in solution. They reported the reactivity ratios for styrene and BA as $0.68 < r_{styrene} < 0.82$ and $0.22 < r_{BA} < 0.26$ respectively (Arehart and Matyjaszewski, 1999). Chambard et al. reported the copolymerization of styrene and BA prepared in bulk via FRP. The reactivity ratios of styrene and BA prepared via FRP at 90°C were reported to be 0.95 and 0.20 respectively (Chambard et al., 1999). Ziegler and Matyjaszewski reported the variation in reactivity ratios of MMA and BA with the change in ligand from 4,4'-di(5-nonyl)-2,2'-bipyridine (dNbpy) ($r_{MMA} = 2.52$, $r_{BA} = 0.26$) to N,N,N',N'',N''-pentamethyldiethylenetriamine (PMDETA) ($r_{MMA} = 3.15$, and $r_{BA} = 0.37$) (Ziegler and Matyjaszewski, 2001). Lessard et al. reported the reactivity ratios of styrene and *tert*-butyl acrylate(*t*-BA) prepared via NMP at 115°C in bulk as $r_{t\text{-}BA} = 0.09$–0.12 and $r_{styrene} = 0.40$–0.49 (Lessard et al., 2007). We reported the ATRP of furfuryl methylacrylate and methyl methacrylate (Kavitha and Singha, 2007), 2-ethylhexyl acrylate and glycidyl methacrylate (Haloi et al., 2009). However, there is no report on the ATRP of styrene and EHA and to determine their reactivity ratios. The objective of this investigation is to study the copolymerization of styrene and EHA via ATRP and to determine their reactivity ratios. The reactivity ratios of styrene and EHA were calculated using Finemann–Ross (FR), inverted Finemann–Ross (IFR), and Kelen–Tudos (KT) methods (Fineman and Ross, 1950; Kelen et al., 1980; Makrikosta et al., 2005).

MATERIALS AND METHODS

The monomers, 2-ethylhexyl acrylate (EHA) (Aldrich, USA) and styrene (Jyoti Chemicals, Mumbai) were purified by vacuum distillation. CuBr (Aldrich, USA) was purified by washing with glacial acetic acid, and then it was washed thoroughly with diethyl ether and was finally dried under vacuum. Phenyl ethylbromide (PEBr) (97%) and N,N,N',N'',N''-pentamethyldiethylenetriamine (PMDETA) (97%) were purchased from Aldrich, USA and were used as received.

CHARACTERIZATION

Number average molecular weight (Mn, GPC) and dispersity (Đ) were determined by Gel Permeation Chromatography (GPC). GPC analysis was carried out at room temperature using a Viscotek GPC equipped with a refractive index detector (Model VE3580), two ViscoGEL GPC columns (model GMHHR-M # 17392) connected in series. GPC analysis was carried out using tetrahydrofuran as eluent at a flow rate of 1 ml/min. Linear and narrow disperse polystyrene was used as calibration standard and Viscotek OMNI-01 software was used for data processing.

^{1}H NMR spectra of the polymers were recorded on a 200 MHz Brucker NMR spectrometer using $CDCl_3$ as solvent which had a small amount of tetramethylsilane (TMS) as an internal standard.

Differential scanning calorimetry (DSC) analysis was carried out by using TA Instrument (DSC Q100 V8.1 Build 251) under nitrogen atmosphere at a heating rate of 10°C/min within a temperature range of −100°C to +150°C. The baseline calibration was done by scanning the temperature domain with the help of an empty pan. The enthalpy was calibrated by using indium standard and the heat capacity was calibrated by using the sapphire disc that was supplied by TA instrument. The glass transition temperature (T_g) was determined from the plot of heat flow vs. temperature in the second heating scan in the DSC analysis.

Thermogravimetric analysis (TGA) was carried out by using a TA Instrument (Q50) at a heating rate of 20°C/min in the temperature range of 30–600°C in nitrogen atmosphere. TGA analyzer consists of high precision balance with a pan which was placed in a small electrically heated oven. The temperature was measured accurately with the help of a thermocouple. From the plot of weight percent vs. temperature the polymer degradation temperature was determined.

SYNTHESIS OF COPOLYMERS OF STYRENE AND EHA VIA ATOM TRANSFER RADICAL COPOLYMERIZATION (ATRcP)

The polymerization reaction was carried out in a Schlenk tube. In a typical ATRP reaction EHA (4.82 g, 26.1 mmol), styrene (0.909 g, 8.7 mmol) and CuBr (0.050 g, 0.35 mmol) were accurately weighed and transferred to the Schlenk tube. The PMDETA ligand (0.0604 g, 0.35 mmol) was then added to the reaction tube. Oxygen was removed from the reaction mixture by passing nitrogen through the reaction tube. The polymerization was started by adding PEBr (0.0646 g, 0.35 mmol) and was carried out at 90°C. Aliquot samples were taken out at different time intervals and were used to calculate the conversion by gravimetric method. The samples were also used to find out the molecular weight by GPC. The final product was diluted with THF and was purified by passing through alumina column to remove the copper catalyst. The same procedure was adopted for other feed ratios of ATRcP (atom transfer radical copolymerization) of EHA and styrene.

HOMOPOLYMERIZATION OF STYRENE VIA ATOM TRANSFER RADICAL POLYMERIZATION (ATRP)

The homopolymerization of styrene was carried out in bulk in a Schlenk tube equipped with silicone septum and magnetic stirring bar. In the Schlenk tube styrene (4.5 g, 43.2 mmol), CuBr (0.031 g, 0.21 mmol) and PMDETA (0.037 g, 0.21 mmol) were weighed and degassed by passing nitrogen gas for 15 min. The reaction was started by adding PEBr (0.040 g, 0.21 mmol) in the mixture. The reaction was carried out at 110°C for 6 h. ^{1}H NMR ($CDCl_3$, 200 MHz): δ (in ppm) = 6.4–7.2 (phenyl protons of polystyrene) and 1.4–2.2 ppm ($-CH_2-$ and $>CH-$ protons).

SYNTHESIS OF DIBLOCK COPOLYMER OF POLYSTYRENE WITH EHA VIA ATRP

The homopolymer of polystyrene (i.e., PS-Br) was used as macroinitiator for the synthesis of diblock copolymer of styrene and EHA. The macroinitiator, PS-Br (1.0 g, 0.066 mmol) was taken in a Schlenk tube and was dissolved in THF solvent into which CuBr (0.009 g, 0.069 mmol) and PMDETA (0.017 g, 0.10 mmol) were added followed by EHA (1 g, 5.4 mmol). The polymerization reaction was carried out at 90°C for 6 h. The polymer obtained was dissolved in THF and was purified by passing through basic alumina column and then precipitated in methanol.

RESULTS AND DISCUSSION

Copolymerization of styrene and 2-ethylhexyl acrylate was carried out via ATRcP (shown in **Scheme 1**) at different feed ratios by using phenylethylbromide (PEBr) as initiator, CuBr as catalyst in combination with PMDETA as a ligand. From the kinetic plot of ln(1/1-X) (where X is the percent conversion of monomer) vs. time, it was observed that the value of ln(1/1-X) was linearly increased with time (**Figure 1**). This linear dependency is the characteristics of the controlled polymerization reaction which follows the first order kinetics. **Figure 1** showed that with increase in styrene content in the feed the rate of polymerization increased (Jianying et al., 2006). **Figure 2** infers that there was linear increase in molecular weight with conversion and dispersity (Đ) was relatively narrow. GPC traces of poly(styrene-co-EHA) (50:50) (sample 3 of **Table 1**) is shown in supplementary section (**Figure S1**). It indicates the controlled nature of the copolymerization reaction. **Table 1** summarizes the feed ratio as well as the copolymer composition of the different copolymerization reactions.

STRUCTURAL CHARACTERIZATION AND COPOLYMER COMPOSITION

The structural characterization and copolymer composition were determined by ^{1}H NMR spectroscopy. **Figure 3** shows the ^{1}H NMR spectra of poly(styrene-co-EHA) of 40:60 feed ratio. The resonance at $\delta = 0.8$ ppm is due to the –CH$_3$ protons of PEHA part. The broad resonances at $\delta = 1.0$ to 2.1 ppm are due to the different –CH$_2$– and >CH– protons of pendant group of PEHA part as well as those of the main chain backbone of the copolymer. The resonances at $\delta = 6.5$–7.1 ppm are due to the different aromatic protons of polystyrene part. The resonances at 3.8 ppm are due to –OCH$_2$– protons of PEHA part. The distinct resonances at 6.5–7.3 for five aromatic protons of styrene and 3.8 ppm for two protons of EHA were used to calculate the composition of the copolymer of styrene and EHA. The copolymer composition

was determined by the equation (1) as shown below

$$F_{EHA} = \frac{5A}{5A + 2B} \times 100 \quad (1)$$

where, A and B represent the integral area at $\delta = 3.7$ and $\delta = 6.5$–7.3 ppm for –OCH$_2$– protons in PEHA and aromatic protons of polystyrene unit respectively.

REACTIVITY RATIO DETERMINATION

In the copolymerization of styrene (1) and EHA (2) the reactivity ratio r_1 is defined as k_{11}/k_{12}, where k_{11} is the rate constant of the reaction between the growing polymer chain carrying free radical of styrene as the terminal unit and styrene (homo propagation) and k_{12} is the rate constant of the reaction between the same reactive chain end and the EHA monomer. Similarly the reactivity ratio r_2 is also defined as k_{22}/k_{21}, where k_{22} is the rate constant of homo propagation reaction between the growing

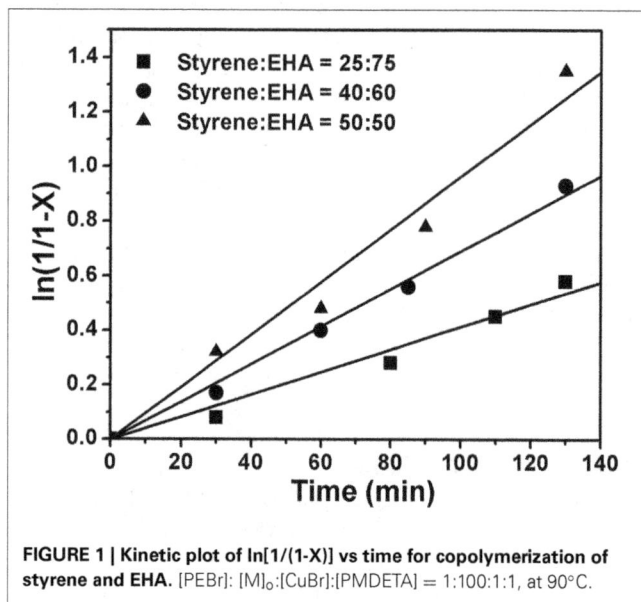

FIGURE 1 | Kinetic plot of ln[1/(1-X)] vs time for copolymerization of styrene and EHA. [PEBr]: [M]$_0$:[CuBr]:[PMDETA] = 1:100:1:1, at 90°C.

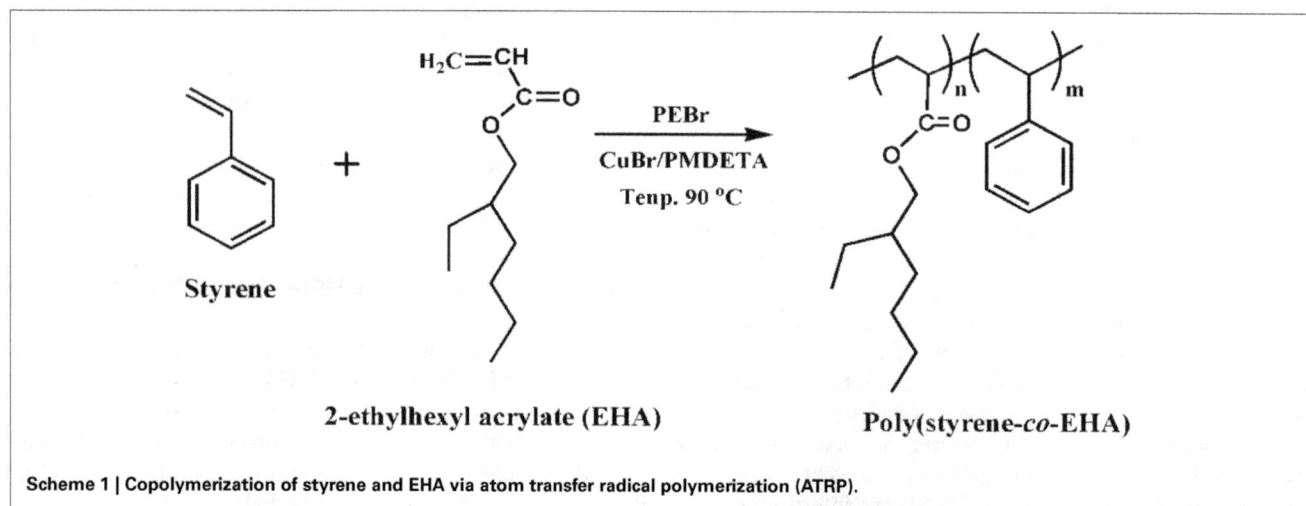

Scheme 1 | Copolymerization of styrene and EHA via atom transfer radical polymerization (ATRP).

macromolecular chain having EHA active radical as the terminal unit and EHA and k_{21} is the rate constant of the reaction between EHA active radical and styrene monomer (cross propagation). For determining the r_1 and r_2, copolymerization of styrene and EHA was carried out at different feed ratio of styrene (1) and EHA

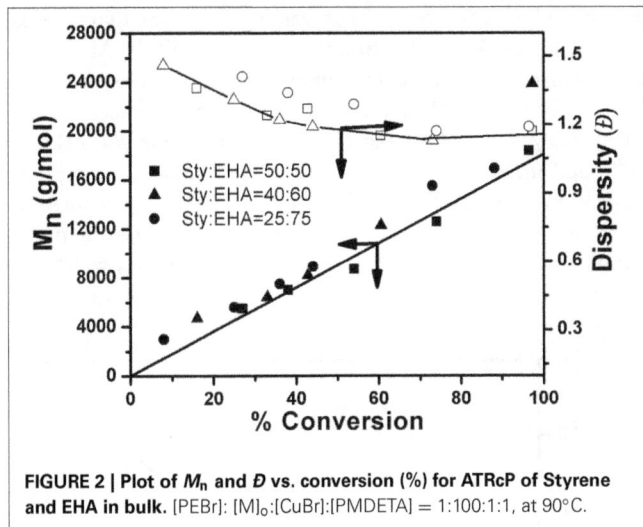

FIGURE 2 | Plot of M_n and $Đ$ vs. conversion (%) for ATRcP of Styrene and EHA in bulk. [PEBr]: [M]$_o$:[CuBr]:[PMDETA] = 1:100:1:1, at 90°C.

Table 1 | Copolymerization of styrene and EHA in bulk at 90°C.

Sl. No.	Feed composition (%) styrene: EHA	Copolymer composition$^{\neq}$ (%) styrene: EHA	Conversion (%)	$M_{n,GPC}$ (g/mol)	$Đ$
1	25:75	27:73	98.1	16,900	1.27
2	40:60	44:56	88.0	24,000	1.15
3	50:50	56.6:43.4	97.8	18,500	1.17

[PEBr]: [M]$_o$:[CuBr]:[PMDETA] = 1:100:1:1. Reaction time = 2 h 20 min.
$^{\neq}$ *Compositions were calculated by ^{1}H NMR spectroscopy.*

(2). In this case copolymerization was carried out at low conversion (~10%) and its molar composition was determined by ^{1}H NMR spectroscopy. Composition of the low conversion copolymer was used for the determination of monomer reactivity ratios. In this case FR, IFR, and KT methods were used to determine the reactivity ratio of the monomers.

In the FR method the following equation was used

$$M - \frac{M}{P} = -r_2 + r_1 \frac{M^2}{P}$$

where, M = molar feed ratio (M_1/M_2) and P = copolymer composition (m_1/m_2)

$$or, \ G = -r_2 + r_1 H$$

where, $G = M - M/P$ and $H = M^2/P$

The plot of G vs. H gives the straight line (**Figure 4**). From this the slope and intercept were calculated to be $r_1 = 1.24$ and $r_2 = 0.71$ respectively.

In the IFR method the equation used is

$$\frac{G}{H} = -r_2 \frac{1}{H} + r_1$$

So, from the plot of G/H and 1/H (**Figure 5**) the reactivity ratios, r_1 and r_2 were calculated as 1.34 and 0.76 from the intercept and the slope respectively.

In KT method the reactivity ratio was determined by using the equation

$$\eta = \left(r_1 + \frac{r_2}{\alpha}\right)\xi - \frac{r_2}{\alpha}$$

where, $\eta = G/(\alpha + H)$ and $\xi = H/(\alpha + H)$ and $\alpha = (H_{min}.H_{max})^{1/2}$, $H_{min} = 0.253$ and $H_{max} = 0.775$. H_{min} and H_{max} values are taken from **Table 2**.

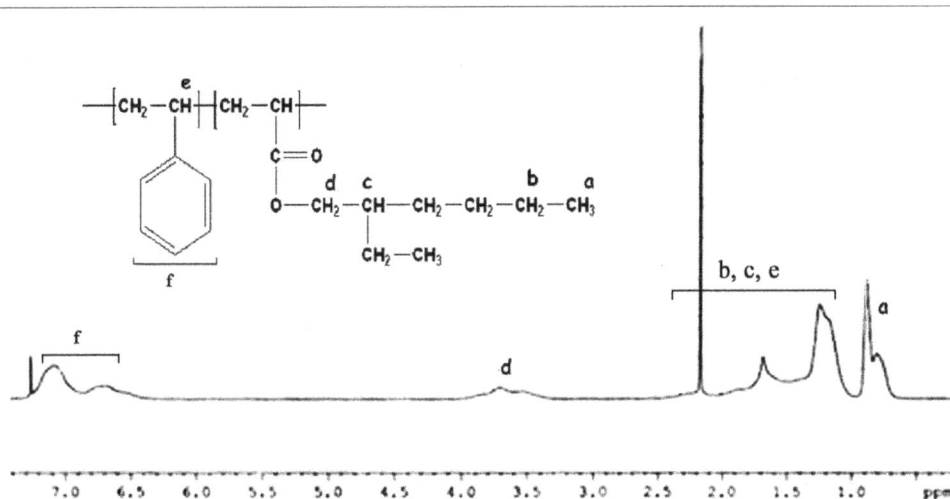

FIGURE 3 | ^{1}H NMR spectrum of poly(styrene-co-EHA) with feed molar composition 50:50.

In the plot of η vs. ξ, (**Figure 6**) the slope gives the value of $(r_1 + r_2/\alpha)$ and intercept provides r_2/α. From these two values r_1 and r_2 were calculated as $r_1 = 1.30$ and $r_2 = 0.73$ respectively. All the parameters used for the three methods are given in **Table 2**.

The reactivity ratios, r_1 and r_2 calculated by the different methods are tabulated in **Table 3**. They are quite comparable. The values; $r_1 > 1$, $r_2 < 1$ and $r_2 < r_1$ indicate that the styrene has much influence on the copolymer formation during the reaction (Jianying et al., 2006). Srivastava et al. (Srivastava and Rai, 1995) reported the copolymerization of styrene and EHA initiated by azobisisobutyronitrile (AIBN) in bulk in the presence of anhydrous $ZnCl_2$. They reported the reactivity ratios for styrene and EHA as 0.10 and 0.175 respectively. Moreover, Kavousian et al. reported the copolymerization of styrene and EHA via conventional radical polymerization (Kavousian et al., 2004). They reported the reactivity ratio of styrene and EHA as 0.926 and 0.238 respectively. There is a difference in the polymerization mechanism of ATRP and non-ATRP processes. In ATRP, atom transfer from an organic halide to a transition metal complex occurs to generate the active radical species, which are

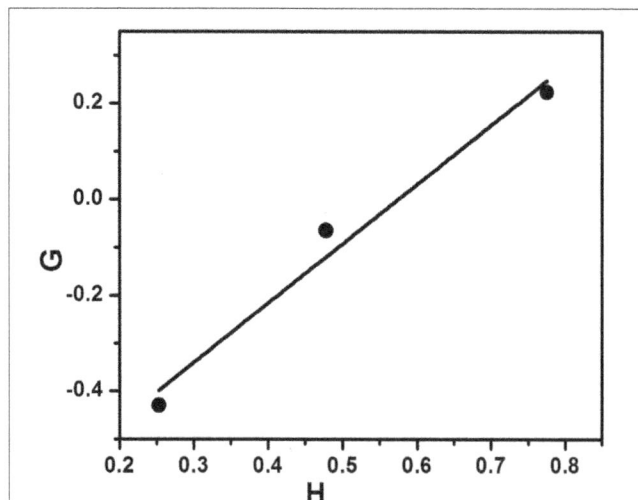

FIGURE 4 | Finemann–Ross plot for copolymerization of styrene with EHA.

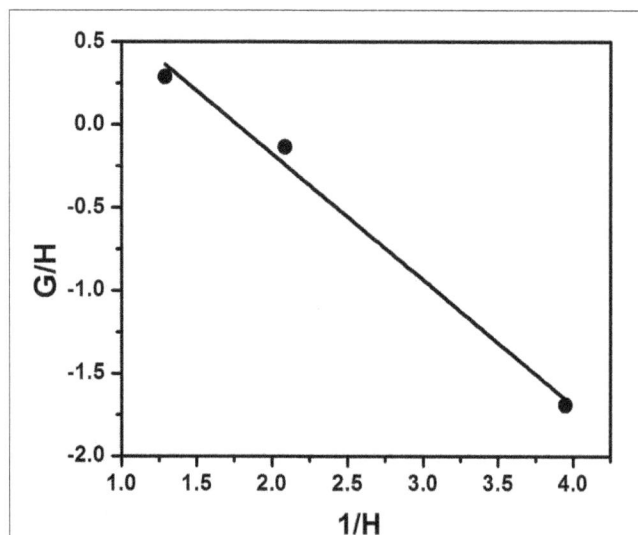

FIGURE 5 | Inverted Finemann–Ross plot for copolymerization of styrene with EHA.

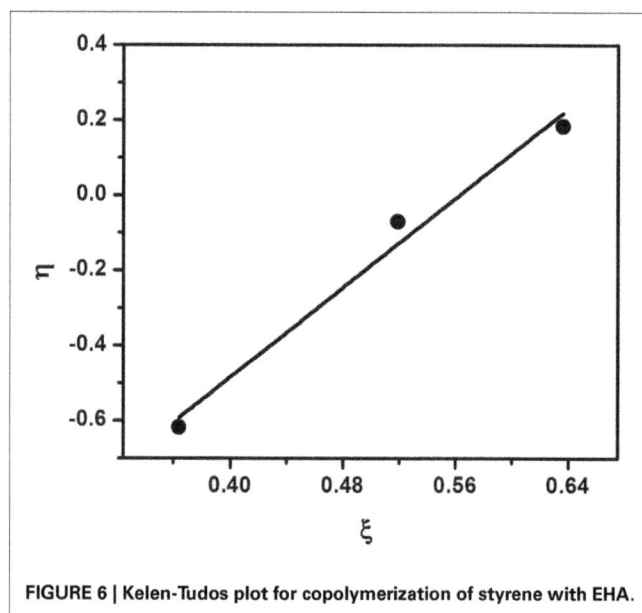

FIGURE 6 | Kelen-Tudos plot for copolymerization of styrene with EHA.

Table 3 | Reactivity ratio of styrene (r_1) and EHA (r_2) determined by three different models.

Methods used	r_1 (styrene)	r_2 (EHA)	$r_1 \cdot r_2$
FR	1.24	0.71	0.88
IFR	1.34	0.76	1.01
KT	1.30	0.73	0.91

Table 2 | Finemann–Ross, Inverted Finemann–Ross and Kelen–Tudos parameter for the copolymer of styrene and EHA in bulk[*].

Monomer feed ratio styrene: EHA	$M = M_1/M_2$	[#]$P = m_1/m_2$	M/P	P-1	G	H	G/H	$1/H$	$\alpha+H$	η	ξ
25:75	0.33	0.43	0.767	−0.57	−0.430	0.253	−1.69	3.95	0.695	−0.618	0.364
40:60	0.66	0.91	0.725	−0.09	−0.065	0.478	−0.135	2.09	0.920	−0.070	0.519
50:50	1.0	1.29	0.775	0.29	0.224	0.775	0.289	1.29	1.217	0.184	0.636

[*][PEBr]: [M]$_o$:[CuBr]:[PMDETA] = 1:100:1:1, at 90°C (where, α= 0.442).

[#] Copolymer composition at 10% conversion.

Table 4 | The monomer composition and sequence length ratio.

Copolymer feed ratio styrene: EHA	M_1/M_2	M_2/M_1	l_1	l_2	$l_1: l_2$
25/75	0.33	3.00	1.42	3.19	4:9
40/60	0.66	1.50	1.85	2.09	4:5
50/50	1.00	1.00	2.29	1.73	4:3

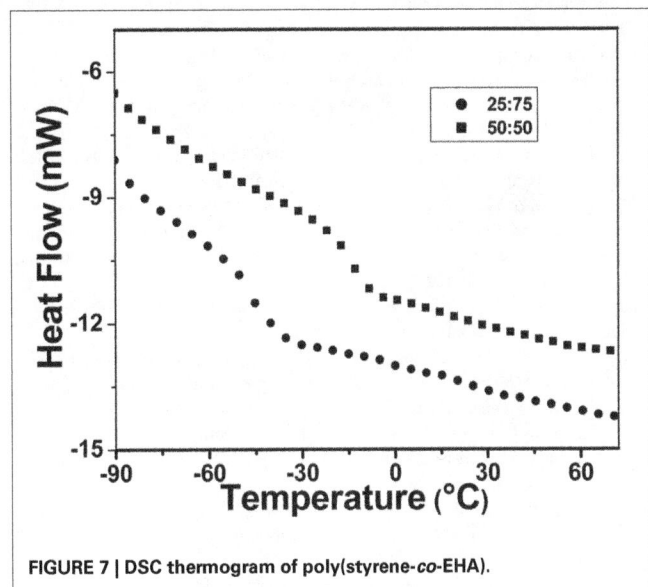

FIGURE 7 | DSC thermogram of poly(styrene-co-EHA).

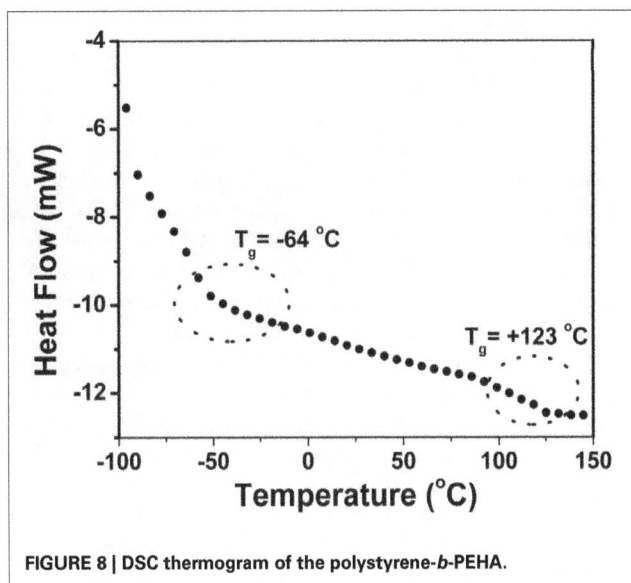

FIGURE 8 | DSC thermogram of the polystyrene-b-PEHA.

Table 5 | Thermal properties of copolymers of styrene and EHA.

SI. no.	Feed composition styrene: EHA	Copolymer composition styrene: EHA	T_g (°C)	*$T_{g,F-F}$ (°C)	#T_{onset} (°C)	T_{max} (°C)
1	25/75	27:73	−47	−36	381	409
2	40/60	44:56	−27	−18	389	414
3	50/50	56.6:43.4	−14	−5	385	415
4	Polystyrene-b-PEHA	85: 15	+123 and −65	–	–	–

*$T_{g,F-F}$ was calculated by using Flory–Fox equation.

#Temperature at 10% weight loss was taken as T_{onset}.

T_{max} represents the maximum degradation temperature at which the polymer back-bone starts degrading. It was determined from the DTG thermogram.

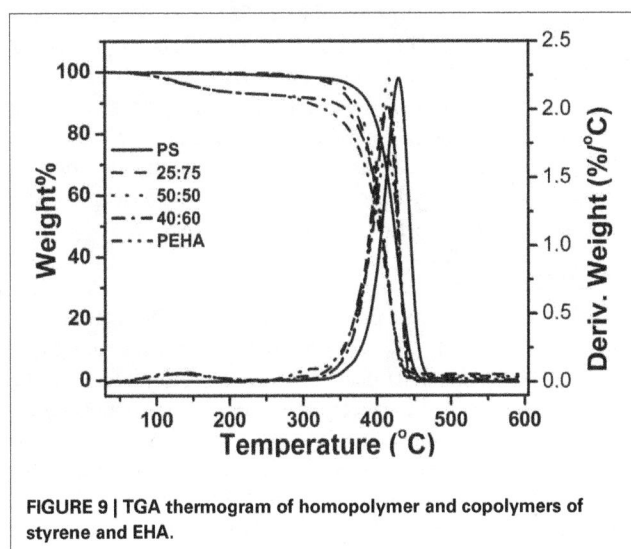

FIGURE 9 | TGA thermogram of homopolymer and copolymers of styrene and EHA.

then quickly "deactivated" by back transfer of the atom from the transition metal to the radical species (Matyjaszewski, 2002; Braunecker and Matyjaszewski, 2007). So, there is difference in reactivity ratio values in comparison to FRP. Barim et al. studied the FRP and ATRP of phenoxycarbonylmethyl methacrylate (PCMMA) and styrene at 110°C. They reported the reactivity ratios of PCMMA and styrene prepared via ATRP were 0.33 and 0.96 respectively and the same prepared via FRP were 0.47 and 1.16 respectively (Barim et al., 2007). The reactivity ratio varies with the polymerization temperature (Chambard et al., 1999; McManus et al., 2002). We did the polymerization reaction via ATRP in bulk at 90°C. However, the reactivity ratios calculated in this work follow the same trend ($r_{EHA} < r_{Styrene}$) as reported by the other authors. In addition, in ATRcP technique the product of

the reactivity ratios is less than one. It shows the tendency of random copolymer formation, where the chances of incorporation of styrene is more in comparison to EHA.

The mean sequence length (l) of the copolymers were determined by using the equations (Pazhanisamy et al., 1997)

$$l_1 = r_1 \frac{M_1}{M_2} + 1$$

and

$$l_2 = r_2 \frac{M_2}{M_1} + 1$$

where, r_1 (styrene) = 1.29 and r_2 (EHA) = 0.73

The results of the mean sequence length in the copolymer are shown in **Table 4**. It indicates that the length of EHA increases as its content in the monomer feed increases.

THERMAL PROPERTIES

The glass transition temperature (T_g) of the copolymer was determined by DSC analysis (shown in **Figure 7**). All the copolymers showed a single T_g and they are shown in **Table 5**. It indicates that as the EHA ($T_g = -60°C$) content increases the T_g decreases. The T_g for the copolymers determined by DSC analysis was compared with the same ($T_{g,F-F}$) determined by Flory-Fox equation (Lijia et al., 1997). There is consistent difference between the T_g values at different content of two comonomers (**Table 5**). **Table 5** shows that there is some discrepancy between the experimental T_g and T_g calculated by Flory-Fox equation. The Flory-Fox model is based on the free volume theory. The discrepancy in the T_g values is due to the fact that the effect of the chemical nature and organization of the monomers on the mobility of a polymer chain was not considered (Fernandez-Garcia et al., 1999).

The block copolymer of styrene and EHA (PS-*b*-PEHA) was prepared by using polystyrene-Br as macroinitiator. The shift of GPC traces of the diblock copolymer toward lower elution volume indicated the successful preparation of block copolymer. (The GPC traces are shown in **Figure S2**). This block copolymer showed two T_gs, PEHA block at $-64°C$ and polystyrene at $+123°C$ (**Figure 8**). However, the copolymers, poly(styrene-*co*-EHA) showed only one T_g indicating the copolymers were not blocky in nature. Thermal stability of the copolymer was studied by TGA (**Figure 9**). It is clear that as the styrene content increases, there is a slight increase in T_{onset}. However, there was no significant change in T_{max} as shown in **Table 5**.

CONCLUSIONS

Copolymers of styrene and 2-ethylhexyl acrylate were synthesized successfully in bulk by using atom transfer radical copolymerization (ATRcP). The chemical composition was studied by ^{1}H NMR spectroscopy and the reactivity ratios of the two monomers were calculated by using FR, IFR, and KT methods. The reactivity ratios of styrene and EHA were somewhat different from the polymerization reaction of styrene and EHA using the FRP system. DSC analysis showed that the T_g of the copolymer increases on increasing styrene content.

ACKNOWLEDGMENT

Bishnu P. Koiry gratefully acknowledges the fellowship from IIT, Kharagpur, India.

REFERENCES

Arehart, S. V., and Matyjaszewski, K. (1999). Atom transfer radical copolymerization of styrene and n-butyl acrylate. *Macromolecules* 32, 2221–2231. doi: 10.1021/ma981693v

Barim, G., Demirelli, K., and Coskun, M. (2007). Conventional and atom transfer radical copolymerization of phenoxycarbonylmethyl methacrylate-styrene and thermal behavior of their copolymers. *Express Polym. Lett.* 1, 535–544. doi: 10.3144/expresspolymlett.2007.76

Barner-Kowollik, C. (2008). *Handbook of RAFT Polymerization.* Weinheim: Wiley-VCH Verlag GmbH and Co. KGaA. doi: 10.1002/9783527622757

Braunecker, W. A., and Matyjaszewski, K. (2007). Controlled/living radical polymerization: features, developments, and perspectives. *Prog. Polym. Sci.* 32, 93–146. doi: 10.1016/j.progpolymsci.2006.11.002

Chambard, G., Klumperman, B., and German, A. L. (1999). Dependence of chemical composition of styrene/butyl acrylate copolymers on temperature and molecular weight. *Polymer* 40, 4459–4463. doi: 10.1016/S0032-3861(98)00690-9

Chiefari, J., Chong, Y. K., Ercole, F., Krstina, J., Jeffery, J., Le, T. P. T., et al. (1998). Living free-radical polymerization by reversible addition-fragmentation chain transfer: the RAFT process. *Macromolecules* 31, 5559–5562. doi: 10.1021/ma9804951

Fernandez-Garcia, M., Cuervo-Rodriguez, R., and Madruga, E. L. (1999). Glass transition temperatures of butyl acrylate–methyl methacrylate copolymers. *J. Polym. Sci. B Polym. Phys.* 37, 2512–2520. doi: 10.1021/ma001182k

Fineman, M., and Ross, S. D. (1950). Liner method for determining monomer reactivity ratios in copolymerization. *J. Polym. Sci.* 5, 259–262. doi: 10.1002/pol.1950.120020210

Haldankar, G. S. (2001). *Compatibilizer for Polymer Blends.* Frankfort, IL: The Sherwin-Williams Co. US Patent 6,541,571.

Haloi, D. J., Roy, S., and Singha, N. K. (2009). Copper catalyzed atom transfer radical copolymerization of glycidyl methacrylate and 2-ethylhexyl acrylate. *J. Polym. Sci. A Polym. Chem.* 47, 6526–6533. doi: 10.1002/pola.23695

Harth, E., Horn, B. V., and Hawker, C. J. (2001). Acceleration in nitroxide mediated 'living' free radical polymerizations. *Chem. Commun.* 9, 823–824. doi: 10.1039/b102145c

Hawker, C. J., Bosman, A. W., and Harth, E. (2001). New polymer synthesis by nitroxide mediated living radical polymerization. *Chem. Rev.* 101, 3661–3688. doi: 10.1021/cr990119u

Jianying, H., Jiayan, C., Jiaming, Z., Yihong, C., Lizong, D., and Yousi, Z. (2006). Some monomer reactivity ratios of styrene and (meth)acrylates in the presence of TEMPO. *J. Appl. Polym. Sci.* 100, 3531–3535. doi: 10.1002/app.22512

Kamigaito, M., Ando, T., and Sawamoto, M. (2001). Metal-catalyzed living radical polymerization. *Chem. Rev.* 101, 3689–3745. doi: 10.1021/cr9901182

Kavitha, A., and Singha, N. K. (2007). Atom-transfer radical copolymerization of furfuryl methacrylate (FMA) and methyl methacrylate (MMA): a thermally-amendable copolymer. *Macromol. Chem. Phys.* 208, 2569–2577. doi: 10.1002/macp.200700239

Kavitha, A., and Singha, N. K. (2008). High temperature resistant tailor-made poly(meth)acrylates bearing adamantyl group via atom transfer radical polymerization. *J. Polym. Sci. A Polym. Chem.* 46, 7101–7113. doi: 10.1002/pola.23015

Kavousian, A., Ziaee, F., Nekoomanesh, H., Leamen, M. J., and Penlidis, A. (2004). Determination of monomer reactivity ratios in styrene/2-ethylhexylacrylate copolymer. *J. Appl. Polym. Sci.* 92, 3368–3370. doi: 10.1002/app.20338

Kelen, T., Tudos, F., and Turcsanyi, B. (1980). Confidence intervals for copolymerization reactivity ratios determined by the Kelen-Tudos method. *Polym. Bull.* 2, 71–76. doi: 10.1007/BF00275556

Lessard, B., Graffe, A., and Maric, M. (2007). Styrene/t-Butyl acrylate random copolymers synthesized by nitroxide-mediated polymerization: effect of free nitroxide on kinetics and copolymer composition. *Macromolecules* 40, 9284–9292. doi: 10.1021/ma071689w

Lijia, A. N., Dayong, H. E., Jlaokai, J., Zhigang, W., Donghong, Y. U., Jiang, B., et al. (1997). Effects of molecular weight and interaction parameter on the glass transition temperature of polystyrene mixture and its blends with polystyrene/poly(2,6-dimethyl-p-phenylene oxide). *Eur. Polym. J.* 33, 1523–1528. doi: 10.1016/S0014-3057(97)00089-X

Makrikosta, G., Georgas, D., Siakali-Kioulafa, E., and Pitsikalis, M. (2005). Statistical copolymers of styrene and 2-vinylpyridine with trimethylsilyl methacrylate and trimethylsilyloxyethyl methacrylate. *Eur. Polym. J.* 41, 47–54. doi: 10.1016/j.eurpolymj.2004.08.001

Matyjaszewski, K. (2000). "Controlled/Living radical polymerisation: progress in ATRP, NMP and RAFT," in *ACS Symposium Series* (New Orleans, LA). doi: 10.1021/bk-2000-0768

Matyjaszewski, K. (2002). Factors affecting rates of comonomer consumption in copolymerization processes with intermittent activation. *Macromolecules* 35, 6773–6781. doi: 10.1021/ma0208180

Matyjaszewski, K., and Jia, J. (2001). Atom transfer radical polymerization. *Chem. Rev.* 101, 2921–2990. doi: 10.1021/cr940534g

McManus, N. T., Penlidis, A., and Dube, M. A. (2002). Copolymerization of alpha-methyl styrene with butyl acrylate in bulk. *Polymer* 43, 1607–1614. doi: 10.1016/S0032-3861(01)00738-8

Mignard, E., Leblanc, T., Bertin, D., Guerret, O., and Reed, W. F. (2004). Online monitoring of controlled radical polymerization: nitroxide-mediated gradient copolymerization. *Macromolecules* 37, 966–975. doi: 10.1021/ma035589b

Moad, G. (2006). The emergence of RAFT polymerization. *Aust. J. Chem.* 59, 661–662. doi: 10.1071/CH06376

Moad, G., Chiefari, J., Chong, Y. K., Krstina, J., Mayadunne, R. T. A., Postma, A., et al. (2000). Living free radical polymerization with reversible addition–fragmentation chain transfer (the life of RAFT). *Polym. Int.* 49, 993–1001. doi: 10.1002/1097-0126(200009)49:9<993::AID-PI506>3.0.CO;2-6

Odian, G. (2004). *Principles of Polymerization*. New Jersey, NJ: Wiley-Interscience publication. doi: 10.1002/047147875X

Pazhanisamy, P., Ariff, M., and Anwaruddin, Q. (1997). Copolymers of α-methylstyrene with *N*-cyclohexylacrylamide: synthesis, monomer reactivity ratios, and mean sequence length. *J. Macromol. Sci. A Pure Appl. Chem.* A34, 1045–1054. doi: 10.1080/10601329708015009

Plessis, C., Arzamendi, G., Alberdi, J. M., Agnely, M., Leiza, J. R., and Asua, J. M. (2001). Intramolecular chain transfer to polymer in the emulsion polymerization of 2-ethylhexyl acrylate. *Macromolecules* 34, 6138–6143. doi: 10.1021/ma0018190

Skeist, I. (1977). *Handbook of Adhesives*. New York, NY: Chapman & Hall.

Solomon, D. H., and Moad, G. (1995). *The Chemistry of Free Radical Polymerization*. Oxford: Elsevier Science Ltd.

Srivastava, N., and Rai, J. S. P. (1995). Kinetics, mechanism, and rheological properties of the copolymers of 2-ethylhexylacrylate and styrene. *J. Macromol. Sci. A Pure Appl. Chem.* A32, 2049–2062. doi: 10.1080/10601329508011044

Webster, D. C., and Crain, A. L. (2002). Synthesis of latexes containing diesters of 3-butene-1,2-diol. *Prog. Org. Coat.* 45, 43–48. doi: 10.1016/S0300-9440(02)00099-1

Ziegler, M. J., and Matyjaszewski, K., (2001). Atom transfer radical copolymerization of methyl, methacrylate and n-butyl acrylate. *Macromolecules* 34, 415–424. doi: 10.1021/ma001182k

Conflict of Interest Statement: The authors declare that the research was conducted in the absence of any commercial or financial relationships that could be construed as a potential conflict of interest.

16

Diffusion and molecular interactions in a methanol/polyimide system probed by coupling *time-resolved* FTIR spectroscopy with gravimetric measurements

Pellegrino Musto[1]*, Michele Galizia[2], Pietro La Manna[1,2], Marianna Pannico[1] and Giuseppe Mensitieri[2]

[1] *Institute of Chemistry and Technology of Polymers, National Research Council of Italy, Naples, Italy*
[2] *Department of Chemical, Materials and Industrial Production Engineering, University of Naples Federico II, Naples, Italy*

Edited by:
Alfonso Jiménez, University of Alicante, Spain

Reviewed by:
Santiago Betancourt, Universidad Pontificia Bolivariana, Colombia
Artur J. M. Valente, University of Coimbra, Portugal

*Correspondence:
Pellegrino Musto, Institute of Chemistry and Technology of Polymers, National Research Council of Italy, Via Campi Flegrei, 34, 80078 Pozzuoli, Naples, Italy
e-mail: pellegrino.musto@ictp.cnr.it

In this contribution the diffusion of methanol in a commercial polyimide (PMDA-ODA) is studied by coupling gravimetric measurements with *in-situ, time-resolved* FTIR spectroscopy. The spectroscopic data have been treated with two complementary techniques, i.e., difference spectroscopy (DS) and least-squares curve fitting (LSCF). These approaches provided information about the overall diffusivity, the nature of the molecular interactions among the system components and the dynamics of the various molecular species. Additional spectroscopic measurements on thin film samples (about $2\,\mu m$) allowed us to identify the interaction site on the polymer backbone and to propose likely structures for the H-bonding aggregates. Molar absorptivity values from a previous literature report allowed us to estimate the population of first-shell and second-shell layers of methanol in the polymer matrix. In terms of diffusion kinetics, the gravimetric and spectroscopic estimates of the diffusion coefficients were found to be in good agreement with each other and with previous literature reports. A Fickian behavior was observed throughout, with diffusivity values markedly affected by the total concentration of sorbed methanol.

Keywords: diffusion, FTIR spectroscopy, polyimide, methanol

INTRODUCTION

Polyimides are high-performance technopolymers characterized by outstanding properties in terms of thermal stability, mechanical performances, high T_g and good resistance to solvents (Bessonov and Zubkov, 1993; Gosh and Mittal, 1996).

These properties make them attractive for numerous applications, among which in microelectronic and opto-electronic devices, or as membranes for separation technologies (Feger et al., 1993; Thompson et al., 1994). Several polyimides have also been employed as polymeric components of high-performance hybrid systems, prepared via the sol-gel route (Mascia, 1995; Hibshman et al., 2003; Musto et al., 2006). The use of diverse polyimide membranes for dehydration of alcohols by pervaporation processes is well documented in the literature (Okamoto et al., 1992; Chen and Martin, 1995). The optimal design of such processes requires a molecular level understanding of the interaction between the penetrant and the polymer matrix. More generally, the theme of H-bonding between low-molecular weight compounds and the polymer substrates within which they diffuse, represents an area of intense research activity, as demonstrated by the increasing number of studies appearing in the literature on the subject. By using solid-state NMR (Jelinski et al., 1985) and *time-resolved* FTIR spectroscopy (Musto et al., 2007, 2012) it has been demonstrated that both water and

alcohols are able to form hydrogen bonds with polymers displaying proton acceptor groups on their backbone, and that such an occurrence strongly affects the transport properties and the separation performances of these systems. The occurrence of self-association of the penetrant to form larger molecular aggregates has been also documented for the water/polyimide system (Musto et al., 2007, 2012), but remains a matter of debate for the majority of the investigated systems with water as penetrant, and is an open issue for the case of methanol. In fact, most of the literature studies report only experimental data on methanol diffusion in polyimides, while there is a lack of information about the molecular mechanisms of these processes, and the role played by the different types of H-bonding interactions.

In the present contribution *time-resolved* FTIR measurements have been performed at different relative pressures of methanol vapor to investigate its diffusion into a commercial polyimide. The scope of the study was to characterize the system at the molecular level in terms of number of H-bonding aggregates, their structure and relative population. Gravimetric measurements in the same experimental conditions were also performed to substantiate the spectroscopic results. In terms of diffusion kinetics, a Fickian behavior was observed with diffusivities markedly affected by the total concentration of sorbed methanol.

EXPERIMENTAL

MATERIALS

The polyimide precursor used in this study was a polyamic acid, Pyre-ML RK 692 from I.S.T (Indian Orchard, MA). It has molecular weights $\bar{M}_w = 1.0 \cdot 10^5$ g/mol, and $\bar{M}_n = 4.6 \cdot 10^4$ g/mol, and is supplied as a 12 wt% solution in a mixture of N-methyl-2-pyrrolidone (NMP) and xylene (weight ratio 80/20). The polyamic acid is obtained by condensation of pyromellitic dianhydride (PMDA) and oxydianiline (ODA). The molecular structure of the PMDA-ODA polyimide is reported in **Scheme 1**.

Methanol used for sorption experiments was purchased from *Sigma-Aldrich* (Milano, I) with purity higher 99.6%: it was further purified and degassed through freezing-thawing cycles.

Free standing films of PMDA-ODA, 20–24 μm thick, were obtained by spreading the polyamic acid solution on a clean glass support with a Gardner knife. The nascent films were then dried 1 h at room temperature and 1 h at 80°C, allowing the solvent to evaporate. The castings were then thermally treated in a stepwise manner at 100, 150, 200, 250, and 290°C for 1 h at each temperature. Finally, the films were removed from the glass support by immersion in distilled water at 80°C. Thinner films (2–4 μm) were prepared by a two-step spin-coating process, by using a Chemat KW-4A apparatus (Northridge, CA). Spinning conditions were 12 s at 700 rpm for the first step and 20 s at 1500 rpm for the second step. After removal of the films from their glass support in distilled water at room temperature, they were treated in the same conditions as for the thicker samples.

FTIR SORPTION EXPERIMENTS

Time-resolved spectra were collected in the transmission mode making use of a vacuum-tight FTIR sorption cell in which a free standing polymer film is exposed to methanol vapor at constant temperature (30°C) and different relative pressures of the penetrant. Full details of the experimental apparatus are reported in Cotugno et al. (2001). Before each sorption measurement, the polymer film was dried overnight under vacuum in the sorption cell to ensure complete removal of absorbed moisture, which was confirmed by the absence of the water bands in the sample spectrum.

The sorption cell was accommodated in the sample compartment of a suitably modified FTIR spectrometer [Spectrum GX from *Perkin-Elmer* (Norwalk, CT)], equipped with a Ge/KBr beam splitter and a wide-band DTGS detector.

Instrumental parameters for data collection were set as follows: resolution = $4\,\text{cm}^{-1}$; Optical Path Difference (OPD) velocity = 0.5 cm/s; spectral range 4000–600 cm^{-1}. Spectra were acquired in the single-beam mode using a dedicated software package for *time-resolved* spectroscopy (*Timebase, Perkin-Elmer*).

Differential sorption tests were performed by increasing stepwise the relative pressures of methanol vapor within the range 0–0.6.

GRAVIMETRIC SORPTION EXPERIMENTS

Methanol sorption experiments were performed at 30°C, using a Quartz Spring Balance equipped with a couple of digital CCD cameras. Experimental sorption isotherms were obtained by increasing stepwise relative pressure of the penetrant (p/p_0, p_0 being the penetrant vapor pressure at the experimental temperature). Before each sorption test, the polymer sample was dried overnight under vacuum at the experimental temperature, up to constant weight. Full details about the experimental procedure are given in Cotugno et al. (2005).

FTIR DATA ANALYSIS

Full absorbance spectra (i.e., polyimide plus absorbed methanol) were obtained using a background collected on the empty cell (i.e., without sample) at the test conditions. The spectra representative of absorbed methanol were obtained by using as background the single-beam spectrum of the cell containing the dry polymer film. The spectrum obtained in this way is equivalent to that resulting from the difference spectroscopy method (subtraction factor, $K = 1$), provided that the sample thickness does not change significantly during the sorption measurement, which was experimentally verified in the present case. The above procedure allows us to eliminate the interference of the polymer spectrum from the methanol peaks located in the 3650–2750 cm^{-1} range [the ν(OH) and ν(CH) vibrations] and at around 1000 cm^{-1} [ν(C-O), vide infra]. To separate the individual components in the case of unresolved bands, a curve fitting algorithm was applied, based on the Levenberg–Marquardt method (Marquardt, 1963; Meier, 2005). The peak functions used for the two components were a log-normal line-shape for the sharp peak at higher frequency and a Gaussian profile for the broader band at lower wavenumbers, which are expressed, respectively, as (Meier, 2005):

$$f(x) = H \exp\left[\frac{-\ln 2}{(\ln \rho)^2} \ln^2\left[\frac{(x - x_0)(\rho^2 - 1)}{w}\right] + 1\right] \quad (1)$$

$$f(x) = H \exp\left[\left(\frac{x - x_0}{w}\right)^2 4\ln 2\right] \quad (2)$$

where x_0 is the peak position; H the peak height; w the full-width at half height (FWHH), and ρ the asymmetry index (half width ratio). In order to keep the number of adjustable parameters to a minimum, the baseline and the number of components were fixed, allowing the curve-fitting algorithm to optimize the height, the FWHH and the position of the peaks.

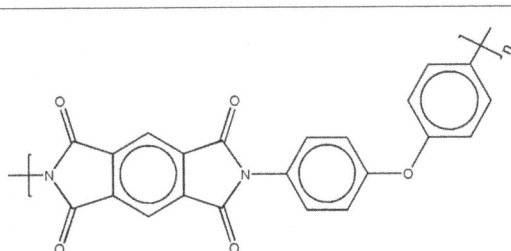

Scheme 1 | Molecular structure of the PMDA-ODA polyimide.

RESULTS AND DISCUSSION
INTERPRETING ABSORBANCE AND DIFFERENCE SPECTRA

In **Figure 1** are reported the FTIR spectra of the fully dried polyimide film (red trace) and of the same film after equilibration with methanol vapor at $p/p_0 = 0.5$ (blue trace). Sorbed methanol displays characteristic bands in three distinct regions of the spectrum, namely, in the 3650–3050 cm^{-1} range [ν(O-H)], in the 3000–2060 cm^{-1} range [ν(C-H)] and at around 1000 cm^{-1} (Shurvell, 2002). The latter peak is a highly coupled vibration, generally quoted as a ν(C-O), but with a significant contribution from the C-O-H bending. The methanol molecule represents an ideal probe to investigate the chemical environment surrounding the penetrant for it has one fragment very sensitive to H-bonding interactions (the hydroxyl group) and the other (the methyl group) completely insensitive. Thus, the ν(O-H) mode is expected to be the most perturbed, while the ν(C-H) vibrations should remain unaffected.

Suppressing the interference of the matrix by difference spectroscopy allows us to isolate the spectrum of sorbed methanol in the regions of interest. These are displayed in **Figures 2A,B**.

In the ν(O-H) range a relatively sharp peak at 3585 cm^{-1} is superimposed onto a much broader band centered at lower wavenumbers. The band at 1023 cm^{-1} does not show evidence of an underlying fine structure. To get information about molecular interactions we analyzed in more detail the most sensitive region, i.e., the ν(O-H) interval which was subjected to a curve fitting analysis. The results obtained on the difference spectrum collected at equilibrium ($p/p_0 = 0.3$) are shown in **Figure 3**. By properly selecting the band-shape of the peaks (a log-normal function and a mixed Guss-Lorentz function for the high- and low-frequency components, respectively) the experimental profile can be very satisfactorily simulated, which suggests the presence of two distinct molecular species.

In order to propose a reasonable interpretation of the spectroscopic results, it is useful to compare the spectrum of methanol sorbed in the polyimide with that of methanol in unperturbed conditions. This reference state cannot be represented by the

isolated molecule *in-vacuo* (i.e., in the gas phase, at low p) because of the superposition of roto-vibrational effects; a more appropriate reference is a dilute solution in a low polarity, non-interacting solvent (CCl_4). In these conditions, at high dilution (below 0.01 M) only the monomer is present, giving rise to a very sharp peak at 3643 cm^{-1}; increasing the concentration, dimers start to appear, producing a well resolved and symmetrical band at 3525 cm^{-1}; At 0.1 M and above, the equilibrium involves also trimers and tetramers, which produce a further broad band centered at 3350 cm^{-1} (Dixon et al., 1997; Galizia et al., 2013). The spectrum of a 0.1 M solution of methanol in CCl_4 is reported in **Figure 4A**. It is noteworthy the considerable increase of full width at half height (FWHH) when passing from the monomer to higher aggregates, which is a direct consequence of the H-bonding interaction in self-associated structures.

The spectrum of methanol sorbed in PMDA-ODA ($p/p_0 = 0.3$) is shown in **Figure 4B**. By comparison with the reference state, the main peak at 3580 cm^{-1} can be safely associated with a monomeric species. However, the larger value of FWHH (from 22 to 90 cm^{-1}) and, especially, the considerable red-shift (68 cm^{-1}) demonstrate that in the polymer, the monomeric species is not free as in the CCl_4 solution, but is involved in a H-bonding interaction of the non-self association type. Instead, self-association is related to the broad band at 3425 cm^{-1}: its symmetrical shape suggests the occurrence of a single type of molecular aggregate (dimers) (Galizia et al., 2013) or, at least, a large prevalence of this species over higher order aggregates.

To identify the active site on the polymer backbone acting as proton acceptor, we ought to investigate the perturbation brought about by methanol to the spectrum of the polyimide. A thin film (in spectroscopic terms, i.e., less than 5.0 μm) is required to perform this analysis, in order to keep the most intense peaks within the limits of absorbance linearity. This sample was prepared *ad-hoc* by a spin-coating process (see experimental); the resulting thickness, as measured by the interference-fringes method (Musto et al., 2007) was 2.4 μm.

Figures 5A,B display the carbonyl range of the PMDA-ODA spectrum in the sample equilibrated at different relative pressures of methanol vapor. In particular, **Figure 5A** shows the absorbance spectra, while **Figure 5B** displays the difference spectra (equilibrated sample–dry sample).

It is observed that the two carbonyl peaks of the imide moiety [ν_s(C = O) at 1778 cm^{-1} and ν_{as}(C = O) at 1727 cm^{-1}] are shifted toward lower frequencies. The effect is very weak in the absorbance spectra, yet well within the detectability limits of interferometric spectroscopy. It is further evidenced in the difference spectra where the typical first-derivative profiles characteristic of a downward shift of the sample peaks with respect to the reference, are readily apparent. The extent of the red-shift, which is reflected in the peak-to-peak height of the difference spectra, is found to increase gradually with increasing methanol concentration in the sample (see **Figure 5B**). It has been verified that the effect is fully reversible upon methanol removal, which further confirms that it originates from the polyimide/methanol interactions.

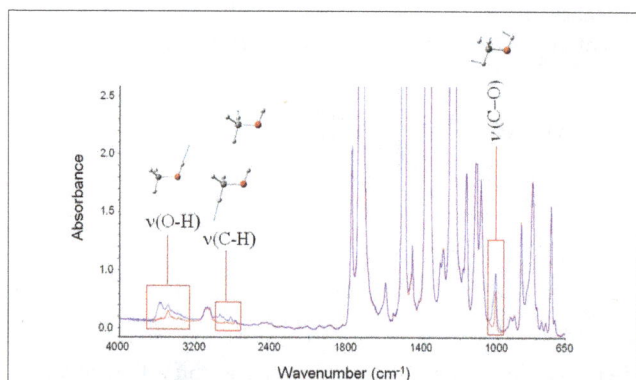

Figure 1 | Absorbance spectra for fully dried PMDA-ODA (red trace) and PMDA-ODA equilibrated at $p/p_0 = 0.5$. The insets represent the bands characteristic of sorbed methanol, and the forms of the respective normal modes.

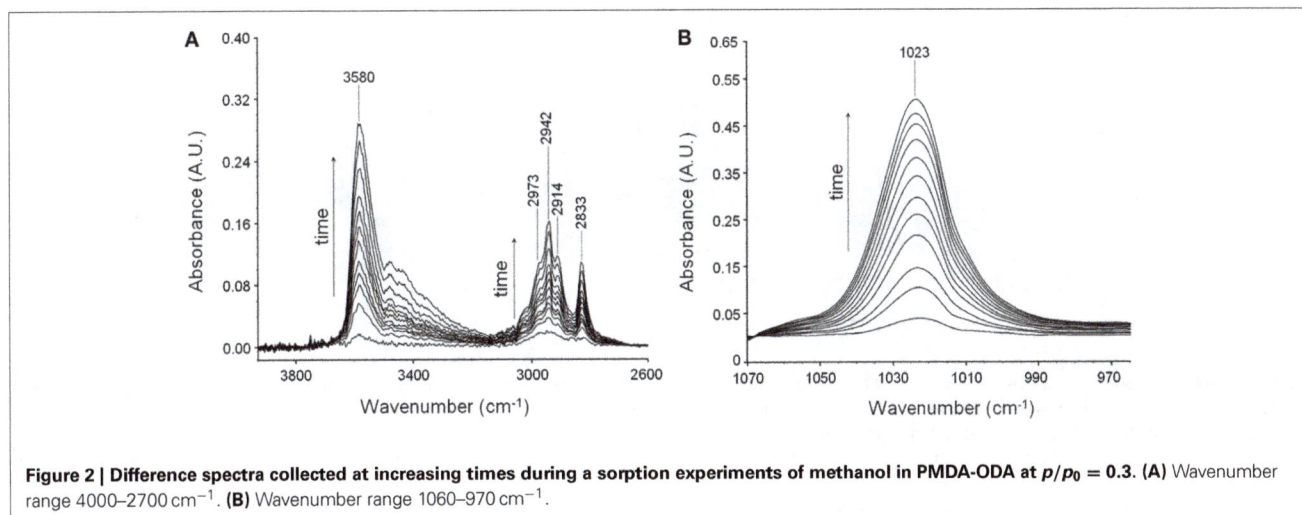

Figure 2 | Difference spectra collected at increasing times during a sorption experiments of methanol in PMDA-ODA at $p/p_0 = 0.3$. (A) Wavenumber range 4000–2700 cm^{-1}. **(B)** Wavenumber range 1060–970 cm^{-1}.

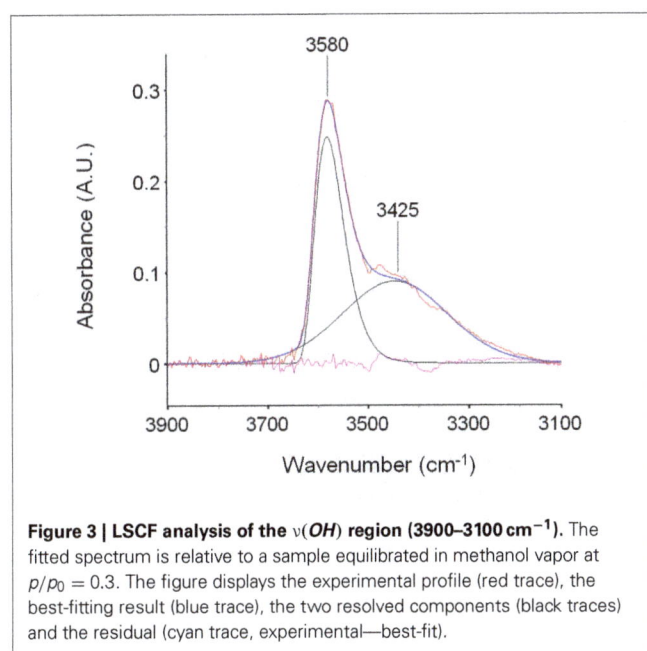

Figure 3 | LSCF analysis of the $\nu(OH)$ region (3900–3100 cm^{-1}). The fitted spectrum is relative to a sample equilibrated in methanol vapor at $p/p_0 = 0.3$. The figure displays the experimental profile (red trace), the best-fitting result (blue trace), the two resolved components (black traces) and the residual (cyan trace, experimental—best-fit).

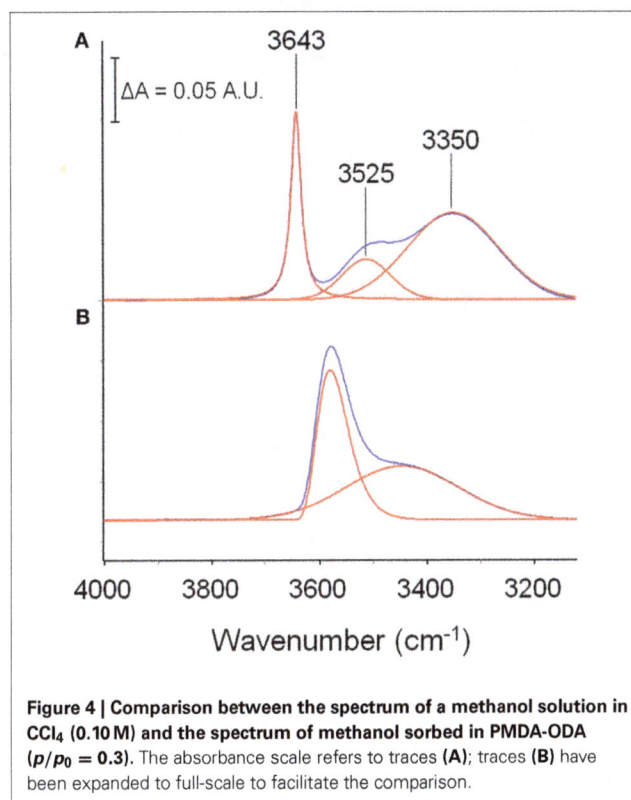

Figure 4 | Comparison between the spectrum of a methanol solution in CCl$_4$ (0.10 M) and the spectrum of methanol sorbed in PMDA-ODA ($p/p_0 = 0.3$). The absorbance scale refers to traces **(A)**; traces **(B)** have been expanded to full-scale to facilitate the comparison.

The above observations can be interpreted assuming that the carbonyl groups of the polyimide act as proton acceptors in the H-bonding interaction with methanol. In fact, the red shift is a direct consequence of the weakening of the C = O force constant due to the redistribution of the electron density caused by the interacting proton.

Taking into consideration the whole of the spectroscopic results, we can assign the 3580 cm^{-1} component to the O-H groups of methanol directly bound to the imide carbonyls, which represent the first shell layer of sorbed methanol in the frame of the multilayer adsorption model of Brunauer-Emmett-Teller (BET) [19]. The band at 3425 cm^{-1} originates from self-associated methanol molecules, as depicted in **Scheme 2**.

Work is in progress to optimize the geometry of the above molecular aggregates by first *principles* computational approaches

so as to calculate theoretically the vibrational spectra and compare these simulations with the experimental results.

DIFFUSION BEHAVIOR

In **Figure 6A** is reported the sorption isotherm as a function of the relative pressure of methanol, as evaluated gravimetrically. S-shaped curves similar to those of **Figure 6A** have been reported in the literature for methanol and other alkyl-alcohols absorbed in high T_g, glassy polyacetylenes (Galizia et al., 2011, 2012). In the case at hand, this behavior is likely due to a dual-mode diffusion

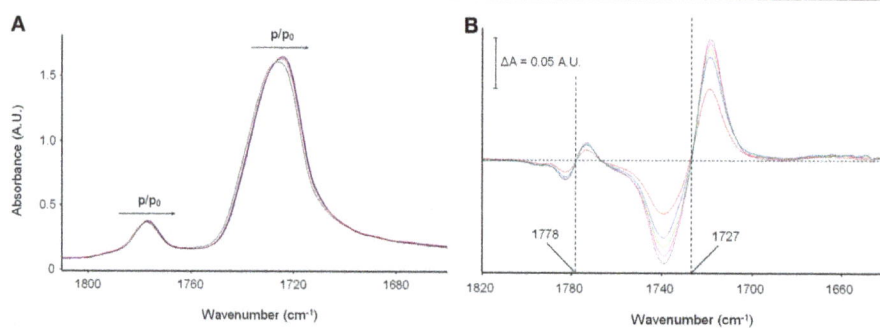

Figure 5 | (A) The Absorbance spectra in the $\nu(C=O)$ frequency range (1810–1640 cm^{-1}) collected on polyimide film equilibrated at different vapor pressures of methanol; **(B)** difference spectra (equilibrated sample–dry

sample). The color code is as follows: black trace: $p/p_0 = 0$ (dry sample); red trace: $p/p_0 = 0.2$; blue trace: $p/p_0 = 0.3$; green trace: $p/p_0 = 0.4$; cyan trace: $p/p_0 = 0.5$; brown trace: $p/p_0 = 0.6$.

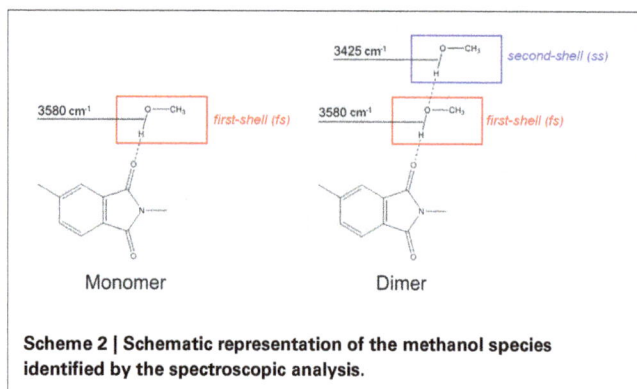

Scheme 2 | Schematic representation of the methanol species identified by the spectroscopic analysis.

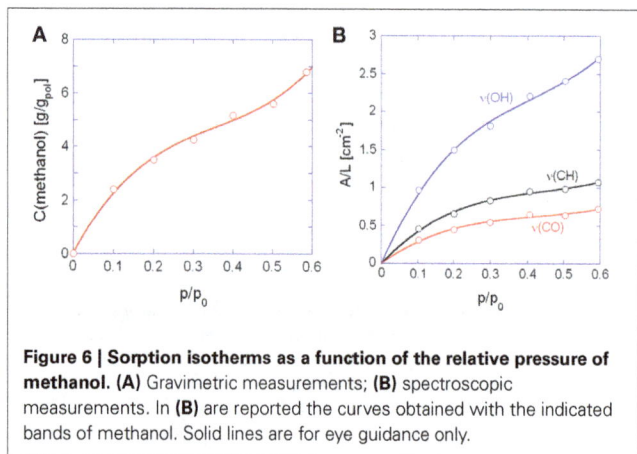

Figure 6 | Sorption isotherms as a function of the relative pressure of methanol. (A) Gravimetric measurements; **(B)** spectroscopic measurements. In **(B)** are reported the curves obtained with the indicated bands of methanol. Solid lines are for eye guidance only.

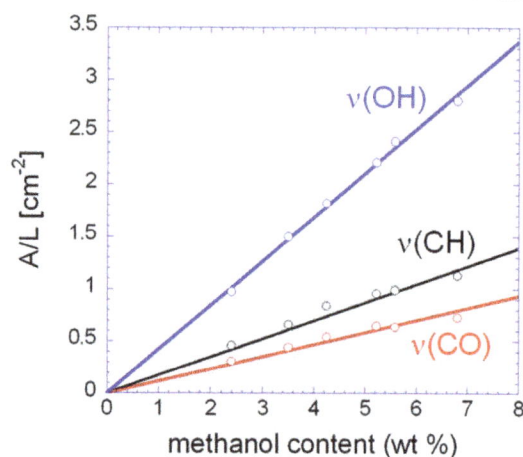

Figure 7 | Normalized absorbance of the three analytical bands of methanol as a function of the methanol concentration in the sample.

regime. The upturn present in sorption isotherms has been associated with two different factors: (1) plasticization of the glassy matrix induced by the penetrant, which promotes the transition from a glassy to a rubbery system, and (2) the gradual onset of penetrant clustering or self-association as the relative pressure increases. In view of the high glass transition temperature of the polyimide (383°C) the second interpretation seems the most likely. The solubility of methanol in PMDA-ODA is conspicuous, close to 7.0 wt% at $p/p_0 = 0.6$, which reflects both the favorable

H-bonding interactions and the contribution of the methyl group in reducing the alcohol polarity.

In **Figure 6B** are shown the isotherms evaluated spectroscopically by considering the absorbance areas (A) normalized for the sample thickness (L) of the three analytical bands of the penetrant. The absorbance-concentration relationships are demonstrated in **Figure 7**: in all cases a Lambert-Beer behavior is observed (i.e., a linear trend through the origin) which allows us to directly transform the absorbance intensities into absolute concentration values.

Figure 8 shows the difference spectra representative of sorbed methanol at the various investigated relative pressures. Normalization with respect to the principal component at 3580 cm^{-1}, evidences that the contribution of the component at 3425 cm^{-1} to the total absorbance area of the $\nu(OH)$ profile increases by increasing p/p_0, that is, by enhancing the concentration of sorbed methanol. This indicates that, at higher concentration of sorbed methanol the equilibrium moves toward self-associated species. To put this observation on a more quantitative basis, we performed a LSCF analysis on the spectral profiles of **Figure 8**; the quality of the fit was in all cases satisfactory, analogous to that reported in **Figure 3**. The absorbance of the curve-resolved components may be converted into concentration values provided that the respective values of molar absorptivity (ε_{fs}, ε_{ss}) are available. In a recent publication on the sorption of methanol in a commercial polyetherimide (Ultem 1000) analogous spectral features were observed and a method was proposed to evaluate ε_{fs} and ε_{ss} based on coupling spectroscopic and gravimetric measurements taken at identical equilibrium conditions (Galizia et al., 2013). The reported values were 76 km/mol for ε_{fs} and 108 km/mol for ε_{ss}. Assuming the above values for the system PMDA-ODA/methanol, we estimated C_{fs} and C_{ss} from the relevant absorbance–concentration relationships, i.e., $C_{fs} = A_{3575}/\varepsilon_{fs}L$ and $C_{ss} = A_{3440}/\varepsilon_{ss}L$. The results of the quantitative analysis are summarized in **Table 1**.

In **Figure 9A** are reported the concentrations of *firs-shell* and *second- (or higher-) shell* methanol in the polyimide as a function of methanol relative pressure. The C_{fs} curve exceeds the C_{ss} curve up to a p/p_0 value of 0.4; in this range it is likely that dimers are the prevailing species, with a slight amount of residual monomers, whose concentration is equal to $C_{fs} - C_{ss}$. At p/p_0 values higher than 0.4 C_{ss} significantly offsets C_{fs} and this may only occur when aggregates comprising more than two methanol

molecules (clusters) are formed. The plot of **Figure 9A** allows us to clearly identify the onset of the clustering phenomenon as the intersection point of the two curves. It is noted that that the absolute spectroscopic evaluation of sorbed methanol as $C_{fs} + C_{ss}$, (that is, with no direct calibration with gravimetry), provides concentration values of the same order of magnitude as those from weight measurements but underestimated by a factor of about 1.6. This could be due to the assumed absorptivity values which may slightly differ in going from a polyetherimide to the PMDA-ODA. **Figure 9B** demonstrates that the concentration ratio C_{fs}/C_{ss} decreases linearly by increasing the relative pressure of methanol vapor.

The evolution of the infrared spectrum with time (see **Figures 2A,B**) can be reliably used to trace the kinetics of the mass transport, both in the sorption and in the desorption regime. This is demonstrated in **Figures 10A–D**, which display the absorbance versus time curves relative to the $\nu(OH)$ and the $\nu(CO)$ bands.

The above data have been analyzed on the basis of the differential equation representing the Fick's second law of diffusion, imposing the initial condition (I.C.) and the boundary conditions (B.C.'s) appropriate for the case of a plane sheet exposed to equal penetrant activity on both surfaces. In terms of concentration of penetrant within the polymer, the I.C. and B.C.'s are as follows:

$$I.C.: \quad C = C_0 \quad 0 < x < L, \quad t = 0$$
$$B.C.'s: \quad C = C_1 \quad \text{at } x = 0, \quad \forall t \geq 0$$
$$C = C_1 \quad \text{at } x = L, \quad \forall t \geq 0$$

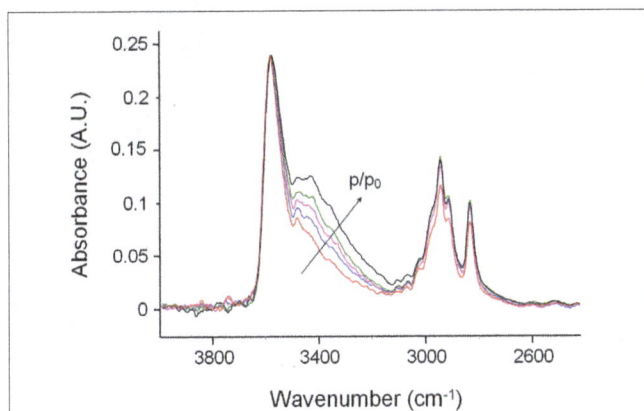

Figure 8 | Difference spectra in the $\nu(OH)-\nu(CH)$ region (4000–2400 cm^{-1}) of PMDA-ODA films equilibrated at different relative pressures of methanol vapor. Color code: red trace: $p/p_0 = 0.2$; blue trace: $p/p_0 = 0.3$; cyan trace: $p/p_0 = 0.4$; green trace: $p/p_0 = 0.5$; black trace: $p/p_0 = 0.6$. The absorbance scale refers to the red trace; the other spectra were normalized with respect to the peak at 3580 cm^{-1} to facilitate the comparison.

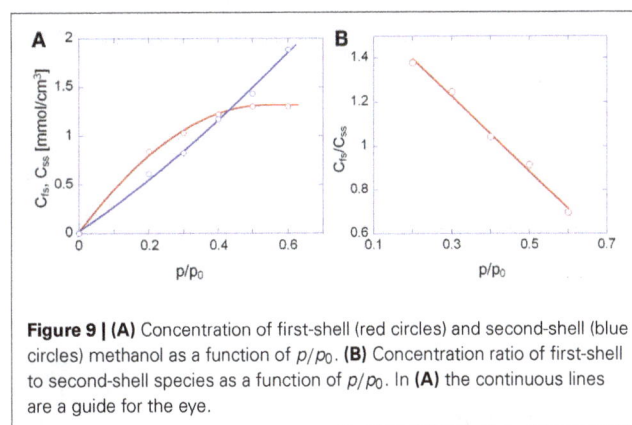

Figure 9 | (A) Concentration of first-shell (red circles) and second-shell (blue circles) methanol as a function of p/p_0. (B) Concentration ratio of first-shell to second-shell species as a function of p/p_0. In (A) the continuous lines are a guide for the eye.

Table 1 | Sample thickness, absorbance of the resolved components, concentration of methanol species and total methanol concentration for the sorption tests performed at different relative pressures of methanol vapor.

p/p_0	L (cm)	A_{3580} (cm^{-1})	A_{3425} (cm^{-1})	C_{fs} (mmol/cm^3)	C_{ss} (mmol/cm^3)	C_{tot}^{spec} (mmol/cm^3)	C_{tot}^{grav} (mmol/cm^3)
0.2	2.4×10^{-3}	14.37	14.79	0.84	0.61	1.45	2.70
0.3	2.2×10^{-3}	18.92	21.56	1.03	0.83	1.86	3.29
0.4	2.4×10^{-3}	18.47	25.23	1.22	1.17	2.39	4.05
0.5	2.0×10^{-3}	20.13	31.27	1.30	1.42	2.72	4.33
0.6	2.0×10^{-3}	23.60	47.14	1.29	1.79	3.08	5.15

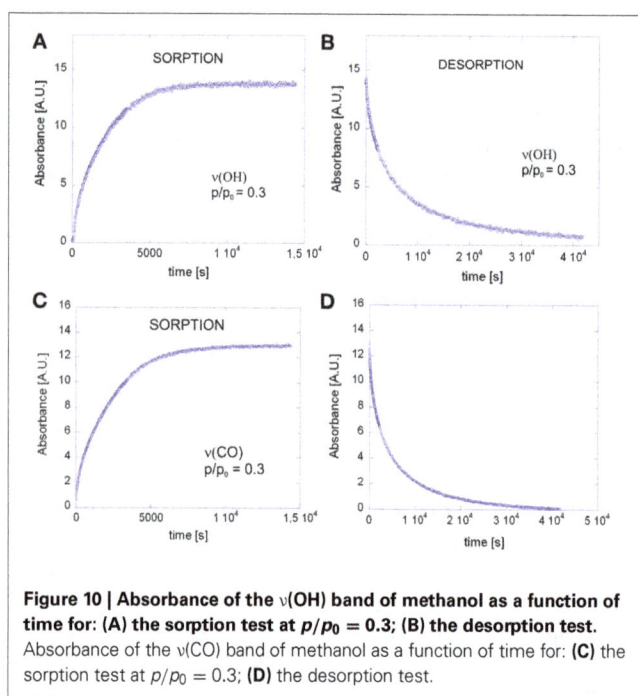

Figure 10 | Absorbance of the v(OH) band of methanol as a function of time for: (A) the sorption test at $p/p_0 = 0.3$; (B) the desorption test. Absorbance of the v(CO) band of methanol as a function of time for: **(C)** the sorption test at $p/p_0 = 0.3$; **(D)** the desorption test.

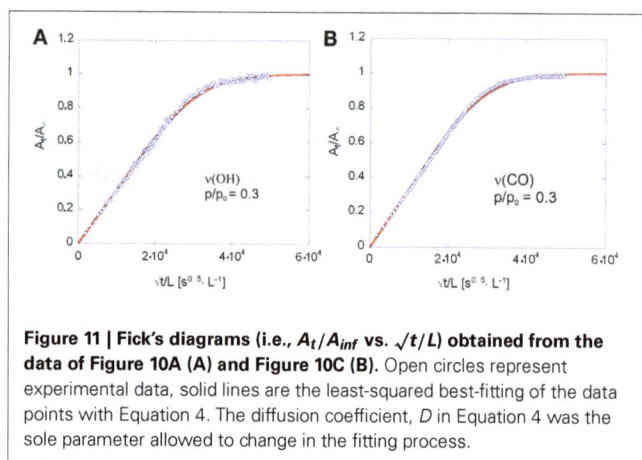

Figure 11 | Fick's diagrams (i.e., A_t/A_{inf} vs. $\sqrt{t}/L$) obtained from the data of Figure 10A (A) and Figure 10C (B). Open circles represent experimental data, solid lines are the least-squared best-fitting of the data points with Equation 4. The diffusion coefficient, D in Equation 4 was the sole parameter allowed to change in the fitting process.

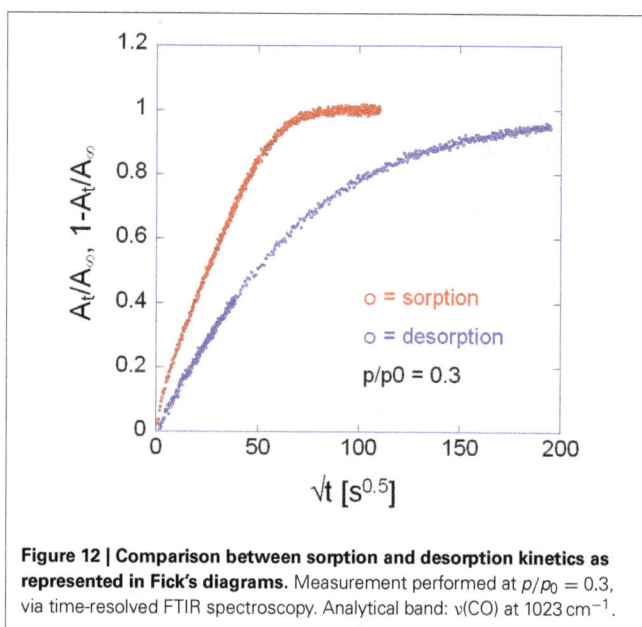

Figure 12 | Comparison between sorption and desorption kinetics as represented in Fick's diagrams. Measurement performed at $p/p_0 = 0.3$, via time-resolved FTIR spectroscopy. Analytical band: v(CO) at 1023 cm^{-1}.

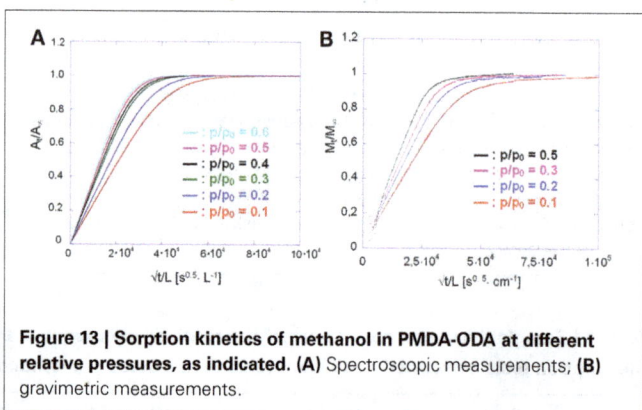

Figure 13 | Sorption kinetics of methanol in PMDA-ODA at different relative pressures, as indicated. (A) Spectroscopic measurements; **(B)** gravimetric measurements.

where L represents the thickness of the plane sheet. For this configuration, the solution of the second Fick's law is (Crank, 1975):

$$\frac{m(t)}{m_{inf}} = 1 - \frac{8}{\pi^2} \sum_{m=0}^{\infty} \frac{1}{(2m+1)^2} \exp\left[\frac{-D(2m+1)^2\pi^2 t}{L^2}\right] \quad (3)$$

where $m(t)$ and m_{inf} are, respectively, the total mass of penetrant absorbed within the sheet at time t and at sorption equilibrium while D is the mutual diffusivity.

Equivalently, in terms of absorbance, the solution reads:

$$\frac{A(t)}{A_{inf}} = 1 - \frac{8}{\pi^2} \sum_{m=0}^{\infty} \frac{1}{(2m+1)^2} \exp\left[\frac{-D(2m+1)^2\pi^2 t}{L^2}\right] \quad (4)$$

where $A(t)$ and A_{inf} are the integrated absorbances at time t and at equilibrium, respectively.

The coincidence of the experimental data with the curves predicted by theory, as well as the linearity of the sorption curves as a function of $t^{0.5}/L$, up to $A(t)/A_{inf}$ values well exceeding 0.6 (see **Figures 11A,B**) demonstrates the Fickian behavior of the system.

Comparing the sorption and desorption kinetics (**Figure 12**) evidences a much slower diffusivity in desorption than in sorption, which is typical of systems characterized by a pronounced dependence of D on the total concentration of penetrant in the sample. In particular, the observation that the desorption curve lies well below the sorption curve, implies that D is an increasing function of concentration. To investigate further this effect the kinetic analysis of the sorption process was performed in the whole range of relative pressures of methanol from 0.1 to 0.6. The results of both the spectroscopic and the gravimetric measurements are reported in **Figures 13A,B**.

The diffusion coefficient can be evaluated directly from Equations 3 or 4 whenever it can be considered constant with penetrant concentration. If this is not the case, a slightly more elaborate procedure is applied (Crank, 1975). The diffusivity from the Fick's plot represents now an average value over the

Table 2 | Sample thickness, equilibrium methanol content and diffusivities for the sorption tests performed at different relative pressures of methanol vapor.

p/p_0	L_{spec} (cm)	C_0 (wt %)	$\bar{D}_{spec}$ (cm^2/s)	L_{grav} (cm)	$\bar{D}_{grav}$ (cm^2/s)	$D(C)$(cm^2/s)
0.1	2.4×10^{-3}	2.40	1.06×10^{-10}	1.6×10^{-3}	1.10×10^{-10}	2.34×10^{-10}
0.2	2.2×10^{-3}	3.51	1.51×10^{-10}	1.5×10^{-3}	1.57×10^{-10}	3.30×10^{-10}
0.3	2.4×10^{-3}	4.25	2.19×10^{-10}	1.5×10^{-3}	1.96×10^{-10}	3.94×10^{-10}
0.4	2.0×10^{-3}	5.23	2.41×10^{-10}	1.6×10^{-3}	2.21×10^{-10}	4.79×10^{-10}
0.5	2.0×10^{-3}	5.59	2.73×10^{-10}	2.0×10^{-3}	2.54×10^{-10}	5.10×10^{-10}
0.6	2.4×10^{-3}	6.77	3.04×10^{-10}	1.3×10^{-3}	3.19×10^{-10}	6.12×10^{-10}

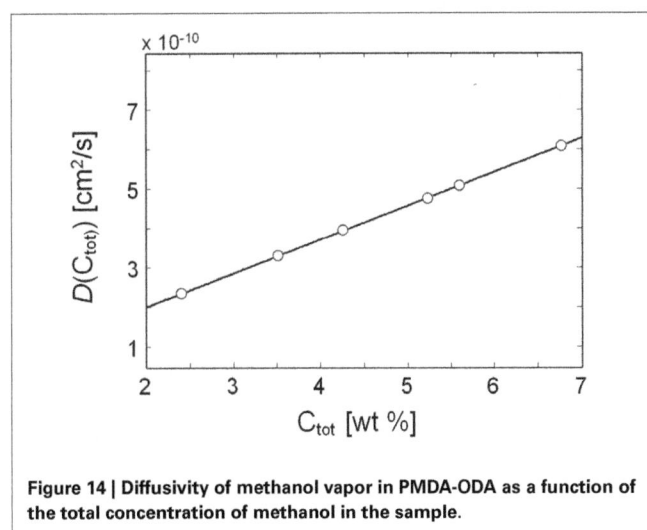

Figure 14 | Diffusivity of methanol vapor in PMDA-ODA as a function of the total concentration of methanol in the sample.

whole range of concentrations encompassed in the diffusion process, i.e.,

$$\bar{D} = \frac{1}{C_0} \int_0^{C_0} D \, dC \qquad (5)$$

where C_0 is the methanol concentration at equilibrium in a specific sorption test. Thus, the $\bar{D}$ values from Equations 3 or 4 are plotted as $\bar{D}C_0$ vs. C_0 (Equation 5) and numerical differentiation of the curve with respect to C_0 yields the sought D vs. C relationship.

The results of such an analysis are summarized in **Table 2**.

From **Table 2** it emerges that the diffusivity values estimated from gravimetric and FTIR data are in excellent agreement with each other; moreover, they also compare well with a previous literature report (Kamaruddin and Koros, 2000). The plot of D vs. C, constructed considering the average between the spectroscopic and the gravimetric values of $\bar{D}$, reveals a linear relationship in the concentration range of interest.(see **Figure 14**).

CONCLUSIONS

In this work the sorption of methanol in PMDA-ODA polyimide has been investigated both kinetically and at equilibrium, coupling gravimetric and *in-situ* FTIR spectroscopy in the transmission mode. This approach allowed a detailed characterization of the investigated system at a molecular level.

The spectral data at equilibrium were analyzed by using two complementary techniques, i.e., difference spectroscopy and least squared curve-fitting analysis: two distinct molecular species were detected in the system: single methanol molecules directly bound to the carbonyls of the imide group via H-bonding (*first-shell* species), and self-associated methanol molecules forming second- and higher shell layers. Up to p/p_0 values of 0.4 the dimer is likely to represent the predominant species; afterwards, higher aggregates (higher shell layers) are formed, giving rise to the clustering phenomenon. A method was proposed to quantify the population of the methanol species, based on literature values of the respective molar absorptivities. The concentration ratio C_{fs}/C_{ss} was found to decrease linearly by increasing the relative pressure of methanol vapor.

Kinetic data from gravimetry and *time-resolved* spectroscopy provided two independent estimations of the diffusion coefficients which were in good agreement with each other. The system was found to behave according to the Fick's second law of diffusion, with the diffusivity displaying a marked dependence on the total concentration of sorbed methanol. The evaluation of the D vs. C curve with the differential method of Crank demonstrated a linear behavior with positive slope.

REFERENCES

Bessonov, M. I., and Zubkov, V. A. (1993). *Polyamic Acids and Polyimides: Synthesis, Transformation and Structure*, Boca Raton, FL, CRC Press.

Chen, W. J., and Martin, C. R. (1995). Highly methanol selective membranes for the pervaporation separation of methyl-t-butyl ether/methanol mixtures. *J. Membr. Sci.* 104, 101–108. doi: 10.1016/0376-7388(95)00017-7

Cotugno, S., Di Maio, E., Mensitieri, G., Iannace, S., Roberts, G. W., Carbonell, R. G., et al. (2005). Characterization of microcellular biodegradable polymeric foams produced from supercritical carbon dioxide solutions. *Ind. Eng. Chem. Res.* 44, 1795–1803. doi: 10.1021/ie049445c

Cotugno, S., Larobina, D., Mensitieri, G., Musto, P., and Ragosta, G. (2001). Novel spectroscopic approach to investigate transport processes in polymers: the case of water/epoxy system. *Polymer* 42, 6431. doi: 10.1016/S0032-3861(01)00096-9

Crank, J. (1975). *The Mathematics of Diffusion, 2nd Edn.* Oxford, UK: Oxford University Press.

Dixon, J. R., George, W. O., Hossain, F., Lewis, R., and Price, J. M. (1997). Hydrogen bonded forms of methanol IR spectra and ab-initio calculations. *J. Chem. Soc. Faraday Trans.* 93 3611–3618. doi: 10.1039/a702176c

Feger, C., Khojasteh, M. M., and Htoo, M. S. (1993). *Advances in Polyimide Science and Technology.* Lancaster, PA: Technomic.

Galizia, M., De Angelis, M. G., Finkelshtein, E., Yampolskii, Y., and Sarti, G. C. (2011). Sorption and transport of hydrocarbons and alcohols in addition type poly trimethylsilyl norbornene. I: Experimental data. *J. Membr. Sci.* 385–386, 141–153. doi: 10.1016/j.memsci.2011.09.032

Galizia, M., De Angelis, M. G., and Sarti, G. C. (2012). Sorption of hydrocarbon and alcohols in addition type poly-trimethylsilyl-norbornene and other high free volume glassy polymers. II. NELF model predictions. *J. Membr. Sci.* 405–406, 201–211. doi: 10.1016/j.memsci.2012.03.009

Galizia, M., La Manna, P., Pannico, M., Mensitieri, G., and Musto, P. (2013). Methanol Diffusion in Polyimides: a Molecular Description. *Polymer.* (in press).

Gosh, M. K., and Mittal, K. L. (1996). *Polyimides: Fundamentals and Applications.* New York, NY: Marcel Dekker.

Hibshman, C., Cornelius, C. J., and Marand, E. (2003). The Gas separation effects of annealing polyimide-organosilicate hybrid membranes. *J. Membr. Sci.* 211, 25–40. doi: 10.1016/S0376-7388(02)00306-X

Jelinski, L. W., Dumais, J. J., Chiolli, A. L., Ellis, T. S., and Karasz, F. E. (1985). Nature of the water-epoxy interaction. *Macromolecules* 18, 1091–1095. doi: 10.1021/ma00148a008

Kamaruddin, H. D., and Koros, W. J. (2000). Sorption of methanol/MTBE and diffusion of methanol in 6FDA-ODA polyimide. *J. Polym. Sci. B Polym. Phys.* 38, 2254–2267. doi: 10.1002/1099-0488(20000901)38:17<2154::AID-POLB60>3.0.CO;2-8

Marquardt, D. W. (1963). Finite difference algorithm for curve fitting. *J. Soc. Ind. Appl. Math.* 11, 431–441. doi: 10.1137/0111030

Mascia, L. (1995). Developments in organic-inorganic polymeric hybrids: Ceramers. *Trends Polym. Sci.* 3, 61–66.

Meier, R. J. (2005). On art and science in curve fitting vibrational spectra. *J. Vib. Spectrosc.* 39, 266–269. doi: 10.1016/j.vibspec.2005.03.003

Musto, P., Abbate, M., Lavorgna, M., Ragosta, G., and Scarinzi, G. (2006). Microstructural features, diffusion and molecular relaxations in poly-imide/silica hybrids. *Polymer* 47, 6172. doi: 10.1016/j.polymer.2006.05.074

Musto, P., Mensitieri, G., Lavorgna, M., Scarinzi, G., and Scherillo, G. (2012). Combining gravimetric and vibrational spectroscopy measurements to quantify first and second shell hydration layers in polyimides with different molecular architectures. *J. Phys. Chem. B* 116, 1209–1220. doi: 10.1021/jp2056943

Musto, P., Ragosta, G., Mensitieri, G., and Lavorgna, M. (2007). On the molecular mechanism of H_2O diffusion into polyimides: a vibrational spectroscopy investigation. *Macromolecules* 40, 9614–9627. doi: 10.1021/ma071385+

Okamoto, K., Tanihara, N., Watanabe, H., Tanaka, K., Kita, H., Nakamura, A., et al. (1992). Vapor permeation and pervaporation separation of water-ethanol through polyimide membranes. *J. Membr. Sci.* 68, 53–63. doi: 10.1016/0376-7388(92)80149-E

Shurvell, H. F. (2002). "Spectra-structure Correlations in the Mid- and Far-infrared," in *Handbook of Vibrational Spectroscopy*, Vol. 3. eds J. M. Chalmers and P. R. Griffiths (Chichester: Wiley), 1783–1816.

Thompson, L. F., Willson, C. G., and Tagawa, S. (1994). *Polymers for Microelectronics: Resists and Dielectrics.* Washington, DC: ACS Symposium Series 537, ACS.

Conflict of Interest Statement: The authors declare that the research was conducted in the absence of any commercial or financial relationships that could be construed as a potential conflict of interest.

17

Sensitive impedimetric biosensor for direct detection of diazinon based on lipases

Nedjla Zehani[1,2], Sergei V. Dzyadevych[3], Rochdi Kherrat[2] and Nicole J. Jaffrezic-Renault[1]*

[1] UMR 5280, Institut des Sciences Analytiques, Université de Lyon, Villeurbanne, France
[2] Laboratory of Environmental Engineering, Faculty of Engineering Sciences, University of Annaba, Annaba, Algeria
[3] Laboratory of Biomolecular Electronics, Institute of Molecular Biology and Genetics, National Academy of Sciences of Ukraine, Kiev, Ukraine

Edited by:
Margarita Stilianova Stoytcheva,
Universidad Autonoma de Baja
California, Mexico

Reviewed by:
Karim Michail, University of Alberta,
Canada
Xiliang Luo, Qingdao University of
Science and Technology, China

*Correspondence:
Nicole J. Jaffrezic-Renault, UMR
5280, Institut des Sciences
Analytiques, Université de Lyon, 5
Rue de la Doua, 69100 Villeurbanne,
France
e-mail: nicole.jaffrezic@univ-lyon1.fr

Two novel impedimetric biosensors for highly sensitive and rapid quantitative detection of diazinon in aqueous medium were developed using two types of lipase, from *Candida Rugosa* (microbial source) (CRL) and from porcine pancreas (animal source) (PPL) immobilized on functionalized gold electrode. Lipase is characterized to specifically catalyze the hydrolysis of ester functions leading to the transformation of diazinon into diethyl phosphorothioic acid (DETP) and 2-isopropyl-4-methyl-6-hydroxypyrimidine (IMHP). The developed biosensors both presented a wide range of linearity up to $50\,\mu M$ with a detection limit of 10 nM for *Candida Rugosa* biosensor and $0.1\,\mu M$ for porcine pancreas biosensor. A comparative study was carried out between the two biosensors and results showed higher efficiency of *Candida Rugosa* sensor. Moreover, it presented good accuracy and reproducibility, had very good storage and multiple use stability for 25 days when stored at 4°C.

Keywords: biosensor, diazinon, lipase, electrochemical impedance spectroscopy

INTRODUCTION

Chemicals released from agriculture or industry may potentially develop toxic effects in the environment and ecological systems. Among them, pesticides are actively applied and globally used for crop control and to prevent damage to plants, animals, humans, and aliments. Organophosphates constitute the most extensive and manifold group of pesticides, they were developed at the beginning of this century by chemical manipulation of nerve gasses that are so toxic compounds (Osterauer and Kohler, 2008). Their mode of toxicity is the inhibition of acetyl cholinesterase, an enzyme responsible for the hydrolysis of the neurotransmitter acetylcholine (Cabello et al., 2001; Gordon and Mack, 2003; Pesando et al., 2003). This inhibition leads to a continuous stimulation of cholinergic neurons and eventually paralyzes the target organs (Wiener and Hoffman, 2004).

Diazinon after malathion is one of the most commonly used organophosphate pesticides (OPs) in the world, it is extensively used as an insecticide in agriculture to control juvenile forms of insects in soil, plants, fruit, vegetable crops and to control external pet parasites (i.e., mites, leaf miner flies, black cherry aphid, and apple maggot) (Karpouzas and Singh, 2006; UAP. Ca, 2007), it can enters the body via skin contact, feeding, and inhalation (Villeneuve et al., 1972). Furthermore, diazinon is one of the substances most responsible for acute poising via insecticides of humans and wildlife (Keizer et al., 1995). Once applied on crops and other plants, it is easily washed by surface waters and enters the ground water. Eventually, it enters the aquatic environment in large quantities as described in a number of studies and thus may affect a wide range of non-target organisms. The high acute toxicity of diazinon to freshwater fish and aquatic invertebrates is reflected by the 96 h LC50 of 2 g/L in *Daphnia magna*, of 1.35 mg/L in *O. mykiss*, and of 8 mg/L in *Danio rerio* (Osterauer and Kohler, 2008). Therefore, the persistence and mobility of diazinon and its metabolites suggest the potential for groundwater contamination, an increasing concentration of diazinon and OPs residues are found to be present in many sampled soils, aquatic eco collected in the United States and Canada and it was the most frequently detected insecticide in surface waters prior to the phase-out of urban uses in 2004 in United States. It degrades in water as a result of hydrolysis, especially under acidic conditions. In sterile water, diazinon was determined to have a half-life of 12 days in acidic water (pH 5), 138 days in neutral water (pH 7), and 77 days (pH 9). Moreover, because diazinon is fat soluble, there is potential for delayed toxicity if significant amounts of diazinon are stored in fatty tissues and causes diseases with long-term (US NPIC, 2009). Diazinon applied to soils can be also absorbed by plant roots and translocated in plants, soil metabolism studies report soil half-lives for diazinon ranging from 21 to 103 days depending on the type of soil. In addition, oxypyrimidine which is the principal metabolite of diazinon hydrolysis is very mobile in the environment and has been measured up to 72 inches below the surface of soils. Oxypyrimidine appears to be more persistent under at least some conditions compared with diazinon (US NPIC, 2009).

Thus, the need to be understood and evaluate the biological effects of pollutants on aquatic ecosystems has generated (Reddy et al., 2013). In this sense, a large number of studies have used biosensors as functional tools to evaluate the toxicity of such compounds for natural populations (Kumar and D'Souza, 2011). Recently, there has been an intense research effort to develop enzymatic biosensor devices for the detection of organophosphorus pesticides. Many enzymes were used for this purpose (Pogacnik and Franco, 1999; Deo et al., 2005; Sajjadi et al., 2009). Among them, we find the lipase which it is an important enzyme in biological systems, where is catalyzes the hydrolysis of ester functions and the transformation of triacylglycerol to glycerol and fatty acids, it is a subclasses of esterases (Bhalchandra et al., 2008).

This paper describes a new biosensor system devoted for environmental application, based on impedimetric transduction incorporating two types of enzymes, lipase from *Candida Rugosa* CRL (microbial source) and lipase from porcine pancreas PPL (animal source) for the detection of diazinon in aqueous medium. This is the first time that lipase is proposed for the design of diazinon biosensor. This enzyme was used to catalyze the hydrolysis of diazinon (O,O-diethylO-(2-isopropyl-6methyl-4-pyrimidyl phosphorothioate) into Diethyl phosphorothioic acid (DETP) and 2-isopropyl-4methyl-6 hydroxypyrimidine (IMHP) (**Figure 1**). In this work, bioselective membranes were prepared by functionalization of microelectrodes with SAMs and enzyme cross-linking using glutaraldehyde vapor and bovine serum albumin (BSA). As a detection mode, we used electrochemical impedance spectroscopy (EIS) which is one of the electrochemical techniques that have been widely employed to study various chemical and biological phenomena on surfaces and to develop sensors (Yagati et al., 2011). EIS has emerged as a powerful tool to study the biomolecular interactions by detecting changes in capacitance and interfacial electron transfer resistance at the surface electrode occurring during these processes (K'Owino and Sadik, 2005; Ferreira et al., 2010). It is a rapidly developing technique for the label free detection of different types of biosensing events occurring at the surface electrode, example: antigen-antibody enzyme-substrate reaction, cell adsorption (Bourigua et al., 2010), it allows such complex recognition events to be probed in a simple, sensitive, label-free, and mediator free strategy (Sadik et al., 2002; Farcas et al., 2010). In this study, the different functionalization steps of gold microelectrodes were first characterized by EIS and cyclic voltammetry (CV), afterward, the analytical characteristics of the developed biosensors were determined. Finally, a comparison study between the two biosensors was carried out.

EXPERIMENTAL
CHEMICAL AND BIOLOGICAL REAGENTS
All reagents used in this study were purchased from Sigma Aldrich (Saint Quentin Fallavier) France, including, lipase from *Candida*

Rugosa enzyme (CRL, type VII, ≥ 700 unit/mg solid), Lipase from porcine pancreas (PPL, type II, 30–90 units/mg protein), bovine serum albumin (BSA), diazinon, parathion methyl, paraoxon methyl, atrazine, sevin, simazine, fenitrothion, sodium phosphate dibasic, sodium phosphate monobasic, KH_2PO_4, K_2HPO_4, glutaraldehyde (grade II, 25% aqueous solution), glycerol ($\geq 99\%$), 6-methyl-5propyl-4 pyrimidinone, N-hydroxysuccinimide (NHS), 1-ethyl-3(3-(dimethyl-amino)propyl)carbodiimide (EDC), acidthiol(16mercacaptohexadecanoicacid). Sulfuric acid (96%), hydrogen peroxide (30%), ethanol (99%) was purchased from Fluka. All solutions were made up with ultrapure water (resistivity no less than 18 MΩ cm and obtained from a Millipore purification system).

ELECTRODE GOLD PREPARATION AND ENZYME IMMOBILIZATION
Gold cleaning
Before analysis, in order to improve the adhesion of the enzymatic membrane on the electrode surface, gold electrode was firstly immersed in acetone for 10 min under sonication to remove the resin layer; then, it was immerged for 3 min in Piranha solution (H_2O_2:H_2SO_4: 3:7 v/v). Finally the electrode was thoroughly rinsed with ultrapure water and dried under a nitrogen flow.

Functionalization and activation process
The gold electrode was incubated overnight in 10 mM solution of thiolacid in ethanol, overnight at a temperature of 4°C, which allowed the formation of a self-assembled monolayer (SAMs) on the surface of the electrode. The thiol functionalized electrode was then rinsed with ethanol in order to remove the unbonded thiol molecules. The resulting monolayer ending with carboxylic acid groups were then activated by an [EDC/NHS] mixture at a concentration of 0.1 mol·l^{-1} for 1 h.

Enzyme attachment
Afterward, the pretreated electrode was rinsed with PBS and dried under nitrogen flow, 10 µL of enzyme based solution containing CRL (5%), BSA (5%), glycerol (10%) in phosphate buffer solution 20 mM, pH = 7.2 (90%) was thoroughly homogenized and deposited onto the surface of the working electrode. Then, the sensor was placed in saturated glutaraldehyde vapor (cross-linker) for 30 min, and dried in air at room temperature for 40 min, the biosensor was stored at 4°C until further use.

ELECTROCHEMICAL SET-UP: CYCLIC VOLTAMMETRY (CV) AND ELECTROCHEMICAL IMPEDANCE SPECTROSCOPY (EIS)
Electrochemical measurements were carried out in a conventional one compartment three electrode cell with an internal volume of 5 ml (Verre equipments Collonges au Mont d'or, France), hermetically closed on one side with a planar gold electrode (300 nm thickness, deposited on insulated silicon with 0.19 cm^2 surface area) used as the working electrode. On the other side, a planar platinum electrode (0.59 cm^2) was used as the counter electrode. A saturated calomel electrode from Hach Lange (France) was used as reference electrode. This electrochemical cell was designed to maintain a fixed distance between the electrodes. It was manufactured with two inlets; one for the positioning of the reference

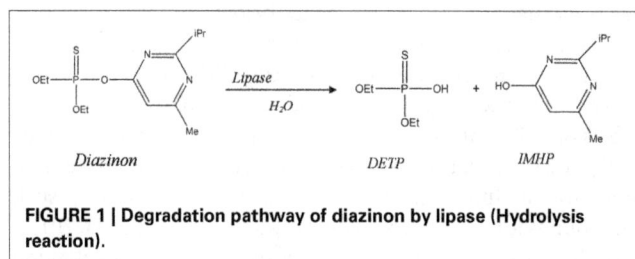

FIGURE 1 | Degradation pathway of diazinon by lipase (Hydrolysis reaction).

electrode and the other for OP injections. This feature prevented further manipulation or movements of the electrodes, fixing the geometry of the cell and also ensuring the reproducibility of measurements.

The electrochemical measurements were performed with an electrochemical impedance analyzer "Voltalab PGZ 402" (Hach Lange, France). Data acquisition and processing via "Voltamaster 4" software was provided by Thesame company. CV measurements were performed in 8 mM solution of $[Fe(CN_6)]^{3-/4-}$ at scan rate of 100 mV/s. All electrochemical measurements were taken at a frequency range of 100 mHz–100 kHz at room temperature. Measurements were performed at room temperature in 20 mM PBS buffer solution, pH 5.2, under magnetic stirring in Faraday cage. A DC potential of -400 mV was applied; as shown in **Figure 2**, applying a DC potential of -400 mV allowed the total impedance of the modified electrode to be minimized before the injection of diazinon. The measured impedance spectra were analyzed in terms of electrical equivalent circuits using the analysis program Zview (Scribner Associates, USA). A classical Randles equivalent circuit presented in **Figure 3** was used to fit Nyquist plots, including the two resistive elements R_1 and R_2, in which the double layer capacitive was replaced by CPE, a constant phase angle element and Warburg impedance Zw (cf. **Figure 3**). R_1 corresponds to the Ohmic resistance of the bulk electrolyte and of electrical contacts and R_2 to the charge transfer resistance between the solution and the modified electrode surface. R_2 is equal to the diameter of semi-circle. Warburg impedance (Zw) is the specific electrochemical element of diffusion and can

be defined as $(Z_w) = (1 - j)^{-0.5}$, where, σ denotes the Warburg coefficient. CPE takes into account the non-homogeneity of the layer and the impedance of such a non-ideal layer that can be expressed as $Z(\omega) = CPE^{-1}(j\omega)^{-n}$, where ω is a circular frequency and n a parameter describing the deviation from an ideal capacitor, varying from 0 to 1 (Katz and Willner, 2003).

RESULTS AND DISCUSSION
ELECTROCHEMICAL CHARACTERIZATION OF THE DIFFERENT STEPS OF BIOSENSOR ELABORATION
Cyclic voltammetry characterization (CV)

Figure 4 shows the electrochemical characterization by CV of the bare and modified electrode in the presence of 8 mM solution of $[Fe(CN)_6]^{4-/3}$. The potential was swept between -0.4 and 0.6 V, at the scan rate of 100 mV/s. As it is clear in the figure, the complete disappearance of the oxidation and reduction peaks after functionalization with the thiol layer confirms the high insulating proprieties of the dense acid thiol layer.

EIS response of the enzymatic biosensor

In **Figure 5**, a 2.5 fold increase of R_2 was observed by assembling the first layer on the electrode surface, which basically reflects the insulating proprieties of SAMs (Chen et al., 2005). Contrariwise, a decrease of interfacial impedance, correlated to 1.5 fold decrease of R_2 was observed following the injection of diazinon, confirming that the hydrolysis of diazinon induces charge redistribution at the functionalized gold electrode/electrolyte interface. The n-value was 0.95 for the bare electrode and was reduced to 0.86–0.91 after modification, reflecting only a slight deviation from ideality and a rather capacitive behavior of the corresponding CPE.

ELECTROCHEMICAL DETECTION OF DIAZINON
Variation of impedance after diazinon injections

As can be seen in **Figure 6**, after increasing concentrations of diazinon contact with CRL biosensor, the total impedance

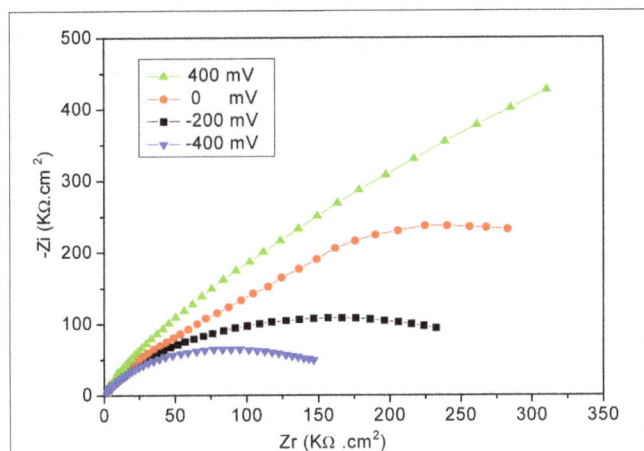

FIGURE 2 | Nyquist plots obtained for the developed biosensor at different applied potentials. Measurements were performed in PBS 20 mM, pH 5.2, for frequencies ranging from 100 mHz to 100 kHz.

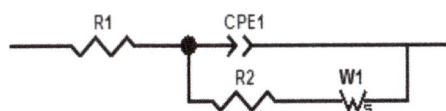

FIGURE 3 | Randles equivalent circuit.

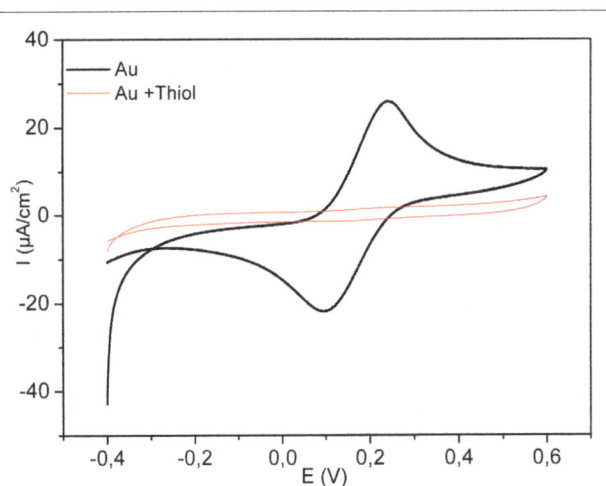

FIGURE 4 | Cyclic Voltammograms for bare electrode (bold plot) and for modified electrode by thiol (thin line). Measurements were performed in 8 mM Fe $(CN_6)^{3-/4-}$, scan rate 100 mV/s.

decreases, from 2 to $50\,\mu$M. For concentrations higher than $50\,\mu$M, a saturation effect is observed. The decrease of total impedance is due to a decrease of R_2, induced by charge redistribution at the functionalized gold electrode/electrolyte interface due to the enzymatic hydrolysis of diazinon, and an associated increase of interface capacitance.

Calibration curves for CRL and PPL biosensors

The relationship between biosensor response and diazinon concentration was examined by recording impedance spectra after injection of different concentrations of diazinon in PBS. For both CRL and PPL biosensors, three measurements were performed at each concentration level. To obtain calibration curves, the values of $\Delta R_2 = R_2 - R_{2(0)}$ were deduced, where $R_{2(0)}$ refers to R_2 for [diazinon] = 0. **Figure 7** presents the calibration plots for both enzymes (a) CRL, (b) PPL, as can be seen, there is a large linear variation of ΔR_2 with diazinon concentration up to $50\,\mu$M for both sensors, with a detection limit of $10\,$nM for the CRL biosensor and $0.1\,\mu$M for the PPL based one.

Besides, the relative variation of R_2 obtained with enzyme CRL is larger (factor 2.5) than that obtained with PPL and as it is shown in **Table 1** that the correlation coefficient and sensitivity of CRL biosensor ($S = 0.78\,$kΩcm^2) are better than those obtained by PPL sensor ($S = 0.49\,$kΩcm^2). Therefore, the analytical performances of CRL based biosensor are the best. This CRL based biosensor had a detection limit of $10\,$nM which is however higher than that biosensors based on inhibition of AchE (Yi et al., 2013), but it presented a large linear range and it was better than enzymatic biosensors already reported like a potentiometric OPH based biosensor (Mulchandani et al., 1999, 2001), an amperometric tyrosinase based biosensor (Tanimoto de Albuquerque and Ferreira, 2007), a tyrosinase based oxygen biosensor (Russell Everett and Rechnitz, 1998), (cf. **Table 2**). Furthermore, this value is sufficient to allow diazinon determination in industrial waste waters and make

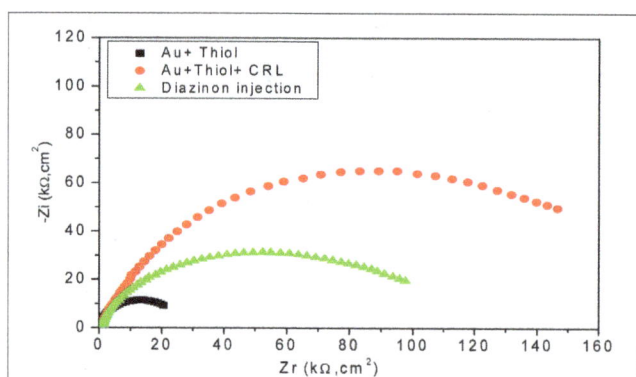

FIGURE 5 | Nyquist diagrams for different layers of developed biosensor with CRL. Measurements were carried out in PBS 20 mM, pH 5.2 with a frequency range of 100 mHz–100 kHz.

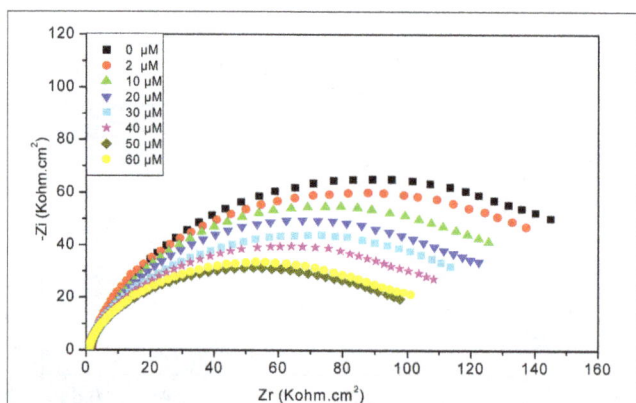

FIGURE 6 | Nyquist diagrams obtained with CRL biosensor, after injection of different concentrations of diazinon. Measurements were carried out in PBS 20 mM, pH 5.2 with a frequency range of 100 mHz–100 kHz.

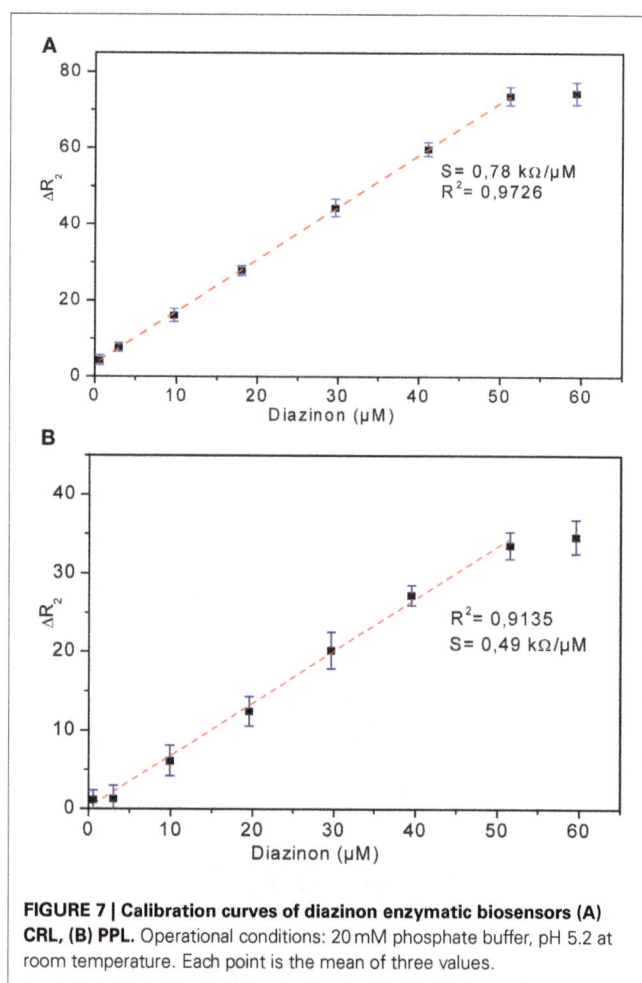

FIGURE 7 | Calibration curves of diazinon enzymatic biosensors (A) CRL, (B) PPL. Operational conditions: 20 mM phosphate buffer, pH 5.2 at room temperature. Each point is the mean of three values.

Table 1 | Analytical characteristics of enzymatic biosensors.

Enzyme	Sensitivity	Linear range (μM)	Detection limit (μM)
CRL	$0.78\,$kΩ/μM	Up to 50	0.01
PPL	$0.49\,$kΩ/μM	Up to 50	0.1

Table 2 | Analytical characteristics of enzymatic biosensors for the detection of diazinon.

Target analyte	Transduction	Linear range	LOD (μM)	Enzyme	References
Diazinon	Amperometric	5–50 μM	5	Tyrosinase	Russell Everett and Rechnitz, 1998
Diazinon	Potentiometric	0.13–2.8 mM	5	OPH	Mulchandani et al., 1999
Diazinon	Amperometric	0.06–016 μM	0.06	Tyrosinase	Tanimoto de Albuquerque and Ferreira, 2007
Diazinon	Amperometric	0.46–8.56 mM	2	OPH	Mulchandani et al., 1998
Diazinon	SiQDs fluorescence	/	2.22×10^{-4}	AchE and ChOx	Yi et al., 2013
Diazinon	Photothermal	/	32.86	AchE	Pogacnik and Franco, 1999
Diazinon	Impedimetric	0.01–50 μM	0.01	CRL	This work
Diazinon	Impedimetric	0.1–50 μM	0.1	PPL	This work

LOD, limit of detection.

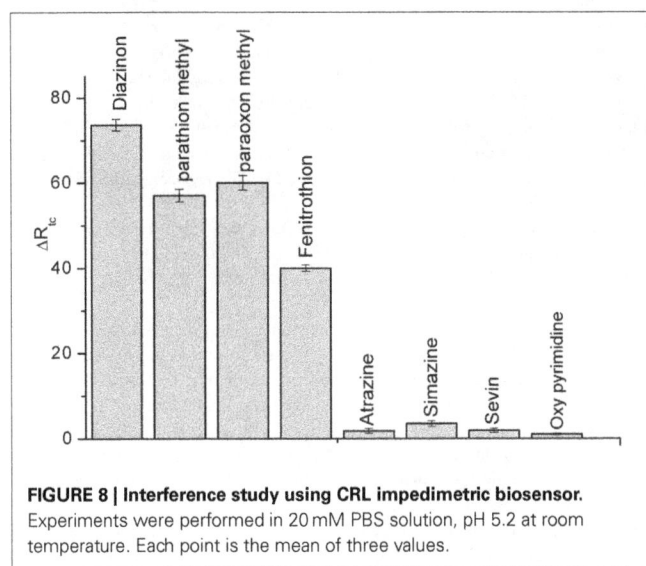

FIGURE 8 | Interference study using CRL impedimetric biosensor. Experiments were performed in 20 mM PBS solution, pH 5.2 at room temperature. Each point is the mean of three values.

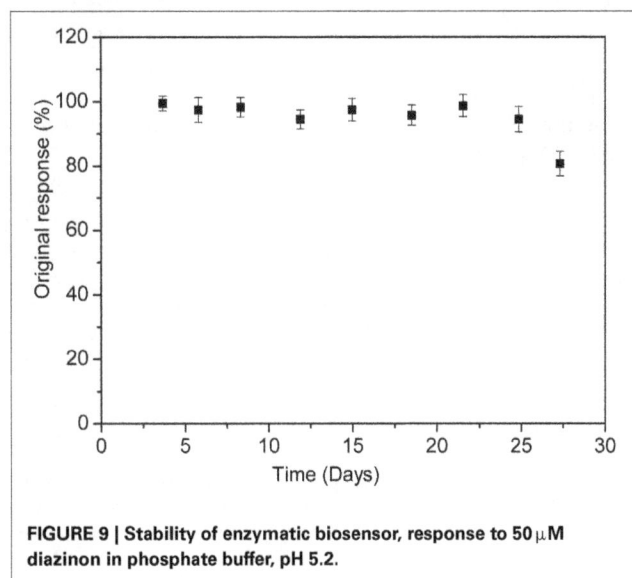

FIGURE 9 | Stability of enzymatic biosensor, response to 50 μM diazinon in phosphate buffer, pH 5.2.

the biosensor suitable for on-line environmental monitoring applications.

The enzymatic biosensor was also evaluated for matrix effect of natural compounds in real samples. Diazinon was spiked in water from La Chaudanne river—Lyon (pH of samples water was adjusted from original 7.5 to 5.2 and analyzed within 24 h after collection). the responses of the biosensor in the river water were almost similar to that in the buffer, validating the potential utility of the present biosensor for detection of OPs contaminated natural waters.

Selectivity of biosensor

Selectivity is a fundamental component of biosensor, therefore, it was tested for the detection of some compounds: parathion methyl, paraoxon methyl, fenitrothion, (organophosphate pesticides), atrazine, sevin, simazine (carbamates) and also for oxypyrimidine; the metabolite of diazinon at the concentration of 50 μM, the results are presented in **Figure 8**. Such as it is clear in the figure, this CRL biosensor is not specific for individual pesticide but to a class of organophosphate pesticides. Conversely, the carbamates and oxypyrimidine did not interfere.

Stability and reproducibility of CRL biosensor

Reproducibility and stability are among the key factors of a sensor's performance, the reproducibility of the enzymatic biosensor was tested for three different sensors in the concentration range from 10 to 50 μM. The variation coefficient obtained from three measurements was very good; it was between 2 and 5% in the concentration range studied.

To investigate the long-term storage stability and multiple use capability, the enzymatic biosensor was used for 1 month to measure the response to 50 μM of diazinon. During that time, the biosensor was stored at 4°C in 20 mM phosphate buffer (pH 7.2). As demonstrated in **Figure 9**, the biosensor was stable for 25 days, after this, only 20% of the initial response was lost. Therefore, life time of this biosensor can be estimated to be one month. This storage time was as stable as other previously reports of organophosphate pesticides detection (32 and 30 days) (Mulchandani et al., 2001; Kumar et al., 2006). However, it exceeds those earlier reported in the literature as amperometric screen printed tyrosinase-modified electrodes (10 days) and microbial Sphingomonas sp biosensor for methyl parathion detection (18 days) (Kumar et al., 2006; Tanimoto de Albuquerque and Ferreira, 2007).

CONCLUSION

In this work, two impedimetric biosensors were developed for the detection of diazinon. Two different types of lipase were immobilized on the surface of functionalized gold electrode. Both biosensors presented a wide linear range and low detection limits, but the performance of CRL based biosensor is better than PPL biosensor (sensitivity and detection limit). This novel CRL biosensor for direct determination of diazinon is simple, one step with rapid response and large dynamic range. Moreover, it is low cost and does not require any expensive measurement apparatus. Unlike the other biosensors based on OPH or AchE, it is specific only for organophosphate pesticides which make it very promising analytical tool for the detection of organophosphate pesticides in real samples. Thus, it will be ideal for on line monitoring of detoxification processes for the treatment of wastewaters generated by the industrial production of organophosphate-pesticides.

ACKNOWLEDGMENT

The authors thank EC for financial support through Marie Curie NANODEV grant N° 318524.

REFERENCES

Bhalchandra, K. V., Ganesh, C. I., Ponrathnam, S., Kulkarni, B. D., and Sanjay, N. N. (2008). Immobilization of Candida Rugosa lipase on poly (allyl glycidyl ether-co-ethylene glycol dimethacrylate) macroporous polymer particles. Bioresour. Technol. 99, 3623–3629. doi: 10.1016/j.biortech.2007.07.035

Bourigua, S., Hnaien, M., Besseuille, F., Lagarde, F., Dzaydevych, S., Maaref, A., et al. (2010). Impedimetric immunosensor based on SWCNT-COOH modified gold microelectrodes for label-free detection of deep venous thrombosis biomarker. Biosens. Bioelectron. 26, 1278–1282. doi: 10.1016/j.bios.2010.07.004

Cabello, G., Valenzuela, M., Vilaxa, M., Duran, A., Rudolph, V., Herpic, I., et al. (2001). A rat mammary tumor model induced by the organophosphorus pesticides parathion and malathion, possibly through acetylcholine inhibition. Environ. Health Perspect. 109, 471–479. doi: 10.1289/ehp.01109471

Chen, H., Heng, C. K., Puiu, P. D., Zhou, X. D., Lee, A. C., Lim, T., et al. (2005). Detection of Saccharomyces cerevisiae immobilized on self-assembled monolayer (SAM) of alkanethiolate using electrochemical impedance spectroscopy. Anal. Chem. Acta 554, 52–59. doi: 10.1016/j.aca.2005.08.086

Deo, R. P., Wang, J., Block, I., Mulchandani, A., Joshi, K. A., Trojanowicz, M., et al. (2005). Determination of organophosphate pesticides at a carbon nanotube/organophosphorus hydrolase electrochemical biosensor. Anal. Chim. Acta 530, 185–189. doi: 10.1016/j.aca.2004.09.072

Farcas, M., Cosman, N., Tings, D., Rscoe, S., and Omanovic, S. (2010). A comparative study of electrochemical techniques in investigating the adsorption behavior fibrinogen on platinum. J. Electroanal. Chem. 649, 206–218. doi: 10.1016/j.jelechem.2010.04.004

Ferreira, D. C., Mendes, R., Kubota, L., and Braz, J. (2010). Kinetic studies of HRP adsorption on ds-DNA immobilized on gold electrode surface by EIS and SPR. Chem. Soc. 21, 1648–1655. doi: 10.1590/S010350532012

Gordon, C. G., and Mack, C. (2003). Influence of gender on thermoregulation and cholinesterase inhibition in the long-evans rat exposed to diazinon. J. Toxicol. Environ. Health 66, 291–304. doi: 10.1080/15287390306371

Karpouzas, D. G., and Singh, K. B. (2006). Microbial degradation of organophosphorus xenobiotics: metabolic pathways and molecular basis. Adv. Microb. Physiol. 51, 185–225. doi: 10.1016/S0065-2911(06)51003-3

Katz, E., and Willner, I. (2003). Probing biomolecular interactions at conductive and semiconductive surfaces by impedance spectroscopy: routes to impedimetric immunosensors, DNA-sensors and enzyme biosensors. Electroanalysis 15, 913–290. doi: 10.1002/elan.200390114

Keizer, J., D'Agostino, G., Nagel, R., Volpe, T., Gnemi, P., and Vittozzi, L. (1995). Enzymological differences of AChE and diazinon hepatic metabolism:

correlation of in vitro data with the selective toxicity of diazinon to fish species. Sci. Total Environ. 171, 213–220. doi: 10.1016/0048-9697(95)04687-0

K'Owino, I., and Sadik, O. (2005). Impedance spectroscopy: a powerful tool for rapid biomolecular screening and cell culture monitoring. Electroanalysis 17, 2101–2113. doi: 10.1002/elan.200503371

Kumar, J., and D'Souza, S. F. (2011). Microbial biosensor for detection of methyl parathion using screen printed carbon electrode and cyclic voltammetry. Biosens. Bioelectron. 25, 4289–4293. doi: 10.1016//j.bios.2011.04.027

Kumar, J., Jha, S. K., and D'Souza, S. F. (2006). Optical microbial biosensor for the detection of methyl parathion pesticide using Flavobacterium sp whole cells adsorbed on glass fiber filters as disposable biocomponent. Biosens. Bioelectron. 21, 2100–2105. doi: 10.1016/j.bios.2005.10.012

Mulchandani, A., Chen, W., Mulchandani, P., Wang, J., and Rogers, K. M. (2001). Biosensor for direct determination of organophosphate pesticides. Biosens. Bioelectron. 16, 225–230. doi: 10.1016/S0956-5663(01)00126-9

Mulchandani, A., Kaneva, I., and Chen, W. (1998). Biosensor for direct determination of organophosphate nerve agents using recombinant Escherichia coli with surface-expressed organophosphorus hydrolase. 2. Fiber-optic microbial biosensor. Anal. Chem. 70, 4140–4145. doi: 10.1021/ac980643l

Mulchandani, P., Mulchandani, A., Kaniva, I., and Chan, W. (1999). Biosensor for direct determination of organophosphate nerve agents. 1. Potentiometric enzyme electrode. Biosens. Bioelectron. 14, 77–85. doi: 10.1016/S0956-5663(98)00096-7

Osterauer, R., and Kohler, H. (2008). Temperature- dependent effects of the pesticides thiacloprid and diazinon on the embryonic development of zebrafish (Danio rerio). Aquat. Toxicol. 86, 485–494. doi: 10.1016/j.aquatox.2007.12.2013

Pesando, D., Huitorel, P., Dolcini, V., Angelini, C., Guidetti, P., and Falugi, C. (2003). Biological targets of neurotoxic pesticides analysed by alteration of developmental events in the Mediterranean sea urchin, Paracentrotus lividus. Mar. Environ. Res. 55, 39–57. doi: 10.1016/S0141-1136(02)00215-5

Pogacnik, L., and Franco, M. (1999). Determination of organophosphate and cabamate pesticides in spiked samples of tap water and fruit juices by a biosensor with photothermal detection. Biosens. Bioelectron. 14, 569–578. doi: 10.1016/S0956-5663(99)00029-9

Reddy, K. G., Madhavi, G., Swamy, B., Reddy, S., Reddy, A. V., and Medhavi, V. (2013). Electrochemical investigations of lipase enzyme activity inhibition by methyl parathion pesticide: voltammetric studies. J. Mol. Liq. 180, 26–30. doi: 10.1016/j.molliq.2012.12.032

Russell Everett, W., and Rechnitz, G. A. (1998). Mediated bioelectrocatalytic determination of organophosphorus pesticides with a tyrosinase- based oxygen biosensor. Anal. Chem. 70, 807–810. doi: 10.1021/ac970958l

Sadik, O. A., Xu, H., Gheorghiu, E., Andreescu, D., Balut, C., Gheorghui, M., et al. (2002). Differential impedance spectroscopy for monitoring protein immobilization and antibody-antigen reactions. Anal. Chem. 74, 3142–3150. doi: 10.1021/ac0156722

Sajjadi, S., Ghourchian, H., and Tavakoli, H. (2009). Choline oxidase as a selective recognition element for determination of paraoxon. Biosens. Bioelectron. 24, 2509–2514. doi: 10.1016/j.bios.2009.01.008

Tanimoto de Albuquerque, D., and Ferreira, L. F. (2007). Amperometric biosensing of carbamate and organophosphate pesticides utilizing screen printed tyrosinase modified electrodes. Anal. Chem. Acta 596, 210–221. doi: 10.1016/j-aca.2007.06.013

UAP. Ca (United Agri Products Canada INC). (2007). Diazinon Technical Fact Sheet. http://www.uap.ca/francais/products/documents/Diazinon50WF_datapak12_FEB_2007_000.pdf

US NPIC (National Pesticide Information Center, Oregon State University Extension Services). (2009). Diazinon Technical Fact Sheet. Available online at: http://npic.orst.edu/factsheets/diazinontech.pdf (Accessed October 12, 2012).

Villeneuve, D. C., Willes, R. F., Lacroix, J. B., and Phillips, W. E. J. (1972). Placental transfer of ^{14}C-parathion administered intravenously to sheep. Toxicol. Appl. Pharmacol. 21, 201–205. doi: 10.1016/0041-008X(72)90010-5

Wiener, S. W., and Hoffman, R. (2004). Nerve agents: a comprehensive review. J. Intensive Care Med. 19, 22–37. doi: 10.1177/0885066603258659

Yagati, A. K., Lee, T., Min, J., and Choi, J. W. (2011). Amperometric sensor for hydrogen peroxide based on direct electron transfer of spinach ferredoxin on Au electrode. *Bioeletrochemistry* 80, 169–174. doi: 10.1016/j.bioelechem.2010.08.002

Yi, Y., Zhu, G., Liu, C., Hang, Y., Zhang, Y., Li, H., et al. (2013). A label-free silicon quantum dots- based photoluminescence sensor for ultrasensitive detection of pesticides. *Anal. Chem.* 85, 11464–11470. doi: 10.1021/ac403257p

Conflict of Interest Statement: The authors declare that the research was conducted in the absence of any commercial or financial relationships that could be construed as a potential conflict of interest.

Permissions

The contributors of this book come from diverse backgrounds, making this book a truly international effort. This book will bring forth new frontiers with its revolutionizing research information and detailed analysis of the nascent developments around the world.

We would like to thank all the contributing authors for lending their expertise to make the book truly unique. They have played a crucial role in the development of this book. Without their invaluable contributions this book wouldn't have been possible. They have made vital efforts to compile up to date information on the varied aspects of this subject to make this book a valuable addition to the collection of many professionals and students.

This book was conceptualized with the vision of imparting up-to-date information and advanced data in this field. To ensure the same, a matchless editorial board was set up. Every individual on the board went through rigorous rounds of assessment to prove their worth. After which they invested a large part of their time researching and compiling the most relevant data for our readers.

The editorial board has been involved in producing this book since its inception. They have spent rigorous hours researching and exploring the diverse topics which have resulted in the successful publishing of this book. They have passed on their knowledge of decades through this book. To expedite this challenging task, the publisher supported the team at every step. A small team of assistant editors was also appointed to further simplify the editing procedure and attain best results for the readers.

Apart from the editorial board, the designing team has also invested a significant amount of their time in understanding the subject and creating the most relevant covers. They scrutinized every image to scout for the most suitable representation of the subject and create an appropriate cover for the book.

The publishing team has been an ardent support to the editorial, designing and production team. Their endless efforts to recruit the best for this project, has resulted in the accomplishment of this book. They are a veteran in the field of academics and their pool of knowledge is as vast as their experience in printing. Their expertise and guidance has proved useful at every step. Their uncompromising quality standards have made this book an exceptional effort. Their encouragement from time to time has been an inspiration for everyone.

The publisher and the editorial board hope that this book will prove to be a valuable piece of knowledge for researchers, students, practitioners and scholars across the globe.

List of Contributors

Georgio K foury
Department of Advanced Materials and Structures, Public Research Center Henri Tudor,Haut charage, Luxembourg Laboratory of Polymeric and Composite Materials, UMONS Research Institute for Materials Science and Engineering, Center for Innovation and Research in Materials and Polymers, University of Mons, Mons, Belgium

Jean-Marie Raquez, Jérémy Odent and Philippe Dubois
Laboratory of Polymeric and Composite Materials, UMONS Research Institute for Materials Science and Engineering, Center for Innovation and Research in Materials and Polymers, University of Mons, Mons, Belgium

Fatima Hassouna, Valérie Toniazzo and David Ruch
Department of Advanced Materials and Structures, Public Research Center Henri Tudor, Haut charage, Luxembourg

Alan Wong and Céline Boutin
CEA Saclay, DSM, IRAMIS, UMRCEA/CNRS3299–NIMBE, Laboratoire Structure et Dynamique par Résonance Magnétique, Gif-sur-Yvette, France

Pedro M.Aguiar
Department of Chemistry, University of York, Heslington, York, UK

Mitra Vasei, Paramita Das, Hayet Cherfouth, Benoît Marsanand Jerome P. Claverie
Department of Chemistry, Nano QAM, Québec Center for Functional Materials, Université du Québec à Montréal, Montreal, QC, Canada

Shigeaki Morita
Department of Engineering Science, Osaka Electro-Communication University, Neyagawa, Japan

Bruno Tota, Tommaso Angel one and Maria C. Cerra
Department of Biology, Ecology and Earth Sciences, University of Calabria, Arcavacata di Rende (CS), Italy

Lorena Pochini, Maria Francesca Scalise, Michele Galluccio and Cesare Indiveri
Department DiBEST (Biologia, Ecologia, Scienzedella Terra) Unit of Biochemistry and Molecular Biotechnology, University of Calabria, Arcavacata di Rende, Italy

Vinod Kumar Gupta
Department of Chemistry, Indian Institute of Technology Roorkee, Roorkee, India

Deepak Pathania, Bhanu Priya and Gaurav Sharma
Department of Chemistry, Shoolini University of Biotechnology and Management Sciences, Solan, India

Amar Singh Singha
Department of Applied Chemistry, National Institute of Technology Hamirpur, Hamirpur, India

Frederik R. Wurm
Physical Chemistry of Polymers, Max Planck Institute for Polymer Research, Mainz, Germany

Clemens K. Weiss
Life Sciences and Engineering, University of Applied Sciences Bingen, Bingen, Germany

Hyder Ali Khan and Bulent Mutus
Chemistry and Biochemistry Department, University of Windsor, Windsor, ON, Canada

Michael Holzinger, Alan Le Goff and Serge Cosnier
Département de Chimie Moléculaire UMR 5250, Biosystèmes Electrochimique and Analytiques, CNRS, University of Grenoble Alpes, Grenoble, France

Venkaiah Betapudi
Department of Cellular and Molecular Medicine, Lerner Research Institute, Cleveland Clinic, Cleveland, OH, USA Department of Physiology and Biophysics, Case Western Reserve University, Cleveland, OH, USA

Edna Bechor, Iris Dahan, Tanya Fradin, Yevgeny Berdichevsky, Anat Zahavi, Aya Federman Gross, Meirav Rafalowski and Edgar Pick
The Julius Friedrich Cohnheim Laboratory of Phagocyte Research, Department of Clinical Microbiology and Immunology, Sackler School of Medicine, Tel Aviv University, Tel Aviv, Israel

Paula M. Brito
URIA-Centro de Patogénese Molecular, Faculdade de Farmácia, Universidade de Lisboa, Lisboa, Portugal Instituto de Medicina Molecular, Faculdade de Medicinada Universidade de Lisboa, Lisboa, Portugal Faculdade de Ciênciasda Saúde, Universida de da Beira Interior, Covilhã, Portugal

Fernando Antunes
Departamento de Químicae Bioquímica and Centro de Químicae Bioquímica, Faculdade de Ciências, Universidade de Lisboa, Lisboa, Portugal

Monica M. Watanabe, Francisco R. M. Laurindo and Denise C. Fernandes
Vascular Biology Laboratory, Heart Institute (InCor), University of São Paulo School of Medicine, São Paulo, Brazil

Bishnu P. Koiry and Nikhil K. Singha
Rubber Technology Centre, Indian Institute of Technology Kharagpur, Kharagpur, India

Pellegrino Musto and Marianna Pannico
Institute of Chemistry and Technology of Polymers, National Research Council of Italy, Naples, Italy

Michele Galizia and Giuseppe Mensitieri
Department of Chemical, Materials and Industrial Production Engineering, University of Naples FedericoII, Naples, Italy

Pietro La Manna
Institute of Chemistry and Technology of Polymers, National Research Council of Italy, Naples, Italy
Department of Chemical, Materials and Industrial Production Engineering, University of Naples FedericoII, Naples, Italy

Nedjla Zehani
UMR 5280, Institut des Sciences Analytiques, Université de Lyon, Villeurbanne, France
Laboratory of Environmental Engineering, Faculty of Engineering Sciences, University of Annaba, Annaba, Algeria

Sergei V. Dzyadevych
Laboratory of Biomolecular Electronics, Institute of Molecular Biology and Genetics, National Academy of Sciences of Ukraine, Kiev, Ukraine

Rochdi Kherrat
Laboratory of Environmental Engineering, Faculty of Engineering Sciences, University of Annaba, Annaba, Algeria

Nicole J. Jaffrezic-Renault
UMR 5280, Institut des Sciences Analytiques, Université de Lyon, Villeurbanne, France

www.ingramcontent.com/pod-product-compliance
Lightning Source LLC
Chambersburg PA
CBHW080536200326
41458CB00012B/4456